D1483652

PRINCIPLES OF ENHANCED HEAT TRANSFER

PRINCIPLES OF ENHANCED HEAT TRANSFER

Ralph L. Webb
Department of Mechanical Engineering
The Pennsylvania State University
University Park, PA

A Wiley-Interscience Publication
JOHN WILEY & SONS, INC.
New York • Chichester • Brisbane • Toronto • Singapore

WIDENER UNIVERSITY
WOLFGRAM LIBRARY
CHESTER, PA.
DISCARDED
WIDENER UNIVERSITY

This text is printed on acid-free paper.

Copyright © 1994 by John Wiley & Sons, Inc.

All rights reserved. Published simultaneously in Canada.

Reproduction or translation of any part of this work beyond
that permitted by Section 107 or 108 of the 1976 United
States Copyright Act without the permission of the copyright
owner is unlawful. Requests for permission or further
information should be addressed to the Permissions Department,
John Wiley & Sons, Inc., 605 Third Avenue, New York, NY
10158-0012.

Library of Congress Cataloging in Publication Data:
Webb, Ralph L., 1934–
    Principles of enhanced heat transfer / by Ralph L. Webb.
      p. cm.
    "Revised December 6, 1992."
    ISBN 0-471-57778-2
    1. Heat—Transmission. 2. Heat exchangers. I. Title.
    TJ260.W36 1994
    621.402′2—dc20   93-4513

Printed in the United States of America

10 9 8 7 6 5 4 3 2

To Professor Ernst Eckert, who trained me in heat transfer and introduced me to the possibilities of analytical modeling. To the memory of Mr. H. Corbyn Rooks, who stimulated and challenged me to provide practical industrial enhanced surfaces. To Professor Richard Goldstein, who has encouraged and supported me in the steps of my career. Finally, to my friend Arthur Bergles, for his many contributions to enhanced heat transfer.

CONTENTS

8   Internally Finned Tubes and Annuli   **200**

9   Integral Roughness   **229**

PREFACE

My involvement in enhanced heat transfer began 30 years ago, when I joined the Applied Research Department of The Trane Co., a manufacturer of air-conditioning and heat exchange equipment. Our objective was to seek new surface geometries for the broad range of corporate products. Over the next 15 years, our group aggressively pursued this goal. We successfully developed shell-side enhancements for nucleate boiling, condensation, and film absorption. Advanced surface geometries for finned tube heat exchangers were also an important objective. We did considerable research on tube-side enhancement, but were limited by commercial manufacturing technology that existed at the time.

After joining Penn State University in 1977, I expanded my interests in enhancement to potential applications and research needs in other industries—automotive, process, and power generation. As an academic, my interests focused on understanding the mechanism of the various enhancement concepts, and we worked to develop predictive methods to account for the effect of fluid properties and surface geometry variables. I interacted with numerous industrial organizations, who were seeking to develop or apply advanced surface geometries. Over the years, I have observed significant advances in manufacturing technology. Among the most significant are high-speed methods to make enhancements inside a tube and to make high-performance boiling surfaces. Enhanced heat transfer surface technology is a multidisciplinary problem, involving both heat transfer and manufacturing technology. Good heat transfer ideas cannot be commercially applied until cost-effective methods are developed.

Industry and academia have shared an informal partnership in advancing the technology of enhanced heat transfer. The advances in the technology of enhanced heat transfer have primarily come from industrial research. Academia has followed along to explain enhancement mechanisms and to provide predictive correlations or models. Although enhancement technology was largely regarded as "empirical" 30 years ago, it has considerably advanced in rational understanding. Much recent work has been done to develop analytically based models to explain the behavior of the complex geometries involved.

This book has two key purposes. First, to summarize the advancement of the technology in an organized manner. The chapters are segregated by "mode of heat transfer." We seek to summarize the data on the various enhancement techniques for each heat transfer mode. Second, we seek to define the "mechanism" by which the enhancement devices work. Where possible, we provide predictive equations and discuss analytically based models for prediction of the heat transfer and friction characteristics of the various enhancement concepts.

In 1983, Webb and Bergles published a paper in *Mechanical Engineering* magazine (June, pp. 60–67) which reviewed the technology. This paper was subtitled "Second Generation Heat Transfer Technology." I believe that this is a good description of the essence of *enhanced heat transfer*. The automotive and refrigeration industries routinely use enhanced surface technology. They do this either to obtain lower-cost, smaller heat exchangers or to obtain higher-energy-efficiency system operation. Other industries are finding practical and economic applications for enhanced surface heat exchangers (e.g., aerospace, process, and power generation). The subject of *enhanced heat transfer* has become so important that there is now a journal devoted to the subject—the *Journal of Enhanced Heat Transfer.*

Problems and special needs associated with different industries are discussed in Chapter 1. Fouling is a barrier to be overcome to open the way for application to fouling fluids. This subject is addressed in Chapter 10. The adoption of this "second-generation technology" has occurred at an accelerating pace, as shown in Figures 1.9 and 1.10 of Chapter 1. It is very likely that most industries will routinely use enhanced surfaces 20 years from now.

I hope that this book will encourage potential users to seriously consider the benefits of enhancement technology in their heat (and mass) transfer applications. The subject offers new research opportunities to academics interested in modeling complex surface geometries. Numerical analysts will find a rich field for application of their powerful tools. Perhaps we will find elements of the subject integrated in future academic courses. I have taught a graduate course on *enhanced heat transfer* since 1984. It provides an excellent opportunity to show students how to apply their understanding of fundamentals to model complex geometries.

My 30-year adventure in *enhanced heat transfer* has been an exciting experience. In my musings, I try to envision what will be said of this technology in a book written 30 years from now!

I am grateful to my friends who have read the various chapters of the book. Their careful proofreading and suggestions contributed to the quality of the manuscript. In particular, I would like to acknowledge my former and present graduate students (Louay Chamra, Neil Gupte, Chris Pais, Bill McQuade, Imam Haider, Eric Dillen, Laing-Han Chien, and Chien-Yuh Yang), my colleagues (Tom Rabas, Horacio Perez-Blanco, Raj Manglik, Ramesh Prasad, and Jerry Taborek), and my secretary

Bonnie Jacobs. Ms. June Leary's excellent drafting skills are responsible for the quality of the figures. Finally, I am very grateful to my wife, Sylvia, who supported my need for the long, quiet periods required to prepare the manuscript.

RALPH L. WEBB

State College, PA
November, 1993

1

INTRODUCTION TO ENHANCED HEAT TRANSFER

1.1 INTRODUCTION

The subject of enhanced heat transfer has developed to the stage that it is of serious interest for heat exchanger application. The refrigeration and automotive industries routinely use enhanced surfaces in their heat exchangers. The process industry is aggressively working to incorporate enhanced heat transfer surfaces in their heat exchangers. Virtually every heat exchanger is a potential candidate for enhanced heat transfer. However, each potential application must be tested to see if enhanced heat transfer "makes sense." The governing criteria will be addressed later.

Heat exchangers were initially developed to use plain (or smooth) heat transfer surfaces. An "enhanced heat transfer surface" has a special surface geometry that provides a higher hA value, per unit base surface area than does a plain surface. The term "enhancement ratio" (E) is the ratio of the hA of an enhanced surface to that of a plain surface. Thus,

$$E_h = \frac{hA}{(hA)_p} \tag{1.1}$$

Consider a two-fluid counterflow heat exchanger. The heat transfer rate for a two-fluid heat exchanger is given by

$$Q = UA \ \Delta T_m \tag{1.2}$$

To illustrate the benefits of enhancement, we will multiply and divide Equation 1.2 by the total tube length, L.

1

$$Q = \frac{UA}{L} L \, \Delta T_m \qquad (1.3)$$

The term L/UA is the overall thermal resistance per unit tube length and is given by

$$\frac{L}{UA} = \frac{L}{\eta_1 h_1 A_1} + \frac{Lt_w}{k_w A_m} + \frac{L}{\eta_2 h_2 A_2} \qquad (1.4)$$

where the subscripts 1 and 2 refer to fluids 1 and 2, respectively. The term η is the surface efficiency, should extended surfaces be employed. For simplicity, Equation 1.4 does not include fouling resistances, which may be important. The performance of the heat exchanger will be enhanced if the term UA/L is increased. An enhanced surface geometry may be used to increase either or both of the hA/L terms, relative to that given by plain surfaces. This will reduce the thermal resistance per unit tube length, L/UA. This reduced L/UA may be used for one of three objectives:

1. *Size Reduction:* If the heat exchange rate (Q) is held constant, the heat exchanger length may be reduced. This will provide a smaller heat exchanger.
2. *Increased UA:* This may be exploited either of two ways:
   a. *Reduced* ΔT_m: If Q and the total tube length (L) are held constant, the ΔT_m may be reduced. This provides increased thermodynamic process efficiency and yields a savings of operating costs.
   b. *Increased Heat Exchange*: Keeping L constant, the increased UA/L will result in increased heat exchange rate for fixed fluid inlet temperatures.
3. *Reduced Pumping Power for Fixed Heat Duty:* Although it may seem surprising that enhanced surfaces can provide reduced pumping power, this is theoretically possible. However, this will typically require that the enhanced heat exchanger operate at a velocity smaller than that of the competing plain surface. This will require increased frontal area, which is normally not desired.

The important principle to be learned is that *an enhanced surface can be used to provide any of three different performance improvements.* Which improvement is obtained depends on the designer's objectives. Thus, designer A may seek a smaller heat exchanger, whereas designer B may want improved thermodynamic process efficiency.

Although the size reduction of objective 1 may be valued, the more important objective may be cost reduction. In many cases, the designer requires that the size reduction be accompanied by cost reduction. Another factor to consider for objective 1 is that the fluid volume in the heat exchanger will also be reduced. This may be an important consideration for a manufacturer of refrigeration equipment, since a smaller volume of expensive refrigerant will be required.

Objectives 2 and 3 are important if "life cycle" costing is of interest. For example, objective 2 for refrigeration condensers and evaporators will result in reduced compressor power costs. Objective 3 is important for upgrading the capacity of an existing heat exchanger. This may allow plant output to be increased.

Pressure drop (or pumping power) is always of concern to the heat exchanger designer. Hence, a practical enhanced surface must provide the desired heat transfer enhancement and meet the required flow rate and pressure drop constraints. Chapters 3 and 4 discuss how one meets these objectives for single-phase (Chapter 3) and two-phase (Chapter 4) flows. A surface geometry that provides a given heat transfer enhancement level with the lowest pressure drop is definitely preferred.

1.2   THE ENHANCEMENT TECHNIQUES

Thirteen enhancement techniques have been identified by Bergles et al. [1983]. These techniques are segregated into two groupings: "passive" and "active" techniques. Passive techniques employ special surface geometries, or fluid additives for enhancement. The active techniques require external power, such as electric or acoustic fields and surface vibration. The techniques are listed in Table 1.1 and are described below.

1.2.1   Passive Techniques

Coated surfaces involve metallic or nonmetallic coating of the surface. Examples include a nonwetting coating, such as Teflon, to promote dropwise condensation. A fine-scale porous coating may be used to enhance nucleate boiling. Figure 1.1a shows the cross section of a sintered porous metal coating for nucleate boiling. The particle size is on the order of 0.005 mm. Figure 1.1b shows larger particles (approximately 0.5 mm) sintered to the surface. These may be used to enhance single-phase convection, or condensation.

Rough surfaces may be integral to the base surface, or they may be made by placing a "roughness" adjacent to the surface. Integral roughness is formed by machining, or "restructuring" the surface. For single-phase flow, the configuration is generally chosen to promote mixing in the boundary layer near the surface, rather than to increase the heat transfer surface area. Figure 1.2a shows two examples of integral roughness. Formation of a rough surface by machining away metal is generally not an economically viable approach. Figure 1.2b shows an enhanced "rough" surface for nucleate boiling. The surface structuring forms artificial nucleation sites, which provide a much higher performance than does a plain surface. Figure 1.2c shows the wire coil insert, which periodically disturbs the boundary layer. A wire coil insert is an example of a nonintegral roughness.

Extended surfaces for liquids are routinely employed in many heat exchangers. As shown in Equation 1.4, the thermal resistance may be reduced by increasing the heat transfer coefficient (h), or the surface area (A), or both h and A. Use of a plain fin may provide only area increase. However, formation of a special shape extended surface may also provide increased h. Current enhancement efforts for gases are directed toward extended surfaces that provide a higher heat transfer coefficient than

TABLE 1.1 Summary of References on Enhancement Techniques as of November 1983

	Single-Phase Natural Convection	Single-Phase Forced Convection	Pool Boiling	Force-Convection Boiling	Condensation	Mass Transfer
Passive Techniques (no external power required)						
Treated surfaces	—	—	149	17	53	—†
Rough surfaces	7	418	62	65	65	29
Extended surface	23	416	75	53	175	33
Displaced enhancement devices	—†	59	4	17	6	15
Swirl flow devices	—†	140	—	83	17	10
Coiled tubes	—†	142	—	50	6	9
Surface tension devices	—	—	12	1	—	—†
Additives for liquids	3	22	61	37	—	6
Additives for gases	—†	211	—	—	5	0
Active Techniques (external power required)						
Mechanical aids	16	60	30	7	23	18
Surface vibration	52	30	11	2	9	11
Fluid vibration	44	127	15	5	2	39
Electric or magnetic fields	50	53	37	10	22	22
Injection or suction	6	25	7	1	6	2
Jet impingement	—	17	2	1	—	2
Compound techniques	2	50	4	4	4	2

— Not applicable.
*Not considered in this survey.
†No citations located.

Source: Bergles and Webb [1985].

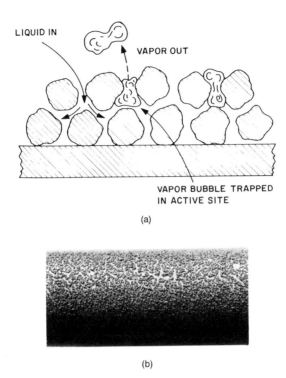

LIQUID IN

VAPOR OUT

VAPOR BUBBLE TRAPPED
IN ACTIVE SITE

(a)

(b)

Figure 1.1   (a) Illustration of cross section of porous boiling surface. (Courtesy of UOP Corporation, Tonawanda, NY.) (b) Attached particles used for film condensation.

that of a plain fin design. Figure 1.3 shows a variety of enhanced extended surfaces used for gases. The surfaces shown in Figures 1.3A through 1.3E involve repeated formation and destruction of thin thermal boundary layers. Extended surfaces for liquids typically use much smaller fin heights than those used for gases. Shorter fin heights are used for liquids, because liquids typically have higher heat transfer coefficients than do gases. Use of high fins with liquids would result in low fin efficiency and result in poor material utilization. Examples of extended surfaces for liquids are shown in Figure 1.4. Figure 1.4a shows an external finned tube, and Figure 1.4b shows an internally finned tube. The internally finned tubes shown in Figure 1.4c are made by multiple, concentric internally finned tubes. The tube shown in Figure 1.4d contains a five-element extruded aluminum insert. The surrounding tube is compressed onto the insert to provide good thermal contact. The geometries shown in Figure 1.4 have also been used for forced convection vaporization and condensation.

   Displaced insert devices are devices inserted into the flow channel to indirectly improve energy transport at the heated surface. They are used with single- and two-phase flows. The devices shown in Figure 1.5a and 1.5b mix the main flow, in addition to that in the wall region. The wire coil insert shown in Figure 1.5c is placed at the edge of the boundary layer, and is intended to promote mixing within the boundary layer, without significantly affecting the main flow.

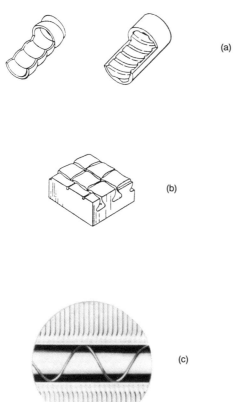

(a)

(b)

(c)

Figure 1.2 (a) Tube-side roughness for single-phase or two-phase flow. (b) "Rough" surface for nucleate boiling. (c) Wire coil insert.

Swirl flow devices include a number of geometrical arrangements or tube inserts for forced flow that create rotating and/or secondary flow. Such devices include full-length twisted-tape inserts (Figure 1.6a), or inlet vortex generators, and axial core inserts with a screw-type winding (Figure 1.6b). Figure 1.6c shows a flow invertor or static mixer intended for laminar flows. This device alternately swirls the flow in clockwise and counterclockwise directions.

Coiled tubes (Figure 1.7) may provide more compact heat exchangers. Secondary flow in the coiled tube produces higher single-phase coefficients and improvement in most boiling regimes. However, a quite small coil diameter is required to obtain moderate enhancement.

Surface tension devices use surface tension forces to drain or transport liquid films. The special "flute" shape of Figure 1.8 promotes condensate drainage from the surface by surface tension forces. The film condensation coefficient is inversely proportional to the condensate film thickness. Heat pipes (not contained in the listing) typically use some form of capillary wicking to transport liquid from the condenser section to the evaporator section.

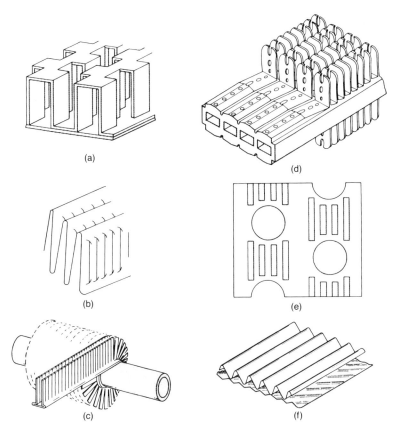

Figure 1.3 Enhanced surfaces for gases. (a) Offset strip fins used in plate-fin exchanger. (b) Louvered fins used in automotive heat exchangers. (c) Segmented fins for circular tubes. (d) Integral aluminum strip finned tube. (e) Louvered tube-and-plate fin. (f) Corrugated plates used in rotary regenerators. (From Webb and Bergles [1983].)

Additives for liquids include (a) solid particles or gas bubbles in single-phase flows and (b) liquid trace additives for boiling systems.

Additives for gases are liquid droplets or solid particles, either dilute phase (gas–solid suspensions) or dense phase (packed tubes and fluidized beds).

1.2.2 Active Techniques

Mechanical aids involve stirring the fluid by mechanical means or rotating the surface. Mechanical surface scrapers widely used for viscous liquids in the chemical process industry can be applied to duct flow of gases. Equipment with rotating heat exchanger ducts is found in commercial practice.

Surface vibration at either low or high frequency has been used primarily to improve single-phase heat transfer.

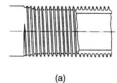

(a)

(b)

(c)

(d)

Figure 1.4 (a) Integral fins on outer tube surface. (b) Internally finned tubes (axial and helical fins). (c) Cross sections of multiply internally finned tubes. (d) Tube with aluminum star insert.

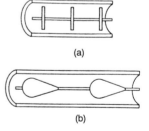

(a)

(b)

Figure 1.5 (a) Spaced disk devices. (b) Spaced streamline-shaped insert devices. (c) Displaced wire coil insert. (From Webb [1987].)

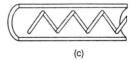

(c)

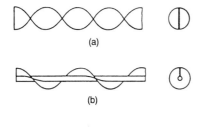

(a)

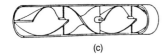

(b)

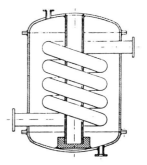

(c)

Figure 1.6   Three types of swirl flow inserts. (a) Twisted-tape insert. (b) Helical vane insert. (c) Static mixer.

Figure 1.7   Helically coiled tube heat exchanger. (From Bergles and Webb [1985].)

Fluid vibration is the more practical type of vibration enhancement because of the mass of most heat exchangers. The vibrations range from pulsations of about 1 Hz to ultrasound. Single-phase fluids are of primary concern.

Electrostatic fields [direct current (dc) or alternating current (ac) are applied in many different ways to dielectric fluids. Generally speaking, electrostatic fields can be directed to cause greater bulk mixing of fluid in the vicinity of the heat transfer surface.

Injection is utilized by supplying gas through a porous heat transfer surface to a flow of liquid or by injecting the same liquid upstream of the heat transfer section. The injected gas augments single-phase flow. Surface degassing of liquids may produce similar effects.

Suction involves vapor removal, in nucleate or film boiling, or fluid withdrawal in single-phase flow through a porous heated surface.

Jet impingement forces a single-phase fluid normally or obliquely toward the surface. Single or multiple jets may be used, and boiling is possible with liquids.

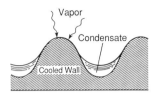

(a)

(b)

Figure 1.8   (a) Fluted tube used for condensation in the vertical orientation. (b) Illustration of surface tension drainage from the flutes into drainage channels.

1.2.3   Technique Versus Mode

Enhancement is applicable to both heat and mass transfer processes, or simultaneous heat and mass transfer. Modes of heat (or mass) transfer of potential interest include:

1. Single-phase flow: natural and forced convection inside or outside tubes
2. Two-phase flow: boiling and condensation inside tubes and tube banks
3. Radiation
4. Convective mass transfer

Compound enhancement results from simultaneous use of two or more of the previously discussed techniques. Such an approach may produce an enhancement that is larger than either of the techniques operating separately.

The majority of commercially interesting enhancement techniques are currently limited to the passive techniques. However, current work on electrohydrodynamic (EHD) enhancement of boiling and condensation suggests significant potential. The lack of use of the active techniques is related to the cost, noise, safety, or reliability concerns associated with the enhancement device.

This book primarily focuses on the passive techniques, derived from special configuration of the surface geometry. Chapters 5 through 14 (except Chapter 10) discuss passive techniques for specific heat transfer modes. Chapter 10 discusses fouling of enhanced surfaces. Chapter 15 surveys EHD enhancement, and Chapter 16 considers fluid additives.

1.3 PUBLISHED LITERATURE

1.3.1 General Remarks

Bergles et al. [1983, 1991] have reported the journal publications on enhanced heat transfer. Webb et al. [1983] provide a similar report on the U.S. patent literature. Bergles and Webb [1985] summarize the information presented in the literature and patent bibliography reports. Figure 1.9, taken from Bergles et al. [1991], shows the journal and conference publications per year between 1900 and 1990, which total 4345, as of December 1990. The period of rapid growth began about 1960. The patent literature is an equally important source for information on enhancement

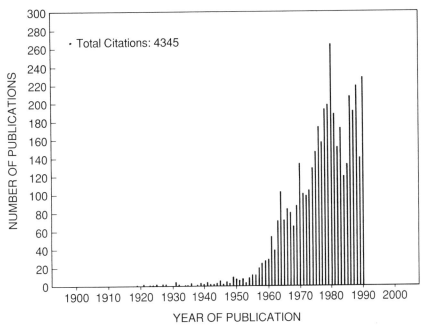

Figure 1.9 Citations on heat transfer augmentation versus year of publication. Status as of December 1990 is illustrated. (From Bergles et al. [1991].)

techniques. Figure 1.10, taken from the Webb et al. [1983] survey of the U.S. patent literature, shows 486 patents issued as of June 1982.

The effectiveness of the technique may depend on the heat (or mass) transfer mode. Table 1.1 classifies the enhancement references according to enhancement technique and heat transfer mode. A dash in the table indicates that the technique is not applicable to that particular heat transfer mode.

Some comments are in order regarding the classifying and cataloging procedures for the report on technical literature. Difficulties in classifying some techniques are acknowledged. For instance, it is somewhat arbitrary whether certain of the new structured boiling surfaces are considered *treated surfaces*, *rough surfaces*, or *extended surfaces*. For example, an enhanced boiling surface may appear to have only a slight surface *roughness*, yet it contains a subsurface microstructure that promotes boiling nucleation. Although the surface may provide a moderate surface area increase, the surface area increase is not responsible for the high enhancement level.

Another example is in the important area of fouling of enhanced surfaces. Papers on this subject are included in the *mass transfer* classification. However, the goal is obviously not to increase the fouling rate!

Several topics are not thoroughly referenced for a variety of reasons. Only a few papers are listed on the theory of dropwise condensation, under *treated surfaces*,

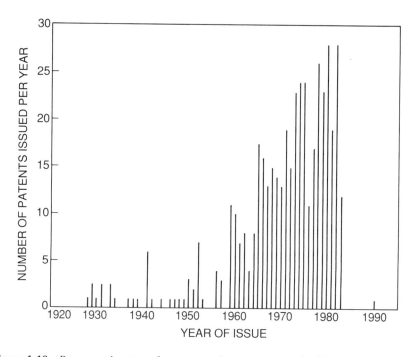

Figure 1.10 Patents on heat transfer augmentation versus year of publication. (From Webb et al. [1983].)

since the emphasis of the classification is on promoters for dropwise condensation. Papers on use of extended surfaces in single-phase natural convection may not be included in the *extended surfaces* listing. Rather, the primary interest is in enhanced extended surfaces. As noted above, the category of *surface tension devices* does not include the vast literature on heat pipes. *Additives for gases* may include some information on fluidized beds. However, the primary interest of the listing is on enhancement of gas-to-wall heat transfer by the particles, rather than toward improvement of gas-to-particle heat transfer. Literature which is "supporting" (e.g., gives details of the flow over a protuberance or pressure drop in a rough tube without accompanying heat transfer area) is not included in the literature file.

The literature reports of Bergles et al. [1983, 1991] are organized according to the following coding:

00 Surveys
01 Treated Surfaces
02 Rough Surfaces
03 Extended Surfaces
04 Displaced Enhancement Devices
05 Swirl Flow Devices
16 Coiled Tubes
06 Surface Tension Devices
07 Additives for Liquids
08 Additives for Gases
09 Mechanical Aids
10 Surface Vibration
11 Fluid Vibration
12 Electric or Magnetic Fields
13 Injection or Suction
17 Jet Impingement
14 Compound Enhancement
15 Performance Evaluation

Each citation includes the following information:

Author(s)
Title
Publication (or source) and date
Several key words describing the work

The English translations of foreign language titles are provided except for French and German works. The key words include reference number, year of publication, journal, technique, and mode of heat transfer. The modes of heat transfer are coded as follows (mass transfer is included for completeness):

1   Single-Phase Natural Convection
2   Single-Phase Forced Convection
3   Pool Boiling
4   Forced-Convection Boiling
5   Condensation
6   Mass Transfer

Occasional difficulty arises when multiple techniques and modes are considered in a single paper, because not every mode may be considered for all the techniques. Because there was no convenient way to couple the mode and technique keywords, many of the ambiguities are left unresolved.

Reference numbers are formed according to technique (first two digits), mode (third digit), and number in that category (last three digits). Citations are set up on a minicomputer, recorded on a disk, and transferred to a mainframe computer that exercises the information retrieval program.

The literature is presented according to enhancement technique and mode of heat transfer, and the citations are listed in alphabetical order within each category. Table 1.1 summarizes the number of citations in each category.

1.3.2   U.S. Patent Literature

The second bibliography described covers U.S. patent literature. This report by Webb et al. [1983] contains 554 references to U.S. patents. Patents with issue dates up to June 30, 1983 are included.

The present report classifies the patents in three ways. The first two are by patent number and by name of the first named inventor. The third compilation is similar to that used in Bergles et al. [1983], in that the patents are classified by enhancement technique, mode of heat transfer, and geometry.

There are several changes in the classification utilized for the first bibliography. Specifically, two new classifications that were primarily subsets of other classifications are introduced below.

There are several differences between the enhancement technique classifications used in patent and journal literature reports. The patent literature report includes "formed channels," which are typified by corrugated plates used in plate-fin heat exchangers and deep spirally fluted tubes. The surface distortion is much greater than that typical of roughness. The surface deformation causes mixing by secondary flows. Surface area increase usually is not a significant factor, although the three-dimensional deformed tubes, also included in this classification, can provide a surface area increase of up to 30%. Formed channels provide enhancement to the fluid flowing on each side of the channel.

The patent report separately identifies "attached promoters," which are discrete roughness elements that are not integral to the heat transfer surface. A good example is the wire coil insert used in circular tubes.

In addition, the categories of Surveys, Coiled Tubes, Jet Impingement, and Performance Evaluation are not included in the patent classification. Coiled Tube concepts are included in the Formed Channels category.

The patents cited in the report are organized according to the following coding system:

Technique

01   Treated Surfaces

02   Rough Surfaces

03   Attached Promoters

04   Extended Surfaces

05   Displaced Enhancement Devices

06   Formed Channels

07   Swirl Flow Devices

08   Surface Tension Devices

09   Additives for Liquids

10   Additives for Gases

11   Mechanical Aids

12   Surface Vibration

13   Fluid Vibration

14   Electrostatic Fields

15   Injection or Suction

16   Compound Enhancement

Mode of Transfer

1   Single-Phase Natural Convection

2   Single-Phase Forced Convection

3   Pool Boiling

4   Flow Boiling

5   Condensation

6   Mass Transfer

Geometry Classification

A   Inside Tubes

B   Outside Tubes—Gases

C   Outside Tubes—Liquids

D   Plate-Type Heat Exchangers

E   Fin-Type Heat Exchangers

F   Other Geometries

For example, a code of 023C indicates a rough surface for pool boiling, with the roughness on the outer surface of a tube. Note that the patent file contains a geometry classification, which is not contained in the literature file.

The report contains patents issued through June 1983. An intensive "hands on" search of patents on file at the U.S. Patent Office was performed for the 1980 edition of the report.

Because there are more than 10,000 patent subclasses, it was necessary to develop a logical approach to establish which subclasses were worthy of search. The patent classifications searched at the U.S. Patent Office included classes 29, 62, 72, 113, 165, 202, 219, 260, 264, 357 and 427. Only certain subclasses were of interest. An illustration of the U.S. Patent Office classification system is shown in Table 1.2. This table lists the subclasses within class 165, "Heat Exchange," that were searched. Study of Table 1.2 shows that the patent literature uses a key wording system that significantly differs from key wording familiar to engineers. Hence, special considerations are required for patent searches.

The patent subclasses selected for search were established by several parallel efforts: First, the *U.S. Patent Office Class Definitions*, which lists and describes the subclasses within each patent class, was reviewed. Study of this document suggested potentially fruitful areas.

The final and most successful method involved an iterative search method. At the beginning of the intensive search, there were approximately 100 patents on file. A "frequency distribution table" that defined the classes/subclasses of these patents was made. The patents themselves also provided some cross-reference information. This work-sheet table, along with the previously indicated fruitful areas, identified the initial classes/subclasses to be searched. In addition, current issues of the biweekly *Official Patent Gazette*, which publishes bimonthly abstracts of new patents, were monitored during the search period. This publication was the primary source of recent patents for the second edition.

Copies of approximately 900 patents were ordered from the Commissioner of Patents and Trademarks, Washington, D.C. The patents were then carefully reviewed for relevance and cataloged by the previously described coding system. Many patents were discarded.

The final list in this report contains 486 entries obtained from 454 separate patents. There are 31 patents for which more than one technique mode is appropriate. One patent actually has three technique-mode entries.

A patent entry, typical of those listed in this bibliography film, is given below.

3,587,730 Milton, R. M.	(patent number and inventor)
1971 June 28	(date of patent issue)
165/110	(patent class/sub-class)
013C	(technique-mode-geometry classification)
Heat Exchange System with Porous Boiling Layer	(title)

TABLE 1.2 Selected Subclass Titles in Patent Class 165, Heat Exchange

Subclass	Subclass Title
1	Process
2	Heating and cooling
68	With external support lets
76	With repair or assembly means
86	Movable heating or cooling surface
109	With agitating or stirring structure
110	With first fluid holder or collector open to second fluid
111	Separate external discharge port for each fluid
115	Trickler
133	With coated, roughened, or polished surface
134	With projector or protective agent
172	Side-by-side tubular structures or tube sections
174	With internal flow director
177	Tubular structure
179	Projecting internal and external heat transfer means
180	Diverse materials
181	With discrete heat transfer means
184	Helical
185	Heat transmitter
DIG. 3	Electrical devices—cooled
DIG. 18	Condensation

Assignees are also identified (opposite the date of issue) for the patents added subsequent to the first edition of the report.

The computer file was used to generate the three output listings contained in Webb et al. (1983). These are:

1. Chronological listing by patent number
2. Alphabetical listing by first listed inventor
3. Technique-mode-geometry classification listing.

Table 1.3 shows the distribution of patents within each technique-mode classification. The 32 multiple listings are reflected in this table. Enhancement techniques for single-phase forced convection comprise 60% of the patents listed in the file; 88% of the patents involve the "passive techniques." Virtually all of the "passive techniques" involve special surface geometries.

Figure 1.10, which shows the number of patents issued by year since 1928, indicates that U.S. patent activity relevant to heat transfer enhancement has continued at a high level since 1965. During the five-year period 1978–1982, more than 20 patents per year were granted.

For those who wish to identify new patents, review of the *Official Patent Gazette* is recommended. The *Official Patent Gazette* is issued biweekly, and it contains

TABLE 1.3   Application of Enhancement Techniques

Geometry	Available	Mode			Materials	Performance Potential
		F	CB	C		
Inside tubes						
Metal coatings	Yes	5	2	5	Al, Cu, St	Hi
Integral fins	Yes	2	1	1	Al, Cu	Hi
Flutes	Yes	4	4	2	Al, Cu	Mod
Integral roughness	Yes	1	2	2	Cu, St	Hi
Wire coil inserts	Yes	1	2	2	Any	Mod
Displaced promoters	Yes	2	2	2	Any	Mod (Lam)
Twisted-tape inserts	Yes	1	2	2	Any	Mod
Outside circular tubes						
Coatings						
Metal	Yes	5	1	3	Al, Cu, St	Hi (B)
Nonmetal	No	5	4	4	"Teflon"	Mod
Roughness (integral)	Yes	3	2	2	Al, Cu	Hi (B)
Roughness (attached)	Yes	2	3	3	Any	Mod (CV)
Axial fins	Yes	1	2	2	Al, St	Hi (CV)
Transverse fins						
Gases	Yes	1	5	5	Al, Cu, St	Hi
Liquids, two-phase	Yes	1	1	1	Any	Hi
Flutes						
Integral	Yes	5	5	1	Al, Cu	Hi
Nonintegral	Yes	5	5	3	Any	Hi
Plate-fin heat exchanger						
Metal coatings	Yes	5	3	5	Al	Hi (B)
Surface roughness	Yes	4	4	4	Al	Hi (B)
Configured channel	Yes	1	1	1	Al, St	Hi
Interrupted fins	Yes	1	1	1	Al, St	
Flutes	No	5	5	4	Al	Mod
Plate-type heat exchanger						
Metal coatings	No	5	4	5	St	Hi (B)
Surface roughness	No	4	4	4	St	Mod (CV)
Configured channel	Yes	1	3	3	St	Hi (CV)

Use Code

1   Common use
2   Limited use
3   Some special cases
4   Essentially no use
5   Not applicable

Heat Transfer Mode

CV   Convection
B   Boiling
C   Condensation

patent abstracts, which are segregated by patent class. The most fruitful classes to review are: 29 (metal working), 62 (refrigeration), 138 (pipes with flow regulators and/or baffles), and 165 (heat exchange). Approximately 60% of the enhancement patents were found in class 165.

Several computer patent search programs are available for search purposes. Engineers must learn keywords favored by patent attorneys to effectively use the patent databases.

1.3.3   Manufacturer's Information

Manufacturers product information is an excellent source for information on enhanced heat transfer. Such sources include manufacturers of enhanced heat transfer tubing, and insert devices. However, much information on enhanced heat transfer products is not reported in manufacturers product brochures. An example is the enhanced heat transfer products used in automobiles. The automotive industry is a prominent user of enhanced heat transfer. However, heat exchanger technology is not discussed in their consumer oriented-product information. The heat exchangers used in the refrigeration industry typically use enhancement. Some of their product information may discuss their enhanced surfaces. However, if the enhanced is not visible in the produce, the manufacturer may choose not to reveal its existence. This is because it is considered as "proprietary information." Chemical process plants also use enhanced surfaces, and these are most certainly proprietary. Hence, an interested person will not gain a very complete perspective of the broad range of "enhanced heat transfer technology" from product information.

An excellent source for industrial information is the patent literature. Industry uses patents to protect their proprietary interests. Unfortunately, you have to consciously seek out the patent information, since much of it is not widely published. Some trade magazines publish current patent information in special industrial areas.

1.4   BENEFITS OF ENHANCEMENT

Special surface geometries provide enhancement by establishing a higher hA per unit base surface area. Three basic methods are employed to increase the hA value:

Method 1: Increase h without an appreciable physical area (A) increase. An example is surface roughness inside the tube shown in Figure 1.2b.

Method 2: Increase A without appreciably changing h. An example is the internally finned tube shown in Figure 1.4b.

Method 3: Increase both h and A. The interrupted fin geometries shown in Figures 1.3A through 1.3E provide a higher heat transfer coefficient than a plain fin. They also provide increased surface area.

For application to a two-fluid tubular heat exchanger, enhancement may be desired for the inner, the outer, or both sides of the tube. If enhancement is applied to the inner and outer tube surfaces, a doubly enhanced tube results. Applications for such doubly enhanced tubes include condenser and evaporator tubes. Depending on the design application, the two-phase heat transfer process may occur on the tube side or shell side. Consider, for example, a shell-and-tube condenser with cooling water on the tube side. The preferred enhancement geometry for the condensing side may be substantially different from that desired for the water side. Hence, one must be aware of possible manufacturing limitations or possibilities that affect independent formation of the enhancement geometry on the inner and outer surfaces of a tube. Figure 1.11 illustrates five basic approaches that may be employed to provide doubly enhanced tubes. Figures 1.11a, 1.11b, and 1.11c allow independent selection of the shell-side and tube-side enhancement geometries. However, the forming process used to make the Figure 1.11d tube-side ridges also deforms the outer tube surface. The tube shown in Figure 1.11a may give good tube-side enhancement but lower shell- side performance than would result from a different shell-side enhancement geometry. For example, in condensation the Figure 1.11a shell-side enhancement would be better.

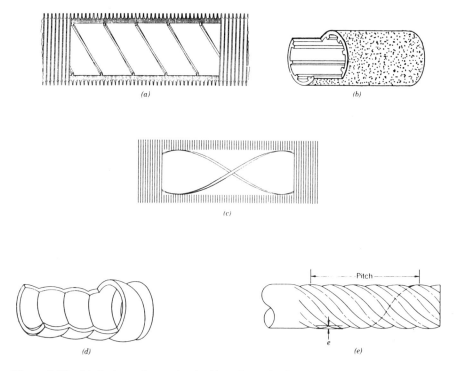

Figure 1.11   Methods used to make doubly enhanced tubes. (a) Helical rib roughness on inner surface and integral fins on outer surface. (b) Internal fins on inner surface and coated (porous boiling surface) on outer surface. (c) Insert device (twisted tape) with integral fins on outer surface. (d) Corrugated inner and outer surfaces. (e) Corrugated strip rolled in tubular form and seam welded. (From Webb [1987].)

Webb [1981] discusses possibilities for manufacture of doubly enhanced condenser tubes. Users should also be aware that it may not be possible to make a given enhancement geometry in all materials of interest. For example, the internally finned tube shown in Figure 1.1b can easily be made as an aluminum extrusion, but it would be extremely difficult to make from a very hard material, such as titanium. Webb [1980] discusses particular enhancement geometries that have been commercially manufactured, and he also discusses their materials of manufacture. More discussion of current manufacturing technology is presented in later chapters. Tables 1.1 and 1.2 show that a variety of enhancement techniques are possible. In a practical sense, the various enhancement techniques may be considered to be in competition with one another. The favored method will provide the highest performance at minimum cost for the material of interest.

1.5 COMMERCIAL APPLICATIONS OF ENHANCED SURFACES

1.5.1 Heat (and Mass) Exchanger Types of Interest

Consider the application of enhancement techniques to four basic heat exchanger types: (1) shell-and-tube or tube banks, (2) fin-and-tube, (3) Plate-fin, and (4) plate-type. The heat transfer modes to be considered are single-phase forced convection of liquids and gases, boiling, and condensation. These heat exchanger types require enhancement for four basic flow geometries:

1. Internal flow in tubes, circular being the most important.
2. External flow along or across tubes, with circular tubes most common.
3. Plate-fin-type heat exchangers made by a stacked construction. This involves flow in noncircular passages.
4. Plate-type heat exchangers which involve flow between formed parallel plates.

Table 1.3 summarizes the present and potential use of spiral surface geometries and other enhancement techniques for the applications of interest. The first part of the table considers application of special surface geometries to the inner and outer surfaces of circular tubes. The latter part of the table considers plate-type and plate-fin heat exchangers. A coding system is employed to define the degree to which an enhancement technique has found application. A dash entry indicates not applicable or not intended to augment that heat transfer mode. The next-to-last column lists typical materials used for commercial manufacture. The last column indicates the author's assessment of the performance potential, independent of whether a commercial manufacturing technique exists. It is evident that a significant number of the available enhancement techniques are commercially utilized.

Enhancement techniques are not limited to heat exchangers. Some types of equipment, such as cooling towers and distillation columns, involve simultaneous heat and mass transfer processes. Enhancement techniques viable for enhancement of heat transfer to gases are equally applicable for convective mass transfer to gases. Convective mass transfer to a gas controls the performance of a cooling tower.

1.5.2   Illustrations of Enhanced Tubular Surfaces

Some of the major commercially used enhanced surfaces are illustrated and discussed. These are broadly representative of commercial technology, but are not meant to be comprehensive.

1.5.2.1   Corrugated Tubes. Figure 1.2a shows corrugated, or "roped," tubes. This tube is formed by pressing a forming tool on the outer surface of the tube. This produces an internal ridge roughness. The tube shown in Figure 1.2a is the popular Wolverine Korodense™ tube.

1.5.2.2   Integral-Fin Tubes. Figure 1.4a shows a standard integral-fin tube. Tubes having 19–26 fins/in. (748–1024 fins/m) and 1.5-mm fin height are routinely used in shell-and-tube exchangers for enhancement of single-phase gases and liquids, for boiling, and for condensation of low-surface-tension fluids, such as refrigerants. A 35-fins/in. (1378-fin/m) integral-fin tube with 0.9-mm fin height is used for condensation of refrigerants. When the outside heat transfer enhancement is sufficiently high, tube-side enhancement is also beneficial. This constitutes a "doubly enhanced" tube. Figure 1.12a shows the Wolverine Turbo-Chil™ tube, which has 1024 fins/m (26 fins/in.) on the outside surface and a 10-start internal rib roughness at 49-degree helix angle.

Jaber and Webb [1993] describe tubes developed by Wolverine and Wieland for steam condensation, a high-surface-tension fluid. High-surface-tension fluids require increased fin pitch. A steam condenser tube may have 11 fins/in. (433 fins/m)

(a)

(b)

Figure 1.12   (a) Wolverine Turbo-Chil™ tube (integral fin outside surface and 10-start helical rib internal rib roughness). (b) Wieland NW™ tube (11 fins/in. on outer surface and a wavy inside roughness).

with 0.80- to 1.0-mm fin height. Figure 1.12b shows the Wieland NW™ tube having 433 fins/m and a wavy inside surface enhancement.

1.5.2.3 Enhanced Condensing Tubes. The tubes shown in Figure 1.13 have a special saw-tooth fin geometry, and they provide higher condensation coefficients than do the standard integral-fin tubes. They were developed for use with refrigerants. Note that the tubes also contain water-side enhancement.

1.5.2.4 Enhanced Boiling Tubes. At least six enhanced boiling tubes are commercially available, three of which are shown in Figure 1.14. One example is the Wolverine Turbo-B™ surface geometry (see Figure 1.14a), which is made by a high-speed thread rolling process. The tubes promote high nucleate boiling at much lower heat fluxes than the integral-fin tubes of Figure 1.4a. The enhanced boiling tubes are typically made with water-side enhancements, as illustrated in Figures 1.13 and 1.14. The tubes shown in Figures 1.1a and Figure 1.14 are also used in process applications. The porous coating shown in Figure 1.1a is commercially known as the UOP High-Flux™ tube, and is frequently applied to a corrugated tube.

(a) 26 FPI

(b) TURBO-C

(c) GEWA-SC

(d) TRED-D

Figure 1.13 Tubes for refrigerant condensation having enhancement on the condensing refrigerant side (outside) and the water side (inside). (a) Standard integral fin (26 fins/in, or 1024 fins/m). (b) Wolverine Turbo-C™. (c) Wieland GEWA-SC™. (d) Sumitomo Tred-19D™.

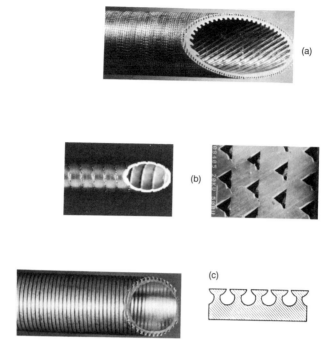

Figure 1.14 The Wolverine Turbo-B™ (a), the Hitachi ThermoExcel-E™ (b), and the Wieland TW™ (c) enhanced boiling tubes.

1.5.2.5 *Tube-Side Enhancements.* A variety of tube-side enhancements exist. Figures 1.13 and 1.14 illustrate several of those commercially used. Internally finned tubes having high fins (e.g., 1 mm or more) are quite expensive to make in copper, but are cheaply made in aluminum by extrusion. Figure 1.15 shows the "microfin" tube, made by Wieland, which is widely used for in direct expansion refrigerant evaporators and condensers. This tube has very short triangular cross-section fins (0.20 mm) high, which are typically made with a helix angle of approximately 15 degrees. Extruded aluminum microfin tubes are also commercially available.

Figure 1.15 The Ripple-fin™ (or "microfin") tube having 0.2-mm-high triangular fins. (Courtesy of Wieland-Werke AG.)

1.5.3 Enhanced Fin Geometries for Gases

Gas-side fins are typically used for heat transfer between a tube-side liquid (or two-phase fluid) and a gas (e.g., air). Air-side fins are used because the air-side heat transfer coefficient is much smaller than that of liquids or two-phase fluids. Plain fins are "old technology." Enhanced surface geometries, such as wavy fins or interrupted fins (Figure 1.3), give a higher performance than do plain fins. Louvered fins (Figure 1.3B) are widely used in automotive heat exchangers having rectangular cross-section tubes. Air-conditioning heat exchangers may use the interrupted fins on round tubes shown in Figure 1.3C or 1.3D. Brazed aluminum, plate-and-fin heat exchangers typically use the fin geometries shown in Figure 1.3A or 1.3B. The interrupted fin geometries shown in Figure 1.3A through 1.3E provide high heat transfer coefficients via the repeated growth and destruction of thin boundary layers.

The geometry shown in Figure 1.3F is typical of enhancements used in rotary regenerators. Use of an interrupted fin geometry would allow leakage between the hot and cold streams, which reduces performance, and allows contamination of the "clean" stream.

1.5.4 Plate-Type Heat Exchangers

These heat exchangers are used to exchange heat between liquids, or between a liquid and a two-phase fluid. Plate exchangers do not use extended surface on the heat transfer plates. Rather, they use corrugated surfaces, which provide secondary flow and mixing to obtain high heat transfer performance (see Figure 1.16).

1.5.5 Cooling Tower Packings

Cooling towers cool a liquid (gravity drained liquid film) by transferring heat from the water film to air passing through the cooling tower "packing." This involves a simultaneous heat and mass transfer process. The water is cooled principally by evaporation of a small fraction of the water. High heat and mass transfer coefficients are provided for the air flow by special packing geometries, such as illustrated in Figure 1.17. The limiting mass transfer resistance is in the gas phase. The packing shown in Figure 1.17a enhances the mass transfer coefficient by the development and destruction of thin gas boundary layers using the same principle employed in the offset strip-fin heat exchanger geometry of Figure 1.3A. The corrugated packing shown in Figure 1.17b provides mixing of the air flow, while providing a large surface area of liquid film. The geometry shown in Figure 1.17b provides interrupted gas boundary layers, quite similar to that of the offset strip fin heat exchanger geometry shown in Figure 1.3a.

1.5.6 Distillation and Column Packings

Distillation columns also involve heat and mass transfer between two fluid mixtures. One fluid exists as a liquid film, and the other fluid is a gas. Typical column

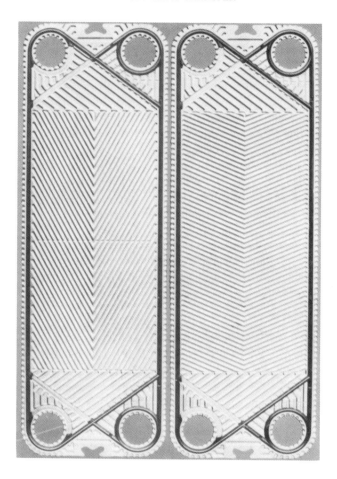

Figure 1.16 Corrugated plate geometry used in plate-type heat exchangers. (Courtesy of Alfa-Laval.)

packings are shown in Figure 1.18. These packings provide enhancement for the gas flow in the same manner as for cooling tower packings.

1.5.7 Factors Affecting Commercial Development

It is appropriate to view enhancement techniques as "second generation" heat transfer technology. Such new technology normally requires several phases of development for successful commercialization. These include the following:

1. Basic performance data for heat transfer and pressure drop, if applicable, must be obtained. General correlations should be developed to predict heat transfer and pressure drop as a function of the geometrical characteristics.

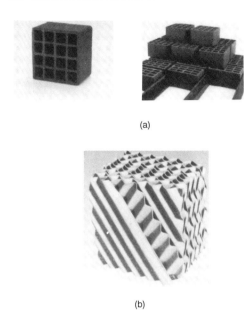

(a)

(b)

Figure 1.17 Enhanced film-type cooling tower packing geometries. (a) Ceramic block type. (Courtesy of Ceramic Cooling Tower Co. affiliate of Baltimore Aircoil Company, Inc.) (b) Plastic corrugated packing. (Courtesy of Munters Corp.)

2. Design methods must be developed to facilitate selection of the optimum surface geometry for the various techniques and particular applications.
3. Manufacturing technology and cost of manufacture must be available for the desired surface geometry and material.
4. Pilot plant tests of the proposed surface are required to permit a complete economic evaluation and to establish the design, including long-term fouling and corrosion characteristics.

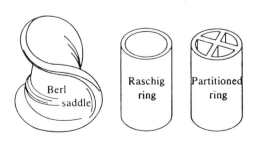

Figure 1.18 Packings for distillation columns. (From Edwards et al. [1973].)

Some enhancement techniques have completed all steps required for commercialization. Examples are (a) extended surface for fin-and-tube heat exchangers and (b) external low-fin tubing. However, techniques such as internally roughened tubes require further development. Table 1.4 attempts to summarize the status of technology development for several of the more important enhancement techniques applied to circular tubes. Commercialization represents the ultimate stage of development; however, even commercial products require additional development work.

It is recognized that several enhancement techniques are in competition with one another. For application to a given heat transfer mode, some techniques will appear more interesting than others. Therefore, a continued high level of interest for all competing techniques cannot be expected.

The development of extended surfaces on the outside of tubes has proceeded on a proprietary basis and has been successfully commercialized. However, additional work directed at developing design methods for optimum surface selection is worthwhile. Extended surface for boiling and condensing applications is similarly well established. Additional work is needed on high-performance boiling and condensing surfaces. Most of the previous work has focused on the outer surface of the tube. Very little has been done to identify high-performance surfaces for boiling and condensing inside tubes. The development of high-performance techniques which allow low manufacturing cost is sorely needed for a variety of materials.

With regard to internally roughened tubes, a considerable amount of experimen-

TABLE 1.4   Status of Technology Development

Geometry	Forced Convection	Boiling	Condensation
Inside tubes			
Coatings		3,5	1
Roughness	2,3,5	1	1
Internal fins	1,2,3,5	1,5	1
Flutes	5	5	1,5
Insert devices	5	1,5	1,5
Outside tubes			
Metal coatings		3,5	1
Nonmetal coatings		1	4
Roughness	1,2	1,2,5	1
Extended surface	5	5	5
Flutes			5

Code
1 Basic performance data
2 Design methods
3 Manufacturing technology
4 Heat exchanger application
5 Commercialization

tal data exist for single-phase forced convection. The major need is development of analytical design methods for optimized surface geometries, with additional work on manufacturing technology.

The further development of any technique will depend on two factors: (1) its performance potential and (2) manufacturing cost. Presently, cost-effective manufacturing technology is probably the most significant barrier to commercial application of high-performance enhancement techniques. Researchers appear intent on identifying surface geometries which offer the highest performance. It is suggested that more communication between researchers and manufacturing engineers is needed to identify the most cost-effective concepts.

1.6 DEFINITION OF HEAT TRANSFER AREA

Enhanced tubes may involve an area increase, relative to that of a plain tube. In this work, we will define the heat transfer coefficient as follows:

1. *Tube Side Enhancement:* The h value is defined on the basis of the *nominal* surface area. The nominal surface area is defined as $A/L = \pi d_i$, where d_i is the tube inside diameter to the base of the enhancement. This definition allows direct comparison of the h value of tube-side enhancements, relative to the h value of a plain tube. Some authors define the h value for an internally finned tube in terms of the total surface area.

2. *Extended Surfaces for Gases:* We define the h value on the basis of the total surface area (A) of the fins and base surface, to which the fins are attached. This is consistent with industrial practice.

1.7 POTENTIAL FOR ENHANCEMENT

For two-fluid heat exchangers, the ratio of the thermal resistances between the two-fluid streams is of primary importance in determining whether enhancement will be beneficial. Enhancement should be considered for the stream that has the controlling thermal resistance. If the thermal resistances of both streams are approximately equal, one should consider enhancing both sides of the heat exchanger. Practical considerations should also include the effect of design fouling resistances on the potential for enhancement.

In addition to two-fluid heat exchangers, cooling is required for equipment or devices having internal heat generation. Examples are nuclear fuel elements, electrical heaters, or electronic components, such as circuit boards. Use of enhancement here will result in either of two benefits. The device may support a greater heat flux for fixed surface temperature. Alternatively, the device may operate at a lower temperature, while operating at the same heat flux. The second benefit is of great importance for electronic equipment, since the component service life is inversely

related to the operating temperature. This problem involves operation with a specified heat flux boundary condition. Because a second fluid is not involved, as in
PEC Example 1.1, any increase of the heat transfer coefficient will be beneficial for
the case of interest.

In Chapter 2, we will discuss performance evaluation criteria used to evaluate the
benefits of enhancement. We will provide a number of examples in the book that
illustrate applications of enhanced surfaces. The example problems will be called
"PEC examples."

1.7.1 PEC Example 1.1

Consider a shell-and-tube heat exchanger that cools engine oil on the shell side with
cooling water on the tube side. The exchanger currently uses 19-mm-outer-diameter
plain copper tubes (0.8-mm wall thickness). Would enhancement be beneficial?
Which side of the tube should be enhanced? What type of enhancement geometry
would offer potential?

We must first determine the heat transfer coefficients for each of the fluid
streams. Using available correlations, the designer has determined that the water-
side heat transfer coefficient is 10 kW/m²-K, and the heat transfer coefficient for the
oil is 1.0 kW/m²-K. For simplicity, the tube wall and fouling thermal resistances are
ignored. The overall thermal resistance is

$$L/UA = L/(h_i A_i) + L/(h_o A_o)$$
$$L/UA = 1/(10 \times 0.019\pi) + 1/(1 \times 0.0174\pi)$$
$$19.96 = 1.68 + 18.29$$

Thus 91.6% (18.29/19.96) of the total thermal resistance is on the shell (oil)
side. There would be no benefit derived from adding enhancement to the tube
(water) side. However, substantial benefit would result from shell-side enhancement. A possible candidate geometry is the low, integral-fin tube. It is likely that the
shell side heat transfer coefficient may be enhanced 200% or more. Assuming a
200% increase of the shell-side heat transfer coefficient, the total thermal resistance
(L/UA) would then be

$$L/UA = 1/(10 \times 0.019\pi) + 1/(3 \times 0.0174\pi)$$
$$L/UA = 1.68 + 6.10 = 7.78$$

The total tubing length in the enhanced tube heat exchanger would be 0.39
(7.78/19.96) that of the original design. No consideration was given to the oil
pressure drop. Pressure drop considerations are invariably part of the design analysis for enhancement considerations.

1.7.2 PEC Example 1.2

Consider a micro-chip on an electronic circuit board. The device is rated for 0.1 W at 60°C. The chip currently dissipates heat to ambient air at 30°C. If the surface temperature can be lowered to 50°C, the expected life of the chip may be increased by 25%. Assume that an enhancement concept will increase the surface heat transfer coefficient by 50%. Calculate the operating surface temperature of the chip.

$$q/A = h_{\text{old}}(60 - 30) = h_{\text{new}}(T_{s,\text{new}} - 30)$$

Using $h_{\text{new}}/h_{\text{old}} = 1.5$ in the above equation gives $T_{s,\text{new}} = 52°C$. The chip surface temperature will be reduced from 60°C to 52°C. By being a little more clever in the enhancement design, we may be able to get the surface temperature down to 50°C!

1.8 REFERENCES

Bergles, A. E., Nirmalan, V., Junkhan, G. H. and Webb, R. L., 1983. "Bibliography on Augmentation of Convective Heat and Mass Transfer II," *Heat Transfer Laboratory Report HTL-31*, ISU-ERI-Ames-84221, Iowa State University, December.

Bergles, A. E., Jensen, M. K., Somerscales, E. F. C., and Manglik, R. M., 1991. "Literature Review of Heat Transfer Enhancement Technology for Heat Exchangers in Gas Fired Applications," GRI Report GRI 91–0146, Gas Research Institute, Chicago, IL.

Bergles, A. E., and Webb, R. L., 1985. "A Guide to the Literature on Convective Heat Transfer Augmentation," in *Advances in Enhanced Heat Transfer—1985*, S. M. Shenkman, J. E. O'Brien, I. S. Habib, and J. A. Kohler, Eds., ASME Symposium, Vol. HTD-Vol. 43, pp. 81–90.

Edwards, D. K., Denny, V. E., and Mills, A. F., 1973. *Transfer Processes—An Introduction to Diffusion, Convection, and Radiation*, Hemisphere Pub. Corp., New York, p. 320.

Jaber, M. H., and Webb, R. L., 1993. "An Experimental Investigation of Enhanced Tubes for Steam Condensers," *Experimental Heat Transfer*, Vol. 6, pp. 35–54.

Webb, R. L., 1980. "Special Surface Geometries for Heat Transfer Augmentation," Chapter 7 in *Developments in Heat Exchanger Technology—1*,. D. Chisholm, Ed., Applied Science Publishers, London.

Webb, R. L., 1981. "The Use of Enhanced Surface Geometries in Condensers," *Power Condenser Heat Transfer Technology*, P. J. Marto and R. H. Nunn, Eds., Hemisphere Publishing Corp., Washington, D.C., pp. 287–324.

Webb, R. L., 1987. "Enhancement of Single-Phase Heat Transfer," Chapter 17 in *Handbook of Single-Phase Heat Transfer*, S. Kakac, R. K. Shah, and W. Aung, Eds., John Wiley & Sons, New York, 62 pages.

Webb, R. L. and Bergles, A. E., 1983. "Heat Transfer Enhancement: Second Generation Technology," *Mechanical Engineering*, Vol. 105, pp. 60–67.

Webb, R.L., Bergles, A. E. and Junkhan, G. H., 1983. "Bibliography of U. S. Patents on Augmentation of Convective Heat and Mass Transfer," *Heat Transfer Laboratory Report HTL-32*, ISU-ERI-Ames-81070, Iowa State University, December.

1.9 NOMENCLATURE

A	Heat transfer surface area, m^2 or ft^2
c_p	Specific heat, J/kg-K or Btu/hr-F
d	Tube diameter, m or ft
E_h	h/h_p at constant Re, dimensionless
e	Roughness height, m or ft
h	Heat transfer coefficient, W/m^2-K or Btu/hr-ft^2-F
L	Length of flow path in heat exchanger, m or ft
ΔT_m	Mean temperature difference, K or F
Q	Heat transfer rate, W or Btu/hr
T	Temperature, K or F
t	Tube wall thickness, m or ft
U	Overall heat transfer coefficient, W/m^2-K or Btu/hr-ft^2-F

Greek Symbols

π	Surface efficiency, dimensionless

Subscripts

1	Fluid 1
2	Fluid 2
i	Tube inner surface
o	Tube outer surface
p	Plain surface
s	Smooth tube
w	Tube wall

2

HEAT TRANSFER FUNDAMENTALS

2.1 INTRODUCTION

This chapter presents a summary of heat transfer fundamentals and heat exchanger design theory, and it introduces the designer to the broad possibilities for the use of enhancement technology. Heat transfer enhancement concepts may be considered for application to virtually any type of heat exchanger or heat exchange device.

The heat exchanger design problem is considerably more complex than the calculation of the heat transfer and pressure drop characteristics of the exchanger. Heat exchanger calculations may be classified by two basic problems:

1. *Rating Problem:* The heat exchanger type, size, and surface geometry are specified. The process conditions (e.g., flow rate, entering fluid conditions, and fouling factors) are also specified. This problem requires calculation of the heat transfer rate and pressure drop of each stream.
2. *Sizing Problem:* This problem involves calculation of the heat exchanger size required for specified process requirements (flow rate, entering fluid conditions, and allowable pressure drops).

Both problems require use of fundamental relations to determine the thermal design and pressure drop characteristics of the exchanger. The sizing problem is more complex than the rating problem because the heat exchanger type and surface geometry must be selected before the thermal analysis can be performed. The preliminary decisions required are (1) heat exchanger materials, (2) heat exchanger type and flow arrangement, and (3) heat transfer surface geometries.

Table 2.1 summarizes the considerations necessary to establish the heat ex-

33

TABLE 2.1 Heat Exchanger Design Considerations

Design Specifications	Design Selection
Process Requirements a. Fluid compositions, and inlet flow conditions (flow rate, temperature and pressure) b. Heat duty or required exit temperatures c. Allowable pressure drops	**Heat Exchanger Materials** a. Fluid temperatures and pressures b. Corrosive characteristics of fluid–material combination
Operating and Maintenance Considerations a. Fouling potential and method of cleaning b. Failure due to corrosion, thermal stress, vibration or freezing c. Repair of leaks d. Part load operating characteristics	**Heat Exchanger Type** a. Design pressure and fluid temperatures b. Corrosion, stress, vibration, and freezing considerations c. Fouling potential and cleaning possibilities d. First cost, operating and maintenance costs
Size and Weight Restrictions a. Frontal area, length, or height b. Possible weight restrictions	**Heat Transfer Surface Geometries** a. Thermal resistance ratio of fluids b. Potential for use of enhanced surfaces c. Fouling potential and cleaning possibilities d. Unit cost of heat transfer surface

changer design concept. These selections result from consideration of (a) given or implied "design specifications" and (b) the "design selections" as influenced by the "design specifications."

Having established the exchanger type, the designer then performs the thermal and hydraulic analysis. However, a single calculation will not directly yield the "optimum" design because different fluid velocities or heat transfer surface geometries will yield different results. Typically, the designer must evaluate several possibilities in the search for an "optimum" design. The selection of an optimum design requires that the designer consciously establish which variables (the objective function) are to be optimized [e.g., first cost, operating cost (or, alternatively "life-cycle costs"), and size dimensions].

The design optimization methods are discussed by Shah et al. [1978]. The optimization process involves a parametric analysis to determine how the specified objective function is influenced by trade-offs among the possible design variables. Shah et al. present a quite detailed discussion of the general methodology of the heat exchanger sizing problem and the selection of an optimum design for a given application.

2.2 HEAT EXCHANGER DESIGN THEORY

This section presents a summary of heat exchanger thermal analysis. The development of the fundamental relations may be found in any heat transfer text.

2.2.1 Thermal Analysis

Consider the design of a two-fluid heat exchanger as illustrated in Figure 2.1. Although a counterflow exchanger is illustrated, our discussion applies to any flow configuration.

The heat exchanged between the two fluids is dependent on three specifications

1. Thermodynamic specifications as defined by the hot and cold fluid flow rates and their entering and exit temperatures. Thus,

$$Q = C_h\,(T_{h1} - T_{h2}) = C_c(T_{c2} - T_{c1}) \tag{2.1}$$

Heat exchanger analysis normally uses the definition $C_h = W_h c_{ph}$ because these two terms occur as a product. The term C_h is defined as the "capacity rate" of the hot fluid.

2. The heat transfer rate equation $dQ = UdA(T_h - T_c)$. The rate of heat exchange is dependent on the overall heat transfer coefficient between the two

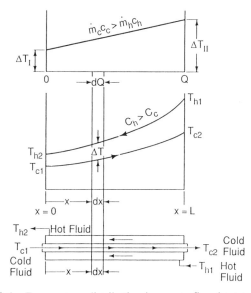

Figure 2.1 Temperature distribution in counterflow heat exchanger.

fluids. The heat exchanger surface area required to satisfy the thermodynamic specifications of Equation 2.1 is obtained by integrating the heat transfer rate equation over the length (or area) of the exchanger. Thus, for constant U,

$$Q = U \int_A (T_h - T_c) \, dA \qquad (2.2)$$

The "effective mean temperature difference," averaged over the total surface area, is

$$\Delta T_m = \frac{1}{A} \int_A (T_h - T_c) \, dA \qquad (2.3)$$

Substitution of Equation 2.3 in Equation 2.2 gives

$$Q = UA\Delta T_m \qquad (2.4)$$

3. The product of the UA terms is the "overall heat conductance" of the heat exchanger. Stated in thermal resistance terms, $1/UA$ is the overall thermal resistance to heat transfer between the two fluids.

The overall resistance, $1/UA$, is calculated by summing the individual thermal resistances given in Equation 2.5 below. Equation 2.5 contains five individual resistances where

$$\frac{1}{UA} = \frac{1}{(\eta hA)_h} + \left(\frac{R_f}{A}\right)_h + \frac{t}{k_w A_w} + \frac{1}{(\eta hA)_c} + \left(\frac{R_f}{A}\right)_c \qquad (2.5)$$

where subscripts h and c refer to the hot and cold streams. The first and fourth terms account for the convective resistance between the flowing fluids and the pipe wall. The third term is the conduction resistance of the solid wall which separates the streams. The second and fifth terms are the fouling resistance on the hot and cold heat transfer surfaces. The overall coefficient, U, may be defined in terms of the surface area of either the hot or cold surface; thus, $UA = U_h A_h = U_c A_c$.

The effective temperature difference ΔT_m is a function of the heat exchanger flow geometry (e.g., counterflow or cross flow) and the degree of fluid mixing within each flow stream. For counterflow and parallel-flow geometries, substitution of Equation 2.1 in equation 2.3 and integration yields the following result for result for ΔT_m, which is commonly known as the *logarithmic mean temperature difference* (LMTD).

$$\text{LMTD} = \frac{\Delta T_{\mathrm{I}} - \Delta T_{\mathrm{II}}}{\ln(\Delta T_{\mathrm{I}} / \Delta T_{\mathrm{II}})} \qquad (2.6)$$

Using the terminology ΔT_{I} and ΔT_{II} as the difference between the fluid temperatures at each end of the exchanger, Equation 2.6 applies to both parallel-flow and counterflow configurations. For more complex geometries such as cross-flow or multipass-flow configurations, integration of Equation 2.3 yields more complex expressions for ΔT_m. For such cases it is customary to define a correction factor

$$F \equiv \frac{\Delta T_m}{\mathrm{LMTD}} \qquad (2.7)$$

where the LMTD is calculated for counterflow. The F correction is presented in graphical terms as illustrated in Figure 2.2 for one fluid "mixed" and the other fluid "unmixed." The term "unmixed" means that a fluid stream (e.g., the hot stream) passes through the heat exchanger in individual flow channels or tubes with no fluid mixing between adjacent flow channels. The F-correction factor for "mixed" flow means that there are no thermal gradients normal to the flow. The parameters P and Z are defined as

$$Z \equiv \frac{C_c}{C_h} = \frac{T_{h1} - T_{h2}}{T_{c2} - T_{c1}} \qquad (2.8)$$

$$P \equiv \frac{T_{c2} - T_{c1}}{T_{h1} - T_{c1}} \qquad (2.9)$$

The term Z is the ratio of the capacity rates of the cold and hot streams. The term P is defined as the "temperature effectiveness of the cold stream," which is the temperature rise of the cold stream divided by the difference between the inlet temperatures of the two fluids. Multipass exchangers have a combination of parallel flow and counterflow in alternate passes. Thus, F will depend on (a) the "pass arrangement," (b) whether fluid mixing occurs within a pass, and (c) the conditions of mixing between passes. Charts to determine F are given in heat transfer textbooks, such as Holman [1986], Incropera and DeWitt [1990], or Kays and London [1984].

2.2.2   Heat Exchanger Design Methods

There are alternate methods for calculating the heat transfer rate between the two fluids. These are the UA–LMTD and the effectiveness–NTU methods.

The UA–LMTD Method: The UA–LMTD method follows directly from Equations 2.4 and 2.9. Substitution of Equation 2.9 in Equation 2.4 gives

$$Q = F \cdot UA \cdot \mathrm{LMTD} \qquad (2.10)$$

For a rating calculation, one calculates UA using Equation 2.5. Or, for a sizing problem, one calculates UA per unit area or length. Second, the LMTD is calcu-

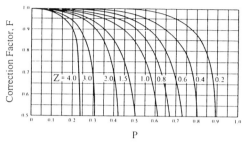

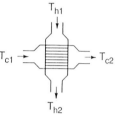

$$P = \frac{T_{c2}\text{-}T_{c1}}{T_{h1}\text{-}T_{c1}} \quad Z = \frac{T_{h1}\text{-}T_{h2}}{T_{c2}\text{-}T_{c1}}$$

When hot fluid flows in tubes
interchange subscripts c and h.

Figure   2.2   Correction factor for crossflow heat exchanger—one fluid mixed and one fluid unmixed.

lated. Third, the geometry correction, F, is determined as noted below Equation 2.7.

Effectiveness–NTU Method: The heat exchanger effectiveness, ϵ, is defined as

$$\epsilon = \frac{\text{actual heat transfer}}{\text{maximum possible heat transfer}} = \frac{Q_{\text{actual}}}{Q_{\text{max}}} \tag{2.11}$$

where Q_{actual} is given by the heat balance equations (Equation 2.1). The maximum possible heat transfer will occur in a counterflow heat exchanger of infinite area if one fluid undergoes a temperature change equal to the maximum temperature difference available, $\Delta T_{\text{max}} = T_{h1} - T_{c1}$.

The calculation for Q_{max} is based on the fluid having the smaller capacity rate, C_{min}, in order to satisfy the first law of thermodynamics. Thus,

$$Q_{\text{max}} = C_{\text{min}}\Delta T_{\text{max}} = C_{\text{min}}(T_{h1} - T_{c1}) \tag{2.12}$$

Substituting Equation 2.12 in Equation 2.11, we obtain

$$Q_{\text{actual}} = \epsilon C_{\min}(T_{h1} - T_{c1}) \tag{2.13}$$

The heat exchanger effectiveness depends on the flow geometry and pass arrangement (e.g., counterflow or cross flow). For a given flow geometry, the effectiveness is a function of two dimensionless quantities, $UA/C_{\min}$ = NTU and $C_{\min}/C_{\max}$ = R. The effectiveness equation is obtained by algebraic manipulation of the equations developed in calculation of the LMTD. The development of the effectiveness equations for parallel flow and counterflow are given in heat transfer texts (e.g., Holman [1986] or Incropera and DeWitt [1990]). The effectiveness for a counterflow heat exchanger is shown in Figure 2.3 and is given by the algebraic relation

$$\epsilon = \frac{1 - e^{-\text{NTU}(1-R)}}{1 - Re^{-\text{NTU}(1-R)}} \tag{2.14}$$

The dimensionless ratio $UA/C_{\min}$ is called the *number of heat transfer units* (NTU). It is indicative of the heat exchanger physical size. The capacity rate ratio, R, is equal to zero for an evaporator or condenser if the fluid remains at a constant temperature during the phase change. If $R = 0$, the ε–NTU relation is independent of the heat exchanger flow geometry. Thus, all flow geometries have the same ε–NTU relation when $R = 0$. Figure 2.3 shows that the required NTU increases for increasing values of effectiveness. The NTU asymptotically approaches infinity as the effectiveness approaches its thermodynamic limit (ε = 1 for counterflow).

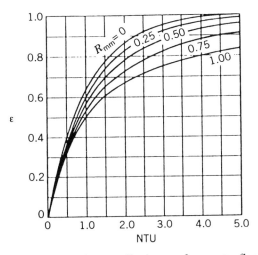

Figure 2.3 Heat exchanger effectiveness for counter-flow exchanger.

For a given NTU, counterflow provides the highest effectiveness whereas parallel flow provides the lowest effectiveness, with cross flow providing intermediate values. Cross flow is frequently used for gas–liquid heat exchangers, because it is impractical to design them for counterflow. Shell-and-tube heat exchangers frequently use some form of multipass cross-counterflow.

The F-correction factors and ϵ–NTU equations for "mixed" and "unmixed" flow are simply limiting thermodynamic definitions. Heat transfer textbooks typically consider banks of tubes as "mixed" flow. It is unlikely that the flow in a tube bank has no temperature gradients normal to the flow. Furthermore, interrupted fin geometries may permit some mixing between lateral flow channels, but it is unlikely that it is fully mixed. Use of the "unmixed" relations will result in overprediction, if some lateral fluid mixing occurs. Webb and DiGiovanni [1989] discuss this concern and provide a method to predict the performance for partially mixed conditions. However, the designer must estimate the mixing fraction.

2.2.3   Comparison of LMTD and NTU Design Methods

The two design methods are equivalent and differ only in the algebraic form of the resulting equations. Which method is used may depend on the designer's preference. Both design methods offer the same relative ease for the sizing problem. However, the NTU is much easier to use for the rating problem. Because at least one fluid exit temperature is unknown in a rating problem, the LMTD is not directly calculable. In this case a trial-and-error solution is required to obtain the LMTD and the correction factor F.

The NTU method allows physical interpretation of the thermodynamic performance of the heat exchanger not provided by the LMTD method. Also, the ϵ–NTU relations are readily available in algebraic form, needed for digital computer calculations.

2.3   FIN EFFICIENCY

Extended surfaces or fins are frequently employed in heat exchangers. Fins will be beneficial if they are applied to the fluid stream having the dominant thermal resistance. The fins provide reduced thermal resistance for this flow stream by providing increased surface area. The heat transfer coefficient on the extended surfaces may be either higher or lower than that which would occur on the unfinned surface. For example, the use of low radial fins on horizontal tubes with condensation provides both an area increase and increased heat transfer coefficient. However, fins used in single-phase forced convection may offer a modest reduction of the coefficient depending on the fin spacing. Because of the temperature gradient in the fin material over its length, a finned surface will not transfer as much heat as a fin of infinite thermal conductivity material. Thus, the heat conductance of a finned surface (hA) must be multiplied by a fin efficiency factor to account for the temperature gradient in the fin. The fin efficiency, η_f, is defined as the ratio of the actual heat transfer from the fin to that which would occur if the entire fin were at its base temperature.

$$\eta_f = \frac{Q}{hA_f(T_w - T_\infty)} \tag{2.15}$$

The efficiency of a fin is a function of its cross-sectional shape, its length, and the geometry of the base surface. Figure 2.4 gives the efficiency of plain and circular fins of uniform cross section. This chart is valid for an adiabatic fin tip boundary condition. Harper and Brown [1922] have shown that fins which have convection from their tips may be determined from Figure 2.4, provided that the fin length is modified to the following value: $L_c = L + t/2$. Fin efficiency equations for a number of specialized fin geometries are discussed in detail by Kern and Kraus [1972].

A finned surface heat exchanger consists of the secondary finned surface and the primary surface to which the fins are attached. A second term, called the "total surface efficiency," is defined to account for the efficiency of the composite structure consisting of the fins and the base surface. The total surface efficiency is defined as

$$\eta \equiv 1 - (1 - \eta_f)\frac{A_f}{A} \tag{2.16}$$

This definition assumes that the heat transfer coefficient on the finned surface is the same as on the base surface.

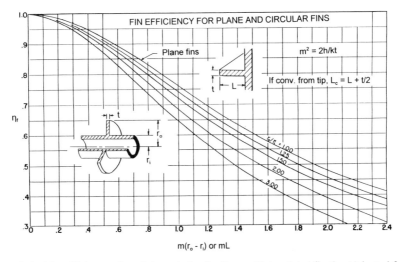

Figure 2.4 Fin efficiency of straight and circular fins, with insulated fin tip. (Adapted from Kays London [1984].)

2.4   HEAT TRANSFER COEFFICIENTS AND FRICTION FACTORS

The performance of enhanced heat transfer surfaces is frequently compared to that of a plain surface. This section provides (a) recommended design equations for heat transfer and friction factor of smooth surfaces in single-phase forced convection and (b) the heat transfer coefficient for boiling and condensation inside and outside tubes. The equations presented are those commonly used in heat exchanger design applications and may be found in Holman [1986] or Incropera and DeWitt [1990].

2.4.1   Laminar Flow Over Flat Plate

The Blasius solution for the average Stanton number with laminar flow over a flat plate of length L, as reported in Chapter 7 of Incropera and DeWitt [1990], is

$$\overline{St} = 0.664 \mathrm{Re}_l^{-0.5} \mathrm{Pr}^{-2/3} \qquad (2.17)$$

The average friction factor for the plate is given by

$$\overline{f} = 1.328\ \mathrm{Re}_l^{-0.5} \qquad (2.18)$$

Writing Equation 2.17 in terms of the j factor ($\mathrm{StPr}^{2/3}$) shows that $j = f/2$, which satisfies Reynolds analogy. Reynolds analogy is discussed in Section 2.6.

2.4.2   Laminar Flow in Ducts

Table 2.2 provides friction factors and Nusselt numbers for circular and noncircular ducts. The Nusselt number is shown for two thermal boundary conditions, namely, constant heat flux (subscript H) and constant wall temperature (subscript T). The Nusselt number values (based on hydraulic diameter for the noncircular ducts) are listed in order of decreasing Nusselt number. By comparing the Nusselt number with the duct cross-sectional shape, one may quickly discern how the duct shape affects Nusselt number. The table also shows that the ratio of j/f decreases with decreasing Nusselt number. High-aspect-ratio rectangular channels yield higher heat transfer coefficients and higher heat transfer per unit of pressure drop (j/f) than do circular tubes and triangular channels.

Figure 7.6 in Chapter 7 shows entrance region solutions (simultaneous velocity and thermal boundary layer development) for laminar flow in circular tubes for q_w = constant and T_w = constant. Figure 5.26 in Chapter 5 shows $\mathrm{Nu}_m/\mathrm{Nu}_{fd}$ versus the entrance region parameter, x_d^* [$= x/(d_i \mathrm{Re}_d \mathrm{Pr}]$, for circular and noncircular ducts with the T_w = constant boundary condition. Figure 5.27 shows the parameter $K/K(\infty)$, which is the pressure drop increment associated with the entrance region. This term is used with Equation 5.25. Bhatti and Shah [1987] give numerous solutions for developing laminar flow for a variety of geometries and for several thermal boundary conditions.

TABLE 2.2 Fully Developed Laminar Flow Solutions

Geometry		Nu_H	Nu_T	$f\,Re$	$K(\infty)$	j_H/f	j_T/f	L_{hy}^{+}
▭	$\dfrac{2b}{2a}=0$	8.235	7.541	24.00	0.686	0.386	0.354	0.0056
$2b$ ▭ $2a$	$\dfrac{2b}{2a}=\dfrac{1}{8}$	6.490	5.597	20.585	0.879	0.355	0.306	0.0094
$2b$ ▭ $2a$	$\dfrac{2b}{2a}=\dfrac{1}{6}$	6.049	5.137	19.702	0.945	0.346	0.294	0.0010
$2b$ ▭ $2a$	$\dfrac{2b}{2a}=\dfrac{1}{4}$	5.331	4.439	18.233	1.076	0.329	0.274	0.0147
◯		4.364	3.657	16.00	1.24	0.307	0.258	0.038
$2b$ ▭ $2a$	$\dfrac{2b}{2a}=\dfrac{1}{2}$	4.123	3.391	15.548	1.383	0.299	0.245	0.0255
$2b$ □ $2a$	$\dfrac{2b}{2a}=1$	3.608	3.091	14.227	1.552	0.286	0.236	0.0324
$2a$ △ $2b$	$\dfrac{2a}{2b}=\dfrac{\sqrt{3}}{2}$	3.111	2.47	13.333	1.818	0.263	0.209	0.0398
$2b$ △ $2a$	$\dfrac{2b}{2a}=\dfrac{\sqrt{3}}{2}$	3.014	2.39	12.630	1.739	0.269	0.214	0.0408
$2b$ △ $2a$	$\dfrac{2b}{2a}=2$	2.880	2.22	13.026	1.991	0.249	0.192	0.0443
△	$\dfrac{2b}{2a}=.25$	2.600	1.99	12.622	2.236	0.232	0.178	0.0515

$$j/f \equiv \frac{Nu\,Pr^{-1/3}}{f\,Re}\,; j_H + j_T \text{ for } Pr = 0.7$$

$H \sim$ heat flux boundary condition
$T \sim$ constant temperature boundary condition

2.4.3 Turbulent Flow in Ducts

The Petukhov [1970] equation is recommended as the most precise available equation. It is

$$\text{St} = \frac{f/2}{1.07 + 12.7(f/2)^{1/2}(\text{Pr}^{2/3} - 1)} \tag{2.19}$$

Webb [1971] compares the ability of Equation 2.19 to predict heat transfer data in smooth tubes. He shows that Equation 2.21 predicts smooth tube, constant property data of four investigators ($0.7 \leq \text{Pr} \leq 75$) within $\pm 8\%$. One may account for the effect of fluid property variations across the boundary layer using the correction factors given in Tables 2.3 and 2.4 and explained in Section 2.5. Petukhov [1970] recommends that the smooth tube friction factor be predicted by

$$f = (1.58 \ln \text{Re}_d - 3.28)^{-2} \tag{2.20}$$

The Dittus–Boelter equation is frequently used for quick, approximate calculations. However, it is not as accurate as the Petukhov equation. It is

$$\text{Nu}_d = 0.023 \, \text{Re}_d^{0.8} \text{Pr}^n \tag{2.21}$$

where the Prandtl number exponent (n) is 0.4 for heating and 0.3 for cooling. The friction factor is frequently predicted by the approximate Blasius equations,

$$f = 0.079 \text{Re}_d^{-0.25}, \qquad \text{Re} < 50{,}000 \tag{2.22}$$

$$f = 0.046 \text{Re}_d^{-0.2}, \qquad \text{Re} > 50{,}000 \tag{2.23}$$

The turbulent duct flow equations are also applicable to noncircular ducts and axial flow in rod bundles by using the hydraulic diameter in place of the tube diameter.

2.4.4 Tube Banks (Single-Phase Flow)

The correlations of Zukauskas [1972] for heat transfer and friction for flow normal to tube banks are recommended. These correlations are given by Incropera and DeWitt [1990] and will not be repeated here.

2.4.5 Film Condensation

The Nusselt analysis as described by Incropera and DeWitt [1990] provides equations for film condensation on vertical plates, inclined plates, and horizontal tubes. The heat transfer coefficient is the average value over the plate length or over the tube diameter. The average condensation coefficient on a vertical plate with a gravity-drained laminar film is given by

$$\bar{h} = 0.943 \left(\frac{k^3 g(\rho_l - \rho_v)\lambda}{v_l \Delta T_{vs} L} \right)^{1/4} \tag{2.24}$$

For an inclined plate, one simply replaces the gravity force by $(g \sin \theta)$, where θ is the plate inclination angle from the vertical direction. Equation 2.24 may be written in terms of the condensate Reynolds number draining from the bottom of the plate, Re_L:

$$\bar{h} = 1.47 \left(\frac{k^3 \rho_l(\rho_l - \rho_v)g}{\mu_l^2} \right)^{1/3} \mathrm{Re}_L^{-1/3} \tag{2.25}$$

The condensate Reynolds number is defined as $\mathrm{Re}_L = 4\Gamma/\mu_l$, where Γ is the condensate flow rate per unit plate width.

The condensation coefficient for laminar film condensation on a single, horizontal tube is given by

$$\bar{h} = 0.728 \left(\frac{k^3 g(\rho_l - \rho_v)\lambda}{v_l \Delta T_{vs} d} \right)^{1/4} \tag{2.26}$$

Equation 2.26 may be written in terms of the condensate Reynolds number draining from the tube.

$$\bar{h} = 1.51 \left(\frac{k^3 \rho_l(\rho_l - \rho_v)g}{\mu_l^2} \right)^{1/3} \mathrm{Re}_L^{-1/3} \tag{2.27}$$

The condensate Reynolds number is again defined as $\mathrm{Re}_L = 4\Gamma/\mu_l$, where Γ is the condensate flow rate per unit tube length.

The Chun and Seban [1971] correlation is recommended for the local turbulent film condensation coefficient on a vertical plate. It is given by

$$h = 0.0038 \left(\frac{k^3 \rho_l(\rho_l - \rho_v)g}{\mu_l^2} \right)^{1/3} \mathrm{Re}_L^{0.4} \mathrm{Pr}^{2/3} \tag{2.28}$$

The nomenclature for Equations 2.25 through 2.30 is given in Chapter 12. Equations 2.24 through 2.28 are also applicable to evaporation of laminar or turbulent films. The equations presented here are for the case of zero interfacial shear stress on the condensate film. Interfacial shear stress on the liquid–vapor interface will increase the condensation coefficient, if the vapor and liquid film flow directions are the same. However, vapor shear will decrease the condensation coefficient if the vapor flow direction is opposite to the condensate flow.

A text on two-phase flow and heat transfer should be consulted for equations appropriate to convective condensation inside tubes. The appropriate equations are sensitive to the geometric orientation of the tube, as well as to the entering and leaving vapor quality. A recommended source is Carey [1992].

2.4.6 Nucleate Boiling

The Cooper [1984] correlation is a well accepted correlation for prediction of the nucleate boiling coefficient on horizontal plain tubes. This correlation is

$$h = 90\, q^{2/3} M^{1/2} p_r^m (-\log_{10} p_r)^{-0.55}, \qquad \text{where } m = 0.12 - 0.2 \log_{10} R_p \quad (2.29)$$

The term R_p is the surface roughness expressed in microns. Webb and Pais [1992] found that the Cooper correlation (with $R_p = 0.3$) predicted their data for five refrigerants boiling at 4.4°C and 26.7°C within ±10%.

Recommended equations for convective vaporization are not given here. As for convective condensation, the equations are sensitive to geometric orientation of the tube, as well as to the entering and leaving vapor quality. A recommended source is Carey [1992]. Webb and Gupte [1992] give a survey of correlations for convective vaporization in tubes (horizontal and vertical) and for tube banks.

2.5 CORRECTION FOR VARIATION OF FLUID PROPERTIES

Simplified heat exchanger analysis assumes the following: (1) The overall heat transfer coefficient is constant along the exchanger length, and (2) the individual heat transfer coefficients are functions only of the mean fluid temperature. In reality, the heat transfer coefficients depend on the physical properties of the fluid, which change with the fluid temperature—both in the flow direction and across the boundary layer thickness. If the heat exchanger is so short that an entrance region exists, the heat transfer coefficient may also vary with length. In cases involving polymers or viscous oils, substantial property variations may exist, thus introducing errors in the calculation. We will consider the effect of changing fluid temperature and local property variation effects on the overall heat transfer coefficient and outline a design method to account for these effects.

2.5.1 Effect of Changing Fluid Temperature

The derivation of the LMTD equation assumes that U is constant. Colburn [1933] has shown that if U varies linearly with temperature difference, $U = a + b(T_h - T_c)$, then

$$\frac{Q}{A} = \frac{U_{\mathrm{I}}\Delta T_{\mathrm{II}} - U_{\mathrm{II}}\Delta T_{\mathrm{I}}}{\ln \dfrac{U_{\mathrm{I}}\Delta T_{\mathrm{II}}}{U_{\mathrm{II}}\Delta T_{\mathrm{I}}}} \qquad (2.30)$$

where the subscripts I and II indicate the inlet and exit ends of the exchanger. Rather than assuming a linear variation of U, one may perform an incremental design which is well adapted to computer methods. The heat exchanger is divided into N increments of equal length. An average value of U is calculated for each increment, and the heat duty (q) is determined for each increment. The summation of q values for all increments will provide an accurate design. This procedure is outlined by Holman [1986] and by others.

2.5.2 Effect Local Property Variation

Certain fluid properties vary significantly with temperature, notably fluid viscosity. Hence the fluid properties at the wall temperature and the local mixed temperature will differ. A method to account for the fluid property variation across the boundary layer is required. Either of two methods are frequently used to correct heat transfer and friction correlations for fluid property variation across the boundary layer. The first is to evaluate the properties at the "film temperature," defined as the average of the local, mixed fluid temperature and the wall temperature. The second is to correct by a fluid property ratio evaluated at the mixed fluid and the wall temperatures. In this method, the fluid properties in the correlation are evaluated at the mixed fluid temperature. We will use second method.

Define $h = h_{cp} F_c$, where h_{cp} is the heat transfer coefficient assuming no temperature difference between the wall and bulk fluid temperature. F_c is a correction factor to account for a finite temperature difference between the wall and the bulk temperature. For liquids, $F_c = (\mu/\mu_w)^n$, where n depends on the flow conditions (flow regime and geometry) and whether the fluid is being heated or cooled. For gases, F_c is assumed to be dependent on $(T/T_w)^n$, where T is the absolute temperature. Summaries of the property correction term have been provided by Petukhov [1970] and by Kays and London [1984]. Our recommendations come from these two sources.

Table 2.3 shows F_c for turbulent flow of gases and liquids for both heating and cooling. The values shown in Table 2.3 are for flow in tubes. If only one entry is provided for F_c, it is from Petukhov [1970]. If a second entry is given (below the first value), it is from Kays and London [1984]. These F_c values are typically used for other flow geometries (e.g., tube banks). Table 2.4 show F_c for laminar flow of gases and liquids for both heating and cooling. These F_c values are taken from Kays and London [1984] and are for flow in tubes. However, these values are also frequently used for other flow geometries.

Figures 2.5 and 2.6 show the fluid property corrections for laminar and turbulent flow of liquids, respectively. Figures 2.5 and 2.6 were prepared using the F_c equations of Petukhov [1970], presented in Tables 2.3 and 2.4, respectively. Note that the abscissa is written as μ_w/μ, rather than μ/μ_w.

To determine the viscosity ratio correction, one must know the wall temperature. This is not explicitly known. It is necessary to iteratively solve for the wall temperature by iteration using

$$Q = h_i A_i (T_i - T_w) = h_o A_o (T_w - T_o) \qquad (2.31)$$

The required calculation procedure is as follows: (1) Calculate h_i and h_o for the constant properties condition (based on the mean fluid temperature). (2) Calculate

TABLE 2.3   Correction for Fluid Properties (Turbulent Flow)

	Heating	Cooling
	Liquids	
$\dfrac{\text{St}}{\text{St}_{cp}} = \left(\dfrac{\mu}{\mu_w}\right)^n$	$n = 0.11$	$n = 0.25$
$\dfrac{f}{f_{cp}} = \left(\dfrac{\mu}{\mu_w}\right)^m$	$m = \dfrac{1}{6}\left(7 - \dfrac{\mu}{\mu_w}\right)$ $m \simeq -0.25$	$m = -0.24$
	Gases	
$\dfrac{\text{St}}{\text{St}_{cp}} = \left(\dfrac{T}{T_w}\right)^n$	$n = 0.3\,\log_{10}\dfrac{T_w}{T} + 0.36$ $n \sim -0.5$	$n = 0.36$
$\dfrac{f}{f_{cp}} = \left(\dfrac{T}{T_w}\right)^m$	$m = 0.6 - 5.6\left(\text{Re}\dfrac{\rho_w}{\rho}\right)^{-0.38}$ $m \simeq 0.1$	$m = 0.6 - 7.9\left(\text{Re}\dfrac{\rho_w}{\rho}\right)^{-0.11}$ $m \simeq 0.1$

Source: Petukhov [1972] and Kays and London [1984].

TABLE 2.4   Correction for Fluid Properties (Laminar Flow)

	Heating	Cooling
	Liquids	
$\dfrac{\text{St}}{\text{St}_{cp}} = \left(\dfrac{\mu}{\mu_w}\right)^n$	$n = 0.14$	$n = 0.14$
$\dfrac{f}{f_{cp}} = \left(\dfrac{\mu}{\mu_w}\right)^m$	$m = -0.58$	$m = -0.50$
	Gases	
$\dfrac{\text{St}}{\text{St}_{cp}} = \left(\dfrac{T}{T_w}\right)^n$	$n = 0.0$	$n = 0.0$
$\dfrac{f}{f_{cp}} = \left(\dfrac{T}{T_w}\right)^m$	$m = -1.0$	$m = -1.0$

Source: Kays and London [1984].

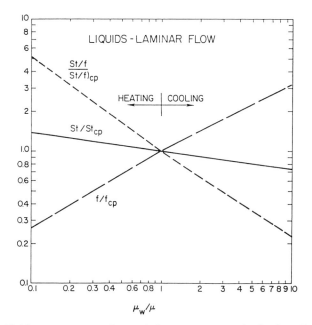

Figure 2.5   Fluid property correction variation versus μ_w/μ for laminar flow of liquids.

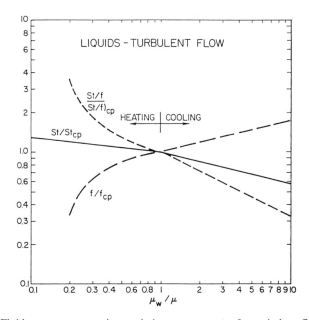

Figure 2.6   Fluid property correction variation versus μ_w/μ for turbulent flow of gases.

T_w using Equation 2.31. (3) Evaluate F_c from Table 2.3 or 2.4. (4) Calculate $h = h_{cp}F_c$. It may be necessary to repeat steps 1 through 4 several times until the assumed and calculated values of T_w agree.

A sophisticated heat exchanger design would account for variations of U due to both (a) changing mean fluid temperature and (b) local property variation effects.

2.6 REYNOLDS ANALOGY

Reynolds analogy provides a useful concept for the evaluating the performance of heat transfer enhancement concepts. Using Nu $\propto$ Pr$^{1/3}$, Colburn's [1933] statement of Reynolds analogy is

$$StPr^{2/3} \equiv j = f/2 \qquad (2.32)$$

Equation 2.32 applies to:

1. Laminar and turbulent flow over flat plates.
2. Turbulent flow in smooth ducts or rod bundles. It may also be applied to noncircular channels using the hydraulic diameter concept.

The equation $2j/f = 1$ establishes a relation between friction due to surface shear and heat transfer. The analogy does not hold when flow separations occur. With flow across a tube bank or over a rough surface, the total drag force will consist of two components, namely, surface shear and form drag. Because form drag is basically a parasitic loss, it contributes little to heat transfer. In this situation $2j/f < 1$, where f includes momentum losses due to flow separations.

Enhanced heat transfer surfaces yield increased heat transfer coefficients, thus $j/j_s > 1$, where the subscript s refers to the smooth surface. For comparison at equal Reynolds numbers, Equation 2.32 implies that the enhanced surface must have a friction increase at least as large as the j/j_s increase. Enhanced surfaces, in general, show $(j/j_s)/(f/f_s) < 1$, with exceptions in special cases.

We may define the relative j/f ratio of enhanced-to-smooth surfaces as an "efficiency index", thus

$$\eta_e = \frac{j/j_s}{f/f_s} \qquad (2.33)$$

This efficiency index is useful to evaluate the quality of the enhancement concept. The goal is to obtain high values for j/j_s—that is, as close as possible to 1.0. Assuming $j/j_s = 2$, a very good enhanced surface may have an η_e value of 0.8 to 0.9. Smaller values of η_e mean increased friction penalty to establish a given enhancement level, j/j_s. In some special cases (turbulent flow of high Prandtl number liquids over roughness) it is possible to obtain $\eta_e > 1$. This is because the Prandtl number dependency of the rough surface flow is greater than that of smooth surfaces. For Pr < 1, enhanced surfaces have not shown η_e values which exceed one.

The $2j/f$ analogy does not hold for laminar flow in ducts. However, the concept may be applied in a qualitative sense. In laminar duct flow, both Nu and f depend on the duct shape as shown in Table 2.2. This table lists values of j_H/f and j_T/f for Pr = 0.7 (gases). The table shows that the parallel plate channel provides the highest j/f, or $2j/f = 0.708$, for constant wall temperature. Compared to $2j/f = 1$ for turbulent flow, the "efficiency index" of this laminar flow geometry is 70.8%. As one moves down the table of laminar flow solutions, the other ducts shapes show smaller values of j/f, and thus they have lower values of the "efficiency index." For laminar flow in a circular tube, j/f is only 51.6% of that for turbulent flow in the same duct shape. This shows that turbulent flow provides greater heat transfer per unit friction than is possible with laminar duct flow. As the Prandtl number increases, the j/f characteristic of laminar flow decreases, since $f \propto Pr^{1/3}$ for laminar duct flows, and j/f is independent of Pr for turbulent flow.

These concepts provide important tools for evaluating the quality of heat transfer enhancement or for selecting laminar flow duct shapes which yield high heat transfer per unit friction.

2.7 FOULING OF HEAT TRANSFER SURFACES

After a period of operation, the heat exchanger surfaces may become coated with a deposit of solid material. The thermal resistance of such deposits cause a reduced overall heat transfer coefficient. Heat exchangers are normally overdesigned to compensate for the anticipated fouling. The fouling factor may be measured by testing the initially clean and fouled conditions. Then, the overall fouling factor, R_f, is calculated as

$$R_f = \frac{1}{U_{\text{fouled}}} - \frac{1}{U_{\text{clean}}} \tag{2.34}$$

A given fouling resistance will usually cause a greater performance reduction with liquids than with gases. There are two reasons for this. First, heat transfer coefficients for liquids are higher than for gases. Second, gases frequently use extended surfaces, so the fouling deposit is apportioned over a larger surface area. To illustrate these effects, consider a finned-tube liquid cooler having $A_o/A_i = 20$, where subscripts o and i refer to the air and liquid sides. Assume that the heat transfer coefficient on the liquid and air sides are $h_i = 6000$ W/m²-K and $h_o = 85$ W/m²-K, and that $R_f = 0.0002$ m²-K/W. The liquid side fouling deposit will reduce the effective liquid side heat transfer coefficient $1/(1/h_i + R_c)$ by 55%, relative to the unfouled condition. However, the effective air-side heat transfer coefficient $1/(1/h_o + R_{fo})$ is reduced only 2.4%. Using Equation 2.5, U for the fouled condition is only 46% as large as for the unfouled surfaces. Furthermore, the air-side fouling accounts for only 0.23% of the reduced U value. This is because of the much smaller air-side heat transfer coefficient and because the air-side fouling deposit is distributed over 20 times the surface area as compared to the liquid-side fouling. Fouling may cause a different problem for gases. The fouling deposit on the heat exchanger surface increases the gas pressure drop. The increased pressure drop may result in a lower gas flow rate, due to the balance point on the fan curve. Hence, the heat transfer rate is reduced, because of the lower gas flow rate through the heat exchanger.

Fouling raises serious implication for enhanced heat transfer surfaces. Assume that internal roughness is employed to obtain a 100% higher tube-side heat transfer coefficient than that of the smooth tube considered in the previous example for $R_f =$ 0.0002 m²-K/W; also assume that the effective liquid-side coefficient, $1/(1/h_i + R_f)$, is only 29% as large as that of the clean tube. Thus, a given fouling factor causes a more severe penalty as the heat transfer coefficient is increased.

For the same fouling factor, an internally finned tube may suffer a smaller penalty due to fouling because the fouling effect is distributed over a larger area. Assume that a clean internally finned tube provides the same heat conductance, $h_i A_i/L$, as the internally roughened tube. If the internal fins provide a 100% surface area increase with $R_f = 0.0002$ m²-K/W, the effective $h_i A_i/L$, including fouling, will be 45.6% as large as that of the unfouled surface.

Fouling deposits may be removed by mechanical, chemical, or thermal cleaning. Various types of mechanical cleaning systems are used. The outside surfaces of tubes may be cleaned with high-pressure air, water, or steam. Inner tube surfaces are cleaned by forcing special plugs or brushes through the tubes. In most cases the exchanger must be taken out of service for cleaning. However, an innovative tube-side cleaning system is described by Leitner [1980] in which a four-way flow valve is used to force cleaning brushes through the tubes. Because the direction of flow may be reversed, the cleaning system is automatic and performed while the exchanger is in service. Chemical cleaning includes flushing the system with chemical cleaning agents selected to dissolve the particular fouling deposit. Thermal cleaning is accomplished by heating the exchanger to a sufficiently high temperature to vaporize the deposits. This method is limited in application and is mainly used for small gas–gas heat exchangers. Closed water-cooling systems (e.g., cooling towers and boilers) use chemical treatment to minimize the fouling characteristics of the water. Fouling of enhanced surfaces is addressed in Chapter 10.

2.8   CONCLUSIONS

This chapter provides basic fundamentals relating to heat transfer from plain surfaces and heat exchanger design fundamentals. The performance of enhanced surfaces is typically compared to that of plain surfaces. The heat exchanger design information is needed to evaluate the performance of enhanced surfaces used in two-fluid heat exchangers. Chapters 3 and 4 apply this knowledge to develop "performance evaluation criteria" for evaluation of the performance of enhanced surfaces used in a heat-exchanger, and compared to a heat-exchanger using plain surfaces.

The Reynolds analogy is an important tool that shows how flow fiction may be related to the heat transfer coefficient. Although enhanced surfaces provide an increased heat transfer coefficient, they may also be expected to demonstrate increased flow friction, especially for single-phase flow. Applied to enhanced surfaces, Reynolds analogy indicates the minimal friction increase that may be expected from an enhanced surface. The "efficiency index" provides a measure of how well one has used increased friction to obtain a given heat transfer enhancement ratio.

Fouling can degrade the performance of a heat-exchanger, particularly for water flow in tubes. Hence, it is important to understand the fundamentals of fouling, and how it may affect heat exchanger performance. Fouling of enhanced surfaces is treated in Chapter 10.

2.9 REFERENCES

Bhatti, M. S., and Shah, R. K., 1987. "Laminar Convective Heat Transfer in Ducts," Chapter 3 in *Handbook of Single-Phase Heat Transfer*, S. Kakaç, R. K. Shah, and W. Aung, Eds., John Wiley & Sons, New York.

Carey, V. P., 1992. *Liquid–Vapor Phase-Change Phenomena*," Hemisphere Publishing Corp., Washington, D.C.

Chun, K. R. and Seban, R. A, 1971. "Heat Transfer to Evaporating Liquid Films." *Journal of Heat Transfer*, Vol. 93, pp. 391–396.

Colburn, A. P., 1933. "A Method of Correlating Forced Convection Heat Transfer Data and a Comparison with Fluid Friction," *Transactions of the AIChE*, Vol. 29, pp. 174–210.

Cooper, M. G., 1984. "Saturation Nucleate, Pool Boiling—A Simple Correlation." *International Chemical Engineering Symposium Series*, No. 86, 785–792.

Harper, W. B., and Brown, D. R., 1922. "Mathematical Equations for Heat Conduction in the Fins of Air-Cooled Engines," NACA Report 158.

Holman, J. P., 1986. *Heat Transfer*, 6th edition, McGraw–Hill, New York.

Incropera, F. P., and DeWitt, D. P., 1990. *Fundamentals of Heat and Mass Transfer*, John Wiley & Sons, New York.

Kays, W. M. and London, A. L., 1984. *Compact Heat Exchangers*, 3rd edition, McGraw–Hill, New York.

Kern, D. Q. and Kraus, A. D., 1972. *Extended Surface Heat Transfer*, McGraw–Hill, New York, p. 168.

Leitner, G. F., 1980. "Controlling Chiller Tube Fouling," *ASHRAE Journal*, Vol. 22, No. 2, pp. 40–43.

Petukhov, B. S., 1970. "Heat Transfer in Turbulent Pipe Flow with Variable Physical Properties," in *Advances in Heat Transfer*, Vol. 6, T. F. Irvine and J. P. Hartnett, Eds., Academic Press, New York, pp. 504–564.

Shah, R. K., Afimiwala, K. A., and Mayne, R. W., 1978. "Heat Exchanger Optimization," *Proceedings of the 6th International Heat Transfer Conference*, Vol. 4, pp. 193–199.

Webb, R. L., 1971. "A Critical Evaluation of Analytical and Reynolds Analogy Equations for Turbulent Heat and Mass Transfer in Smooth Tubes, *Wärme-und-Stoffübertragung*, Vol. 4, pp. 197–204.

Webb, R. L. and DiGiovanni, M. A., 1989. "Uncertainty in Effectiveness–NTU Calculations for Crossflow Heat Exchangers," *Heat Transfer Engineering*, Vol. 10, No. 3, pp. 61–70.

Webb, R. L., and Gupte, N. S., 1992. "A Critical Review of Correlations for Convective Vaporization in Tubes and Tube Banks." *Heat Transfer Engineering*, Vol. 13, No. 3, pp. 58–81.

Webb, R. L., and Pais, C., 1992. "Nucleate Boiling Data for Five Refrigerants on Plain, Integral-Fin and Enhanced Tube Geometries," *International Journal Heat and Mass Transfer*, Vol. 35, No. 8, pp. 1893–1904.

Zukauskas, A., 1972. "Heat Transfer from Tubes in Crossflow," in *Advances in Heat Transfer*, Vol. 8, J. P. Hartnett and T. F. Irvine, Eds., Academic Press, New York, pp. 93–160.

2.10   NOMENCLATURE

A	Heat transfer surface area, m² or ft²
A_f	Area of extended surface, m² or ft²
c_p	Specific heat, kg/kg-K or Btu/lbm-°F
C	Capacity rate (Wc_p), W/K or Btu/hr-°F
d	Tube diameter, m or ft
D_h	Hydraulic diameter, m or ft
f	Fanning friction factor, $\Delta p = (4fL/d)(\rho u^2/2)$; dimensionless
F	Correction factor for logarithmic mean temperature difference, dimensionless
F_c	Correction for fluid property variation across boundary layer, dimensionless
h	Heat transfer coefficient, W/m²-K or Btu/hr²-°F
k	Thermal conductivity, W/m-K or Btu/hr-°F
K	Pressure drop increment to account for flow development, $K(\infty)$ (full entrance region) or $K(x)$ (over length x), shown in Figure 5.27; dimensionless
L	Fin length, m or ft
LMTD	Logarithmic mean temperature difference, K or ft
M	Molecular weight, kg/mol
Nu	Nusselt number (hd/k), dimensionless
p_{cr}	Critical pressure, kPa or lbf/ft²
p_r	p/p_{cr}, dimensionless
Pr	Prandtl number, dimensionless
q	Heat flux, W/m² or Btu/hr-ft²
Q	Heat transfer rate, W or Btu/hr
Re	Reynolds number, Re_d (based on tube diameter) or Re_{Dh} (based on hydraulic diameter); dimensionless
Re_L	Condensate film Reynolds number, $4\Gamma/\mu_l$; dimensionless.

R	Capacity rate ratio (C_{min}/C_{max}), dimensionless
R_f	Fouling resistance, m²-K/W or hr-ft²-°F/Btu
R_p	Surface roughness (Equation 2.29), μm
St	Stanton number, dimensionless
t	Tube wall thickness, m or ft
T_c	Cold fluid temperature, T_{c1} (entering) and T_{c2} (leaving); °C or °F
T_h	Hot fluid temperature, T_{h1} (entering) and T_{h2} (leaving); °C or ft
T_w	Wall temperature, °C or ft
$T\infty$	Fluid temperature, °C or ft
U	Overall heat transfer coefficient, W/m²-K or Btu/hr²-°F
x_d^*	$x/(d_t Re_d Pr)$, dimensionless
W	Mass flow rate of fluid, kg/s or lbm/s

Greek Symbols

Γ	Condensate flow rate per unit plate width (or tube length), dimensionless.
ΔT_m	Mean temperature difference; K or °F
ϵ	Heat exchanger effectiveness; dimensionless
η_f	Fin efficiency, surface efficiency (Equation 2.15); dimensionless
η	Surface efficiency (Equation 2.16), dimensionless
μ	Dynamic viscosity, kg/s-m² or lbm/s-ft²

Subscripts

c	Refers to cold fluid in heat exchanger
fd	Fully developed flow
h	Refers to hot fluid in heat exchanger
i	Refers to conditions inside tube
m	Average value over flow length
o	Refers to conditions external to tube
H	Constant heat flux thermal boundary condition
T	Constant wall temperature thermal boundary condition
w	At wall

3

PERFORMANCE EVALUATION CRITERIA FOR SINGLE-PHASE FLOWS

3.1 PERFORMANCE EVALUATION CRITERIA (PEC)

A quantitative method is required to evaluate the performance improvement provided by a given enhancement. Normally, one compares the performance of an enhanced surface with that of the corresponding plain (smooth) surface.

There are three considerations in surface performance:

1. The performance objective: The four basic objectives of interest were discussed in Section 1.1.
2. The operating conditions: This includes the fluid flow rate and the entering fluid temperature.
3. The constraints: This includes the allowable pressure drop (or fan or pumping power) and the frontal area (or velocity) constraints.

The "basic performance characteristic" of an enhanced surface for single-phase heat transfer is defined by the j factor ($= \mathrm{StPr}^{2/3}$) and f versus Reynolds number curves. One possibility to quantify the performance improvement is to calculate the ratios j/j_s and f/f_s, where the subscript s (or p) is for a smooth (or plain) surface at the same Re. Generally, the friction factor of an enhanced surface in single-phase flow is higher than that of the smooth surface, when operated at the same velocity (or Reynolds number). However, this method is not recommended because it does not define the actual performance improvement, subject to specific operating constraints. If one simply calculated the enhancement ratio, j/j_s at equal velocities, an unfair comparison may result. This is because the enhanced surface would be allowed to operate at a higher pressure drop. The plain surface would give a higher h

value if it were allowed to operate at a higher velocity, giving the same pressure drop as the enhanced surface. Thus, the pressure drop constraint is a very important consideration for calculating the performance benefits of an enhanced surface in single-phase flow.

Several measures of "performance" are possible, and they have been proposed in the literature. The first may compare the "performance" of the enhanced surface with that of a smooth surface. This may be described as "surface performance comparison," since it does not account for the effects of the second fluid stream in the heat exchanger. The second may determine the effect of the enhanced surface on the overall performance of a two fluid heat exchanger. We will describe and discuss the merits of several possible comparison methods.

PEC analysis for enhancement of a two-phase flow (boiling or condensation) should be handled differently from that for single-phase flow. This is because pressure drop of a two-phase fluid also reduces the local saturation temperature of the fluid. Thus, the driving potential for heat transfer is also affected. The special PEC for two-phase heat transfer is treated in Chapter 4.

The PEC presented here are based on those previously described by Webb and Eckert [1972], Webb [1981a], Bergles et al. [1974a, 1974b], Bergles [1981], and Webb and Bergles [1983]. The algebraic formulations of the PEC are applicable to single-phase laminar or turbulent flows in tubes or normal to tube banks.

3.2 PEC FOR HEAT EXCHANGERS

The primary interest is to determine how the enhanced surface (or doubly enhanced surfaces) will affect the performance of the heat exchanger. The preferred evaluation method sets a performance objective (e.g., reduced surface area) and calculates the performance improvement relative to a reference design (e.g., smooth tubes) for a given set of operating conditions and design constraints (e.g., constant pumping power). Hence, this method defines the improvement of the objective function (e.g., percent surface area decrease) relative to a smooth surface heat exchanger. Possible performance objectives of interest were summarized in Section 1.1. These objectives assume fixed heat exchanger flow rate and entering fluid temperature. The four objectives stated in Section 1.1 are as follows:

1. Reduced heat transfer surface material for fixed heat duty and pressure drop.
2. Reduced logarithmic mean temperature difference (LMTD) for fixed heat duty and surface area.
3. Increased heat duty for fixed surface area.
4. Reduced pumping power for fixed heat duty and surface area.

Objective 1 allows a smaller heat exchanger size and, hopefully, reduced capital cost. Objective 2 offers reduced operating cost. The reduced LMTD of Objective 2 will affect improved system thermodynamic efficiency, yielding lower system operating cost. Objective 3 is important if the heat exchange capacity of a given heat exchanger is to be increased. Objectives 2 and 3 will typically result in a more expensive heat exchanger, since the enhanced and smooth tube exchangers have the same total tubing length. A more costly enhanced surface will be justified if the operating cost savings are sufficiently high.

The major operational variables include the heat transfer rate, the pumping power (or pressure drop), the heat exchanger flow rate, and the fluid velocity (or the flow frontal area). A performance evaluation criterion, (PEC) is established by selecting one of the operational variables for the performance objective, subject to design constraints on the remaining variables.

The design constraints placed on the exchanger flow rate and velocity cause key differences among the possible PEC relations. The increased friction factor of enhanced surfaces may require reduced velocity to satisfy a fixed pumping power (or pressure drop) constraint. If the exchanger flow rate is held constant, it may be necessary to increase the flow frontal area to satisfy the pumping power constraint. However, if the mass flow rate is reduced, it is possible to maintain constant flow frontal area at reduced velocity. When the exchanger flow rate is reduced, it must operate at higher thermal effectiveness to provide the required heat duty. This may significantly reduce the performance potential of the enhanced surface if the design thermal effectiveness is sufficiently high. In many cases the heat exchanger flow rate is specified and thus flow rate reduction is not permitted.

3.3 PEC FOR SINGLE-PHASE FLOW

3.3.1 Objective Function and Constraints

Table 3.1 defines PEC for 12 cases of interest with flow of a single phase fluid inside enhanced and smooth tubes of the same envelope diameter. The PEC are segregated by three different geometry constraints.

1. *FG Criteria:* The cross-sectional flow area and tube length are held constant. The FG criteria may be thought of as a retrofit situation in which there is a one-for-one replacement of smooth surfaces with enhanced surfaces of the same basic geometry (e.g., tube envelope diameter, tube length, and number of tubes for in-tube flow). The FG-2 criteria have the same objective functions as the FG-1 criteria but require the enhanced surface design to operate at the same pumping power as the reference smooth tube design. In most cases, this will require the enhanced exchanger to operate at reduced flow rate. The FG-3 criteria seek reduced pumping power for fixed heat duty.

2. *FN Criteria:* These criteria maintain fixed cross-sectional flow area and allow the length of the heat exchanger to be a variable. These criteria seek reduced surface area (FN-1, -2) or reduced pumping power (FN-2) for constant heat duty.

3. *VG Criteria:* In many cases, a heat exchanger is sized for a required thermal duty with a specified flow rate. In these situations, the FG and FN criteria are not applicable. Because the tube-side velocity must be reduced to accommodate the

TABLE 3.1 Performance Evaluation Criteria for Single-Phase Heat Exchange System with d_i = Constant

Case	Geometry	Fixed				Objective
		W	P	Q	ΔT_i	
FG-1a	N, L^a	×			×	$\uparrow Q$
FG-1b	N, L	×		×		$\downarrow \Delta T_i$
FG-2a	N, L		×		×	$\uparrow Q$
FG-2b	N, L		×	×		$\downarrow \Delta T_i$
FG-3	N, L			×	×	$\downarrow P$
FN-1	N		×	×	×	$\downarrow L$
FN-2	N	×		×	×	$\downarrow L$
FN-3	N	×		×	×	$\downarrow P$
VG-1		×	×	×	×	$\downarrow NL$
VG-2a	$N L^b$	×	×		×	$\uparrow Q$
VG-2b	$N L^b$	×	×	×		$\downarrow \Delta T_i$
VG-3	$N L^b$	×		×	×	$\downarrow P$

$^a N$ and L are constant in all FG cases.

^{b}The product NL is constant for VG-2 and VG-3

higher friction characteristics of the enhanced surface, it is necessary to increase the flow area to maintain constant flow rate. This is accomplished by using a greater number of parallel flow circuits. Maintenance of a constant exchanger flow rate avoids the penalty encountered in the previous FG and FN cases of operating at higher thermal effectiveness.

Calculation of the performance improvement for any of the 12 cases in Table 3.1 requires algebraic relations which quantify the objective function and constraints. The concept of the PEC analysis will be explained for the case of a prescribed wall temperature or a prescribed heat flux. It is convenient to develop the algebraic relations relative to a smooth surface operating at the same fluid temperature. This allows cancellation of the fluid properties from the equations. The equations applicable to the PEC cases of Table 3.1 are presented below.

3.3.2 Algebraic Formulation of the PEC

The different cases listed in Table 3.1 are derived for flow inside enhanced and smooth tubes of the same inside diameter. Consider a shell-and-tube heat exchanger of length L, having N tubes in each pass and having N_p passes. The total tube-side surface area in the heat exchanger is $A = \pi d_i L N N_p$, where N is the number of tubes in the flow passage, and N_p is the number of tube passes. The cross sectional flow area in the tube is A_c. The basic heat transfer and friction performance characteristics of the enhanced and smooth tubes are normally presented as j ($= \text{StPr}^{2/3}$) and f versus Re $= d_i G / \mu$. Because the tube inside diameter is held constant, we may write

$$h = c_p \text{Pr}^{-2/3} jG \qquad (3.1)$$

We are interested in the hA value of the enhanced surface, relative to that of the competing smooth surface. Writing Equation 3.1 as the ratio, relative to a smooth surface gives

$$\frac{hA}{h_s A_s} = \frac{j}{j_s} \frac{A}{A_s} \frac{G}{G_s} \tag{3.2}$$

The pumping power is calculated as

$$P = \left(\frac{fA}{A_c} \frac{G^2}{2\rho} \right) \left(\frac{GA_c}{\rho} \right) \tag{3.3}$$

Writing Equation 3.3 as the ratio relative to the smooth surface gives

$$\frac{P}{P_s} = \frac{f}{f_s} \frac{A}{A_s} \left(\frac{G}{G_s} \right)^3 \tag{3.4}$$

Elimination of the term G/G_s from Equations 3.2 and 3.4 gives

$$\frac{hA/h_s A_s}{(P/P_s)^{1/3}(A/A_s)^{2/3}} = \frac{j/j_s}{(f/f_s)^{1/3}} \tag{3.5}$$

The variables on the left side of Equation 3.5 are $hA/h_s A_s$, P/P_s, and A/A_s. One of these variables is set as the objective function, and the remaining two are set as operating constraints with the value 1.0. It is necessary to determine the G/G_s ratio that satisfies Equation 3.5. We assume that the Reynolds (Re_s) number and the j_s and f_s of the smooth tube heat exchanger design are known. We will also assume that the equations for enhanced surface j and f as functions of Re are known.

Continuation of the solution for the general case requires specification of (1) the heat exchanger flow rate, (2) the flow frontal area, and (3) the heat transfer coefficient on the outer and inner tube surfaces for the smooth surface heat exchanger. Solution of the general problem involves performing a complete heat exchanger analysis. Before discussing the general problem, we will first discuss a simple case, which avoids much of the complexity of the general case.

3.3.3   Simple Surface Performance Comparison

Assume that enhancement is applied to the tube side of a shell-and-tube heat exchanger. The simplest case exists if the total thermal resistance is on the tube side. Thus, $UA = hA$. Two cases will be illustrated. The first is for constant total flow rate, and the second will be for a tube-for-tube replacement (the same number of tubes in the enhanced and smooth tube exchangers). The tube-for-tube replacement corresponds to fixed flow frontal area. The PEC analysis is developed in detail for case VG-1 in Table 3.1.

3.3.4 Constant Flow Rate

For case VG-1, Q/Q_s = constant. Because the flow rate is assumed constant, and all of the thermal resistance is on the tube side, the Q/Q_s constraint is satisfied by $hA/h_sA_s = 1$. The pumping power constraint of case VG-1 is $P/P_s = 1$. Substitution of these ratios in Equation 3.5 gives

$$\frac{A}{A_s} = \left(\frac{f}{f_s}\right)^{1/2} \left(\frac{j_s}{j}\right)^{3/2} \tag{3.6}$$

The G/G_s ratio required to meet both the heat transfer and pressure drop constraints is obtained by substituting A/A_s from Equation 3.6 in Equation 3.4 and solving for G/G_s. The result is

$$\frac{G}{G_s} = \left(\frac{j}{j_s}\frac{f_s}{f}\right)^{1/2} \tag{3.7}$$

It is necessary to iteratively solve Equations 3.6 and 3.7 for the ratios G/G_s and A/A_s. Using the assumed known expressions for $f(\text{Re})$, and $j(\text{Re})$, one may guess Re/Re_s (= G/G_s), solve for j/j_s and f/f_s and then calculate G/G_s and A/A_s. This iterative process is repeated until the solutions converge. With G/G_s, j/j_s and f/f_s known, A/A_s is calculated using Equation 3.2 or 3.6. For fixed internal tube diameter, the flow rate constraint $W/W_s = 1$ implies

$$\frac{W}{W_s} = 1 = \frac{G}{G_s}\frac{N}{N_s} \tag{3.8}$$

Solution of Equation 3.8 gives $N/N_s = G_s/G$, which defines the number of tubes required to satisfy the pumping power constraint. Typically, the enhanced tube exchanger must operate at a lower mass velocity, which means it will have more tubes per pass and, thus, a larger shell diameter. The tube pass length, L, is obtained from the equation

$$\frac{A}{A_s} = \frac{N}{N_s}\frac{L}{L_s} \tag{3.9}$$

3.3.5 Fixed Flow Area

If the number of tubes in the exchanger is fixed, the flow frontal area remains constant. The reduced velocity in the enhanced tube exchanger will result in reduced flow rate. The previous solution for G/G_s still applies. Because $N/N_s = 1$, Equation 3.8 gives $W/W_s = G/G_s$, which defines the flow rate in the enhanced tube exchanger. This heat exchanger must operate at higher thermal effectiveness to provide $Q/Q_s = 1$, which will require greater surface area than for the previous case. This shows that if the heat exchanger flow rate is reduced, the benefits of enhancement will be decreased. Webb [1981a] provides details on the methodology to account for the effect of reduced flow rate on the performance improvement. The designer should avoid flow rate reduction, if possible.

3.4   THERMAL RESISTANCE ON BOTH SIDES

The simple example analyzed above assumed that the total thermal resistance was on the side to which enhancement is applied. This is an unrealistic situation. Webb [1981a] extends the PEC analysis to account for the thermal resistance of both fluid streams, the wall resistance, and fouling resistances. Equations 3.10 and 3.11 below are the generalized relations necessary for the performance improvement provided by an enhanced surface used in an actual heat exchanger. These equations include the possibility that enhancement may be provided to both fluid streams. The dimensionless variables β and β_s are defined in Table 3.2.

$$\frac{P}{P_s} = \frac{f}{f_s}\frac{A}{A_s}\left(\frac{G}{G_s}\right)^3 \tag{3.10}$$

$$\frac{UA}{U_s A_s} = \frac{1 + \beta_s}{\dfrac{\text{St}_s}{\text{St}}\left[\dfrac{f}{f_s}\dfrac{P_s}{P}\left(\dfrac{A_s}{A}\right)^2\right]^{1/3} + \beta\left(\dfrac{A_s}{A}\right)} \tag{3.11}$$

Equations 3.10 and 3.11 were developed for tube-side enhancement. However, they are equally applicable to shell-side enhancement or air-side enhancement of a fin-and-tube heat exchanger, if the various terms are appropriately defined. This extension is discussed in Section 3.8.

Consider the application of Equations 3.10 and 3.11 to tube-side enhancement in a shell-and-tube or fin-and-tube exchanger. The equations give the UA and pumping power (P) ratios of the enhanced exchanger relative to a smooth tube exchanger that operates at specified G. The thermal resistance terms β_s and β contain known or specified information. Equation 3.11 includes the possibility that the enhanced exchanger may also contain an enhanced outer tube surface, $E_{ho} = h_o/h_{op}$. The parameters of interest in Equation 3.11 are A/A_s, P/P_s, and $UA/U_s A_s$. For each case of interest in Table 3.1, two of these parameters are set as constraints and the third is the objective function.

TABLE 3.2   Dimensionless Ratios to be Used in Equations 3.10–3.12

Definition	Reference Exchanger	Enhanced Exchanger
Surface area	$B = A_{os}/A_s$	$B = A_o/A$
Outer surface conductance	$r_{os} = h_s/B_s h_{os}$	$r_o = r_{os}B_s/B$
Metal resistance	$r_{wp} = h_s t A_s/kA_m$	$r_w = htA/kA_m$
Fouling resistance	$r_{fp} = h_s R_f$	$r_f = hR_f$
Composite resistance	$\beta_s = r_{os} + r_{wp} + r_{fs}$	$\beta = r_o/E_{ho} + r_w A/A_s + r_f$

The tube diameter of the enhanced exchanger may be different from that of the reference smooth tube exchanger. The definition of geometrical parameters for air-side extended surfaces is more complicated, as will be discussed subsequently.

As an example, assume tube-side enhancement in a shell-and-tube exchanger for case FG-2. Here $P/P_s = A/A_s = 1$ and the objective is to achieve $UA/U_sA_s > 1$. Equation 3.11 becomes

$$\frac{UA}{U_sA_s} = \frac{1 + \beta_s}{\dfrac{St_s}{St}\left(\dfrac{f}{f_s}\right)^{1/3} + \beta} \tag{3.12}$$

and from Equation 3.10, we obtain

$$\frac{G}{G_s} = \left(\frac{f_s}{f}\right)^{1/3} \tag{3.13}$$

These equations may be solved for specified G_s and E_{ho} once appropriate equations are established for the St and f of the enhanced surface.

Figure 3.1 shows the performance benefits of tube-side transverse-rib roughness in a condenser, relative to a condenser having a smooth inside tube surface as reported by Webb [1981b]. The tube-side operating conditions are Pr = 3.0 and Re_s = 40,000. The curves are prepared for case VG-1 (constant total tube-side flow rate, pumping power, heat duty, and entering fluid temperature). To satisfy the

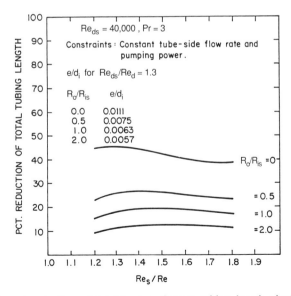

Figure 3.1 Percent reduction of total heat exchanger tubing length obtained by use of transverse ribs ($p/e = 10$), as reported by Webb [1981b]. (From Webb [1981b].)

constant pumping power constraint, it is necessary to operate the rough tube at reduced velocity, or $Re/Re_s < 1$. This means the enhanced tube exchanger will have a greater number of tubes of shorter length. The parameter r_o/r_{ip} is the outside-to-inside thermal resistance ratio for the exchanger having smooth inner tube surface. The line labeled $R_o/R_{ip} = 0$ is for all of the thermal resistance on the tube-side. For $R_o/R_{ip} = 0$, 45% reduction of total tubing length is possible. However, a practical heat exchanger would not be expected to operate with $R_o/R_{ip} \leq 0$. The curves show that the tube-side enhancement is less beneficial as $R_o/R_{ip} \geq 0$. For $R_o/R_{ip} = 0.5$, the enhanced tube provides 27% reduction of total tubing length. Figure 3.1 suggests that $Re_s/Re = 1.3$ provides a good compromise between heat exchanger flow frontal area and material savings. The table in Figure 3.1 shows the dimensionless roughness height (e/d_i) which satisfies the design constraints at $Re_s/Re = 1.3$. The roughness height was selected using methods described in Chapter 9.

3.5 RELATIONS FOR St AND f

Most of the available data for enhanced surfaces are given in graphic form, rather than equation form. However, this is only a minor inconvenience in the implicit solution of the above equations. With these equations, caution should be exercised regarding generality. Because a given enhancement type will usually involve several possible characteristic dimensions, the nominal tube diameter or hydraulic diameter is insufficient for a generalized correlation in the form of St and f versus Re. Therefore, the resulting "correlation" may be limited to a specific surface geometry and tube diameter (or fin spacing). Furthermore, the data may have been taken with a single fluid, so application to other fluids may be questionable unless the Prandtl number dependency is known. Bergles [1981] also discusses problems associated with applying single tube data to a "scaled-up" situation which involves flow outside a large number of tubes. This situation is probably a greater problem for boiling and condensing than for single-phase heat transfer.

3.6 HEAT EXCHANGER EFFECTIVENESS

The enhanced and smooth exchangers may not operate at the same effectiveness (ϵ). For those cases when the objective is increased heat duty, the ϵ–NTU design method gives

$$\frac{Q}{Q_s} = \frac{W}{W_s} \frac{\epsilon}{\epsilon_s} \frac{\Delta T_i}{\Delta T_{is}} \tag{3.14}$$

where ΔT_i is the temperature difference between the two inlet streams. For fixed inlet temperatures ($\Delta T_i = \Delta T_{is}$), Equation 3.14 yields

$$\frac{Q}{Q_s} = \frac{W}{W_s} \frac{\epsilon}{\epsilon_s} \qquad (3.15)$$

Because the operating conditions of the smooth tube exchanger are known, NTU_s ($= A_s/W_s c_{ps}$) is known and ϵ can be calculated. Once $UA/U_s A_s$ and W/W_s for the enhanced exchanger are known, its NTU is calculated from

$$NTU = NTU_s \frac{UA}{U_s A_p} \frac{W_s}{W} \qquad (3.16)$$

Then the ϵ of the enhanced exchanger may be calculated, and Q/Q_s obtained from Equation 3.15.

When the objective function is increased UA with $Q/Q_s = 1$, Equation 3.14 shows that $\Delta T_i/\Delta T_{is} < 1$.

3.7   EFFECT OF REDUCED EXCHANGER FLOW RATE

The FN and FG cases maintain $N/N_s = 1$. For $P/P_s = 1$, it is necessary to operate at reduced exchanger flow rate ($W/W_s < 1$). The reduced flow rate of these cases may penalize the enhanced exchanger. Such a penalty does not occur for the VG cases, which maintain $W/W_s = 1$. Let us compare this effect for the FN-1 and VG-1 cases which seek reduced surface area for $Q/Q_s = P/P_s = 1$. Because the FN exchanger operates at higher thermal effectiveness, additional surface area is required to compensate for the reduced LMTD.

Figure 3.2 may be used to illustrate the penalty associated with the $W/W_s < 1$ exchanger. This figure is constructed for shell-side condensation ($C_{min}/C_{max} = 0$) which operates with constant UA and ΔT_i. The ϵ is given by $\epsilon = 1 - \exp(-NTU)$. To obtain $Q/Q_s = 1$ for $UA = $ constant, the surface area of the $W/W_s < 1$ exchanger must be increased. The area penalty increases with larger NTU of the reference smooth tube exchanger. The $W/W_s < 1$ and $W/W_s = 1$ enhanced exchangers have the same U value if $\Delta p/\Delta p_s = 1$. If $P/P_s = 1$, the $W/W_s = 1$ exchanger will have a slightly higher U value.

To summarize the effect, assume that the smooth tube exchanger is designed for $\epsilon_s = 0.632$ (NTU = 1.0) and that the enhanced exchanger provides a 50% larger U value. For the VG case ($W/W_s = 1$), $U/U_s = 1.5$ thus yields $A/A_s = 1/1.5$.

Figure 3.2 shows $Q/Q_s = 0.903$ for the FN case ($W/W_s = 0.80$). The FN exchanger will require $A/A_s = 0.667/0.903 = 0.74$ to satisfy $Q/Q_s = 1$. Thus, the VG exchanger provides 10% greater surface area reduction than that allowed by the FN exchanger. Of course, there may be benefits with reduced flow rate, such as a smaller pump, reduced pressure drop in the supply pipe and header, or smaller supply pipe size.

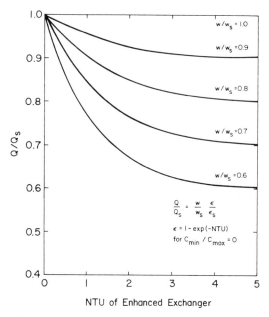

Figure 3.2   Effect of reduced flow rate on Q/Q_s for $UA/U_s A_s = 1$. (From Webb [1981a].)

3.8   FLOW NORMAL TO FINNED TUBE BANKS

The PEC defined in Table 3.1 may also be interpreted for flow normal to tube banks (bare or finned) and for plate-fin heat exchangers. The FN and FG cases maintain constant cross-sectional flow area ($N/N_s = 1$). Cases FG-2 and FN-1 constrain $P/P_s = 1$, for which a reduced flow rate ($W/W_s < 1$) may be required to satisfy the $P/P_s = 1$ constraint. It is possible that the potential benefits of an enhanced surface will be lost if the heat exchanger is required to operate with $W/W_s < 1$ as discussed by Webb [1981a]. When the flow rate is reduced, the LMTD is reduced (for constant ΔT_i), and additional surface is required to compensate for the reduced LMTD. The VG cases avoid this problem by increasing the flow cross-sectional area sufficiently to maintain $W/W_s = 1$; this is accomplished by adding tubes in parallel. Hence, when a pumping power constraint is applied, the greatest performance benefit of an enhanced surface will be realized by use of the VG criteria. This implies that enhanced surfaces will yield smaller benefits in retrofit applications with fixed flow area than in new designs, for which the flow frontal area may be increased over that of the corresponding smooth tube design.

The PEC defined in Table 3.1 may also be interpreted for flow normal to tube banks (bare or finned) and for plate-fin heat exchangers which are discussed in

TABLE 3.3   Interpretation of Table 3.1 PEC for Finned Tube Banks

Variable	Flow in Tubes	Tube Banks
Flow area	$A_c N$	A_{fr}
Mass Flow Rate (W)	$A_c N G$	$A_{fr} G_{fr}$
Surface area	$\pi d_i N L$	βV
Flow rate ratio (W/W_s)	$NG/(N_s G_s)$	$A_{fr} G_{fr}/(A_{fr} G_{fr})s$

Source: Webb [1982].

Chapter 5 and illustrated by Figure 5.1. Table 3.3, taken from Webb [1982], defines how the variables used in Table 3.2 (flow in tubes) should be interpreted for the geometries shown in Figure 5.1.

3.9   VARIANTS OF THE PEC

The flow resistance constraint used in Table 3.1 and in the previous PEC equations is the flow friction power. One may choose to use different constraints, as is appropriate to the situation. For example, one may choose to fix the pressure drop, rather than the pumping power. An important third possible constraint is the balance point on the fan or pump curve.

Consider case FG-1 involving a tube-for-tube retrofit of a shell-and-tube condenser. The analysis for this problem is described by Webb et al. [1984]. Cooling water from a lake or cooling tower supplies the condenser cooling water. The enhanced tubes have a higher friction factor than that of the smooth tubes they are replacing. Hence, the circulating water pump must work against a higher system friction resistance characteristic. The flow rate in the condenser will be determined by the balance point between the pump and system characteristics. Figure 3.3 shows this situation, as analyzed by Webb et al. [1984]. The system resistance consists of the sum of the friction loss in the condenser tubes and the piping and fittings external to the condenser. Using the known friction characteristics of the external piping and the condenser tubes, the designer constructs polynomial functions, which represent the system characteristic curves. Assume that $H_e = f_e(W)$ and $H_s = f_s(W_s)$ represent the system characteristics of the enhanced and smooth tubes, respectively. The pump curve is represented by $H_p = f_p(W)$. The smooth tube flow rate at the balance point on the pump curve is given by iterative solution of $f_s(W_s) = f_p(W_s)$. Similarly, the balance point for the enhanced tubes is given by iterative solution of $f_e(W) = f_p(W)$. Because the number of tubes is constant, $G/G_s = W/W_s$. Using the known values of G and G_s, one continues the solution for the h and UA of the enhanced surface using Equation 3.11.

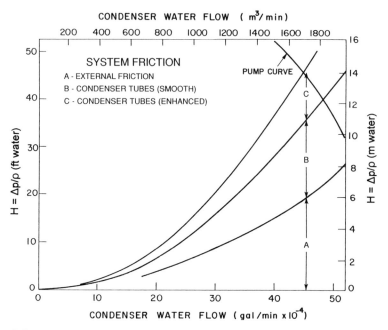

Figure 3.3 Illustration of balance point between pump curve and system resistance for smooth and enhanced tubes in steam condenser of a nuclear plant. (Adapted from Webb et al. [1984].)

3.10   COMMENTS ON OTHER PERFORMANCE INDICATORS

Many authors have proposed alternate performance comparison methods. Shah [1978] surveys more than 30 comparison methods, and more have been published since 1978. Most of the methods assume that the total thermal resistance is on one side of the heat exchanger. A recent offering by Cowell [1990] describes a methodology to compare compact heat exchanger surfaces, which is basically equivalent to that presented in Section 3.3. However, he allows the hydraulic diameter to be either fixed or variable. Two popular methods are discussed below and are compared to the methodology of Section 3.3.

3.10.1   Shah

Kays and London [1984] and Shah [1978] describe a PEC method that they term a "goodness factor" comparison. This dimensional method allows direct comparison of the air-side performance ($r_{tot} = r_{tot,s} = 0$) of different surface geometries. Rather than writing Equations 3.1 and 3.3 in the form of a ratio, the heat transfer coefficient-surface efficiency product ($h\eta_o$) and friction power (P) are written in dimensional terms as

$$h\eta_o = \frac{c_p\mu}{\mathrm{Pr}^{2/3}} \frac{\eta_o j \mathrm{Re}}{D_h} \qquad (3.17)$$

$$\frac{P}{A} = \frac{\mu^3}{2\rho^2} \frac{f\mathrm{Re}^3}{D_h^3} \qquad (3.18)$$

The fluid properties in Equations 3.17 and 3.18 are evaluated at any preferred "standard condition," and Equation 3.17 is plotted versus Equation 3.18, giving $(h\eta_o)_{\mathrm{STD}}$ versus $(P/A)_{\mathrm{STD}}$. For fixed $(P/A)_{\mathrm{STD}}$, the surface having the highest h will require the least heat transfer surface area (A) for equal $h\eta_o A$. Alternatively, one may modify the equations to allow comparison on a volume basis. Multiplying each equation by $\beta\,(=A/V)$ allows comparison of $(h\eta_o A/V)_{\mathrm{STD}}$ versus $(P/V)_{\mathrm{STD}}$. Thus, at fixed P/V, one compares the $h\eta_o A$ per unit volume. Geometries having large $h\eta_o A/V$ will require the least core volume for a given air-side performance ($h\eta_o A$). Equations 3.17 and 3.18 may be written in terms of j, f, and Re to allow direct substitution from a graph or table of the dimensionless surface performance data. The author suggests that the effects of the geometric parameters can be better visualized by writing the P/V and $h\eta_o A/V$ equations in forms containing the mass velocity:

$$\frac{h\eta_o A}{V} = \frac{4\sigma c_p}{\mathrm{Pr}^{2/3}} \frac{jG}{D_h} \qquad (3.19)$$

$$\frac{P}{V} = \frac{2\sigma}{\rho^2} \frac{fG^3}{D_h} \qquad (3.20)$$

The above form clearly shows the effect of fin pitch, which is influenced by D_h. As the fin pitch decreases, D_h decreases. Thus, both $h\eta_o A/V$ and P/V will increase in direct proportion to fin pitch. We feel that comparison methods should not confuse the effect of fin pitch and surface geometry. One way to avoid this is to evaluate Equations 3.17 and 3.18 at a fixed D_h for all surfaces. Although the surfaces being compared may have different hydraulic diameters, the j and f versus Re values of all surface geometries can be evaluated at a common D_h. The PEC equations presented in Section 3.3 use fixed tube diameter to show only the effect of the surface configuration on the performance.

The goodness factor method may be compared to the cases given in Table 3.1:

1. The equations supporting Table 3.1 are written in the form of a ratio, relative to a reference surface. Although this is not necessary, it provides simplicity because the fluid property terms drop out.
2. The Table 3.1 equations allow inclusion of the total resistance between the hot and cold streams.
3. The surfaces are compared at equal D_h in Table 3.1.

Case VG-2a with $r_{tot} = r_{tot,s} = 0$ directly corresponds to the surface area goodness factor comparison method. This case yields h/h_s for $P/P_s = A/A_s = 1$ with $W/W_s = 1$. This is equivalent to $P/A = P_s/A_s$.

3.10.2 Soland et al.

Soland et al. [1976, 1978] have compared the air-side performance of 17 parallel plate-fin surface geometries given in Kays and London [1984]. Their analysis is for $r = r_s = 0$ and corresponds approximately to cases FG-1, VG-la, VG-2a, and VG-3 of Table 3.1. They redefine the j and f factors in terms of the area of the primary surface, and the characteristic dimension in f and Re is given by the spacing between the parting sheets, which contain the extended surface. Figure 3.4, taken from Figure 11 of Soland et al. [1976], shows V_s/V for the VG-1 case. The reference surface is a plain channel having 6.4 mm between parting sheets, with no fins between the parting sheets. The VG-1 comparison used by Soland et al. [1976, 1978] seeks minimum heat exchanger volume (V/V_s), and geometries having small fin pitches and greater distance between parting sheets will be favored in the compari-

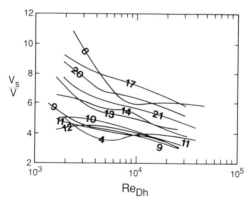

Plate-and-Fin Surfaces Compared

Surface Geometry	Plate Spacing b (mm)	Surface Code	Surface Number
Plain	6.4	11.1	4
	6.4	19.86	8
Louvered	6.4	3/8-6.06	9
	6.4	3/8(a)-6.06	10
	6.4	1/2-6.06	11
	6.4	1/2(a)-6.06	12
	6.4	3/8-8.7	13
	6.4	3/8(a)-8.7	14
	6.4	1/4(b)-11.1	17
	6.4	1/2-11.1	20
	6.4	3/4-11.1	21

Figure 3.4 V/V_s for constant hA and P using case VG-1. All geometries are the parallel plate-fin heat exchanger geometry having 6.35-mm plate spacing. (From Soland et al. [1976].)

son. Thus, the V/V_s ratio is strongly influenced by the basic geometric parameters, independent of the heat transfer coefficient. Although this method does indicate compactness, it gives no indication of the relative heat transfer surface area or fin material requirements.

3.11 CONCLUSIONS

The PEC described in this chapter have wide application to single-phase heat exchangers. The PEC equations allow one to make a simple surface performance comparison, or to account for the thermal resistance of both fluid streams. The PEC analysis allows the designer to quickly identify the most effective surfaces or tube inserts of a variety of available choices. Use of the criteria equations with consideration of external thermal resistances and changes in temperature difference along the flow path allows the designer to assess the performance of actual heat exchangers.

The designer should be careful in applying PEC methods that compare two surface geometries having different diameter (or hydraulic diameter). The resulting comparison shows the effect of changing both variables, rather than the effect of surface geometry alone. The effect of surface geometry alone is determined, if the diameter is maintained constant.

3.12 REFERENCES

Bergles, A. E., 1981. "Applications of Heat Transfer Augmentation," in *Heat Exchangers: Thermal Hydraulic Fundamentals and Design*, S. Kakaç, A. E. Bergles, and F. Mayinger, Eds., Hemisphere Publishing Co., Washington, D.C.

Bergles, A. E., Blumenkrantz, A. R., and Taborek, J., 1974a. "Performance Evaluation Criteria for Enhanced Heat Transfer Surfaces," in *Heat Transfer 1974, Proceedings of the 5th International Heat Transfer Conference*, The Japan Society of Mechanical Engineers, Tokyo, Japan, Vol. 2, pp. 234–238.

Bergles, A. E., Bunn, R. L., and Junkhan, G. K., 1974b. "Extended Performance Evaluation Criteria for Enhanced Heat Transfer Surfaces," *Letters in Heat and Mass Transfer*, Vol. 1, pp. 113–120.

Cowell, T. A., 1990. "A General Method for the Comparison of Compact Heat Transfer Surfaces," *Journal of Heat Transfer*, Vol. 112, pp. 288–294.

Kays, W. M., and London, A. L., 1984. *Compact Heat Exchangers*, 3rd edition, McGraw–Hill, New York.

Shah, R. K., 1978. "Compact Heat Exchanger Surface Selection Methods," in *Heat Transfer 1974, Proceedings of the 5th International Heat Transfer Conference*, Hemisphere Publishing Co., New York, Vol. 4, pp. 193–199.

Soland, J. G., Mack, W. M., Jr., and Rohsenow, W. B., 1976. "Performance Ranking of Plate-Pin Heat Exchanger Surfaces," ASME paper 76-WA/HT-31.

Soland, J. G., Mack, W. M., Jr., and Rohsenow, W. B., 1978. "Performance Ranking of Plate-Pin Heat Exchanger Surfaces," *Journal of Heat Transfer*, Vol. 100, pp. 514–519.

Webb, R. L., 1981a. "Performance Evaluation Criteria for Use of Enhanced Heat Transfer Surfaces in Heat Exchanger Design," *International Journal of Heat and Mass Transfer*, Vol. 24, pp. 715–726.

Webb, R. L., 1981b. "The Use of Enhanced Heat Transfer Surface Geometries in Condensers," in *Power Condenser Heat Transfer Technology: Computer Modeling, Design, Fouling*, P. J. Marto, and R. H. Nunn, Eds., Hemisphere Publishing Corp., pp. 287–324.

Webb, R. L., 1982. "Performance Evaluation Criteria for Air-Cooled Finned Tube Heat Exchanger Surface Geometries," *International Journal of Heat and Mass Transfer*, Vol. 25, p. 1770.

Webb, R. L., and Bergles, A. E., 1983. "Performance Evaluation Criteria for Selection of Heat Transfer Surface Geometries Used in Low Reynolds Number Heat Exchangers," in *Low Reynolds Number Flow Heat Exchangers*, S. Kakaç, R. K. Shah, and A. E. Bergles, Eds., Hemisphere Publishing Corp., Washington, D.C., pp. 735–752.

Webb, R. L., and Eckert, E. R. G., 1972. "Application of Rough Surfaces to Heat Exchanger Design," *International Journal of Heat and Mass Transfer*, Vol. 15, pp. 1647–1658.

Webb, R. L., Haman, L. L., and Hui, T. S., 1984. "Enhanced Tubes in Electric Utility Steam Condensers," in *Heat Transfer in Heat Rejection Systems*, S. Sengupta and Y. G. Mussalli, Eds., ASME Symposium, Vol. HTD–Vol. 37, ASME, New York, pp. 17–26.

3.13 NOMENCLATURE

A	Heat transfer surface area, m^2 or ft^2
A_c	Flow area at minimum cross section, m^2 or ft^2
A_{fr}	Frontal area, m^2 or ft^2
A_m	Mean conduction surface area, m^2 or ft^2
B	External-to-internal surface area ratio, dimensionless
b	Distance between parting sheets, m or ft
c_p	Specific heat, J/kg-K or Btu/lbm-°F
d_i	Tube inside diameter, m or ft
D_h	Hydraulic diameter, m or ft
E_h	h/h_s at constant Re, dimensionless
e	Roughness height, m or ft
f	Fanning friction factor, dimensionless
G	Mass velocity, kg/s-m^2 or lbm/s-ft^2
G_{fr}	Mass velocity based on heat exchanger frontal area, kg/s-m^2 or lbm/s-ft^2
g	Gravitational acceleration, m/s^2 or ft/s^2
h	Heat transfer coefficient, W/m^2-K or Btu/hr-ft^2
j	Heat transfer factor, StPr$^{2/3}$; dimensionless
k	Thermal conductivity, W/m-K or Btu/hr-°F
L	Length of flow path in heat exchanger, m or ft
LMTD	Logarithmic mean temperature difference, K or °F
N	Number of tubes in each pass of heat exchanger
NTU	Number of heat transfer units, $UA/c_p W_{min}$; dimensionless
p	Roughness axial pitch, m or ft

Δp	Pressure drop, Pa or lbf/ft^2
P	Pumping power, W or hp
Pr	Prandtl number, $c_p\mu/k$, dimensionless
Q	Heat transfer rate, W or Btu/hr
q	Heat flux, W/m^2 or $Btu/hr\text{-}ft^2$
r	Resistance ratios defined in Table 3.2, dimensionless
r_{tot}	Sum of resistances defined in Table 3.2, $m^2\text{-}K/W$ or $ft^2\text{-}hr\text{-}°F/Btu$
Re	Reynolds number: smooth or rough tube (du/ν); internally finned tube $(D_h u/\nu)$; dimensionless
R_f	Fouling factor, $m^2\text{-}K/W$ or $ft^2\text{-}hr\text{-}°F/Btu$
R_i	$1/h_i A_i$, $m^2\text{-}K/W$ or $ft^2\text{-}hr\text{-}°F/Btu$
R_o	$1/h_o A_o$, $m^2\text{-}K/W$ or $ft^2\text{-}hr\text{-}°F/Btu$
St	Stanton number, $h/c_p G$; dimensionless
T	Temperature, K or °F
ΔT	Average temperature difference between hot and cold streams, K or °F
ΔT_i	Temperature difference between inlet hot and cold streams, K or °F
t	Tube wall thickness, m or ft
u	Average flow velocity, m/s or ft/s
U	Overall heat transfer coefficient, $W/m^2\text{-}K$ or $Btu/hr\text{-}ft^2\text{-}°F$
V	Heat exchanger volume, $A_{fr}L$; m^3 or ft^3
W	Flow rate, kg/s or lbm/s

Greek Symbols

β	Composite resistance ratio defined in Table 3.2, dimensionless
β	Ratio of surface area to exchanger volume, m^{-1} or ft^{-1}
β_T	Volume coefficient of thermal expansion, $-(1/p)(\partial p/\partial T)p$; K^{-1} or R^{-1}
ϵ	Heat exchanger thermal effectiveness, dimensionless
η_o	Surface efficiency, diimensionless
μ	Dynamic viscosity, $kg/s\text{-}m^2$ or $lbm/s\text{-}ft^2$
ν	Kinematic viscosity, m^2/s or ft^2/s
ρ	Density, kg/m^3 or lbm/ft^3
σ	Area ratio, A_{fr}/A_c; dimensionless

Subscripts

f	Designates fouling condition
o	Tube outer surface
p	Plain surface
s	Smooth tube
ref	Reference geometry
STD	Reference condition used in Eqs. 3.17 and 3.18
w	Wall condition

Note: Properties are evaluated at the bulk fluid temperature unless otherwise noted.

4

PERFORMANCE EVALUATION CRITERIA FOR TWO-PHASE HEAT EXCHANGERS

4.1   INTRODUCTION

This chapter describes quantitative methods to define the performance benefits of enhanced heat transfer surfaces in a heat exchanger, for which at least one of the fluids is evaporating or condensing. Pressure drop of a two-phase fluid may reduce the mean temperature difference for heat exchange. This situation does not exist for heat exchange between single-phase fluids. This chapter shows how the performance evaluation criteria performance evaluation criteria (PEC) previously defined for single phase flow may be modified to account for the effect of pressure drop. Generalized PEC are defined for three basic applications of two-phase heat exchangers. Two examples of PEC analysis are presented. The PEC analysis for two-phase heat transfer was originally formulated by Webb [1988]. It is the only PEC analysis advanced to date for two-phase heat transfer.

4.2   OPERATING CHARACTERISTICS OF TWO-PHASE HEAT EXCHANGERS

A "two-phase heat exchanger" is defined here as one for which at least one of the fluids undergoes a phase change. The heat exchangers of interest here are vaporizers (evaporators, boilers, and reboilers) and condensers. Use of two-phase heat exchangers include three general application areas:

1. Work-producing systems, such as a Rankine power cycle.
2. Work-consuming systems, such as a vapor compression refrigeration system

or a vacuum distillation process, which use a compressor to support the process.

3. Heat-actuated systems, such as an absorption refrigeration cycle or a distillation process. As used here, a turbine or compressor would not be in this system.

The quantitative relations given in Chapter 3 assume that the logarithmic mean temperature difference (LMTD) is not affected by the pressure drop, if the total flow rate is held constant. This is not true for two-phase flow. In two-phase flow, pressure drop causes a reduction of the saturation temperature, causing reduced LMTD for fixed entering condenser temperature or for fixed leaving evaporator temperature. This is illustrated in Figure 4.1. Hence the previously proposed quantitative relations must be modified to account for the effect of two-phase pressure drop on the LMTD.

The effect of a given pressure drop on the decrease of saturation temperature (ΔT_{sat}) depends on the value of dT/dp at the operating pressure. The Clausius–Clapeyron equation may be employed to establish the magnitude of dT/dp:

$$\frac{dT}{dp} = \frac{T(v_v - v_l)}{\lambda} \tag{4.1}$$

Figure 4.2 shows dT/dp versus p/p_{cr} for several fluids. This figure shows that dT/dp increases as p/p_{cr} decreases. Typical operating points for condensers and

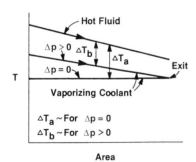

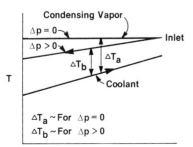

Figure 4.1 Illustration of the effect of two-phase pressure drop on the LMTD. (From Webb [1988].)

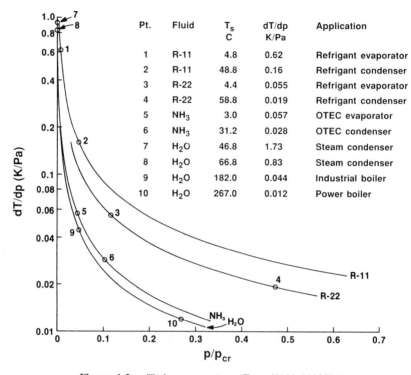

Pt.	Fluid	T_s C	dT/dp K/Pa	Application
1	R-11	4.8	0.62	Refrigant evaporator
2	R-11	48.8	0.16	Refrigant condenser
3	R-22	4.4	0.055	Refrigant evaporator
4	R-22	58.8	0.019	Refrigant condenser
5	NH_3	3.0	0.057	OTEC evaporator
6	NH_3	31.2	0.028	OTEC condenser
7	H_2O	46.8	1.73	Steam condenser
8	H_2O	66.8	0.83	Steam condenser
9	H_2O	182.0	0.044	Industrial boiler
10	H_2O	267.0	0.012	Power boiler

Figure 4.2 dT/dp versus p/p_{cr}. (From Webb [1988].)

evaporators of refrigeration and power cycles are noted in Figure 4.2. Examination of the figure shows that refrigerant evaporators and power cycle condensers will suffer the greatest LMTD reduction due to two-phase pressure drop.

For single-phase flow in heat exchangers, the pump or fan power requirement (P) is directly proportional to the fluid pressure drop for a fixed flow rate. This is why the pumping power is included as an objective function (or constraint) in Table 3.1. However, the same relationship between pumping power and pressure drop does not exist for two-phase heat exchanger applications.

4.3   ENHANCEMENT IN TWO-PHASE HEAT EXCHANGE SYSTEMS

This section describes how enhanced heat transfer may be employed in the previously defined "work-producing," "work-consuming," and "heat-actuated" systems.

4.3.1   Work-Consuming Systems

The purpose of a refrigeration cycle is to cool air, water, or a process fluid. Mechanical power is consumed by the compressor, which "pumps" the evaporated refrigerant from the evaporator pressure to the condenser pressure.

Performance improvements that may be affected by the use of enhanced surfaces are as follows:

1. Reduced heat transfer surface area for fixed compressor power (P_c).
2. Increased evaporator heat duty for fixed compressor lift (pressure difference between condenser and evaporator).
3. Reduced compressor power for fixed evaporator heat duty. This would be affected by reducing the LMTD of the evaporator and/or condenser. This would increase the suction pressure to the compressor and reduce the inlet pressure to the condenser.

For the first case, one would seek reduced surface area in the evaporator and condenser for fixed compressor inlet and outlet conditions. The compressor power is not influenced by pressure drop in either the evaporator or condenser. However, pressure drop in either heat exchanger will reduce the LMTD and reduce the net performance improvement. Figure 4.3 illustrates a refrigeration cycle for this case and shows the effect of two-phase pressure drop on the LMTD.

The second case would maintain constant heat transfer surface area and take advantage of the increased *UA* to obtain increased heat duty. Again, pressure drop would act to reduce the increase of evaporator heat duty. A larger compressor

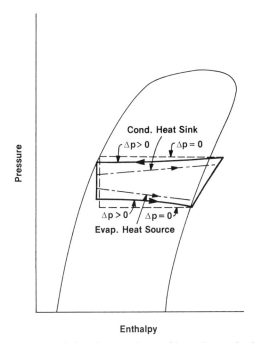

Figure 4.3   Pressure versus enthalpy diagram for a refrigeration cycle showing the effect of pressure drop on the LMTD in each heat exchanger. (From Webb [1988].)

capacity would be required, since the refrigerant flow rate would increase in proportion to the heat transfer increase.

The surface area is maintained constant for the third case, and the UA increase is employed to reduce the LMTD. This allows the compressor lift to be reduced for the fixed evaporator heat duty.

A "vacuum distillation" process may occur at a sufficiently low temperature that a refrigeration system is required to provide condenser coolant (an evaporating liquid) at the required low condensing temperature. Use of an enhanced heat transfer surface in the distillation reflux condenser may allow the compressor of the refrigeration system to operate at a higher suction pressure and thereby save compressor power.

4.3.2   Work-Producing Systems

Next, consider the case of a Rankine power cycle. The boiler feed pump raises the pressure of the working fluid from the condenser pressure to the boiler pressure. This pumping power is calculated by

$$P = (p_b - p_c)Wv_l \tag{4.2}$$

The pressure drop in the boiler is negligible compared to the pressure difference between the boiler and the condenser ($p_b - p_c$). Hence the mechanical energy consumed by frictional pressure drop in the boiler is of negligible concern. However, pressure drop in the boiler or condenser will decrease the thermodynamic efficiency of the power cycle, because of the effect of dT/dp on the LMTD.

Performance improvements that may be affected by the use of enhanced surfaces are as follows:

1. Reduced boiler and/or condenser surface area for constant turbine output (P_t).
2. Increased turbine output for fixed boiler heat input, or fixed condenser heat rejection. This would increase the boiler pressure or reduce the condensing temperature.

The first case maintains fixed boiler exit pressure and/or fixed condenser inlet pressure. Increased pressure drop in either heat exchanger will reduce the LMTD and hence reduce the performance improvement. Based on the information presented in Figure 4.2, one sees that the pressure drop in the boiler will have very small effect on the boiler performance improvement. However, the same pressure drop in the condenser may significantly reduce the LMTD.

The second case maintains constant heat transfer surface area, and it employs the increased UA to allow reduction of the LMTD. If the LMTD in the condenser is reduced, the turbine can expand to a lower back pressure and produce more work. It is unlikely that the LMTD reduction in the boiler would have a significant effect on the turbine power output.

4.3.3 Heat-Actuated Systems

These systems may have a pump to transport liquids, but they do not use compressors or turbines in the process operations. A good example of a heat-actuated system is the absorption refrigeration cycle. This cycle has four heat exchangers that involve condensation or boiling process (evaporator, concentrator, condenser, and absorber). A pump is used to transport the working fluid between the absorber and the concentrator, but a compressor is not employed. In the LiBr–water absorption cycle, water is the refrigerant. Heat, in the form of steam or combustion products, is added in the concentrator to boil off water and concentrate the LiBr–water solution. If the LMTD in any of the four heat exchangers can be reduced, the thermodynamic irreversibility of the heat exchanger processes will be reduced.

A second example of a heat actuated system is the oil-refining process. A number of cascaded process condensers may be employed in a refinery "crude train." The heat rejected by one condenser may provide heat input to an evaporator. Energy, in the form of combustion products, is employed to provide heat input to the top crude. Energy consumption will be decreased if the LMTD to each cascaded heat exchanger is reduced. This is probably more important than using the higher performance of the enhanced surface to reduce the capital cost of the heat exchanger.

A third example is distillation processes, in which heat rejection occurs to cooling water. Heat, in the form of steam, is added to a reboiler and heat is rejected in a water-cooled condenser. Enhanced surfaces may be employed for the same purposes as in the first two examples.

Mechanical energy (e.g., compressors or turbines) is not used in the above examples. Performance improvements may be affected by use of enhanced surfaces for three different purposes:

1. Reduced heat transfer surface area for fixed operating temperatures.
2. Increased heat exchange capacity for fixed amount of heat exchange surface area.
3. Reduced LMTD for fixed amount of heat exchange surface area. This will increase the thermodynamic efficiency of the process or cycle.

Pressure drop of either two-phase stream will act to reduce the LMTD. The smaller LMTD will decrease the surface area reduction possible or will decrease the efficiency improvement.

The above discussion shows that the different operating conditions of two-phase exchangers used in refrigeration and power cycles establishes a need for a different set of PEC than those of Table 3.1, established for single-phase flow. The basic differences are as follows:

1. The LMTD is affected by the two-phase pressure drop.
2. The mechanical power of interest in work-producing/consuming cycles is not a pump or fan; rather it is a turbine or compressor. The power consumed by

the circulation pump of a power cycle is negligible compared to that produced by the turbine. Hence the effect of the heat exchanger performance on the work produced by the turbine, or on the work input to the compressor, is the key point of concern.

3. In heat-actuated systems, the heat exchanger size or the thermodynamic efficiency of the system is influenced by the LMTD in the heat exchangers. Hence the effect of the two-phase fluid pressure drop on the LMTD is the point of concern.

4.4   PEC FOR TWO-PHASE HEAT EXCHANGE SYSTEMS

Having defined the performance improvements that may be affected in two-phase heat exchange systems, it is now possible to prepare a table similar to Table 3.1. Table 4.1 shows PEC corresponding to the FG, FN, and VG cases for the previously discussed two-phase heat exchanger applications.

Table 4.1 is laid out using the same cases as are listed in Table 3.1. However there are several key differences between Table 3.1 and Table 4.1. These differences are as follows:

1. The pumping power variable (P) of Table 3.1 is not used in Table 4.1. This is replaced in work-consuming (or work-producing) systems by the variable P_w, which means work performed by (or on) the system. For heat-actuated systems, one may wish to constrain or reduce the pressure drop, because of its effect on the LMTD. The variable Δp applies for heat actuated systems.

TABLE 4.1   Performance Evaluation Criteria for Two-Phase Heat Exchange System

| Case | Geometry | Fixed | | | | Objective |
		W	P_w	Q	ΔT_i^*	
FG-1a	N, L			$\times$		$\uparrow Q$
FG-1b	N, L	$\times$	$\times$	$\times$		$\downarrow \Delta T_i^*$
FG-3	N, L	$\times$		$\times$	$\times$	$\downarrow P_w^{**}$
FN-1	N	$\times$		$\times$	$\times$	$\downarrow L$
FN-2	N	$\times$		$\times$	$\times$	$\downarrow L$
FN-3	N	$\times$		$\times$		$\downarrow P_w^{**}$
VG-1		$\times$	$\times$	$\times$		$\downarrow NL$
VG-2a	$N L$		$\times$		$\times$	$\uparrow Q$
VG-2b	$N L$	$\times$		$\times$		$\downarrow \Delta T_i$
VG-3	$N L$	$\times$		$\times$		$\downarrow P_w^{**}$

*ΔT_i is defined as the temperature difference between the leaving boiling fluid and the entering process fluid for vaporizers. For condensers, it is the difference between the entering vapor temperature and the entering coolant temperature.

**Replace by Δp for heat-actuated systems.

2. The condensers and evaporators considered here are for complete evaporation or condensation. In this case, the heat duty is given by $q = W\Delta x$. Because Δx will be a constant for the evaporator or condenser, $q \propto W$. Hence W and q are not independent variables, as they are treated in Table 3.1.

3. Several of the cases in Table 3.1 do not apply for the present considerations. Hence they do not appear in Table 4.1.

4. Because the variable P_w applies to work-producing or work-consuming systems and Δp applies to heat-actuated systems, separate tables should apply to these two system types. In order to avoid this complication, Table 4.1 is prepared using an asterisk code (*) to differentiate between the two system types.

Table 4.1 is equally applicable for vaporization or condensation inside tubes, or on the outside of tubes in a bundle.

4.5 PEC CALCULATION METHOD

An algebraic calculation must be performed to determine the value of the objective function for the case of interest in Table 4.1. Before any PEC calculations can be made, it is necessary to know the heat transfer coefficient and friction factor (or Δp characteristics) of the enhanced tube as a function of the flow and geometry conditions. The heat transfer and friction characteristics for vaporization or condensation on enhanced surfaces depend on the following variables:

1. Geometry: The tube diameter and the specific geometric features of the enhanced surface.

2. The operating pressure (or saturation temperature).

3. The mass velocity (G), the vapor quality (x), and the heat flux (q) or the wall-to-fluid temperature difference $(T_w - T_s)$.

The data may be measured as a function of the local vapor quality, or as average values over a given inlet and exit vapor quality. Jensen [1988] provides a review of correlations for forced convection vaporization and condensation inside tubes (plain and enhanced). Webb and Gupte [1992] provide a more recent review for forced convection vaporization inside tubes and across tube bundles. Chapters 12 and 13 treat convective vaporization and condensation in enhanced tubes, respectively.

With the basic heat transfer and friction correlations in hand, for the particular enhanced geometry of interest, one is prepared to perform the PEC analysis. Two examples of the PEC calculation will be presented.

4.5.1 PEC Example 4.1

Consider case FN-2 for a refrigeration evaporator with tube-side vaporization. Assume that an evaporator design for vaporization in smooth tubes has been completed. Figure 4.4 illustrates the operating conditions for the smooth tube design.

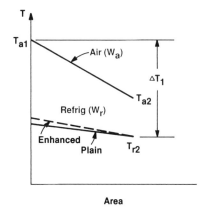

Figure 4.4 Illustration of operating conditions for an air cooled refrigerant evaporator with tube-side vaporization. (From Webb [1988].)

Air enters the evaporator at flow rate W_{a1} and T_{a1} and is cooled to T_{a2}. The total heat duty of the evaporator is Q_{tot}. Assume that the plain tube design has N refrigerant circuits in parallel. The heat duty per circuit is given by $Q_c = Q_{tot}/N$ and the refrigerant mass flow rate per circuit is given by $W_{rc} = W_r/N$. The refrigerant mass velocity is given by $G = W_{rc}/A_c$, where A_c is the flow cross-section area for the circuit.

The enhanced tube design operates at the conditions shown on Figure 4.4, except for the entering refrigerant temperature (T_{r1}), which is a dependent variable. The objective is to reduce the evaporator size (circuit length) for fixed compressor suction pressure, which fixes T_{r2}. It is assumed that the frontal area of the finned tube heat exchanger is held constant, and that the evaporator length reduction is accomplished by removing tube rows in the air-flow direction. If the pressure drop in the enhanced evaporator tube is greater than that of the plain tube, the LMTD will be reduced, as is qualitatively illustrated by the dashed line of Figure 4.4. The evaporator is designed to operate between entering vapor quality x_1 and 1.0 exit vapor quality. Assume that a correlation exists to calculate the "circuit average" refrigerant heat transfer coefficient (h_i) at the given operating pressure. This is given by

$$h_i = fcn(G,q) \tag{4.3}$$

Similarly, the pressure gradient is assumed to be known in the form

$$\Delta p = fcn(G,L) \tag{4.4}$$

The PEC calculation proceeds as follows:

1. Guess the required circuit length and calculate the circuit averaged heat flux from the required heat load (Q_c) as

$$q = \frac{Q_c}{\pi d_i L} \tag{4.5}$$

2. Calculate the refrigerant-side h and Δp using the known refrigerant mass velocity (G), heat flux (q), and the assumed circuit length (L) using Equations 4.3 and 4.4.

3. Calculate the UA value of the enhanced tube exchanger from

$$\frac{1}{UA} = \frac{1}{h_i A_i} + \frac{t_w}{k_w A_w} + \frac{1}{h_o A_o} \tag{4.6}$$

4. From the calculated refrigerant pressure drop, calculate the entering refrigerant temperature (T_{r1}). Then calculate the LMTD for the heat exchanger.

5. Calculate the "available heat transfer rate" (Q_{av}) by

$$Q_{av} = UA \cdot \text{LMTD} \tag{4.7}$$

6. If $Q_{av} > Q_c$ (the required value), one assumes a shorter circuit length and repeats steps 1–5. The procedure is repeated until $Q_{av} = Q_c$. The material savings offered by the enhanced surface is L/L_p, where L_p is the evaporator circuit length of the plain tube design.

It is possible that the enhanced tube design may provide better performance characteristics if it were designed to operate at a different flow rate per circuit than the one used for the plain tube design. This question must be resolved by performing a parametric study of the enhanced tube design as a function of the number of circuits. This calculation procedure would follow the previously outlined steps.

4.5.2   PEC Example 4.2

Assume the same plain tube heat exchanger design employed in PEC Example 4.1. For the present problem, the enhanced evaporator tube will be used to increase the compressor suction pressure and, hence, reduce the compressor power. This calculation involves case FN-3 of Table 4.1. The same enhanced tube operating conditions are the same as for PEC Example 1.1, except now the refrigerant temperature leaving the evaporator (T_{r2}) is not fixed. One may be tempted to fix the entering evaporator temperature (T_{r1}) to the value of the plain tube design. However, this has little meaning, since the entering temperature inherently adjusts to meet the compressor suction pressure and the evaporator circuit pressure drop. A more direct approach is to specify the desired compressor suction pressure which fixes T_{r2} and to calculate the circuit length of the enhanced tube that gives the required heat duty and the desired T_{r2}. If the desired T_{r2} is set too high, one will find that a solution does not exist for the smaller LMTD corresponding to the increased T_{r2}. As T_{r2} increases, a longer evaporator circuit will be required, and the pressure drop will increase with increasing evaporator circuit length. It is possible that the increased UA of the enhanced tube design will not be sufficiently high to overcome the smaller LMTD resulting from the increased T_{r2} and the possible larger pressure drop of the enhanced tube design.

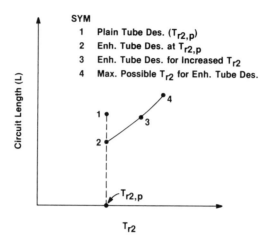

Figure 4.5   Illustration of the effects of enhanced evaporator tubes on the evaporator performance. (From Webb [1988].)

The recommended approach for this problem is then to specify the desired T_{r2} and proceed to calculate the required evaporator circuit length. This calculation would proceed following the same procedures outlined for PEC Example 1.1. For a specified T_{r2} one calculates the circuit length (L) required to satisfy the specified evaporator load. One may repeat the calculations for several specified values of T_{r2} and plot a graph as indicated by Figure 4.5. This graph provides information to evaluate the trade-offs between evaporator cost and compressor power.

4.6   CONCLUSIONS

PEC have been defined for application to heat exchangers in which at least one of the fluids is a two-phase fluid. Because pressure drop of the two-phase fluid may affect the mean temperature difference in the heat exchanger, the PEC equations must account for the effect of this situation. An assessment of two-phase heat exchanger applications showed that the "reduced pumping power" objective function used for single-phase flow is not generally applicable to the two-phase flow situation. Three generic applications for two- phase heat exchangers were defined and provided a basis for establishing relevant objective functions and operating constraints. The 11 PEC cases defined for two-phase heat exchangers are closely related to the 12 single-phase PEC cases. A new objective function is defined for two-phase exchangers and relates to the work done on or by the system in which the two-phase heat exchanger is installed. This objective function replaces the "reduced pumping power" objective function for single-phase flow. Two examples using the two-phase PEC were presented.

4.7 REFERENCES

Jensen, M. K., 1988. "Enhanced Forced Convective Vaporization and Condensation Inside Tubes," in *Heat Transfer Equipment Design*, R. K. Shah, E. C. Subbarao, and R. A. Mashelkar, Eds., Hemisphere Publishing Corp., Washington, D.C., pp. 681–696.

Webb, R. L. and Gupte, N. S., 1992. "A Critical Review of Correlations for Convective Vaporization in Tubes and Tube Banks," *Heat Transfer Engineering*, Vol. 13, No. 3, pp. 58–81.

Webb, R. L., 1988. "Performance Evaluation Criteria for Enhanced Surface Geometries Used in Two-Phase Heat Exchangers," in *Heat Transfer Equipment*, R. K. Shah, E. C. Subbarao and R. A. Mashelkar, Eds., Hemisphere Publishing Corp., Washington, D.C., pp. 697–706.

4.8 NOMENCLATURE

A_c	Cross-sectional flow area, m^2 or ft^2
A_i	Inside tube surface area, m^2 or ft^2
A_o	Outside tube surface area, m^2 or ft^2
d_i	Tube inside diameter, m or ft
FG	FG criteria used in Table 4.1
FN	FN criteria used in Table 4.1
G	Mass velocity, kg/s m^2 or lbm/s-ft^2
H	Head ($\Delta p/\rho$), H$_p$ (pump), H$_s$ (smooth tube), H$_e$ (enhanced tube)
h	Heat transfer coefficient, W/m K or Btu/hr-ft-°F
k_w	Thermal conductivity of tube, W/m-K or Btu/hr-ft-°F
L	Circuit length, m or ft
LMTD	Logarithmic mean temperature difference, K or °F
N	Number of flow circuits in parallel, dimensionless
P	Power: P_c (compressor), P_t (turbine), or P_w (work done on/by system); W or hp
p	Pressure: p_b (boiler), p_c (condenser), or p_{cr} (critical); Pa
Δp	Pressure difference, Pa or lbf/in.^{2}
Q	Heat load, W or Btu/hr
Q_{av}	Available heat transfer rate, W or Btu/hr
Q_c	Heat load per circuit, W or Btu/hr
Q_{tot}	Heat load on all circuits, W or Btu/hr
q	Heat flux, W/m^2 or Btu/hr ft^2
R_w	Thermal resistance of tube wall, m^2-K/W or hr-ft^2-°F/Btu
T	Temperature, T_r (refrigerant), T_w (wall), or T_{sat} (saturation); K or °F
ΔT_i	Difference between entering fluid temperatures, K or °F
T_s	Saturation temperature, K or °F
ΔT_{sat}	$T_w - T_s$, K or °F
T_w	Wall temperature, K or °F
U	Overall heat transfer coefficient, W/m^2-K or Btu/hr-ft^2-°F
VG	VG Criteria used in Table 4.1

v_l	Specific volume of liquid, m³/kg
v_g	Saturated vapor specific volume, m³/kg or ft³/lbm
W	Mass flow rate, kg/s or lbm/s
x_i	Inlet vapor quality to heat exchanger, dimensionless
Δx	Vapor quality change, dimensionless

Greek Symbol

λ	Latent heat of vaporization, J/kg or Btu/lb

Subscripts

a	Air
c	Per refrigerant circuit
o	Refers to value on the outside of the tube
p	Plain or smooth surface geometry
r	Refrigerant
t	Refers to tube side value
1	Inlet to heat exchanger
2	Exit of heat exchanger

5

PLATE-AND-FIN EXTENDED SURFACES

5.1 INTRODUCTION

This chapter discusses enhanced extended surface geometries for the plate-and-fin heat exchanger geometry, which is shown in Figure 5.1. Normally, at least one of the fluids used in the plate-and-fin geometry is a gas. In forced convection heat transfer between a gas and a liquid, the heat transfer coefficient of the gas is typically 5–20% that of the liquid. The use of extended surfaces will reduce the gas-side thermal resistance. However, the resulting gas-side resistance may still exceed that of the liquid. In this case, it will be advantageous to use specially configured extended surfaces which provide increased heat transfer coefficients. Such special surface geometries may provide heat transfer coefficients 50–150% higher than those given by plain extended surfaces. For heat transfer between gases, such enhanced surfaces will provide a substantial heat exchanger size reduction. There is a trend toward using enhanced surface geometries with liquids for cooling electronic equipment. Data taken for gases may be applied to liquids if the Prandtl number dependency is known. Brinkman et al. [1988] provide data for water and a dielectric fluid (FC-77), but they do not provide a Prandtl number dependency for their data. In the absence of specific data on Prandtl number dependency, one may assume St $\propto$ Pr$^{-2/3}$.

Figure 5.2 shows six enhanced surface geometries. Typical fin spacings are 300–800 fins/m. Because of the small hydraulic diameter and low density of gases, these surfaces are usually operated with $500 < \mathrm{Re_{Dh}} < 1500$ (hydraulic diameter basis). To be effective, the enhancement technique must be applicable to low-Reynolds-number flows. The use of surface roughness will not provide appreciable enhance-

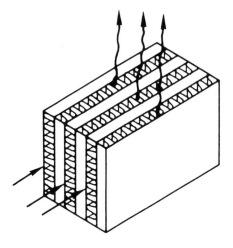

Figure 5.1   A cross-flow plate-and-fin heat exchanger geometry.

ment for such low-Reynolds-number flows. Two basic concepts have been extensively employed to provide enhancement:

1. Special channel shapes, such as the wavy channel, which provide mixing due to secondary flows or boundary layer separation within the channel.
2. Repeated growth and wake destruction of boundary layers. This concept is

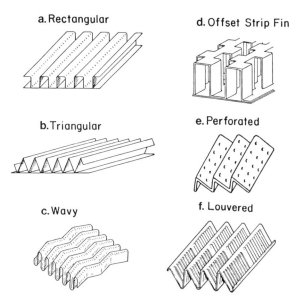

Figure 5.2   Plate-fin exchanger surface geometries. (a) Plain rectangular fins. (b) Plain triangular fins. (c) Wavy fins. (d) Offset strip fins. (e) Perforated fins. (f) Louvered fins. (From Webb [1987].)

employed in the offset strip fin, in the louvered fin, and, to some extent, in the perforated fin.

Sources or summaries of data on plate-and-fin surface geometries include Kays and London [1984], Creswick et al. [1964], and Tishchenko and Bondarenko [1983] (plain fins). Other sources are noted in the following sections.

5.2 OFFSET STRIP FIN

Figure 5.3 illustrates the enhancement principle of the offset strip fin (OSF) shown in Figure 5.2d. A laminar boundary layer is developed on the short strip length, followed by its dissipation in the wake region between strips. Typical strip lengths are 3–6 mm, and the Reynolds number (based on strip length) is well within the laminar region. The enhancement provided by an OSF of $L_p/D_h = 1.88$ is shown by Figure 5.4. The OSF and plain fins are taken from Figures 10–58 and 10–27, respectively, of Kays and London [1984]. The table in Figure 5.4 compares the dimensions with the OSF scaled to the same hydraulic diameter as that of the plain fin. The scaled dimensions (fin height and fin thickness) of the OSF are approximately equal to those of the plain fin surface. Therefore, the performance difference is due to the enhancement provided by the short strip lengths (6.6 mm) of the OSF. At $Re_{Dh} = 1,000$, the j factor of the OSF is 2.5 times higher than that of the plain fin, and the relative friction factor increase is 3.0. Considering the j/f ratio as an "efficiency index" ratio between heat transfer and friction, the OSF yields 150% increased heat transfer coefficient with a heat transfer to friction ratio (j/f) 83% as high as that of the plain fin. Greater enhancement will be obtained by using shorter strip lengths.

5.2.1 PEC Example 5.1

A more realistic comparison of the OSF and plain fin performance is obtained using case VG-1 of Table 3.1. Assume that the plain fin (subscript p) operates at $Re_{Dh,p} =$

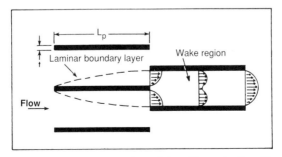

Figure 5.3 Boundary layer and wake region of the offset strip fin. (From Webb [1987].)

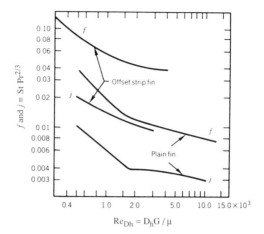

Surface Geometry

	Plain	Offset Strip
Surface designation	10–27	10–58[a]
Fins/m	437	437
Plate spacing (mm)	12.2	11.9
Hydraulic diam (mm)	3.51	3.51
Fin thickness (mm)	0.20	0.24
Offset strip length in flow dir. (mm)	—	6.6

[a] Dimensions geometrically scaled to give same hydraulic diameter as plain fin.

Figure 5.4   Comparison of j and f for the offset strip fin and the plain fin surface geometries. (From Webb [1987].)

834. Equation (3.7), which applies to case VG-1 of Table 3.1, defines the value of G/G_p at which both surfaces satisfy $1 = hA/h_pA_p = P/P_p = W/W_p$. Hence, we must satisfy

$$\frac{G}{G_p} = \left(\frac{j/j_p}{f/f_p} \right)^{1/2} \tag{5.1}$$

It is necessary to iteratively solve Equation 5.1 for G/G_p, reading the j and f values for the OSF surface from Figure 5.4. The final solution of Equation 5.1 yields $G/G_p = 0.923$. Thus, Re $= 0.923 \times 834 = 770$. The j and f values of the OSF at $\text{Re}_{Dh} = 770$ are now known.

The surface area ratio (A/A_p) is obtained using Equation 3.2, repeated here as Equation 5.2.

$$\frac{hA}{h_pA_p} = \frac{j}{j_p} \frac{A}{A_p} \frac{G}{G_p} \tag{5.2}$$

Substituting the known values of j/j_p and G/G_p in Equation 5.2, one calculates $A/A_p = 0.446$. The OSF geometry requires only 44.6% as much surface area to provide the same hA as the plain-fin geometry. Using Equation 3.3, one may show that the flow frontal area must be increased 10% to maintain $P/P_p = 1$.

5.2.2   Prediction of j and f Factors of Offset Strip Fin

Kays [1972] proposed a simple, approximate model to predict the j and f versus Re curves for the OSF. The model assumes that there are laminar boundary layers on the fins and that the boundary layers developed on the fin are totally dissipated in the wake region between fins. Using the equations for laminar flow over a flat plate in a free stream, Kays' analysis gives

$$j = 0.664 \, \mathrm{Re}_L^{-0.5} \tag{5.3}$$

$$f = \frac{C_D t}{2L_p} + 1.328 \mathrm{Re}_L^{-0.5} \tag{5.4}$$

The first term in Equation 5.4 accounts for the form drag on the plate. The form drag contribution is proportional to the fin thickness (t), and it has a negligible effect on the heat transfer coefficient. Kays suggests use of $C_D = 0.88$ based on potential flow normal to a thin plate. This value is derived by Milne-Thomson (1960). Although this model is only approximate, it will allow the designer to predict the effect of strip length and thickness.

Multiple regression, power law correlations have been developed by Wieting [1975] and by Joshi and Webb [1987] to predict the j and f versus $\mathrm{Re}_{\mathrm{Dh}}$ characteristics for the OSF array; these correlations were based on data for 21 OSF geometries. Both references present separate correlations for the laminar and nonlaminar wake regions. Transition from the laminar region occurs at the Reynolds number where the j factor departs from log-linear form as the Reynolds number is increased. This transition point is defined on Figure 5.5. Joshi and Webb [1987] determined this point from the j versus $\mathrm{Re}_{\mathrm{Dh}}$ data for 21 core geometries, and they developed a semiempirical correlation to define the transition point. The Reynolds number (GD_h/μ) at which the wake departs from laminar flow is defined as $\mathrm{Re}_{\mathrm{Dh,tr}}$ and is given by

$$\mathrm{Re}_{\mathrm{Dh,tr}} = 257 \left(\frac{L_p}{s} \right)^{1.23} \left(\frac{t}{L_p} \right)^{0.58} \left(\frac{D_h}{\delta_{\mathrm{mom}}} \right) \tag{5.5}$$

where $\delta_{\mathrm{mom}} \equiv t + 1.328 \, L_p/\mathrm{Re}_L^{0.5}$ is fin thickness, plus twice the momentum thickness of the boundary layer at the end of the strip length. The flow visualization measurements of Joshi and Webb show that this transition point occurs before eddies are shed in the wake region.

The Joshi and Webb [1987] correlation predicts the j factor in the transition region using log-linear interpolation between j_L (at $\mathrm{Re}_{\mathrm{Dh,tr}}$) and j_T (at $\mathrm{Re}_{\mathrm{Dh,tr}} + 1000$). A similar equation is used for the friction factor. The correlations predicted 82% of the f data and 91% of the j data within $\pm 15\%$. The geometric parameters of the test arrays on which Equations 5.7 and 5.8 were based are $0.13 < s/b < 1.0$,

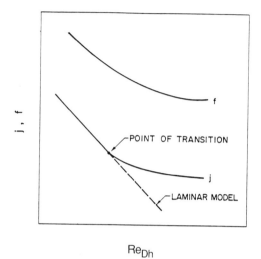

$$Re_{Dh}$$

Figure 5.5 Illustration of transition of the j factor from the laminar region. (From Joshi and Webb [1987].)

$0.012 < t/L_p < 0.048$ and $0.04 < t/s < 0.20$. The Wieting [1975] correlation assumes $Re_{Dh,tr} = 1000$ for all geometries.

Manglik and Bergles [1990] used an asymptotic correlation method developed by Churchill and Ursagi [1972] to provide an alternate correlation for the OSF. This method requires that one know the asymptotic values for very small and very high Reynolds number values. The advantage of the method is that it gives a single, smooth curve for the entire Reynolds number range. However, the equation is more complex, and may not be worth the complication, unless a prediction is required in the transition region, whose lower limit is given by Equation 5.5. Their correlation is based on 18 of the 21 surface geometries in the Joshi and Webb [1987] data set. They give no reason for excluding three of the geometries. Manglik and Bergles first developed their own linear multiple regression correlations for the laminar and nonlaminar regions. They used Equation 5.5 to define the upper limit of the laminar correlation. The laminar region correlations ($Re_{Dh} \leq Re_{Dh,tr}$) are given by

$$f = 9.624Re_{Dh}^{-0.742} \left(\frac{s}{b} \right)^{-0.186} \left(\frac{t}{L_p} \right)^{0.305} \left(\frac{t}{s} \right)^{-0.266} \tag{5.6}$$

$$j = 0.652Re_{Dh}^{-0.540} \left(\frac{s}{b} \right)^{-0.154} \left(\frac{t}{L_p} \right)^{0.150} \left(\frac{t}{s} \right)^{-0.068} \tag{5.7}$$

The nonlaminar region correlations ($Re_{Dh} \geq Re_{Dh,tr} + 1000$) are given by

$$f = 1.870Re_{Dh}^{-0.299} \left(\frac{s}{b} \right)^{-0.094} \left(\frac{t}{L_p} \right)^{0.682} \left(\frac{t}{s} \right)^{-0.242} \tag{5.8}$$

$$j = 0.244Re_{Dh}^{-0.406} \left(\frac{s}{b} \right)^{-0.104} \left(\frac{t}{L_p} \right)^{0.196} \left(\frac{t}{s} \right)^{-0.173} \tag{5.9}$$

The general friction factor correlation for the laminar, transition, and turbulent regions was developed using the asymptotes given by Equations 5.6 and 5.8, for Re → 0 and Re → ∞, respectively. This method, developed by Churchill and Ursagi [1972] , yields the final friction correlation. A similar procedure was used to obtain the final j-factor correlation,

$$f = 9.624 \text{Re}_{\text{Dh}}^{-0.742} \left(\frac{s}{b} \right)^{-0.186} \left(\frac{t}{L_p} \right)^{0.305} \left(\frac{t}{s} \right)^{-0.266}$$ (5.10)

$$\left[1 + 7.669E - 8\text{Re}_{\text{Dh}}^{4.43} \left(\frac{s}{b} \right)^{0.92} \left(\frac{t}{L_p} \right)^{3.77} \left(\frac{t}{s} \right)^{0.236} \right]^{0.1}$$

which is given by

$$j = 0.652 \text{Re}_{\text{Dh}}^{-0.540} \left(\frac{s}{b} \right)^{-0.154} \left(\frac{t}{L_p} \right)^{0.150} \left(\frac{t}{s} \right)^{-0.068}$$ (5.11)

$$\left[1 + 5.63E - 5\text{Re}_{\text{Dh}}^{1.34} \left(\frac{s}{b} \right)^{0.51} \left(\frac{t}{L_p} \right)^{0.46} \left(\frac{t}{s} \right)^{-1.06} \right]^{0.1}$$

The hydraulic diameter definition applicable to Equations 5.6 through 5.11 is defined as

$$D_h = \frac{4sbL_p}{[2(sL_p + bL_p + tb) + ts]}$$ (5.12)

Notice that as Re_{Dh} → 0, Equations 5.10 and 5.11 asymptotically approach Equations 5.6 and 5.8, respectively. As Re_{Dh} → ∞, Equations 5.10 and 5.11 approach Equations 5.7 and 5.9, respectively. The 0.1 exponent on Equations 5.10 and 5.11 is empirically chosen to give the best fit of the data set. Equations 5.10 and 5.11 predict all of the heat transfer data and approximately 90% of the friction data within ±20%.

Webb and Joshi [1983] also used the Churchill and Ursagi [1972] method to correlate their friction data on eight scaled-up OSF model geometries. They correlated 99% of their data within ±15%. Usami [1991] performed a very comprehensive test program using 48 scaled up OSF geometries. He varied the fin thickness (t), the louver pitch (L_p), the fin pitch (p_f), and the number of louvers in the flow depth. Some flow visualization was done and friction data were taken for all geometries.

Note that the multiple regression methods, including the Churchill and Ursagi [1972] variant, are strictly empirical. A rationally based method would numerically solve the momentum and energy equations or would use an analytical model. Patankar and co-workers have published several papers in which the numerical method is used. A summary of this work is given by Patankar [1990]. This approach is generally limited to two-dimensional laminar flow. Thus, the analysis assumes laminar wakes and $s/h = 0$.

Joshi and Webb [1987] have developed a rationally based model to predict the j

and f versus Re_{Dh} characteristics of the OSF array. This model properly accounts for (1) all geometric factors of the array ($\alpha \equiv s/b$, $\delta \equiv t/L_p$, $\gamma \equiv t/s$), (2) heat transfer to the base surface area to which the fins are attached, and (3) the nonlaminar region. Writing an energy balance on the cell shown in Figure 5.6, one obtains

$$\eta Nu = \frac{1 - \gamma}{(1 + \alpha + \delta)^2} \left(\frac{1 + \delta}{1 + \alpha + \delta} \eta_f Nu_p + \frac{\alpha}{1 + \alpha + \delta} Nu_e \right) \quad (5.13)$$

Nu_p ($= h_p 2s/k$) involves the heat transfer coefficient on the strip fin, and Nu_e ($= h_e D_h/k$) involves the heat transfer coefficient on the top and bottom channel walls. Similarly, a force–momentum balance on the Figure 5.6 cell gives

$$f = \frac{1 - \gamma}{(1 - \gamma)(1 + \alpha + \delta)} \left(f_p + \alpha f_e + \frac{C_D t}{2L_p} \right) \quad (5.14)$$

where f_p and f_e are the friction factors on the strip fin and on the top and bottom walls. Comparison of Equation 5.14 with Equation 5.4 shows that the simple Kays [1972] model neglects the friction on the top and bottom walls, and does not include the expression of geometric factors α, δ, and γ which multiplies Equation 5.14. The model is segregated into two submodels—one for the laminar region and one for the nonlaminar (or "turbulent") region. Equation 5.5 is used as the criterion to select the laminar or nonlaminar region models. Table 5.1 shows the equations used to predict Nu_p, Nu_e, f_p, and f_e for the laminar region.

The laminar Nu_p and f_p terms use the numerical solution of Sparrow and Liu [1979] for $\alpha = 0$ and for zero-thickness plates. That analysis assumes laminar flow on the fin and wake regions. The term $F_{h,\alpha}$ and $F_{f,\alpha}$ correct for the effect of channel aspect ratio (α) and are taken from Figure 5.7. The $F_{h,\alpha}$ is the ratio of the Nu for an entrance length region of aspect ratio α divided by the entrance length Nu for a parallel plate channel ($\alpha = 0$). The analytical solutions used to plot Figure 5.7 were taken from Shah and London [1978]. Figure 5.7 shows that $F_{f,\alpha}$ is greater than 0.95 and may be neglected for f_p. The terms Nu_e and f_e are predicted using curve fits for

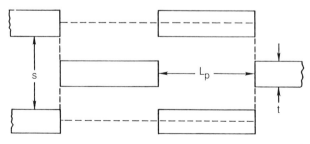

Figure 5.6   Unit cell used to derive the Joshi and Webb [1987] analytical model for the OSF. (From Joshi and Webb [1987].)

TABLE 5.1 Equations Used in the Joshi and Webb [1987] Model for the OSF

Term	Equation
Laminar-Wake Region	
Nu_p	$F_{h,\alpha}[24.2 - 3690L_s^+ + 37.0E + 4(L_s^+)^2]$
Nu_e	$7.45 - 16.9\alpha + 22.1\alpha^2 - 9.75\alpha^3$
f_p	$F_{f,\alpha}[65.5 - 11.63E + 3(L_s^+) + 13.38E + 5(L_s^+)^2]/Re_s$
f_e	$(23.94 - 30.05\alpha + 32.37\alpha^2 - 12.08\alpha^3)/Re_{sh}$
C_D	0.88
Turbulent-Wake Region	
Nu_p	$0.36(Re_{Dh})^{0.067}Pr^{1/3}(L_p/D_h)^{-0.174}$
Nu_e	$0.023(Re_{Dh})^{0.8}Pr^{1/3}$
f_p	$15.33(Re_{Dh}) - 0.785(L_p/D_h)^{-0.324}$
f_e	$0.079(Re_{sh})^{-0.25}$
C_D	0.88

$$Re_{Dh} = \frac{D_h \nu}{2} \quad Re_{sn} = \frac{D_{sh} u}{2} \quad Re_s = \frac{2su}{2}$$

$$D_h = \frac{2s(1-\gamma)}{1+\alpha+\delta} \quad L_s^+ = \frac{L_p}{2s}\frac{1}{Re_s}$$

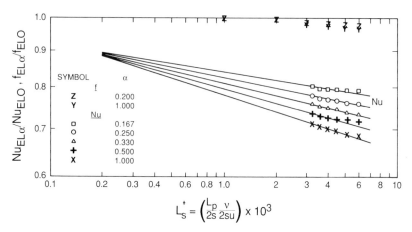

Figure 5.7   Correction factors for Nu_p in finite aspect channels from Joshi and Webb [1987].

fully developed laminar flow in rectangular channels from Shah and London [1978]. The Reynolds number, Re_{sh}, is for an s by h channel shape and is related to Re_{Dh} by

$$Re_{sh} = \frac{1 + \alpha}{1 + \alpha + \delta} Re_{Dh} \qquad (5.15)$$

The terms Nu_e and f_e for the turbulent region are predicted by standard correlations for fully developed turbulent flow in smooth channels. A semiempirical approach was used to develop Nu_p and f_p for the turbulent region. Using the predicted values for the second term in Equation 5.13 and the second and third terms in Equation 5.14, the terms Nu_p and f_p were calculated for each data point. Then a multiple regression analysis was performed to obtain the correlations for Nu_p and f_p shown in Table 5.1. The value 0.88 is used for the drag coefficient (C_D) as was used in the Kays friction model (Equation 5.4).

Figure 5.8 shows ratio of the predicted-to-experimental j and f factors for the 21 cores in the database. The root mean square (rms) error of the j-factor prediction is 11.5%, and all except two cores are predicted within ±20%. The rms error of the f-factor correlation is 16.8%, and 16 of the 21 cores are predicted within ±20%. This is fully competitive with the Manglik and Bergles [1990] empirical correlation. Note that Manglik and Bergles excluded three of the 21 cores in the database.

There is a possibility that burrs may be formed on the upstream and downstream fin edges during the shearing operation to make the OSF geometry. The existence of such burrs was not established by the original experimenters for the 21 cores in the database. In order to investigate this possibility, Webb and Joshi [1983] used their asymptotic correlation for the eight scaled-up, "burr free" geometries to predict the friction factor of the 21-actual-core database. The friction factors of the actual cores were underpredicted by 10–20%. Hence, it appears that the some degree of burrs existed in the 21 actual cores.

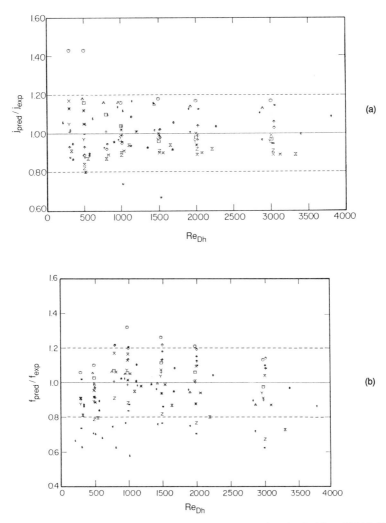

Figure 5.8 Error plots for prediction of the 21 OSF cores using the Joshi and Webb [1987] model. (a) The heat transfer predictions. (b) The friction factor predictions.

5.2.3 Degree of Fin Offset

As conceived, the strips of the OSF fin are offset one-half fin spacing. In this configuration, the wake length is equal to the strip width in the flow direction. However, this may not be the optimum displacement.

It is typically assumed that the thermal and velocity boundary layers are fully dissipated in the wake region. However, there is no rational basis for assuming that full dissipation will occur for a wake length equal to the strip width. Kurosake et al.

[1988] have investigated the benefits of offsetting the strips 30% of the fin pitch, rather than the usual 50%. This provides a wake length of three strip widths. Their work was performed using an array having a total of five rows of strips in the flow direction. They measured the average Nusselt number on each row of strips. Figure 5.9a shows their results for the standard OSF arrangement. This figure shows that the Nu of row 1 equals that of row 2, and the Nu of row 3 equals that of row 4. Row 5 has a smaller Nu than that of rows 3 or 4. Figure 5.9b shows that the results for row 3 offset 30% of the fin pitch. In this array, the wake length between rows 1 and 5 is three strip widths, rather than one strip width. Figure 5.9b shows that the Nu for row 5 is 10–20% higher than that for row 5 of the array shown in Figure 5.9a. Hatada and Senshu [1984] have also studied the effects of fin offset for the OSF. Their heat transfer measurements are discussed in Section 5.4.

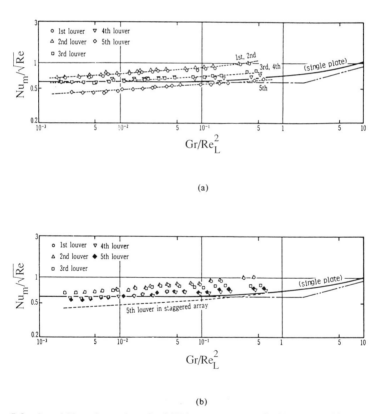

(a)

(b)

Figure 5.9 Local Nusselt numbers in OSF louver array studied by Kurosaki et al. [1988]. (a) Standard array with 50% offset in rows 2 and 4. (b) Array with 50% offset in rows 2 and 4, and 30% offset in row 3. (From Kurosaki et al. [1988].)

5.3 LOUVERED FIN

The louvered fin geometry of Figure 5.2f bears a similarity to the OSF. Rather than offsetting the slit strips, the entire slit fin is rotated 20–45 degrees relative to the air-flow direction. The louvered surface is the standard geometry for automotive "radiators." Current radiators use a louver strip width (in the air-flow direction) of 1.0–1.5 mm. For the same strip width, the louver-fin geometry provides heat transfer coefficients comparable to those of the OSF. Shah and Webb [1982] provide descriptions of modified louver fin geometries and give references to several data sources. Davenport [1983a, 1983b] provides data on 32 one-row cores. Davenport systematically varied the louver dimensions for two louver heights (12.7 and 7.8 mm). Sunden and Svantesson [1990] investigated louvers that are not oriented 90 degrees to the air flow and found that they give lower j factors than the standard 90-degree orientation. Aoki et al. [1989] and Fujikake [1983] measured the effect of louver angle and louver pitch in specially instrumented laboratory test sections. Kajino and Hiramatsu [1987] and Achaichia and Cowell [1988] provide numerical simulations for laminar flow in a variety of louver arrays. Suga and Aoki [1991] performed an extensive numerical study to determine the optimum value of p_f/L_p for fixed louver angle. Their analysis was performed for louver angles of 20, 26, and 30 degrees and $p_f/L_p \le 1.125$. They concluded that the optimum design is given by $p_f/L_p = 1.5 \tan \theta$. The optimum designs provide a good balance between high heat transfer performance and pressure drop.

5.3.1 Heat Transfer and Friction Correlations

Davenport [1983b] tested 32 one-row louver fin geometries (Figure 5.10a) and developed multiple regression correlations for the j and f versus Re_{Dh}. Their correlations are given below.

$$j = 0.249 Re_L^{-0.42} L_h^{0.33} H^{0.26} \left(\frac{L_L}{H} \right)^{1.1} \qquad (300 < Re_{Dh} < 4000) \quad (5.16)$$

$$f = 5.47 Re_L^{-0.72} L_h^{0.37} L_p^{0.2} H^{0.23} \left(\frac{L_L}{H} \right)^{0.89} \qquad (70 < Re_{Dh} < 1000) \quad (5.17)$$

$$f = 0.494 Re_L^{-0.39} H^{0.46} \left(\frac{L_h}{L_p} \right)^{0.33} \left(\frac{L_L}{H} \right)^{1.1} \qquad (1000 < Re_{Dh} < 4000) \quad (5.18)$$

Equations 5.16–5.18 contain dimensional terms; the required dimensions are expressed in millimeters. Figure 5.10a illustrates the one-row cores tested, and Figure 5.10b defines the terms in the equations. The correlation contains the louver height ($2L_h = L_p \sin \theta$) rather than the louver angle (θ). Note that the louvers were made with a wide fin base for better solder contact with the tube. Typical commercial louver arrays are formed as Figure 5.2f, with the fin tips soldered to the tube (or separator plate). Also notice that the louver length (L_L) is shorter than the fin height

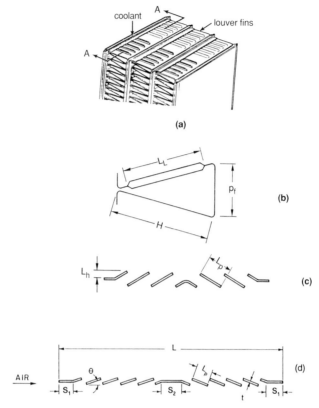

Figure 5.10   One-row louver-fin core geometry tested by Davenport [1983b]. (a) Illustration of core geometry. (b) View of fin in air-flow direction. (c) Cross section A-A of Figure 5.10a as used by Davenport. (d) Precise illustration of cross section A-A of Figure 5.10a. (From Webb [1987].)

(H), which is typical for the louver fin geometry. The characteristic dimension in the Reynolds number limits is D_h, not L_L as used in the equations. The author has used Equations 5.16-to-5.18 to predict the heat transfer performance of automotive radiators and has found agreement to within ±15% of the test data. The Davenport correlation is somewhat flawed, because it does not fully account for all the dimensional variables in the louver fin geometry. Figure 5.10c shows a precise depiction of the cross section A-A on Figure 5.10a. Dimension S_1 defines the length of the inlet louver, and dimension S_2 defines the length of the flow re-direction louver. The Davenport correlation does not contain dimensions S_1 and S_2, nor does it contain the number of louvers. Dillen and Webb [1993] have developed a rationally based correlation, based on the analytical model of Sahnoun and Webb [1992], which includes the S_1, S_2 dimensions, and the number of louvers.

5.3.2 Flow Structure in the Louver Fin Array

Although louvered surfaces have been in existence since the 1950s, it has only been within the past 20 years that serious attempts have been made to understand the flow phenomena and performance characteristics of the louvered fin. Until the flow visualization studies of Davenport [1980], it was assumed that the flow is parallel to the louvers. At very low values of Re_L, Davenport observed that the main flow stream did not pass through the louvers. However, at high values of Re_L the flow became nearly parallel to the louvers. Davenport speculated that at low air velocities the developing boundary layers on adjacent louvers became thick enough to effectively block the passage, resulting in nearly axial flow through the array. The Davenport results are also discussed by Achaichia and Cowell [1988].

Figure 5.11 shows the Stanton number data of Achaichia and Cowell [1988] for three louver-fin geometries used on plate fins (Figure 6.16) with 11-mm tube pitch. At the highest Re_L, the data are parallel (but lower than) those for laminar boundary flow over a flat plate (the "flat plate" line). At low Re_L, the data show characteristics similar to laminar duct flow. The "duct flow" line is for fully developed laminar flow in a rectangular channel with a 0.30 aspect ratio. The aspect ratio for the finned tubes is between 0.25 and 0.37. This behavior is consistent with Davenport's flow observations. Because the heat transfer performance is closely related to the flow structure, we may infer that two types of flow structure exist within the louvered plate-fin array:

1. Duct flow, in which the fluid travels axially through the array, essentially bypassing the louvers.
2. Boundary layer flow, in which the fluid travels parallel to the louvers.

Kajino and Hiramatsu [1987] conducted a flow visualization study of the louvered fin array using dye injection and hydrogen bubble techniques. Figure 5.12

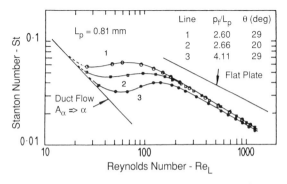

Figure 5.11 Heat transfer characteristics of several louvered plate fin geometries tested by Aichichia and Cowell [1988]. (From Aichichia and Cowell [1988].)

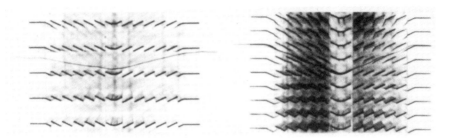

Figure 5.12   Steaklines at $Re_L = 500$ in 160-mm-deep flow visualization model of louver fin array ($l = 10$ mm, $\theta = 26$ degrees) tested by Kajino and Hiramatsu [1987]. (a) $p_f = 10$ mm. (b) $p_f = 10$ mm. (From Kajino and Hiramatsu [1987].)

shows two louver arrays which have the same louver pitch but different fin pitches. In the left-hand portion of Figure 5.12a, a significant fraction of the flow bypasses the louvers. This is because the hydraulic resistance of the "duct" flow region is substantially smaller than that for boundary layer flow across the louvers. When the fin pitch is reduced, as in the right-hand part of Figure 5.12b, the hydraulic resistance of the "duct" is increased, so that most of the flow passes through the louvers.

Webb and Trauger [1991] conducted detailed flow visualization experiments, using large scale models, to accurately simulate the flow in louver fin arrays. The louvered arrays selected permitted study of the effect of louver angle (θ) and the louver-to-fin-pitch ratio (L_p/p_f) on the flow structure within the array. They defined the term "flow efficiency" (η) to quantify their observations. Figure 5.13 provides a visual definition of the flow efficiency. The flow efficiency (η) is defined using Figure 5.13 as

$$\eta = \frac{N}{D} = \frac{\text{Actual transverse distance}}{\text{Ideal transverse distance}} \qquad (5.19)$$

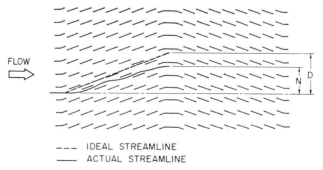

Figure 5.13   Definition of method used to define the flow efficiency used by Webb and Trauger [1991]. (From Webb and Trauger [1991].)

where $D = (L_p/2 - S) \tan \theta$. The flow efficiency is equal to one when the flow is parallel to and through the louvers. It is equal to zero when the flow is axial through the array (100% duct flow). Data were taken for six louver-to-fin-pitch ratios (L_p/p_f = 0.49, 0.56, 0.66, 0.79, 0.98, and 1.31) and two louver angles (θ = 20 and 30 degrees). Data were taken over a Reynolds number range of $400 < \mathrm{Re}_L < 4000$. Figure 5.14 shows the flow efficiency for a 20-degree louver angle. Figure 5.14 shows that the flow efficiency increases with increasing Re_L. This occurs up to a particular value of Reynolds number, defined as the critical Reynolds number, Re_L^*. Above, Re_L^*, the flow efficiency becomes independent of Reynolds number for fixed L_p/p_f. The flow efficiency is given by the following equations developed by Sahnoun and Webb [1992]:

$$\eta = 0.95 \left(\frac{L_p}{p_f} \right)^{0.23} \qquad (\mathrm{Re}_L > \mathrm{Re}_L^*) \qquad (5.20)$$

$$\eta = \eta^* - 37.17 \times 10^{-6}(\mathrm{Re}_L - \mathrm{Re}_L^*)^{1.1} \left(\frac{L_p}{p_f} \right)^{-1.35} \left(\frac{2\theta}{\pi} \right)^{-0.61}$$
$$(\mathrm{Re}_L < \mathrm{Re}_L^*) \qquad (5.21)$$

where $\eta = \eta^*$ at $\mathrm{Re}_L = \mathrm{Re}_L^*$ and $\mathrm{Re}_L^* = 828(\theta/90)^{-0.34}$ with θ in degrees.

Aoki et al. [1989] measured the heat transfer coefficient on the individual fins in the two-dimensional array illustrated in Figure 5.15a. The louvers were 0.080 mm thick and coated with an electrical insulation layer (0.005 mm Al$_2$O$_3$) on which a 0.001-mm-thick layer of nickel was deposited. Electric current in the 0.001-mm nickel coating provided a constant heat flux boundary condition. The heat transfer

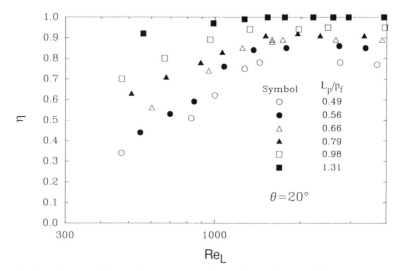

Figure 5.14 Measured flow efficiency versus Reynolds number for 20-degree louver angle as reported by Webb and Trauger [1991]. (From Webb and Trauger [1991].)

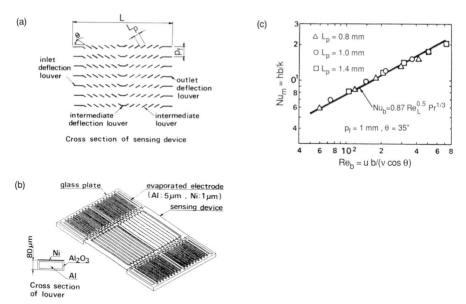

Figure 5.15   Louver-fin tests by Aoki et al. [1989]. (a) Louvered heat transfer plate. (b) Array of louvered plates tested. (c) Average experimental Nusselt numbers. (From Aoki et al. [1989].)

data were taken with six layers of the Figure 5.15a assembly, which is shown in Figure 5.15b. The data points on Figure 5.15c show the average Nusselt numbers obtained for three louver pitches. The equation for the curve fit of the data is shown in Figure 5.15c. Note that the velocity in the Reynolds number is the local velocity over the louvers ($\mu/\cos\theta$), where θ is the louver angle. The Pohlhausen solution as given by Kays and Crawford [1980] for laminar flow over a flat plate with constant heat flux gives

$$\frac{h_{av}L}{k} = 0.906\mathrm{Re}_L^{1/2}\mathrm{Pr}^{1/3} \tag{5.22}$$

The curve fit of the data shown in Figure 5.15c is 4% lower than the theoretical solution given by Equation 5.22. Hence, the average heat transfer coefficient on the louver is well approximated by the Pohlhausen solution.

5.3.3   Analytical Model for Heat Transfer and Friction

Sahnoun and Webb [1992] developed an analytical model to predict the j and f factor versus Re_L for the louver fin array. The development is similar to that of the previously discussed Joshi and Webb [1987] model for the OSF array. However, flow in the louver fin is more complex for several reasons. In addition to the internal louvers, Figure 5.10b shows inlet and redirection louvers having louver pitches S_1

and S_2, respectively. The heat transfer coefficient on all louvers is predicted using the Pohlhausen solution for laminar flow over a flat plate (Equation 5.22). As shown on Figure 5.15c, the Pohlhausen solution predicts the heat transfer coefficient on the louvers with good accuracy. The velocity is adjusted to account for the effect of the louver angle and for the flow efficiency factor. Figure 5.10 shows that the louver height (L_h) is less than the fin height (H). The heat transfer coefficient in the unlouvered end regions is predicted using a fully developed laminar flow solution. Similar procedures are used for the friction model. The friction model accounts for pressure drag caused by the finite fin thickness.

The model was validated by predicting the j and f versus Re data for the 32 one-row cores tested by Davenport [1983b]. The Davenport database are for two fin heights: 19 cores with 12.7-mm fin height, and 11 cores with 7.8-mm fin height. At 10.7-m/s air frontal velocity, the model predicted the j and f factors within $\pm 20\%$.

5.3.4 PEC Example 5.2

Automotive radiators typically use the louver-fin geometry, rather than the OSF geometry. Compare the performance of the OSF with that of the louver fin. Assume the geometric design parameters shown in Table 5.2. The radiator is designed to operate at 27°C with 20-mi/hr (9.5-m/sec) air frontal velocity. Compare the j and f factors for the two fin designs at the same air velocity. This corresponds to the simple PEC case FN-2 of Table 2.1.

The j and f factors are calculated using (a) the Davenport correlations for the louver fin and (b) the Joshi and Webb correlations for the OSF geometries. Use of the correlations gives:

1. OSF: $j = 0.0124, f = 0.0574$
2. Louver: $j = 0.0132, f = 0.045$

Using PEC case FN-2, $A_{osf}/A_{louv} = j_{louv}/j_{osf} = 0.0132/0.0124 = 1.065$. The friction power ratio is proportional to the product fA for $G = $ constant. Hence, $P_{osf}/P_{louv} = (f_{osf}/f_{louv})(A_{osf}/A_{louv}) = (0.0574/0.045)(1.065) = 1.35$.

TABLE 5.2 Geometric Design Parameters for PEC Example 5.2

OSF	Louver	Item
472	Same	Fins/m
0.10	Same	Fin thickness (t), mm
7.62	Same	Height of fin array (H), mm
2.03	Same	Width of louver (L_p), mm
NA	26-degree angle	Louver angle (for louver fin only)
1.0	0.8	Ratio of louver length to fin height
3.18	Same	Hydraulic diameter (D_h), mm
0.940	Same	Contraction ratio (σ)

^{a}NA, not applicable.

This example shows that the louver fin has a marginally higher j factor (6.5%) which provides a 6.5% surface area reduction, assuming negligible tube-side thermal resistance. However, the 28% higher friction factor of the OSF, combined with its 6.5% lower j factor results in 35% higher friction power (and pressure drop).

5.4   CONVEX LOUVER FIN

Hatada and Senshu [1984] have developed a variant of the OSF, which they call the "convex louver fin" geometry. This is a variant of the OSF geometry, which we will describe as the "offset convex louver fin" (OCLF). Figure 5.16 shows cross section details of the plate-and-fin geometries they tested, which are numbered 1 through 8.

No.	θ (deg)	% Offset
1	0	0
2	12.8	23
3	17.4	33
4	24.6	53
5	9.7	20
6	17.4	33
7	20.7	42
8	24.6	53

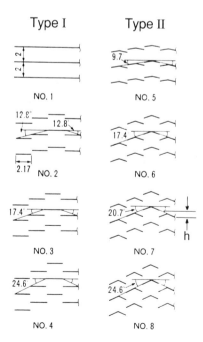

Figure 5.16   Fin geometries tested by Hatada and Senshu [1984]. (From Hatada and Senshu [1984].)

Hatada and Senshu [1984] describe the OSF and OCLF geometries as "Type I" and "Type II," respectively. All of the geometries have 500 fins/m, 2.0-mm fin pitch (p_f), and 0.115-mm fin thickness (t), and the louver width (L_p) of arrays 2 to 8 are 2.17 mm. Array 1 has plain fins, and arrays 2–4 are of the OSF type. Array 2 is the "standard" OSF, in which the louver is offset 50% of the fin spacing. Arrays 5 to 8 are of the convex louver geometry. The several OSF and convex louver geometries differ in the amount of fin offset. Hatada and Senshu use the angle θ shown in Figure 5.16 to define the louver offset. One may translate this to the physical offset using the table shown in Figure 5.16 using "% offset" = h/p_f, where h is the offset dimension shown in Figure 5.16 and p_f is the fin pitch.

The heat transfer and friction characteristics of the arrays shown in Figure 5.16 was measured with air flowing in a test section consisting of 38 fin channels, with each plate 43.3 mm wide by 150 mm long. The heat transfer performance was measured using a step change transient method.

Figure 5.17a shows the effect of the angle θ on the j and f factors for the OSF geometry. This figure shows that the 50% offset (θ = 24.6 degrees) does not

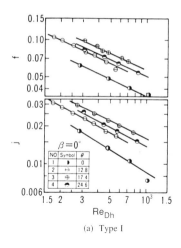

(a) Type I

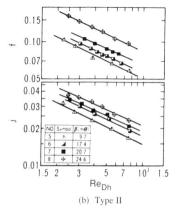

(b) Type II

Figure 5.17 j and f versus Re_D for the louver geometries tested by Hatada and Senshu [1984]. (a) OSF (Type I). (b) OCLF (Type II). (From Hatada and Senshu [1984].)

provide the highest j factor. At $Re_{Dh} = 500$, the $\theta = 17.4$-degree (33% offset) array provides a 22% higher j factor than does the $\theta = 24.6$-degree (50% offset) array. This agrees well with the findings of Kurosake et al. [1988].

For the same angle θ, Figure 5.17b shows that the convex louver array (OCLF) provides higher j factor than does the OSF geometry. For $\theta = 24.6$ degrees, the j factor of the OCLF array is 48% higher than that for the OSF array at $Re_{Dh} = 500$. However, the friction factor increase is approximately 70%. Figure 5.18 is a plot of j and f versus θ for the OSF and OCLF geometries for $Re_{Dh} = 500$. It shows that the maximum values of j and f occur at $\theta \simeq 20$ degrees for the OSF geometry. The j factor for the OCLF array increases with θ, at least up to the highest θ tested (24.6 degrees). Figure 5.18 shows that the higher j factor of the OCLF is accompanied by a higher friction factor than that of the OSF array. Furthermore, the OCLF friction factor drastically increases above 20 degrees in contrast to the decreasing friction factor of the OSF array. For the same friction power, the j factor of the $\theta = 20$-degree OCLF geometry is 22% higher than that of the OSF geometry. Figure 5.18 indicates that the j/f ratio of the OCLF array is not favorable for $\theta > 20$ degrees. The most surprising feature of the OCLF array is the small friction increase of the OCLF array, compared to the OSF array, for $\theta \leq 20.7$ degrees. As shown by Figure 5.18, the friction factor for $\theta = 17.4$ degrees in the OCLF geometry is only 7% higher than that of the OSF geometry ($\theta = 17.4$ degrees). For $\theta > 20$ degrees, Figure 5.18 shows that a large friction increase occurs in the OCLF geometry, whereas the friction factor decreases in the OSF geometry.

Hatada and Senshu also performed flow visualization experiments in scaled-up OSF and OCLF arrays, without tubes present. Figure 5.19 shows observed flow patterns for the OSF array ($\theta = 17.4$ degrees) and the OCLF array ($\theta = 17.4$ and 24.6 degrees). Examination of Figure 5.19c suggests that flow separation on the convex louvers causes significant profile drag for the $\theta = 24.6$-degree louver angle.

Hitachi [1984] uses the convex louver fin geometry in their commercial plate fin-and-tube heat exchangers. Figure 5.20 is taken from their product brochure.

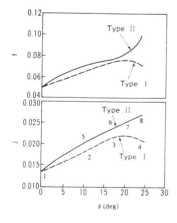

Figure 5.18   Effect of angle θ on j and f of the OSF (Type I) and OCLF (Type II) geometries tested by Hatada and Senshu [1984]. (From Hatada and Senshu [1984].)

No Powder Region   Powder Trace Line

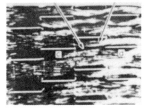

(a) Type I (θ = 17.4 deg)

(b) Type II (θ = 17.4 deg)

(c) Type II (θ = 24.6 deg)

Figure 5.19 Streamline flow patterns observed by Hatada and Senshu [1984] in finned arrays without tubes. (From Hatada and Senshu [1984].)

Figure 5.20 The convex louver fin geometry in the Hitachi fin-and-tube heat exchanger. (Courtesy of Hitachi Cable, Ltd.)

The results of the Hatada and Senshu study indicate that the convex louver fin may provide higher performance than is yielded by the standard OSF with 50% louver offset. The Hatada and Senshu data suggest that a 23% higher j factor with a 7% friction factor increase is possible. No j and f data have been reported for the OCLF geometry in steady-state tests of plate-and-fin heat exchanger configurations. Hence, confirmation of this speculative possibility must await experimental evaluation.

5.5   WAVY FIN

The term "wavy" or "corrugated" is used to describe the geometry shown in Figure 5.2c. For a corrugated geometry having constant corrugation angles and sharp wave tips, the key parameters that affect the performance are the wave pitch (p_w), the corrugation angle (θ), and channel spacing (s). Whether the wave geometry has smooth or sharp corners will affect the performance. Goldstein and Sparrow [1977] used a mass transfer technique to measure the local mass transfer coefficient distribution for a herringbone wave configuration. They propose that the enhancement results from Goertler vortices that form as the flow passes over the concave wave surfaces. These are counter-rotating vortices which have a corkscrew-like flow pattern. Flow visualization studies by this author show local zones of flow separation and reattachment on the concave surfaces for a true "wavy" channel shape. The redeveloping boundary layer from the reattachment point also contributes to heat transfer enhancement. Ali and Ramadhyani [1992] provide a detailed flow visualization study of one corrugated geometry, including local heat transfer and friction factors for $150 \leq \mathrm{Re}_{\mathrm{Dh}} \leq 4000$ which is within the typical operating range. Figure 5.21 shows their data with water at Pr = 7. At $\mathrm{Re}_{\mathrm{Dh}} = 2000$, the Nusselt number enhancement ratio is 2.3 and 3.2 for the narrow and wide channels, respectively. The corresponding friction increases are 2.3 and 3.8, respectively. The solid line shows the correlation of O'Brien and Sparrow [1982], who tested a channel having $\theta = 30$ degrees and $b/p_w = 0.29$ using water with $4 \leq \mathrm{Pr} \leq 8$. The O'Brien and Sparrow correlation accounts for Pr, but is not expected to account for the effect of θ and b/p_w, since these were not varied in their work. Sparrow and Hossfeld [1984] studied the effect of rounding the corners of the corrugation, as opposed to a true wave shape. Several other studies have been performed, as discussed by Ali and Ramadhyani [1992]; these studies are in the $1500 \leq \mathrm{Re}_{\mathrm{Dh}} \leq 25,000$ range and are typically outside the usual heat exchanger operating range.

Kays and London [1984] provide j and f versus Re curves for two wavy fin geometries (Figure 5.2c). Their performance is competitive with that of the OSF. No specific correlations exist for the j and f characteristics of wavy or herringbone fins. Two studies (see Rosenblad and Kullendorf [1975] and Okada et al. [1972]) on wavy-channel geometries used in plate-type heat exchangers provide additional data for small-aspect-ratio channels.

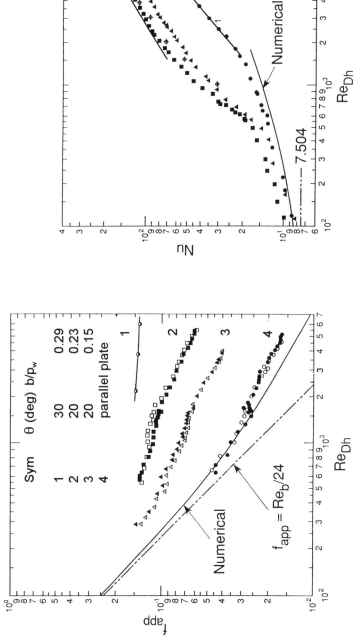

Figure 5.21 Wavy channel data of Ali and Ramadhyani [1992]. (a) Friction factor. (b) Nusselt number for Pr = 7. (From Ali and Ramadhyani [1992]). #1, θ = 30 deg, b/p_w = 0.29; #2, θ = 20 deg, b/p_w = 0.23; #3, θ = 20 deg, b/p_w = 0.15, #4, parallel plate.

5.6   PERFORATED FIN

This surface geometry shown in Figure 5.2e is made by punching a pattern of spaced holes in the fin material before it is folded to form the U-shaped flow channels. If the porosity of the resulting surface is sufficiently high, enhancement can occur due to boundary layer dissipation in the wake region formed by the holes. Shah [1975] provides a detailed evaluation of the perforated fin based on his study of test data on 68 perforated fin geometries. Shah concludes that little enhancement occurs for $Re_{Dh} < 2000$, if the heat transfer coefficient is based on the plate area before the holes were punched. Moderate enhancement may occur in the transition and turbulent flow regimes, $Re_{Dh} > 2000$, depending on the hole size and the plate porosity. Shen et al. [1987] tested flat plate-and-fin channels having round holes or rectangular cutouts. They found no benefit in the laminar range, but holes promote earlier transition and they observed moderate j increase in the turbulent range. Substantially higher performance was obtained with the rectangular slots.

The performance of the perforated fin is less than that of a good offset strip fin, and thus the perforated fin is rarely used today. Furthermore, the perforated fin it represents a wasteful way of making an enhanced surface, since the material removed in making the perforated hole is relegated to the scrap barrel.

Figure 5.22a shows an interesting variant of the perforated fin, which was tested

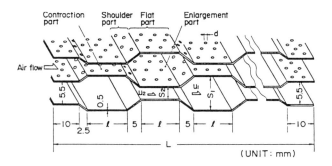

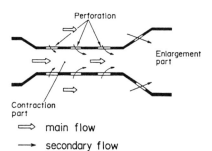

Figure 5.22   (a) Perforated plate fin geometry tested by Fujii et al. [1988]. (b) Illustration of secondary flows through perforations. (From Fujii et al. [1988].)

by Fujii et al. [1988]. This geometry has 2.0-mm-diameter holes in corrugated plates. The plates are aligned so that the channels have expanding and contracting flow areas. As shown in Figure 5.22b, secondary flow through the perforations is speculated to be an important contributor to the enhancement level. Figure 5.23a shows the geometry of the six surfaces, for which Nu and f versus Re_{Dh} curves are shown in Figure 5.23b. The fins are 0.5 mm and the fin pitch is 5.5 mm for all geometries. The dimensions of the corrugated plate geometries (NP-1, P-2, P-2a, and P-2b) defined in Figure 5.23a are L = 15 mm, S_1 = 8.6 m, S_2 = 2.4 mm, and 0.145 porosity (fractional hole area of fin plate). The performance of the different geometry variants, relative to that of the plain fin (NP-1), are as follows: (1) Perforations alone (P-1) provide no enhancement over NP-1; (2) the NP-2 geometry, without perforations, provides j/j_p = 1.90 and f/f_p = 4.7; (3) the corrugated/perforated P-2 geometry provides an impressive j/j_p = 2.7 and f/f_p = 6.1; and (4) the P-2a and P-2b geometries provide performance intermediate between P-2 and NP-2.

Although the P-2 geometry provides high heat transfer enhancement, the resulting efficiency index (η = 0.44) is small compared to other high-performance enhanced surfaces. The authors provide data on 10 other geometry variants. However, these variants provide negligible performance advantage over P-2. They also provide an empirical correlation Nu and f-correlations for the various geometries.

Fujii et al. [1991] also tested the fin geometry in Figure 5.22 as a plate fin-and-tube heat exchanger. These data are discussed in Section 6.5.4.

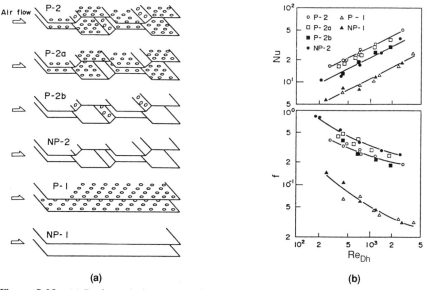

(a)   (b)

Figure 5.23   (a) Perforated plate geometries tested by Fujii et al. [1988]. (b) Test results on illustrated surfaces. (From Fujii et al. [1988].)

5.7   PIN FINS AND WIRE MESH

A pin fin surface geometry may be made of a special weave of "screen wire." Forming this screening into the shape of U-shaped channels results in the "pin fin" surface geometry. The wire may have a round, elliptical, or square cross-section shape. Although such pin fin geometries may have high performance, they are not widely used because the cost of such surfaces is significantly higher that the cost of thin sheet used to make geometries such as the OSF and louver fins. Performance data (j and f versus Re) on five round-pin geometries are given by Kays and London [1984], and those on nine square-pin geometries are given by Theoclitus [1966]. The Theoclitus data were for inline arrangements ($2 < S_t/d < 4$) having fin height (e) to pin diameter (d) ratios of $2 < e/d < 12$. Theoclitus [1966] showed that performance of the square pin fin geometries were approximately predicted using correlations for tube banks with round tubes. The measured heat transfer coefficients were typically 20% below the predicted values. This prediction is reasonably good considering that the pin fins were of square cross section, and that end effects existed on the short e/d pin lengths.

Hamaguchi et al. [1983] tested a regenerator matrix made of multiple layers of woven screen wire made with circular wires. The flow was normal to the wires. They tested a variety of matrix configurations having wire diameters between 0.04 and 0.5 mm, with 20–80 screen layers. Torikoshi and Kawabata [1989] developed a "wire mesh" fin geometry. The matrix is better described as "expanded metal mesh." This was made of 0.2-mm-thick sheet, in which short, parallel slits are made. The sheet is then pulled in one direction, and the slits open to form the expanded metal mesh fin shown in Figure 5.24. Boundary layers form on the square wires of thickness d, and dissipation occurs in the void region between the wire attachment points. By increasing the slit length, l, the porosity of the matrix is increased (dimension m). It would be possible to corrugate this matrix material to form the packing of a plate-and-fin heat exchanger. However, their tests were

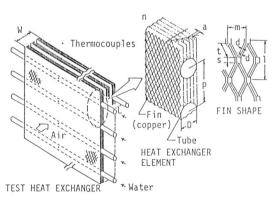

Figure 5.24   Expanded metal (copper) matrix geometry tested by Torikoshi and Kawabata [1989]. (From Torikoshi and Kawabata [1989].)

performed by compressing layers of the expanded metal to conform to the tube shape and then soldering the metal to parallel tubes, as shown in Figure 5.24. For six layers of mesh having dimensions d = 0.2-mm, m = 2.3-mm, l = 5.0-mm mesh joined to 4-mm-diameter tubes at 24 mm pitch, their test results are represented by

$$\frac{hd}{k} = 0.84 \left(\frac{dG}{\mu} \right)^{0.5} \tag{5.23}$$

Hamaguchi's woven screen wire matrix provided an approximately 20% higher heat transfer coefficient than the one given by Equation 5.23. Note that the matrix shown in Figure 5.24 will have less surface area per unit volume (A/V) than the offset-strip fin, or the louver-fin array, for the same fin pitch.

5.8   PLAIN FIN

If plain fins are used (e.g., Figures 5.23a and 5.23b), the flow channel will have a rectangular or triangular cross section. If the flow is turbulent, standard equations for turbulent flow in circular tubes may be used to calculate j and f, provided that Re is based on the hydraulic diameter (D_h). If $\mathrm{Re_{Dh}}$ < 2000, one may use theoretical laminar flow solutions for j and f. Values of j and f for developing and fully developed laminar flow are given in Section 2.4 for a variety of duct shapes. However, many of these solutions assume constant properties. Criteria to establish the value of x/D_h at which the flow becomes fully developed are also given in Table 2.5. Shah and London [1978] provide detailed results of many of the published solutions.

Sources of published data on plain fin geometries include Kays and London [1984] and Tischenko and Bondarenko [1983]. The latter reference provides data on 19 plain fin geometries.

The Reynolds number ($\mathrm{Re_{Dh}}$) is typically below 2200 in the channels of plate-and-fin heat exchangers. Theoretical solutions for laminar have been developed for many channel shapes. The cross-sectional geometry of the plain channel can have a very significant effect of the Nu and f. This is illustrated by the PEC Example 5.3 given below.

5.8.1   PEC Example 5.3

This example will be presented using the FN-3 criterion. Two plain fin geometries having the same fin pitch are compared: a triangular geometry (T) and a rectangular geometry (R). Figure 5.25 shows the flow cross section of the two geometries. The FN-3 criterion constrains the two geometries to operate at the same mass flow rate with equal frontal velocity and hA values. Because the velocities are known, the j and f values are directly calculable. The calculations are made for air flow at 27°C and 4-m/s frontal velocity. The j and f factors are given by the fully developed

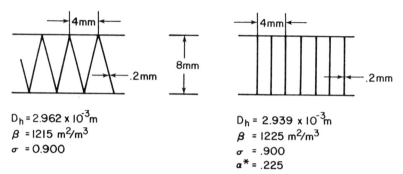

$D_h = 2.962 \times 10^{-3} m$
$\beta = 1215\ m^2/m^3$
$\sigma = 0.900$

$D_h = 2.939 \times 10^{-3} m$
$\beta = 1225\ m^2/m^3$
$\sigma = .900$
$a^* = .225$

Figure 5.25 Rectangular and triangular fin geometries used for PEC Example 5.3. (From Webb [1983].)

laminar solution for constant wall temperature and are taken from Table 2.5. Using the calculated h and f values, we obtain

$$\frac{A_T}{A_R} = \frac{h_R}{h_T} = \frac{39.80}{18.53} = 2.148$$

$$\frac{V_T}{V_R} = \frac{A_T}{A_R}\frac{\beta_R}{\beta_T} = 2.148 \times \frac{1225}{1215} = 2.30$$

$$\frac{P_T}{P_R} = \frac{f_T}{f_R}\frac{A_T}{A_R}\left(\frac{D_{h,T}}{D_{h,R}}\right)^3$$

$$= \frac{0.0167}{0.024} \times 2.148 \times \left(\frac{2.94 \times 781}{2.96 \times 776}\right)^3 = 1.47$$

(5.24)

For the same frontal area, flow rate, and heat transfer rate, this case FN-3 example shows that the triangular geometry requires 115% more surface area and 49% greater pumping power.

5.9   ENTRANCE LENGTH EFFECTS

Extended surface geometries for gases are typically designed for operation at Re $<$ 2000. If the flow length is sufficiently short, it is possible that the average j and f over the flow length are higher than the fully developed values, which are given in Table 2.5 for a number of channel shapes. Plain fin geometries are more susceptible to entrance length effects than are the enhanced fin geometries of Figures 5.2c and 5.2f. Because of the periodic flow interruptions, it is unlikely that entrance region effects would exist for interrupted fin heat exchangers. However, this may not be a good assumption for plain fin geometries. If the developing flow exists over more than, say, 20% of the flow length, it is possible that the average Nu and f are moderately higher than the fully developed values. In this case, one should determine the average Nu and f over the air-flow length.

The entrance region effect on heat transfer for several plain fin channel geometries is illustrated in Figure 5.26. This figure shows the ratio of the average Nu to the fully developed Nu for a constant wall temperature boundary condition with a developed velocity profile. Consider, for example, air flow (Pr = 0.7) in an equilateral triangular-shaped channel whose dimensionless flow length is $X^* = 0.10$. The mean Nu over the $X^* = 0.10$ flow length is 42% greater than the fully developed value. If $X^* \leq 0.2$, it appears that the entrance region solutions should be used. The figure shows that triangular channels have a longer entrance region than do circular or rectangular channels.

The entrance region pressure drop is also higher than the fully developed value. It is calculated by the equation

$$\Delta p = \left[\frac{4f_{fd}L}{D_h} + K(\infty) \right] \frac{G^2}{2\rho} \qquad (5.25)$$

where $K(\infty)$ is the pressure drop increment to be added to account for the increased friction in the development region. Figure 5.27 shows the ratio $K(x)/K(\infty)$ for several duct shapes. $K(x)$ is the increment to be added for a duct whose profile is less

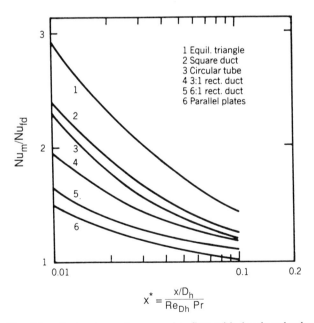

Figure 5.26 Nu_m/Nu_{fd} for laminar entrance region flow with developed velocity profile in different channel shapes for constant wall temperature boundary condition. (From Webb [1987].)

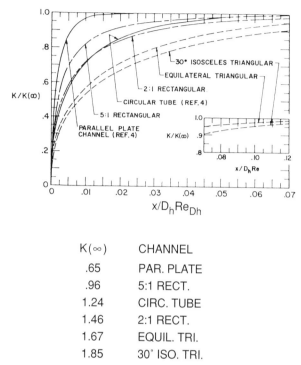

$K(\infty)$	CHANNEL
.65	PAR. PLATE
.96	5:1 RECT.
1.24	CIRC. TUBE
1.46	2:1 RECT.
1.67	EQUIL. TRI.
1.85	30° ISO. TRI.

Figure 5.27 $K(x)/K(\infty)$ for laminar flow in different duct shapes. (From Webb [1987].)

than fully developed at the duct exit. The hydraulic entrance length (L_{hy}^+) may be determined from Figure 5.27 and is the value of $x/(D_h Re)$ at which $K/K(\infty) \simeq 1$. The $K(\infty)$ and L_{hy}^+ values for the channel shapes on Figure 5.27 are listed in Table 2.5. To calculate the pressure drop in an entrance region duct, use Equation 5.25 with $K(\infty)$ replaced by K from Figure 5.27.

5.10 PACKINGS FOR GAS–GAS REGENERATORS

Regenerators, either rotary or valved, are commonly used to transfer heat from combustion products to inlet combustion air. The valved-type regenerator has two identical packings which alternately serve the hot and cold streams and are switched by quick operating valves. Any of the corrugated plate fin geometries discussed in this chapter may conceivably be applied to regenerators. Because the hot and cold streams are normally of different pressures, any packing geometry that allows significant transverse flow leakage will reduce a rotary regenerator performance. The louvered and offset strip fin geometries would be susceptible to this problem. However, one may design around this problem by including full-height continuous

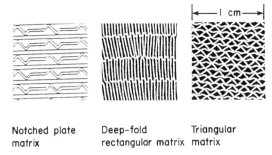

|← 1 cm →|

Notched plate   Deep-fold   Triangular
matrix   rectangular matrix matrix

Figure 5.28 Matrix geometries for rotary regenerators. (a) Illustration of plate stacking, (b) Cross section of stacked plates. (Courtesy of Combustion Engineering Air Preheater Division, Wellsville, NY.)

fins aligned with the flow at discrete spacings. The "brick checkers" geometry commonly used in the valved-type glass furnace regenerator is essentially an offset strip fin (see Figure 5.2d) having a large fin thickness.

Packings having small hydraulic diameter (small s) will provide a higher heat transfer coefficient than will a packing having a larger hydraulic diameter; for laminar flow, $h \propto 1/D_h$. However, the fouling characteristics of the hot gas may limit the size of the flow passage. Considerably smaller passage size may be used in building ventilation heat recovery regenerators than in building those used for heat recovery from coal-fired exhaust gases or glass furnace exhausts. Coal-fired electric utility plants frequently use notched plate packings, such as that illustrated in Figure 5.28. Conceivably, corrugated plate geometries similar to those used in plate-type heat exchangers may also be used. However, the offset strip fin is not a totally viable candidate. This is because it would permit mixing and contamination between the hot and cold streams.

5.11 CONCLUSIONS

The plate-and-fin heat exchanger geometry has become an increasingly important design. The high-performance offset strip and louver fins provide quite high heat transfer coefficients for gases and two-phase applications. It offers significant advantages over the traditional fin-and-round tube geometry. Key advantages are (a) lower gas pressure drop than circular tube designs and (b) the ability to have the fins normal to the gas flow over the full gas flow depth.

Early variants were applied to gas-to-gas applications. It is now used for gases, liquids, or two-phase fluids on either side. Designs using extruded aluminum tubes with internal membranes allow quite high tube-side design pressure (e.g. 150 atm). Further innovative designs, additional applications, and advanced fin geometries are expected. It is currently made in aluminum, steel, and even ceramics.

5.12 REFERENCES

Achaichia, A., and Cowell, T. A., 1988. "Heat Transfer and Pressure Drop Characteristics of Flat Tube and Louvered Plate Fin Surfaces," *Experimental Thermal and Fluid Science*, Vol. 1, pp. 147–157.

Ali, M. M., and Ramadhyani, S., 1992. "Experiments on Convective Heat Transfer in Corrugated Channels," *Experimental Heat Transfer*, Vol. 5, pp. 175–193.

Aoki, H., Shinagawa, and T. Suga, K., 1989. "An Experimental Study of the Local Heat Transfer Characteristics in Automotive Louvered Fins," *Experimental Thermal and Fluid Science*, Vol. 2, pp. 293–300.

Brinkman, R., Ramadhyani, S., and Incropera, F. P., 1988. "Enhancement of Convective Heat Transfer from Small Heat Sources to Liquid Coolants Using Strip Fins," *Experimental Heat Transfer*, Vol. 1, pp. 315–330.

Churchill, S. W., and Ursagi, R., 1972. "A General Expression for the Correlation of Rates of Transfer and Other Phenomena," *AIChE Journal*, Vol. 18, No. 6, pp. 1121–1128.

Creswick, F. A., Talbert, S. G., and Bloemer, J. W., 1964. "Compact Heat Exchanger Study," Battelle Memorial Institute Report, Columbus, OH, April 15.

Davenport, C. J., 1980. "Heat Transfer and Fluid Flow in the Louvred-Fin Heat Exchanger," Ph.D. Thesis, Lanchester Polytechnic, Lanchester, England.

Davenport, C. J., 1983a. "Heat Transfer and Flow Friction Characteristics of Louvered Heat Exchanger Surfaces," in *Heat Exchangers: Theory and Practice*, J. Taborek, G. F. Hewitt, and N. Afgan, Eds., Hemisphere Publishing Corp., Washington, D.C., pp. 387–412.

Davenport, C. J., 1983b. "Correlations for Heat Transfer and Flow Friction Characteristics of Louvered Fin," in *Heat Transfer—Seattle 1983*, AIChE Symposium Series, No. 225, Vol. 79, pp. 19–27.

Fujii, M., Seshimo, Y., and Yamananaka, G., 1988. "Heat Transfer and Pressure Drop of the Perforated Surface Heat Exchanger with Passage Enlargement and Contraction," *International Journal of Heat and Mass Transfer*, Vol. 31, pp. 135–142.

Fujii, M., Seshimo, Y., and Yoshida, T., 1991. "Heat Transfer and Pressure Drop of Tube-Fin Heat Exchanger with Trapezoidal Perforated Fins," in *Proceedings of the 1991 ASME-JSME Joint Thermal Engineering Conference.*, Vol. 4, J. R. Lloyd and Y. Kurosake, Eds., pp. 355–360.

Fujikake, K., Aoki, H., and Mitui, H., 1983. "An Apparatus for Measuring the Heat Transfer coefficients of Finned Heat Exchangers by Use of a Transient Method," *Proc. Japan 20th Symposium on Heat Transfer*, pp. 466–468.

Goldstein, L. J., and Sparrow, E. M., 1977. "Heat/Mass Transfer Characteristics for Flow in a Corrugated Wall Channel," *Journal of Heat Transfer*, Vol. 99 pp. 187–195.

Hamaguchi, K., Takahashi, S., and Miyabe, H., 1983. "Heat Transfer Characteristics of a Regenerator Matrix (Case of Packed Wire Gauzes)," *Transactions of the JSME*, Vol. 49, B, No. 445, pp. 2001–2009.

Hatada, T., and Senshu, T., 1984. "Experimental Study on Heat Transfer Characteristics of Convex Louver Fins for Air Conditioning Heat Exchangers," ASME paper ASME 84-HT-74.

Hitachi, 1984. *Hitachi High-Performance Heat Transfer Tubes*, Cat. No. EA-500, Hitachi Cable Co., Tokyo, Japan.

Joshi, H. M., and Webb, R. L., 1987. "Prediction of Heat Transfer and Friction in the Offset Strip Fin Array," *International Journal of Heat and Mass Transfer*, Vol. 30, pp. 69–84.

Kajino, M., and Hiramatsu, M. 1987. "Research and Development of Automotive Heat Exchangers," in *Heat Transfer in High Technology and Power Engineering*, W. J. Yang and Y. Mori, Eds., Hemisphere Publishing Corp., Washington, D.C., pp. 420–432.

Kays, W. M., 1972. *Compact Heat Exchangers, AGARD Lecture Series on Heat Exchangers*, No. 57, J. J. Ginoux, Ed., AGARD-LS-57–72, January 1972.

Kays, W. M., and Crawford, M. E., 1980. *Convective Heat and Mass Transfer*, McGraw–Hill, New York, p. 151.

Kays, W. M., and London, A. L., 1984. *Compact Heat Exchangers*, 3rd edition, McGraw–Hill, New York.

Kurosaki, Y., Kashiwagi, T., Kobayashi, H., Uzuhashi, H., and Tang, S-C., 1988. "Experimental Study on Heat Transfer from Parallel Louvered Fins by Laser Holographic Interferometry," *Experimental Thermal and Fluid Science*, Vol. 1, pp. 59–67.

Manglik, R. M., and Bergles, A. E., 1990. "The Thermal–Hydraulic Design of the Rectangular Offset-Strip-Fin Compact Heat Exchanger," in *Compact Heat Exchangers*, R. K. Shah, A. D. Kraus and D. Metzger, Eds., Hemisphere Publishing Corp., Washington, D.C., pp. 123–150.

Milne-Thomson, L. M., 1960. *Theoretical Hydrodynamics*, 4th edition, Macmillan, New York, p. 319.

O'Brien, J. E., and Sparrow, E. M., 1982. "Corrugated-Duct Heat Transfer, Pressure Drop and Flow Visualization," *Journal of Heat Transfer*, Vol. 104, pp. 410–416.

Okada, K., Ono, M., Tomimura, T., Okuma, T., Konno, H., and Ohtani, S., 1972. "Design and Heat Transfer Characteristics of New Plate Type Heat Exchanger," *Heat Transfer—Japanese Research*, Vol. 1, No. 1, pp. 90–95.

Patankar, S.V., 1990. "Numerical Prediction of Flow and Heat Transfer in Compact Heat Exchanger Passages," in *Compact Heat Exchangers*, R. K. Shah, A. D. Kraus, and D. Metzger, Eds., Hemisphere Publishing Corp., Washington, D.C., pp. 191–204.

Rosenblad, G., and Kullendorf, A., 1975. "Estimating Heat Transfer Rates from Mass Transfer Studies on Plate Heat Exchanger Surfaces," *Wärme-und-Stoffübertragung*, Vol. 8, pp. 187–191.

Sahnoun, A., and Webb, R. L., 1992. "Prediction of Heat Transfer and Friction for the Louver Fin Geometry," *Journal of Heat Transfer*, Vol. 114, pp. 893–900.

Shah, R. K., 1975. "Perforated Heat Exchanger Surfaces: Part 2—Heat Transfer and Flow Friction Characteristics," ASME paper 75-WA/HT-9.

Shah, R. K., and London, A. L., 1978. *Laminar Flow Forced Convection in Ducts*, Supplement 1 to *Advances in Heat Transfer*, Academic Press, New York.

Shah, R. K., and Webb, R. L., 1982. "Compact and Enhanced Heat Exchangers," in *Heat Exchangers: Theory and Practice*, J. Taborek, G. F. Hewitt and N. H. Afgan, Eds., Hemisphere Publishing Corp., Washington, D. C., pp. 425–468.

Shen, J., Gu, W., and Zhang, Y., 1987. "An Investigation on the Heat Transfer Augmentation and Friction Loss Performances of Plate-Perforated Fin Surfaces," in *Heat Transfer Science and Technology*, B-X. Wang, Ed., Hemisphere Publishing Corp., Washington, D.C., pp. 798–804.

Sparrow, E. M., and Hossfeld, M., 1984. "Effect of Rounding Protruding Edges on Heat

Transfer and Pressure Drop in a Duct," *International Journal of Heat and Mass Transfer*, Vol. 27, pp. 1715–1723.

Sparrow, E. M., and Liu, C. H., 1979. "Heat Transfer, Pressure Drop and Performance Relationships for In-Line, Staggered, and Continuous Plate Heat Exchangers," *International Journal of Heat and Mass Transfer*, Vol. 22, pp. 1613–1625.

Suga, T., and Aoki, H., 1991. "Numerical Study on Heat Transfer and Pressure Drop in Multilouvered Fins," in *Proceedings of the 1991 ASME/JSME Joint Thermal Engineering Conference*, Vol. 4, J. R. Lloyd and Y. Kurosake, Eds., pp. 361–368.

Sunden, B., and Svantesson, J. 1990. "Thermal Hydraulic Performance of New Multilouvered Fins," *Proceedings of the 9th International Heat Transfer Conference*, Vol. 5, pp. 91–96.

Theoclitus, G., 1966. "Heat Transfer and Flow-Friction Characteristics of Nine Pin-Fin Surfaces," *Journal of Heat Transfer*, Vol. 88, pp. 383–390.

Tischenko, Z. V., and Bondarenko, V. N., 1983. "Comparison of the Efficiency of Smooth-Finned Plate Heat Exchangers," *International Chemical Engineering*, Vol. 23, No. 3, pp. 550–557.

Torikoshi, K., and Kawabata, K., 1989. "Heat Transfer and Flow Friction Characteristics of a Mesh Finned Air-Cooled Heat Exchanger," in *Convection Heat Transfer and Transport Processes*, R. S. Figliola, M. Kaviany, and M. A. Ebadian, Eds., ASME Symposium, Vol. HTD-Vol. 116, ASME, New York, pp. 71–77.

Usami, H., 1991. "Pressure Drop Characteristics of Offset Strip Fin Surfaces," in *Proceedings of the 1991 ASME/JSME Joint Thermal Engineering Conference,* Vol. 4, J. R. Lloyd and Y. Kurosaki, Eds., ASME, New York, pp. 425–432.

Webb, R. L., and Joshi, H. M., 1983. "Prediction of the Friction Factor for the Offset Strip-Fin Matrix," *ASME-JSME Thermal Engineering Joint Conference*, Vol. 1, ASME, New York, pp. 461–470.

Webb, R. L., 1983. "Enhancement for Extended Surface Geometries Used in Air-Cooled Heat Exchangers," in *Low Reynolds Number Flow Heat Exchangers*, Hemisphere Publishing Corp., Washington, D.C., pp. 721–734.

Webb, R. L., 1987. Chapter 17 in *Handbook of Single-Phase Heat Transfer*, S. Kakaç, R. K. Shah, and W. Aung, Eds., John Wiley & Sons, New York, pp. 17.1–17.62.

Webb, R. L., and Trauger, P., 1991. "The Flow Structure in the Louver Fin Heat Exchanger Geometry," *Experimental Thermal and Fluid Science*, Vol. 4, pp. 205–217.

Wieting, A. R., 1975. "Empirical Correlations for Heat Transfer and Flow Friction Characteristics of Rectangular Offset Fin Heat Exchangers," *Journal of Heat Transfer*, Vol. 97, pp. 488–490.

5.13   NOMENCLATURE

A	Total heat transfer surface area (both primary and secondary, if any) on one side of a direct-transfer-type exchanger; total heat transfer surface area of a regenerator, m^2 or ft^2
A_c	Flow cross-sectional area in minimum flow area, m^2 or ft^2
b	Distance between plates in a plate-fin exchanger, or channel height, m or ft^2
c_p	Specific heat of fluid at constant pressure, J/kg-K or Btu/lbm-°F
C_D	Drag coefficient, dimensionless

d	Diameter of pin fins, m
D_h	Hydraulic diameter, m or ft
D_{sh}	Hydraulic diameter based on $A_c = sh$, m or ft
f	Fanning friction factor, $\Delta p_f D_h / 2LG^2$, dimensionless
$F_{f,\alpha}$	$f_{EL\alpha}/f_{EL0}$ as shown on Figure 5.7, dimensionless
$F_{h,\alpha}$	$Nu_{EL\alpha}/Nu_{EL0}$ as shown on Figure 5.7, dimensionless
G	Mass velocity based on the minimum flow area, kg/m²-s or lbm/ft²-s
h	Heat transfer coefficient based on A, W/m²-K or Btu/hr-ft²-°F
h	Fin height ($h = b - t$), m
H	Louver fin height, defined in Figure 5.10, m or ft
j	$StPr^{2/3}$, dimensionless
k	Thermal conductivity of fluid, W/m-K or Btu/hr-ft-°F
K	Pressure drop increment to account for flow development: $K(\infty)$ (full entrance region) or $K(x)$ (over length x); shown in Figure 5.25
L	Fluid flow (core) length on one side of the exchanger, m or ft
L_p	Strip flow length of OSF or louver pitch of louver fin, m or ft
L_h	Louver height shown in Figure 5.10, m or ft
L_L	Louver length shown in Figure 5.10, m or ft
L_s^+	$(L_p/2s)/(2su/v)$, dimensionless
Nu	Nusselt number ($= hD_h/k$), dimensionless
p_f	Fin pitch, center-to-center spacing, m or ft
p_w	Axial wave pitch of wavy fin, m or ft
P	Fluid pumping power, W or hp
Pr	Prandtl number ($= c_p\mu/k$), dimensionless
Re_{Dh}	Reynolds number based on hydraulic diameter, $Re_{Dh} = D_h G/\mu$, $Re_{sh} = D_{sh}G/\mu$, dimensionless
Re_s	Reynolds number used in Table 5.1, $Re_s = 2su/v$, dimensionless
$Re_{Dh,tr}$	Transition from laminar to turbulent flow, dimensionless
Re_{sh}	Reynolds number based on D_{sh} hydraulic diameter, dimensionless
Re_{tr}	Transition Reynolds number , $Re_{Dh}^* (\delta_{mom}/D_h) + 1000$, dimensionless
Re_L	Reynolds number based on the interruption length $= GL_p/\mu$, Gx/μ, dimensionless
Re_S	Reynolds number based on S_l, dimensionless
s	Spacing between two fins ($= p_f - t$), m or ft
S_t	Transverse tube pitch, m or ft
St	Stanton number $= h/Gc_p$, dimensionless
t	Fin thickness, m
u	Velocity based on $A_c = s(b - t)$, m² or ft²
v	Velocity based on $A_c = (s - t)(b - t)$, m² or ft²
x	Cartesian coordinate along the flow direction, m or ft
x^*	$x/(D_h RePr)$, dimensionless

Greek Letters

α	Aspect ratio of rectangular duct (s/b), dimensionless
γ	t/s, dimensionless

δ	t/l, dimensionless
δ_{mom}	$t + 1.328\, L_p/(\mathrm{Re}_L)^{0.5}$
η	Surface efficiency of finned surface $[= 1 - (1 - \eta_f)A_f/A]$, dimensionless
η_f	Fin efficiency or temperature effectiveness of the fin, dimensionless
θ	Corrugation angle in corrugated fin geometry; also included half-angle of internal fin cross section normal to flow, radians or degrees
μ	Fluid dynamic viscosity coefficient, Pa-s or lbm/s-ft
ν	Kinematic viscosity, m/s^2 or ft/s^2
ρ	Fluid density, kg/m^3 or lbm/ft^3

Subscripts

fd	Fully developed flow
L	Laminar region
m	Average value over flow length
p	Plain tube or surface
T	Turbulent region
w	Evaluated at wall temperature
x	Local value

6

EXTERNALLY FINNED TUBES

6.1 INTRODUCTION

Finned-tube heat exchangers have been used for heat exchange between gases and liquids (single- or two-phase) for many years. Figures 6.1a and 6.1b show two important finned-tube heat exchanger construction types. Figure 6.1a shows the plate fin-and-tube geometry, and Figure 6.1b shows individually finned tubes. Although round tubes are shown in Figure 6.1, oval or flat tubes are also used—for example, in automotive radiators. A plain air-side geometry is shown in the Figure 6.1 geometries. Figures 6.1a and 6.1b show a staggered tube arrangement, which provides higher performance than an inline tube arrangement. Externally finned tubes are also frequently used for liquids. Figure 6.1c shows an integral-fin tube used for liquids. However, extended surfaces for liquids typically use lower fin height than that used for gases. Because liquids have higher heat transfer coefficients than do gases, fin efficiency considerations require shorter fins with liquids than with gases. When used with liquids, the fin height is typically in the 1.5- to 3-mm range. The dominant amount of material in this chapter is applicable to high fins, which are used for gases.

Because the gas-side heat transfer coefficient is typically much smaller than the tube-side value, it is important to increase the air-side hA value. A plain surface geometry will increase the air-side hA value by increasing the area (A). Use of enhanced fin surface geometries will provide higher heat transfer coefficients than will the use of a plain surface. In order to maintain reasonable friction power with low-density gases, the gas velocity is usually less than 5 m/s.

Important basic enhancement geometries include wavy and interrupted fins. Variants of the interrupted strip fin are also used with finned-tube heat exchangers,

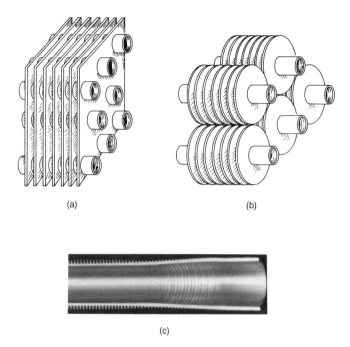

(a)   (b)

(c)

Figure 6.1 Finned-tube geometries used with circular tubes. (a) Plate fin-and-tube used for gases. (From Webb [1987].) (b) Individually finned tube having high fins, used for gases. (From Webb [1982].) (c) Low, integral-fin tube.

for heat exchange to a tube-side fluid. Figure 6.2 shows four variants of the interrupted strip fin applied to gas–fluid heat exchangers. Figures 6.2a through 6.2d are applied to circular tubes, and Figures 6.2e and 6.2f are used with flat, extruded aluminum tubes. These extruded tubes have internal membranes for pressure containment. Figures 6.2a through 6.2d are commonly used in commercial air-conditioning equipment. In Figure 6.2a, the segmented aluminum fin (or "spine fin") is spiral wound on the tube, and is affixed using an adhesive as described by Abbott et al. [1980] and by Webb [1983, 1987]. In Figures 6.2b through 6.2d, a copper or aluminum tube is mechanically expanded on aluminum plate fins. The version shown in Figure 6.2e is typically made with aluminum fins brazed on flat, extruded aluminum tubes. Figure 6.2f shows the "skive fin" design, as described by O'Connor and Pasternak [1976], for which the fins are slit from the thick wall of an aluminum extrusion and bent upward. These constructions are described by Webb [1983], Shah and Webb [1982], and Webb [1987]. The Figure 6.2e and 6.2f geometries on flat aluminum tubes have not found the wide commercial acceptance as have the Figure 6.2a through 6.2d geometries for residential air-conditioning applications. This may be in part due to the cost of brazing, or of the extruded tubes. However, the Figure 6.2e and 6.2f designs have been recently introduced for use in automotive air-conditioning evaporators and condensers. Recent developments in automotive

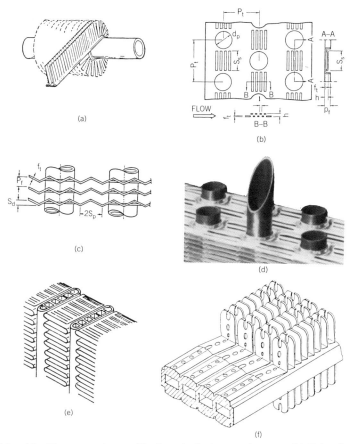

Figure 6.2   Air-side geometries used in finned tube heat exchangers. (a) Spine fin. (b) Slit-type OSF. (c) Wavy fins. (d) Convex louver fin. (e) OSF fins brazed to extruded aluminum tube. (f) Interrupted skive fin integral to extruded aluminum tube.

brazed aluminum manufacturing technology have made the costs of the Figure 6.2e heat exchanger construction more favorable.

Because the gas-side heat transfer coefficient may be 5–20% that of the tube-side fluid, the use of closely spaced, high fins is desirable. High fin efficiency can be obtained, if the fin material has high thermal conductivity (e.g., aluminum or copper). If steel fins are required, fin efficiency considerations will dictate shorter or thicker fins. Operational constraints, such as gas-side fouling, may limit the fin density. Air-conditioning applications use 500–800 fins/m, whereas process air coolers are usually limited to 400 fins/m. Dirty, soot-laden gases may limit the fin density to 200 fins/m. Different correlations are required for the Figure 6.1a and 6.1b geometries.

The fin material used also depends on the operating temperature and the corrosion potential. Listed below are the fin-and-tube materials used in a variety of applications:

1. Residential air conditioning: Aluminum fins and copper or aluminum tubes.
2. Automotive air conditioning: Aluminum fins and aluminum tubes.
3. Automotive radiators: Aluminum fins brazed to aluminum tubes, or copper fins soldered to brass tubes.
4. Process industry heat exchangers: Air-cooled condensers may use aluminum fins on copper or steel tubes.
5. Boiler economizers and heat recovery exchangers: The higher operating temperature requires steel fins on steel tubes.

In addition to describing the various fin geometries and their performance characteristics, we will compare the performance of alternative heat exchanger and fin configurations. We will then identify heat exchanger and enhanced fin geometries that will yield the highest performance per unit heat exchanger core weight. Finally, we will consider possible improvements in the air-side surface geometry.

6.2   THE GEOMETRIC PARAMETERS AND THE REYNOLDS NUMBER

6.2.1   Dimensionless Variables

The flow pattern in finned-tube heat exchangers is very complex, due to its three-dimensional nature and flow separations. The use of enhanced-fin geometries introduces further complications. Little progress has been made in attempts to analytically or numerically predict the heat transfer coefficient and friction factor. Some progress has been made in geometries that use "flat" tubes, but no attempts have been made to theoretically model finned-tube geometries having round tubes.

Equations to predict the heat transfer coefficient and friction factor are usually based on power law correlations using multiple regression techniques. Use of this method requires that one know the geometric and flow variables involved. The geometric and flow variables that affect the heat transfer coefficient and friction factor are as follows:

1. Flow variables: Air velocity (u), viscosity (μ), density (ρ), thermal conductivity (k), and specific heat (c_p).
2. Tube bank variables: Tube root diameter (d_o), transverse tube pitch (S_t), row pitch (S_l), tube layout (staggered or inline), and the number of rows (N).
3. Fin geometry variables: For a plain fin, these are the fin pitch (p_f), fin height (e), and fin thickness (t). If, for example, an enhanced wavy fin geometry is used, the added variables are the wave height (e_w), the wave pitch (p_w), and the wave shape.

Thus, there are seven geometry variables for a plain fin (excluding the tube layout) and five flow variables. Two additional variables are introduced to account for the wavy fin geometry. Dimensional analysis specifies that the number of possibly important dimensionless groups are the number of variables minus the number of dimensions. Because four dimensions are involved for heat transfer (mass, length, time, and temperature), there are eight dimensionless variables for the plain fin and 10 dimensionless variables for the wavy fin. The dimensionless flow variables typically used in correlations are the Reynolds number and the Prandtl number. For heat transfer, one has the Nusselt number or the Stanton number. For pressure drop, one uses the friction factor.

There are no "rules" for selecting the appropriate dimensionless geometric variables. This is simply "cut-and-try" to select the ones that give the best correlation of the data set. Furthermore, power law correlations have no rational basis and provide only an empirical correlation of the data set. It is dangerous to extrapolate such correlations beyond the range of the variables used to develop the correlation.

6.2.2 Definition of Reynolds Number

The basic definition of the Reynolds number is $L_c G/\nu$, where L_c is a characteristic dimension and G is usually defined as the mass velocity in the minimum flow area. For fully developed flow inside a plain tube, there is only one possible characteristic dimension, namely, the tube diameter. We have previously stated that there are seven dimensions associated with a plain fin geometry, and nine for the wavy-fin geometry. Hence, there is no unique characteristic dimension. Thus, there are eight possible values of L_c for definition of the Reynolds number of the plain fin geometry. One approach to defining Reynolds number is to identify a characteristic dimension that appears to dominate over the other possible choices. For a bare tube bank, the possible choices are S_t, S_l, and d_o. However, there is no uniform agreement on the characteristic dimension used to define the Reynolds number. Two different characteristic dimensions have been used to define L_c used in the Reynolds number. They are the tube diameter (d_o) or the hydraulic diameter (D_h). Kays and London [1984] choose to use the hydraulic diameter for the characteristic dimension for all situations, including bare and finned tube banks. There is no evidence to suggest that hydraulic diameter is a better choice. In fact, there is evidence that the tube diameter may be a better choice for finned tube banks. This will be shown later. We conclude that the choice of characteristic dimension is arbitrary.

For fully developed flow in tubes, one defines laminar and turbulent regimes. Do such regimes also exist for finned tube banks? To evaluate this, consider the case of the Figure 6.1a geometry. Assume that the geometry uses $d_o = 19$ mm with an equilateral triangular pitch of $S_t = 44.45$ mm, and 472 fins/m with 0.2-mm thickness. Assume air enters the exchange at 3 m/s and 20°C. The mass velocity in the minimum flow area (G) is 7.34 kg/m²s, and the hydraulic diameter is 3.68 mm. The Reynolds numbers based on d_o and D_h are 8710 and 1640, respectively. Is the flow laminar or turbulent? Based on Re_{Dh}, one would say it is laminar. However, based on tube diameter (Re_d), one would say it is turbulent. In reality, it exhibits some of both characteristics. If the tubes were not present, the flow geometry would be a parallel plate channel, for which $D_h = 3.82$ mm. The Reynolds number is 1588, which is clearly laminar. However, the tubes shed eddies, which wash over the fin surface and provide mixing of the flow.

If the Reynolds number based on hydraulic diameter were dominant over the Reynolds number based on tube diameter, one would expect that the Nu and f data for different fin pitches would tend to fall on one line. We will show that this is not the case.

6.2.3   Definition of the Friction Factor

We strive to use only the Fanning friction factor (f), defined in the nomenclature section. However, other friction factor definitions are frequently used for tube banks (bare and finned). A common definition for tube banks is given the symbol f_{tb}. It is related to the Fanning friction factor by the equation

$$f_{tb}N = \frac{fL}{D_h} \qquad (6.1)$$

where N is the number of tube rows in the flow direction, and L is the flow depth. For bare or finned tubes, $L = S_l(N - 1) + d_e$, where d_e is diameter over the fins. For a bare tube bank, $d_e = d_o$.

6.2.4   Sources of Data

Much of the data on finned-tube heat exchangers were developed by industrial organizations and are therefore proprietary. However, some have been published in the open literature. There are two sources for compilations of published data. One is the book by Kays and London [1984], and the other is a report by Rozenman [1976a]. Both are relatively old and do not contain state-of-the-art data on enhanced surfaces. Kays and London present data for 22 geometries, and Rozenman provides data for 161 geometries. The data of both authors are presented in the format of j and f versus Re_{Dh}, and complete geometry details are provided. Rozenman [1976b] provides empirical correlations for j and f versus Re_{Dh}. Since publication of these two references, various journal and conference publications have provided additional data on numerous geometries, including enhanced fin designs. These data are discussed in the following sections.

6.3   PLAIN PLATE FINS ON ROUND TUBES

Figure 6.1a shows the finned-tube geometry with continuous, plain plate fins in a staggered tube layout. An inline tube geometry is seldom used because it provides substantially lower performance than the staggered tube geometry. The performance difference between inline and staggered tube layouts is discussed in Section 6.8.

6.3.1   Effect of Fin Spacing

Rich [1973] measured heat transfer and friction data for the Figure 6.1a geometry having plain fins, four rows deep, on 12.7-mm-diameter tubes equilateral spaced on 32-mm centers. The tubes and fins were made of copper, and the fins were solder-bonded to minimize contact resistance. The geometry of all heat exchangers were identical, except the fin density $(1/p_f)$, which was varied from 114 to 811 fins/m. All fins were 0.25 mm thick.

Figure 6.3 shows the friction factor and the Colburn j-factor $(StPr^{2/3})$ data (smoothed curve fit) as a function of Reynolds number (based on D_h) for the eight fin spacings tested. Entrance and exit losses were subtracted from the pressure drop, and are not included in the friction factor. Figure 6.3 clearly shows that the hydraulic-diameter-based Re (Re_{Dh}) does not correlate either the j or f data.

Rich proposed that the friction drag force is the sum of the drag force on a bare tube bank (Δp_t) and the drag caused by the fins (Δp_f). The difference between the total drag force and the drag force associated with the corresponding bare tube bank is the drag force on the fins. Thus, the friction component resulting from the fins is given by

$$f_F = (\Delta p - \Delta p_t)\,\frac{2A_o\rho}{G^2 A_f} \tag{6.2}$$

The term Δp_t is that measured for a bare tube bank of the same geometry, without fins. Both Δp drop contributions are evaluated at the same minimum area mass velocity. Figure 6.4 shows the same j-factor data and the fin friction factor calcu-

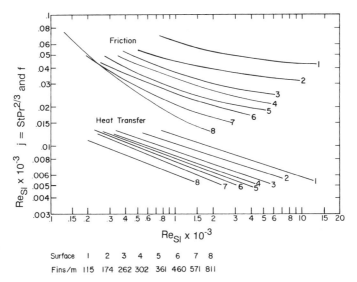

Figure 6.3   Heat transfer and friction characteristics of a four-row plain plate fin heat exchanger for different fin spacings as reported by Rich [1973].

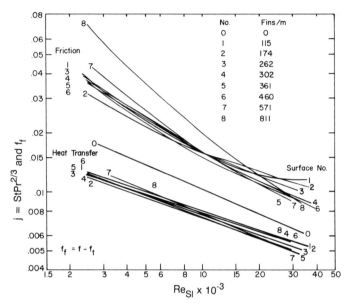

Figure 6.4   Plot of the j factor and the fin friction versus Re_{Sl} based on data reported by Rich [1973].

lated by Equation 6.2 plotted versus the Reynolds number based on the longitudinal row pitch (S_l). The row pitch (S_l) is constant for all of the test geometries. Figure 6.4 shows that the j factor is a function of velocity in the minimum flow area (G_c) and is essentially independent of fin spacing. At the same mass velocity (G_c), the heat transfer coefficient of the bare tube bank is 40% larger than that of the finned tube bank. Figure 6.4 shows that the resulting friction correlation is reasonably good, except for the closest fin spacings. The friction factor data of surfaces 7 and 8 may be questionable, since these surfaces show smaller j/f values than for the other fin spacings. This behavior is unexpected. Normally, the j/f ratio will increase as the fin spacing is reduced, since the fractional parasitic drag associated with the tube is reduced. Use of the Reynolds number based on S_l has no real significance, since all geometries tested had the same S_l. The same degree of correlation would result from use of a Reynolds number based on the tube diameter (d_o), which was also constant. Figure 6.4 may be regarded as evidence that the Reynolds number based on hydraulic diameter will not correlate the effect of fin pitch.

In a later study, Rich [1975] used the same heat exchanger geometry with 551 fins/m to determine the effect of the number of tube rows on the j factor. Figure 6.5 shows the average j factor (smoothed data fit) for each exchanger as a function of Re_{Sl}. The numbers on the figure indicate the number of rows in each coil. The row effect is greatest at low Reynolds numbers and becomes negligible at $Re_{Sl} >$ 15,000.

6.3.2   Correlations for Staggered Tube Geometries

Correlations to predict the j and f factors versus Reynolds number for plain fins on staggered tube arrangements were developed by McQuiston [1978] and by Gray and Webb [1986]. The McQuiston correlation is based on the data of Rich [1973, 1975] shown in Figures 6.3 and 6.5, as well as the data of three other investigators. Gray and Webb used the same data set as McQuiston, plus two additional investigators. The McQuiston and the Gray and Webb heat transfer correlations are comparable in accuracy. However, the Gray and Webb friction factor correlation is much more accurate than that of McQuiston.

The Gray and Webb [1986] heat transfer correlation (for four or more tube rows of a staggered tube geometry) is

$$j_4 = 0.14 \, \text{Re}_d^{-0.328} \left(\frac{S_t}{S_l} \right)^{-0.502} \left(\frac{s}{d_o} \right)^{0.031} \tag{6.3}$$

Equation 6.3 assumes that the heat transfer coefficient is stabilized by the fourth tube row, and hence the j factor for more than four tube rows is the same as that for a four-row exchanger. The correction for rows less than four is based on correlation of the data in Figure 6.5 and is given by

$$\frac{j_N}{j_4} = 0.991 \left[2.24 \text{Re}_d^{-0.092} \left(\frac{N}{4} \right)^{-0.031} \right]^{0.607(4-N)} \tag{6.4}$$

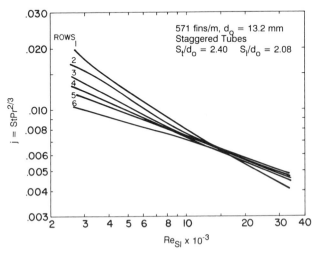

Figure 6.5   Average heat transfer coefficients for plain plate-finned tubes (571 fins/m) having 1–6 rows as reported by Rich [1975]. Same geometry dimensions as in Figure 6.4.

Equations 6.3 and 6.4 correlated 89% of the data for 16 heat exchangers within 10%. The McQuiston [1978] correlation gives comparable results.

Gray and Webb [1986] friction correlation assumes that the pressure drop is composed of two terms. The first term accounts for the drag force on the fins, and the second term accounts for the drag force on the tubes. The validity of this model was previously established in the discussion of Figure 6.4. The friction factor of the heat exchanger is given by

$$f = f_f \frac{A_f}{A} + f_t \left(1 - \frac{A_f}{A} \right) \left(1 - \frac{t}{p_f} \right)$$

(6.5)

The friction factor associated with the fins (f_f) is given by Equation 6.6:

$$f_f = 0.508 \, \text{Re}_d^{-0.521} \left(\frac{S_t}{d_o} \right)^{1.318}$$

(6.6)

The friction factor associated with the tubes (f_t) is obtained from a correlation for flow normal to a staggered bank of plain tubes. Gray and Webb used the Zukauskas [1972] tube bank correlation, also given in Incropera and DeWitt [1990], to calculate the tube bank contribution, Δp_t. f_t is calculated at the same mass velocity (G) that exists in the finned tube exchanger. Equation 6.5 correlated 95% of the data for 19 heat exchangers within $\pm 13\%$. The equation is valid for any number of tube rows. McQuiston [1978] also developed a friction correlation using the same data set; however, his friction correlation has quite high error limits, namely, $+167\%/-21\%$ for the same data.

The range of dimensionless variables used in the development of the Gray and Webb correlations are $500 \le \text{Re}_d \le 24{,}700$, $1.97 \le S_t/d_o \le 2.55$, $1.7 \le S_l/d_o \le 2.58$ and $0.08 \le s/d_o \le 0.64$.

Recent work by Seshimo and Fujii [1991] provides more generalized correlations for staggered banks of plain fins having 1–5 tube rows. They tested 35 heat exchangers, having systematically changed geometric parameters. They used three tube diameters (6.35, 7.94, and 9.52 mm) with the multi-row designs using an equilateral triangular pitch. Data were obtained for four fin densities, from 454 to 1000 fins/m. Eleven one-row designs having different transverse tube pitch and fin depth were tested. They show that the one- and two-row data may be separately correlated using an entrance length parameter. Their data were correlated using a Reynolds number (Re_{Dv}) defined in terms of the volumetric hydraulic diameter (D_{Dv}). The D_v is given by

$$D_v = \frac{4 A_m L}{A}$$

(6.7)

where $A_m L$ is the defined as the total volume of the exchanger, less the volume of the tube bank. The one- and two-row data were correlated in terms of the entrance length parameter, $x_{\text{Dv}}^+ \equiv \text{Re}_{\text{Dv}} \text{Pr} D_v/L$. The correlations are

$$\mathrm{Nu} = 2.1(X_{\mathrm{Dv}}^{+})^{n} \qquad\qquad (6.8)$$

$$fLD_{v} = c_{1} + c_{2}(X_{\mathrm{Dv}}^{+})^{-m} \qquad\qquad (6.9)$$

where then constants and exponents differ for one- or two-row exchangers. They are as follows:

1. One-row: $n = 0.38$, $m = 1.07$, $c_1 = 0.43$, and $c_2 = 35.1$.

2. Two-row: $n = 0.47$, $m = 0.89$, $c_1 = 0.83$, and $c_2 = 24.7$.

For three or more rows, these entrance-length-based correlation did not work very well over the entire Reynolds number range ($200 < \mathrm{Re}_{\mathrm{Dh}} < 800$), because vortex shedding from the tubes seems to be an important factor. For $\mathrm{Re}_{\mathrm{Dh}} > 400$, the data were correlated using the conventional Nusselt number and Reynolds number ($\mathrm{Re}_{\mathrm{Dh}}$) and flow based on the minimum flow area. For $\mathrm{Re}_{\mathrm{Dh}} < 400$, the one-row variant of Equations 6.8 and 6.9 correlated the data for 1-to-5 rows.

6.3.3 Correlations for Inline Tube Geometries

Schmidt [1963] reports data and a correlation for the inline geometry. However, little use exists for an inline tube arrangement. This is because tube bypass effects substantially degrade the performance of an inline tube arrangement. The degree of performance degradation for inline circular fins is discussed in Section 6.4.1.

6.4 PLAIN INDIVIDUALLY FINNED TUBES

6.4.1 Circular Fins with Staggered Tubes

Extruded fins or helically wrapped fins on circular tubes, as shown by Figure 6.1b, are frequently used in the process industries and in combustion heat recovery equipment. Both plain and enhanced fin geometries are used. A staggered tube layout is used, especially for high fins ($e/d_o > 0.2$). A substantial amount of performance data has been published, and several heat transfer and pressure drop correlations have been proposed. The dominant amount of data were taken with a staggered tube arrangement, six or more tube rows deep. The correlations must account for the three tube bank variables (d_o, S_t, and S_l), the fin geometry variables (t, e, and s), and the number of tube rows. Webb [1987] provides a survey of the published data and correlations.

The recommended correlations for a staggered tube layout are the Briggs and Young [1963] correlation for heat transfer and the Robinson and Briggs [1966] correlation for pressure drop. Both correlations are empirically based and are valid for four or more tube rows. The heat transfer correlation is given by

$$j = 0.134\mathrm{Re}_d^{-0.319} \left(\frac{s}{e} \right)^{0.2} \left(\frac{s}{t} \right)^{0.11} \qquad (6.10)$$

Equation 6.10 is based on air flow over 14 equilateral triangular tube banks and covers the following ranges: $1100 \leq \mathrm{Re}_d \leq 18{,}000$, $0.13 \leq s/e \leq 0.63$, $1.0 \leq s/t \leq 6.6$, $0.09 \leq e/d_o \leq 0.69$, $0.01 \leq t/d_o \leq 0.15$, $1.5 \leq S_t/d_o \leq 8.2$. The standard deviation was 5.1%.

The isothermal friction correlation of Robinson and Briggs [1966], which we have rewritten in terms of the tube bank friction factor, is given by

$$f_{tb} = 9.47\mathrm{Re}_d^{-0.316} \left(\frac{S_t}{d_o} \right)^{-0.927} \left(\frac{S_t}{S_d} \right)^{0.515} \qquad (6.11)$$

Equation 6.11 is based on isothermal air-flow data over 17 triangular-pitch tube banks (15 equilateral and two isosceles). The data span the ranges $2000 \leq \mathrm{Re}_d \leq 50{,}000$, $0.15 \leq s/e \leq 0.19$, $3.8 \leq s/t \leq 6.0$, $0.35 \leq e/d_o \leq 0.56$, $0.01 \leq t/d_o \leq 0.03$, $1.9 \leq S_t/d_o \leq 4.6$. The standard deviation of the correlated data was 7.8%. Equation 6.11 is recommended with strong reservations, because it does not contain any of the fin geometry variables (e, s, or t). Because only a small range of s/e was covered in the tests, it is probable that the correlation will fail outside the s/e range used for developing the correlation. Gianolio and Cuti [1981] compared their data for 17 tube bank geometries containing 1–6 rows with the Briggs and Young [1963] and the Robinson and Briggs [1966] correlations. For induced draft, their 6-row data are 0–10% above that of Briggs and Young. Their data for $N < 6$ were increasingly underpredicted as the number of rows decreases. For $N < 6$, they recommended that the Briggs and Young value be multiplied by the factor $(1 + G/\rho N^2)^{-0.14}$, where G is in kg/m²-s units. Gianolio and Cuti [1981] also state that their induced draft h values are 10–40% higher than those for forced draft. No explanation is provided for this unexpected result. The Robinson and Briggs [1965] correlation did not predict their data very well.

Although the data on which Equation 6.10 is based included low fin data (e.g., $e/d_o < 0.1$), Rabas et al. [1981] developed more accurate j and f correlations for low fin heights and small fin spacings. The correlations are given below with the exponents rounded off to two significant digits.

$$j = 0.292 \left(\frac{d_o G_c}{\mu} \right)^n \left(\frac{s}{d_o} \right)^{1.12} \left(\frac{s}{e} \right)^{0.26} \left(\frac{t}{s} \right)^{0.67} \left(\frac{d_e}{d_o} \right)^{0.47} \left(\frac{d_e}{t} \right)^{0.77} \qquad (6.12)$$

where $n = -0.415 + 0.0346(d_e/s)$. The friction correlation is given by

$$f = 3.805 \left(\frac{d_o G_c}{\mu} \right)^{-0.234} \left(\frac{s}{d_e} \right)^{0.25} \left(\frac{e}{s} \right)^{0.76} \left(\frac{d_o}{d_e} \right)^{0.73} \left(\frac{d_o}{S_t} \right)^{0.71} \left(\frac{S_t}{S_1} \right)^{0.38} \qquad (6.13)$$

The equations are valid for staggered tubes with $N \geq 6$, $5000 \leq \text{Re}_d \leq 25{,}000$, $1.3 \leq s/e \leq 1.5$, $0.01 \leq s/t \leq 0.06$, $e/d_o \leq 0.10$, $0.01 \leq t/d_o \leq 0.02$, and $1.3 \leq S_t/d_o \leq 1.5$. The equations predicted 94% of the j data and 90% of the f data within $+15\%$. Rabas and Taborek [1987] present a survey of correlations, row correction factors, and other issues concerning low, integral-fin tube banks. They compare the ability of Equations 6.12 and 6.13 to predict other data sets (e.g., that of Groehn [1977]). Other correlations have been developed by Groehn [1977] and ESDU [1985], which were tested to higher Re_d. Apparently, no correlations exist for $\text{Re}_d \leq 1000$.

A staggered tube layout gives higher values for j at the same Re, especially for high fins ($e/d_o > 0.3$). Hence the inline tube layout is not recommended for $e/d_o > 0.3$. Rabas and Huber [1989] discuss the reduction of the j factor with increased number of tube rows. Designers interested in inline finned tube banks should refer to Schmidt [1963], who developed a heat transfer correlation based on data from 11 sources.

6.4.2   Low Integral-Fin Tubes

The Rabas et al. [1981] correlations given by Equations 6.12 and 6.13 are recommended for a staggered layout of low integral fins. Corresponding equations have not been developed for inline tube layouts. However, Brauer's [1964] j data for the $e/d_o = 0.07$ inline tube bank shown in Figure 6.16 is within 20% of that of a staggered bank having the same e/d_o, S_t/d_o and S_l/d_o. The inline friction factor was approximately 35% smaller than that of the staggered bank. It appears that the performance decrement for inline banks having low fins (e.g., $e/d_o = 0.1$) is not nearly as severe as that for high fins (e.g. $e/d_o = 0.4$).

6.5   ENHANCED PLATE FIN GEOMETRIES WITH ROUND TUBES

The wavy (or herringbone) fin and the offset strip fin (also referred to as parallel louver) geometries are the major enhanced surface geometries used on circular tubes. Figure 6.2c shows the wavy fin geometry applied to circular tubes. This figure shows the geometrical dimensions which influence the heat transfer and friction characteristics. The combination of tubes plus a special surface geometry establishes a very complex flow geometry. The heat transfer coefficient of the wavy fin is typically 50–70% greater than that of a plain (flat) fin.

6.5.1   Wavy Fin

Beecher and Fagan [1987] published heat transfer data for 20 three-row plate fin-and-tube geometries having the wavy fin geometry shown in Figure 6.2c. Figure 6.2c defines the geometric parameters of the wavy fin. All cores had staggered tubes with three rows with $S_t/S_l = 1.15$. Two tube diameters were tested, namely, $d_o =$

9.53 mm and 12.7 mm. The fin pitch was varied from 244 fins/m (6.2 fins/in) to 510 fins/m (13 fins/in) with dimensionless wave heights of $0.076 \leq S_d/d_o \leq 0.25$ and wave pitches of $0.025 \leq e_w/p_w \leq 0.18$. The wavy fins have a 3.18-mm-wide (0.125-in.-wide) flat region around the fin collar. The Nusselt number data were presented as $Nu_a (= h_a D_h/k)$ versus the Graetz number, $Gz = Re_{Dh} Pr D_H/L$. Nu_a is based on the arithmetic mean temperature difference (AMTD) rather than on the LMTD.

Webb [1990] developed a multiple regression correlation of the Beecher and Fagan [1987] wavy fin data. Because the curve of Nu_a versus Gz was not a straight line on log–log coordinates, a two region correlation was used. The correlations are given by:

$$Nu_a = 0.5Gz \left(\frac{S_t}{d_o} \right)^{0.11} \left(\frac{s}{d_o} \right)^{-0.09} \left(\frac{S_d}{S_t} \right)^{0.12} \left(\frac{2S_p}{S_t} \right)^{0.34} \qquad \text{for Gz} > 25$$

$$(6.14)$$

$$Nu_a = 0.83Gz \left(\frac{S_t}{d_o} \right)^{0.13} \left(\frac{s}{d_o} \right)^{-0.16} \left(\frac{S_d}{S_t} \right)^{0.25} \left(\frac{2S_p}{S_t} \right)^{0.43} \qquad \text{for Gz} < 25$$

$$(6.15)$$

For $5 \leq Gz \leq 180$, 96% of the data were correlated to within $\pm 10\%$ The Nusselt number is traditionally based on the logarithmic mean temperature difference (LMTD) rather than on the arithmetic mean temperature difference; however, Beecher and Fagan discovered that at low air velocities, small errors in air temperature measurement led to large errors in calculation of the LMTD and chose to base the Nusselt number on the AMTD. The Nu based on the AMTD may be converted to the LMTD-based Nusselt number (Nu_l), by using the following equation.

$$Nu_t = \frac{Gz(1 + 2N_a/Gz)}{4(1 - 2N_a/Gz)} \qquad (6.16)$$

6.5.2 Offset Strip Fins

The OSF concept (also known as "slit fins") illustrated in Figure 6.2b has been applied to finned-tube heat exchangers with plain fins for dry cooling towers and for refrigerant condensers. Figure 6.2b shows one such geometry, which was studied by Nakayama and Xu [1983]. Figure 6.6 shows the heat transfer coefficients of the OSF and a plain fin used in a two-row staggered tube heat exchanger having 966 fins/m on 10-mm-diameter tubes. At 3-m/s air velocity, the OSF provides a heat transfer coefficient which is 78% higher than that of the plain fin. For the same louver geometry, the OSF will provide a higher heat transfer coefficient when used in the plate-and-fin-type heat exchanger. The OSF shown in Figure 5.4 provides a heat transfer coefficient which is 150% higher than that of the plain fin at the same velocity. Comparison of plain fin geometries on Figures 5.4 and 6.6 shows that the heat transfer coefficient of the plain fin shown in Figure 6.6 is 90% greater than that

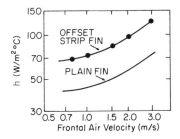

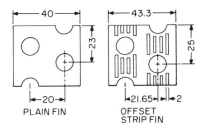

Figure 6.6 Comparison of the heat transfer coefficient for the OSF and plain fin geometries for 9.5-mm-diameter tubes, 525 fins/m, and 0.2-mm fin thickness as reported by Nakayama and Xu [1983]. (From Nakayama and Xu [1983].)

of the plain fin geometry shown in Figure 5.4. Thus, the flow acceleration and fluid mixing in the wake of the tube provide a substantial enhancement for plain fins on tubes.

Generalized empirical correlations for j and f versus Re have not been developed for OSF geometry on round tubes. However, Nakayama and Xu [1983] propose an empirical correlation to define the enhancement level (h/h_p) of an OSF geometry having 2.0-mm strip width (in the flow direction) and 0.2-mm fin thickness.

6.5.3 Convex Louver Fins

Hitachi [1984] uses the convex louver fin geometry in their commercial plate fin-and-tube heat exchangers. Figure 6.2d is taken from the Hitachi [1984] product brochure. The performance of the convex louver fin plate-and-fin surface geometry was compared to the OSF geometry in Section 5.4. Hatada et al. [1989] report performance data of the Figure 6.2d-type fin geometry for a one-row heat exchanger. Figure 6.7 illustrates the finned tube geometry, and Figure 6.8 shows the air-side h and Δp values versus air velocity for three geometries tested. The geometry details of the three geometries shown in Figure 6.8 are defined on Figure 6.7. Figure 6.8 shows data for two variants of the convex louver fin geometry. Fin number 2 has a uniform convex louver shape (louver angle $\theta_1 = 12.5$ degrees). Fin number 1 has $\theta_1 = 17.5$ degrees in the regions between the tubes (between sections B-B and C-C in Figure 6.7) and has $\theta_1 = 4.0$ degrees adjacent to the tubes. The reduced louver angle near the tubes allows more air flow in the vicinity of the tubes. The 17.5-degree louver angle in the fin region between the tubes was found to give

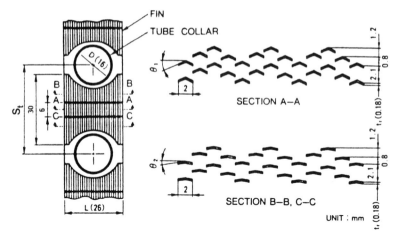

Figure 6.7   Convex louver plate fin-and-tube geometry tested by Hatada et al. [1989]. (From Hatada et al. [1989].)

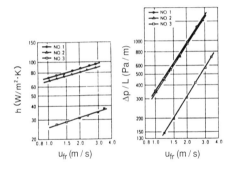

Heat Exchanger Dimensions

Feature	Convex Strip Fin		Plain Plate Fin
	Fin No 1	Fin No 2	Fin No 3
Transverse Tube Pitch S_t (mm)	38	38	36
Fin Depth L (mm)	26	26	42
Number of Rows	1	1	1
Tube Diameter d (mm)	16	16	16
Fin Pitch p_f (mm)	2.2	2.2	2.1
Fin Thickness t (mm)	1.8	1.8	1.8
Ramp Angle θ_1 (Degrees)	17.5	12.5	--
Ramp Angle θ_2 (Degrees)	4.0	12.5	--

Figure 6.8   Performance data for the convex louver surface geometries of Hatada et al. [1989] shown in Figure 6.7. (From Hatada et al. [1989].)

high j and j/f by Hatada and Senshu's [1984] studies of the plate-and-fin geometry, as discussed in Section 5.4. Figure 6.8 shows that the number 1 fin geometry gives an approximately 10% higher h value than does the number 2 fin geometry. The h value of the number 1 louver fin geometry is 2.85 times that of the plain fin (number 3) at the same air velocity. When compared at the same air friction power, the h value of the number 1 fin is 2.3 times that of the number 3 plain fin.

6.5.4 Perforated Fins

Fujii et al. [1991] tested a plate-fin geometry made of corrugated, perforated plates whose fin geometry is illustrated in Figure 5.22. Section 5.6 discusses the performance of this geometry as a plate-and-fin heat exchanger configuration. Fujii et al. [1991] applied the surface illustrated in Figure 5.22 to a one-row plate-fin heat exchanger having 0.5-mm-thick copper fins and obtained experimental results. Their heat exchanger was made of 28-mm-diameter tubes at 76-mm tube pitch and 66-mm fin depth. Figure 6.9 defines the geometry details and presents the test results for the two fin geometry variants, each for 6.0 and 8.0 mm fin pitch. Fin

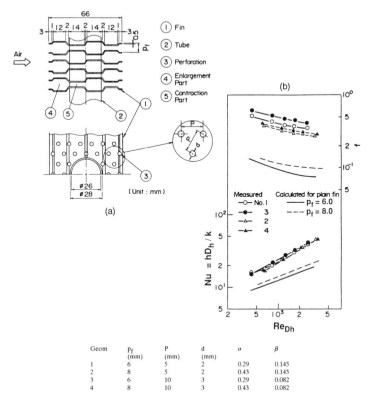

Geom	P_f (mm)	P (mm)	d (mm)	σ	β
1	6	5	2	0.29	0.145
2	8	5	2	0.43	0.145
3	6	10	3	0.29	0.082
4	8	10	3	0.43	0.082

Figure 6.9 (a) Illustration of one-row finned-tube heat exchanger tested by Fujii et al. [1991]. (b) Air-side test results. (From Fujii et al. [1991].)

geometries number 1 and number 3 are the same, as are number 2 and number 4. Both fin geometry variants provide approximately equal Nu. However, geometry number 3 provides a lower friction factor than does geometry number 4 (6-mm fin pitch). Although the Nu is increased approximately 100%, the friction factor of geometry number 3 is increased by a factor of 2.3, relative to a plain fin. This friction performance is not competitive with other high-performance fin geometries discussed in this chapter. Note that the data in Figure 6.9 may be scaled to other tube diameters by scaling all the dimensions in the ratio of the new and original tube diameters.

6.6   ENHANCED CIRCULAR FIN GEOMETRIES

6.6.1   Illustrations of Enhanced Fin Geometries

Figure 6.10 shows some of the enhanced fin geometries that have been used on circular tubes. In Table 2 of Webb [1980], references are provided for information on performance of the fin geometries shown in Figure 6.10. All of the geometries provide enhancement by the periodic development of thin boundary layers on small-diameter wires or flat strips, followed by their dissipation in the wake region between elements. Perhaps the most popular enhancement geometry is the "segmented fin" or "spine fin" (Figure 6.10d), which is similar in concept to the offset strip fin shown in Figure 6.6. The segmented fin is used in a wide range of applications, from air-conditioning to boiler economizers. Figure 6.11 shows two versions of the segmented fin used in air-conditioning applications. In Figure 6.11b, a 0.15-mm-thick aluminum fin strip is tension-wound on the tube and bonded to the tube with an epoxy resin. The fin segment width is 0.75 mm. Figure 6.12 shows the j and f performance of the fin geometries shown in Figures 6.11a and 6.11b and compares their performance with that of plain fins.

6.6.2   Spine or Segmented Fins

The data of Figure 6.12 are for eight rows on a triangular pitch with $d_o = 12.7$ mm and $t = 0.51$ mm. Note that the configurations shown in Figure 6.12 have a negative fin tip clearance. Figure 6.12 shows that the geometries illustrated in Figures 6.11a and 6.11b have approximately the same j and f for the same fin spacing. At $\text{Re}_d =$

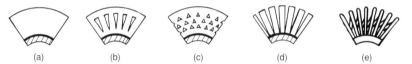

Figure 6.10   Enhanced circular fin geometries. (a) Plain circular fin. (b) Slotted fin. (c) Punched and bent triangular projections. (d) Segmented fin. (e) Wire loop extended surface. (From Webb [1987].)

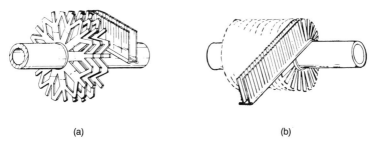

(a)   (b)

Figure 6.11 "Segmented" or "spine" fin geometries used in air-conditioning applications. (a) Geometry studied by LaPorte et al. [1979]. (b) Geometry described by Abbot et al. [1980] and tested by Eckels and Rabas [1985].

4000, $j/j_p \simeq 2.1$ and $f/f_p \simeq 4.2$, where the subscript p refers to the plain fin. The substantial heat transfer enhancement is accompanied by a relatively high friction factor increase. One is interested in increasing the value of hA/L by the use of an enhanced fin geometry. The fin shown in Figure 6.11a has only 43% as much surface area as a plain fin of the same d_e. With $j/j_p = 2.1$, one would obtain $hA/h_p A_p = 2.1 \times 0.43 = 0.90$. For the same hA, the geometry shown in Figure 6.11a would yield a 53% saving of fin material $(1 - 0.43/0.90)$.

A steel segmented fin geometry has been used for boiler economizers and waste heat recovery boilers. Figure 6.13 shows the j and f versus Re_d curves for a four-row staggered and a seven-row inline tube segmented fin geometry as reported by

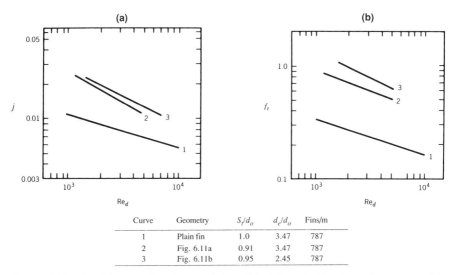

Curve	Geometry	S_f/d_o	d_e/d_o	Fins/m
1	Plain fin	1.0	3.47	787
2	Fig. 6.11a	0.91	3.47	787
3	Fig. 6.11b	0.95	2.45	787

Figure 6.12 j and f versus Re_d characteristics of the Figure 6.11 enhanced fin geometries compared with a plain, circular fin geometry as reported by Eckels and Rabas [1985]. (From Webb [1987].)

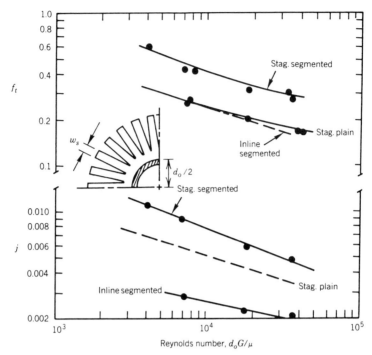

Figure 6.13 Comparison of segmented fins (staggered and inline tube layouts) with plain, staggered fin tube geometry as reported by Weierman et al. [1978]. $S_t/d_o = 2.25$, $e/d_o = 0.51$, $s/e = 0.12$, $w_s/e = 0.17$. (From Webb [1987].)

Weierman et al. [1978]. Also shown are the j and f curves for a staggered plain fin geometry having the same geometrical parameters as the staggered segmented geometry. The plain fin j and f values were calculated using Equations 6.10 and 6.11, respectively. For a staggered tube geometry, Figure 6.13 shows that the j factor of the segmented fin is 40% greater than that of the plain fin geometry. Note that the heat transfer performance of the inline segmented fin geometry is much lower than that of the staggered, segmented fin geometry. The poor performance of the inline geometry was previously noted, and is further discussed in Section 6.8.

Steel fin geometries are used for boiler economizers and heat recovery boilers. Steel is preferred because of the high gas temperature and also because of the corrosive potential of the combustion products. Data on several types of fin geometries are reported. Weierman [1976] gives empirical design correlations for steel segmented and plain fin geometries for staggered and inline tube layouts. Rabas et al. [1986] present additional data on steel segmented fin tubes. Rabas et al. [1986] provide data for staggered and inline arrangements. Breber [1991] provides a review of data and new data on "stud fin" circular tubes. This geometry consists of steel studs welded to a base tube. The stud cross-sectional shape may be round, rectangular, or elliptical. Breber recommends appropriate correlations to predict the heat transfer coefficient and friction factor.

Holtzapple and Carranza [1990] and Holtzapple et al. [1990] provide friction and heat transfer data, respectively, on a spine fin tube made of copper tubes and fins. The spine fins are cut from the thick wall tube, and they are bent so that they project from the tube wall. Their tests were performed on 9.4- and 12.6-mm-diameter tubes with 6.8- and 9.3-mm spine lengths, respectively. There were 26 spines around the circumference, with an axial spacing of 4.76 mm. This geometry provides considerably less total finned surface area than typical of the Figure 6.11 geometries. An advantage of the Holtzapple et al. spine fin is that the fins are integral to the tube wall. However, the geometry is relatively expensive to manufacture. Data are provided on several tube pitch layouts. Carranza [1991] provides anempirical pressure drop correlation.

6.6.3 Wire-Loop Fins

Benforado and Palmer [1964] provide data on the wire loop fin geometry shown in Figure 6.10e. The data were taken for equilateral triangular and inline arrangements on 25.4-mm-diameter tubes on approximately 66-mm transverse tube pitch. The wire loops were made of 0.71-mm-diameter wire, with 12.9-mm fin height. They also tested a plain circular fin geometry having the same fin pitch and height. The wire loop geometry gave approximately 50% higher heat transfer coefficient (based on actual surface area) and the same pressure drop when compared to the plain fin.

6.7 OVAL- AND FLAT-TUBE GEOMETRIES

6.7.1 Oval Versus Circular Individually Finned Tubes

Oval and flat cross-sectional tube shapes are also applied to individually finned tubes. Figure 6.14 compares the performance of staggered banks of oval and circular finned tubes tested by Brauer [1964]. Both banks have 312-fin/m, 10-mm-high fins on approximately the same transverse and longitudinal pitches. The oval tubes gave 15% higher heat transfer coefficient and 25% less pressure drop than did the circular tubes. The performance advantage of the oval tubes results from lower form drag on the tubes and the smaller wake region on the fin behind the tube. The use of oval tubes may not be practical unless the tube-side design pressure is sufficiently low.

6.7.2 Flat, Extruded Aluminum Tubes with Internal Membranes

Higher design pressures are possible using flattened aluminum tubes made by an extrusion process. Figure 6.2f shows a patented finned tube concept made from an aluminum extrusion, as described by O'Connor and Pasternak [1976]. Such tubes can be made with internal membranes which strengthen the tube and allow for a high tube-side design pressure. A variety of fin and tube shapes may be made in aluminum extrusions of different shapes. These "skive fins" are formed from the thick wall using a modified high-speed punch press without creation of scrap mate-

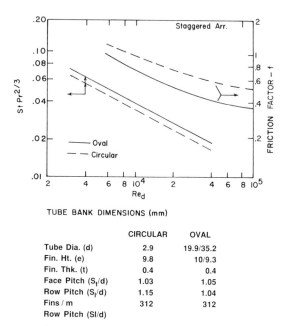

TUBE BANK DIMENSIONS (mm)

	CIRCULAR	OVAL
Tube Dia. (d)	2.9	19.9/35.2
Fin. Ht. (e)	9.8	10/9.3
Fin. Thk. (t)	0.4	0.4
Face Pitch (S_t/d)	1.03	1.05
Row Pitch (S_l/d)	1.15	1.04
Fins / m	312	312
Row Pitch (Sl/d)		

Figure 6.14   Heat transfer and friction characteristics of circular and oval finned tubes in a staggered tube layout as reported by Brauer [1964]. (From Webb [1987].)

rial. The punch press slits the thick aluminum wall and simultaneously bends the chip outward to form the fin. The process is applicable to virtually any cross-section geometry of the extrusion. Designs have been made with circular and flattened tube geometries. Haberski and Raco [1976] show photographs of a number of geometries which have been fabricated. Cox [1973] provides test data on a circular tube geometry, and Cox and Jallouk [1973] give test results on the geometry shown in Figure 6.2f.

The flat tube geometry shown in Figure 6.2e offers significant advantages over the strip fin geometries on round tubes shown in Figures 6.2a and 6.2b.

1. The air flow is normal to all of the narrow strips on the Figure 6.2e geometry, which is not the case for the Figure 6.2a spine fin design. Furthermore, the wake dissipation length decreases in the direction of the fin base in the Figure 6.2a geometry.

2. A low-velocity wake region does not occur behind the tubes of the Figure 6.2e flat tube geometry. The low wake velocity of the Figure 6.2a and 6.2b geometries causes a substantial reduction of the heat transfer coefficient, as documented by Webb [1980].

3. The fraction of the Figure 6.2e surface that is louvered is substantially greater than that in the Figure 6.2b geometry. If a greater area distribution of louvers were provided in the Figure 6.2b geometry, the fin efficiency would substan-

tially decrease. This is because the slits would cut the heat conduction path from the base tube.

4. Furthermore, the low projected area of the Figure 6.2e flat tube will result in lower profile drag.

Some automotive air-conditioning condensers and evaporators use brazed aluminum heat exchangers having flat extruded aluminum tubes with internal membranes. Figure 14.7 shows such heat exchangers. The internal membranes are required to meet the 11,000-kPa (1600-psi) burst pressure required for R-12 or R-134a used in automotive air conditioners. These heat exchangers typically use the louver fin geometry shown in Figure 5.2f. Webb and Gupte [1990] compare the performance of this heat exchanger construction with that of (a) the wavy plate fin-and-tube geometry and (b) the spine fin geometry. These comparisons are presented in Section 6.10.1.

6.7.3 Plate-and-Fin Automotive Radiators

Figure 6.15 shows an automotive radiator geometry having louvered plate fins on flat tubes. The automotive radiator is designed to operate at low pressure, so internal membranes are not required in the tubes. The outer tube cross-section dimensions of brass tubes are typically 1.0–1.5 mm × 12.0–19.0 mm. Achaichia and Cowell [1988] developed a correlation for the Figure 6.15 inline louver fin geometry, based on their tests of 16 core geometries. Their test data span $1.7 \leq p_f \leq 3.44$ mm, $0.81 \leq L_p \leq 1.4$ mm, $8 \leq S_t \leq 14$ mm, and $22 \leq \theta \leq 30$ degrees. Their j- and f-factor correlations are

$$j = 1.234\gamma Re_L^{-0.59} \left(\frac{S_t}{L_p} \right)^{-0.09} \left(\frac{p_f}{L_p} \right)^{-0.04} \tag{6.17}$$

$$\gamma = \frac{1}{\alpha} \left(0.936 - \frac{243}{Re_L} - 1.76 \frac{p_f}{L_p} + 0.995\alpha \right) \tag{6.18}$$

$$f = 533 p_f^{-0.22} L_p^{0.25} S_t^{0.26} H^{0.33} [Re_L^{(0.318 \ \log_{10} Re_L - 2.25)}]^{1.07} \tag{6.19}$$

where all dimensions in Equation 6.19 are in mm.

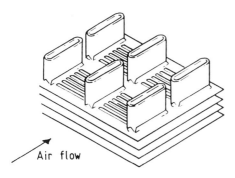

Air flow

Figure 6.15 Illustration of the louvered plate fin automotive radiator with inline tubes. (From Achaichia and Cowell [1988].)

6.8   ROW EFFECTS—STAGGERED AND INLINE LAYOUTS

The published correlations are generally for deep tube banks and do not account for row effects. The heat transfer coefficient will decrease with rows in an inline bank due to the bypass effects. However, the coefficient increases with number of tube rows in a staggered bank. This is because the turbulent eddies shed from the tubes cause good mixing in the downstream fin region. As an approximate rule, one may assume that the heat transfer coefficient for a staggered tube bank of the finned tubes shown in Figures 6.2a and 6.2b finned-tubes has attained its asymptotic value at the fourth tube row.

Inline tube banks generally have a smaller heat transfer coefficient than do staggered tube banks. At low Re_d ($Re_d < 1000$) with deep tube banks ($N \geq 8$), Rabas and Huber [1989] show that the heat transfer coefficient may be as small as 60% of the staggered tube value. The heat transfer coefficient of the inline bank increases as the Re_d is increased; at $Re_d = 50,000$ with $N \geq 8$, the inline-to-staggered ratio may approach 0.80.

There is a basic difference in the flow phenomena in staggered and inline finned tube banks. Figure 6.16 compares the performance of inline and staggered banks of plain, circular finned tubes (Figure 6.1b) as reported by Brauer [1964]. If the fins are short ($e/d_o = 0.07$), the staggered bank gives 30% higher heat transfer coefficient. However, when the fin height is increased to $e/d_o = 0.53$, the heat transfer coefficient for the staggered tube arrangement is as much as 100% higher than that for the inline arrangement. Brauer [1964] argues that bypass effects in the inline

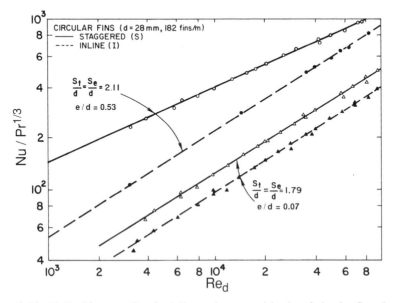

Figure 6.16 $NuPr^{-1/3}$ versus Re_d for inline and staggered banks of circular finned tubes with plain fins, as reported by Brauer [1964]. (From Webb [1983].)

arrangement are responsible for the poor performance. Figure 6.17, taken from Brauer [1964], shows the flow patterns in staggered and inline tube arrangements. The streamlines are shown by the dashed lines. The low-velocity wake regions, or "dead spaces," are shown by the shaded area with fine dots. A much greater fraction of the fin surface area is contained in the low-velocity wake region for the inline arrangement as compared to the staggered arrangement. Consequently, the inline arrangement will have a lower surface average heat transfer coefficient. Outside the shaded zone, particularly between the fin tips, a strong bypass stream exists. Because of poor mixing between the wake stream and bypass stream, the weaker wake stream is quickly heated; its mixed temperature is greater than that of the bypass stream. Thus, the actual temperature difference between the surface and the wake stream is much less than indicated by an overall LMTD based on the mixed outlet temperature. The staggered arrangement provides a good mixing of the wake and bypass streams after each tube row.

The superiority of the staggered fin geometry is shown in Figure 6.13 for $e/d_o =$ 0.51 segmented fins tested by Weierman et al. [1978]. The staggered bank (st) has four rows on an equilateral triangular pitch ($S_t/d_o = 2.25$), and the seven row inline (il) bank has $S_t/d_o = S_l/d_o = 2.25$. At $Re_d = 10,000$, $j_{st}/j_{il} = 2.17$ and $f_{st}/f_{il} = 1.73$. Figure 6.18 shows the row effect of the inline segmented fin geometry. It shows that the bypass effect reduces the performance of the inline geometry as the number of rows increases. Rabas and Huber [1989] have performed tests of the performance differences of staggered and inline banks of tubes having plain fins. Their work shows that the inline bank attains asymptotic values of j and f. In general, their work shows that plain, inline finned tube banks yield the lowest performance for (1) low Re_d, (2) large S_t/d_o and (3) small s/d_o. Rabas et al. [1986] provide additional data on inline versus staggered layout for plain and the Figure 6.11a segmented fin geometry.

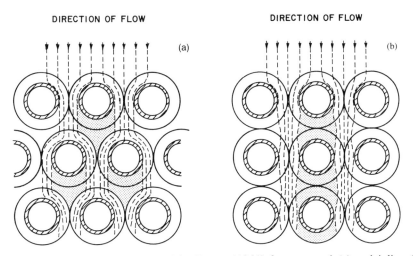

DIRECTION OF FLOW (a)   DIRECTION OF FLOW (b)

Figure 6.17 Flow patterns observed by Brauer [1964] for staggered (a) and inline (b) finned tube banks. (From Webb [1983].)

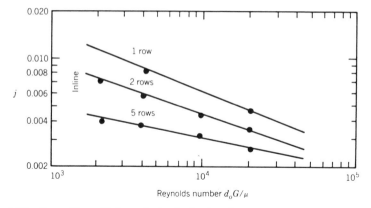

Figure 6.18   Row effect of inline tube banks for the Figure 6.13 segmented fin tubes (S_t/d_o = S_l/d_o = 2.25). (From Webb [1987].)

6.9   HEAT TRANSFER COEFFICIENT DISTRIBUTION (PLAIN FINS)

6.9.1   Experimental Methods

Although the standard fin efficiency calculation assumes that the heat transfer coefficient is constant over the fin surface area, local measurements have shown that this is not the case. The flow accelerates around the tube and forms a wake region behind the tube. This causes local variations of the heat transfer coefficient. Several researchers have measured the distribution of the local heat transfer coefficient on plain fins. Such measurements under steady-state heat transfer conditions are very difficult to perform, since the local fin temperature and local heat flux are required.

   Neal and Hitchcock [1966] used the steady-state method to determine the local values of the heat transfer coefficient. Jones and Russell [1980] used a transient heating method for the circular fin geometry. The local heat flux is calculated from the slope of the temperature versus time curve at each local fin temperature. Jones and Russell also used a different transient method, which avoided measurement of the local fin temperature distribution. They measured the time required to melt a thermal paint of known melting temperature. Others have used mass transfer techniques to infer the local heat transfer coefficients from the heat–mass transfer analogy. Saboya and Sparrow [1974] cast solid naphthalene plates in the form of a plate-fin-and-tube flow passage. The local mass transfer coefficients were defined by measuring the thickness of naphthalene lost by sublimation, during a timed test run. Krückels and Kottke [1970] used another mass transfer technique. Their method involves a chemical reaction between a surface coating and ammonia added to the air stream. The reaction caused permanent discoloration of the coated surface. A photometric method is used to establish the local mass transfer coefficient, which is related to the degree of discoloration.

6.9.2 Plate Fin-and-Tube Measurements

Saboya and Sparrow [1974, 1976a, 1976b] used the naphthalene mass transfer method to measure the local coefficients for one-, two-, and three-row plate fin-and-tube geometries. Figure 6.19 shows the geometry of the two-row layout. All geometries had the same tube size, pitches, and fin spacing (1.65 mm). Figure 6.20 shows the local Sherwood numbers for the one-row layout at Re = 648. The Sherwood number corresponds to the Nusselt number for heat transfer. The tube is located at $0.27 \leq x/L \leq 0.73$. Near the leading edge of the fin ($x/L \leq 0.2$) the developing boundary layer produces a relatively high Sh (12–18). Just upstream of the tube at $x/L = 0.221$, St shows a peak beginning to form. This is caused by the vortex which develops on the front of the tube and which is then swept around the tube. This vortex produces a natural heat transfer augmentation. Following the vortex around the tube, Sh attains a peak (34) at $x/L = 0.257$ and decreases to 14 at the tube centerline, where the boundary layer separates from the tube. The curves do not extend to $y = 0$, because of the presence of the tube. Away from the tube-induced vortex, $8 < Sh < 10$ for $0.283 \leq x/L \leq 1.0$, except for the wake region behind the tube. Aside from the peak Sh caused by the tube vortex, the maximum Sh does not occur at the tube centerline ($x/L = 0.5$), where the minimum flow area exists. A very-low-performance wake region exists behind the tube ($x/L > 0.73$). For $x/L > 0.744$, $1 \leq Sh \leq 4$ in the wake region, compared with $Sh = 8$ outside of the wake. Data taken at 100% higher velocities showed no appreciable increase of Sh in the tube wake region. However, the peak values of Sh caused by the tube vortex were 50% larger and persisted until the end of the fin, where $Sh = 16$. The Re = 648 data show that the local coefficient varies by a 3:1 factor when the wake region is excluded, and by 28:1 when the wake regions are included. The variations are even larger at high Re.

Figure 6.21 shows the local mass transfer coefficient measured by Krückels and Kottke [1970] for a two-row inline plate-fin geometry. The upstream air Reynolds

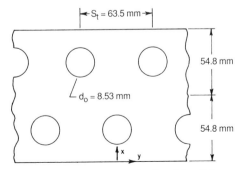

Figure 6.19 Two-row finned tube geometry simulated by Saboya and Sparrow [1974]. (From Saboya and Sparrow [1974].)

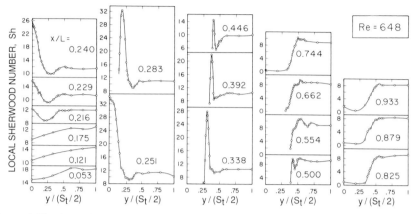

Figure 6.20   Local Sherwood number distribution measured by Saboya and Sparrow [1974] for a one-row plate fin geometry (see Figure 6.20). (From Saboya and Sparrow [1974].)

numbers (Re$_d$) are 1160 and 5800 for the upper and lower portions of the figure, respectively. The results are in qualitative agreement with the observations of Saboya and Sparrow.

6.9.3   Circular Fin-and-Tube Measurements

The local heat (or mass) transfer coefficients on this geometry have been measured by Neal and Hitchcock [1966] and Krückels and Kottke [1970] for single circular finned tubes and for banks of staggered arrangement. Figure 6.22 shows the mass transfer results of Krückels and Kottke for a single circular finned tube. Neal and Hitchcock's results for row 2 and row 6 of a staggered bank show qualitatively

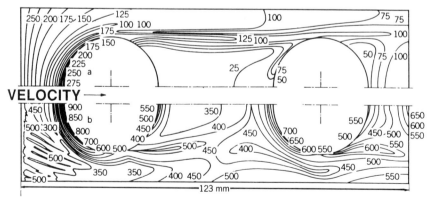

Figure 6.21   Distribution of mass transfer coefficients (m^3/m^2-hr) on a two-row plate finned tube measured by Krückels and Kottke [1970]. (a) Re$_d$ = 1160. (b) Re$_d$ = 5800. (From Krückels and Kottke [1970].)

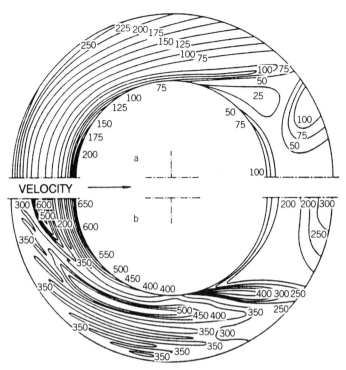

Figure 6.22   Distribution of mass transfer coefficients (m³/m²-hr) on a single circular finned tube (212 fins/m) measured by Krückels and Kottke [1970]. (a) Re_d = 1940. (b) Re_d = 9700. (From Krückels and Kottke [1970].)

similar coefficient distributions as indicated by Figure 6.22. Salient among the findings of these researchers are the following:

1. There is markedly higher heat transfer on the upstream area of the fin than on the downstream area. Maximum coefficients on the fin occur 70–90 degrees from the forward stagnation point.

2. There are higher heat transfer coefficients near the fin tip than near the base of the fin.

3. Stagnation flow on the front of the tube produces high heat transfer at the fin root. Slightly smaller coefficients near the fin tip, at the front, are caused by flow separation.

4. Radial heat flow occurs only near the front of the fin. There is a general migration of heat toward the front of the tube, especially as the fin tip is approached.

5. The measured fin efficiency is less than that calculated, assuming uniform heat transfer coefficients and radial heat flow (0.6 versus 0.7 for the case studied).

6.10   PERFORMANCE COMPARISON OF DIFFERENT GEOMETRIES

6.10.1   Geometries Compared

Webb and Gupte [1990] compared the performance of the six enhanced surface geometries described in Table 6.1. The first four geometries have fins on 9.52-mm-diameter round tubes. The last two geometries (OSF and louver) are brazed aluminum heat exchangers and are illustrated in Figure 6.23. The extruded aluminum tube used in the brazed aluminum exchanger shown in Figure 6.23a has 4.0-mm minor tube diameter with 0.78-mm wall thickness and internal membranes at 5.0-mm pitch.

The analysis was performed to compare candidate geometries for residential air-conditioning condensers. A key objective was to compare the performance of enhanced fin geometries using round tubes with brazed aluminum exchangers that use flat aluminum tubes. Aluminum fin and tube material was used in all of the heat exchangers.

The first four geometries in Table 6.1 are commercially used and employ round tubes. The louver fin geometry is used in automotive and air conditioning heat exchangers. A fin diameter (diameter over the fins) of 25.4 mm was used for the Figure 6.11b spine fins. The extruded aluminum tube used with the OSF and louver geometries (Figure 6.23) had 3.45-mm minor diameter with 0.75-m wall thickness, and can be made with any tube depth. Ideally, one would calculate the heat exchanger depth (in the air flow direction) needed to meet the required heat-exchange duty, and then form the aluminum extrusion having this depth. We define row 1 of the plate-and-fin geometry as the tube depth required to provide the same value of $A/A_{fr}N$ as that of the spine fin, with both having the same fin pitch.

The potential advantages of the Figure 6.23 flat tube OSF geometry, relative to the Figure 6.2a and 6.2b strip fin geometries, were discussed in Section 6.6.2. The 21.84-mm transverse tube used for the Figure 6.23 geometries resulted from an optimization study.

TABLE 6.1   Heat Exchanger Geometries Compared by Webb and Gupte [1990]

	Spine	Wavy	Slit	CLF	OSF	Louver
Figure	6.2a	6.2c	6.2b	6.2D	6.23b	6.23c
d_o (mm)	9.52	9.52	9.52	9.52		
$d_{o,min}$ (mm)					3.46	3.46
$d_{o,max}$ (mm)					7.87	8.38
t (mm)	0.76	0.76	0.76	0.76	0.76	0.76
P_t (mm)	25.4	23.62	25.4	25.4	21.84	21.84
P_l (mm)	25.4	20.60	21.60	21.60		
L_p (mm)			1.98		1.59	1.59
n_L			4		5	5
θ (degrees)					20	20

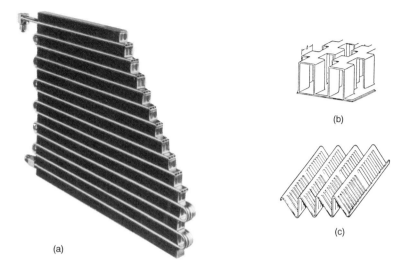

Figure 6.23   Brazed aluminum heat exchanger having (a) 4.0-mm-minor diameter extruded aluminum tube (Courtesy of Modine Manufacturing Co.) used with either (b) offset strip fins (OSF) or (c) louver fins.

6.10.2   Analysis Method

The sources of the various correlations, or data, used to predict the air-side heat transfer and friction characteristics are as follows:

1. Figure 6.2a spine fin: Scaled data of Eckels and Rabas [1985].
2. Figure 6.2b slit fin: Mori and Nakayama [1980] data.
3. Figure 6.2c wavy fin: Webb [1990] wavy fin correlation.
4. Figure 6.2d convex louver fin (CLF): Hatada and Senshu [1984] data.
5. Figure 6.23b offset strip fin: Wieting [1975] correlation.
6. Figure 6.23c louver fin: Davenport [1984] correlation.

The Kandlikar [1987] correlation was used to predict the tube-side heat transfer coefficient for vaporization of R-22 in plain tubes. The same correlation is used to calculate the tube-side heat transfer coefficient for the flat tubes using the hydraulic diameter in the Reynolds number definition.

Figures 6.24a and 6.24b show h_o and Δp versus air frontal velocity (u_{fr}) for the six geometries. The row depth of the OSF and louver plate-and-fin geometry is arbitrarily defined as 15.24 mm, which provides the same $A_o/A_{fr}N$ as the spine fin, finned-tube geometry. Figure 6.24a shows that the OSF and louver fins have much higher heat transfer coefficients than any other fin geometries, and that the heat transfer coefficient of the louver fins is slightly above that of the OSF at 1.3-m/s

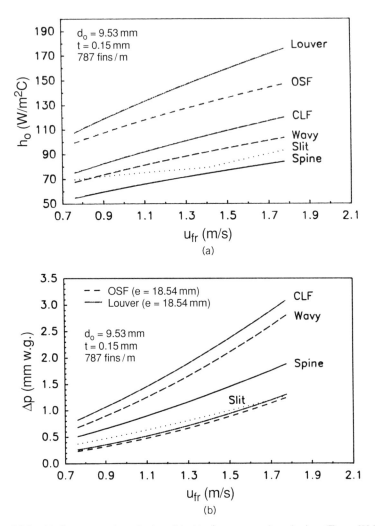

Figure 6.24 (a) h_o versus air velocity. (b) Air Δp versus air velocity. (From Webb and Gupte [1990].)

frontal velocity. Figure 6.24b also shows that the Δp for the convex louver fins is higher than that for the other geometries.

The calculation methodology used the VG-1 criterion described in Chapter 3. The objective is to reduce the heat exchanger size and weight. Reduced-air-flow frontal area is also desirable. The performance evaluation criterion (PEC) determined the heat exchanger frontal area and fin pitch needed to meet the required heat transfer rate for fixed fan power and constant air mass flow rate, with a given number of tube rows in the air-flow direction. For the case of the flat tubes shown in Figure 6.24, the tube depth (in the air-flow direction) may be made any desired value. Hence the tube depth was taken as an independent variable. The frontal area and fin pitch were adjusted to meet the required heat duty and friction power constraints.

The case of an R-22 evaporator is used to compare the performance of the various geometries. The evaporator operates at the following conditions: heat duty (2408 W), air-flow rate (0.53 m³/s), refrigeration saturation temperature (2.77°C), inlet air temperature (10°C), fins/m for the reference spine fin (787), and air Δp (0.74-mm w.g.). The air frontal velocity was allowed to vary from 0.76 to 1.78 m/s, and the fins/m value was allowed to vary from 314 to 867 for each air-side surface geometry. The calculation procedure used to size the heat exchanger for a specified frontal velocity is as follows:

1. The required UA is calculated for the specified air-flow rate, heat duty, inlet air temperature, and refrigerant saturation temperature.
2. Set the fins per meter at the lowest value (314 fins/m) and set the number of tube rows in the air-flow direction (N) to one.
3. Calculate the heat exchanger frontal area, and the air- and tube-side heat transfer coefficients for the specified air frontal velocity.
4. Calculate the available UA from

$$\frac{A_{fr}N}{UA} = \frac{A_{fr}N}{h_iA_i} + \frac{A_{fr}N}{\eta h_oA_o} \tag{6.20}$$

5. Calculate the air pressure drop.
6. If the available UA is less than the required value, increase the fins/m value by 20, and repeat steps 2 through 4. If the available $UA \geq$ required UA, then the pressure drop is checked. If this is within the specified 0.74-mm w.g. limit, an acceptable solution exists. If no design is obtained, the number of rows is increased by one and the calculations are repeated from step 2. This methodology results in selecting the heat exchanger having the minimum number of tube rows, which should yield the minimum material cost.

The heat exchanger calculation for each geometry was repeated for 0.025-m/s increments of frontal velocities between 0.76 and 1.78 m/s. For all the designs, the number of refrigerant circuits was set to one. The refrigerant mass velocity in the round tubes was 336 kg/s-m². This refrigerant mass velocity is typical of that used in commercial heat pump evaporators.

6.10.3   Calculated Results

Table 6.2 compares the various performance parameters of the six heat exchangers configurations. The Figure 6.11b spine fin is taken as the reference for comparison. Each design listed in Table 6.2 provides the same heat duty and operates at the same air-side pressure drop and air-flow rate. The higher pressure drops of the finned-tube

TABLE 6.2   Comparison of All Aluminum Heat Exchangers (Plain Tubes with Two Refrigerant Circuits)

Feature	Spine	Wavy	Slit	CLF	OSF	Louver
Rows	1	1	1	1	0.59	0.59
fins/m	728	433	590	433	866	866
u_{fr} (m/s)	0.96	1.07	1.32	1.19	1.39	1.44
h_o (W/m^2 $-$°C)	64.8	81.7	77.7	93.1	130.0	133.1
η	0.93	0.91	0.90	0.89	0.83	0.83
G_{ref} (kg/m^2 $-$s)	336	336	336	336	1127	1107
h_i (W/m^2 $-$°C)	3351	3554	3690	3667	10288	9732
w_{fin} (kg)	1.96	1.58	1.85	1.49	0.97	0.93
w_{tub}	1.23	1.19	0.96	0.99	0.84	0.84
w_{tot} (kg)	3.19	2.77	2.81	2.48	1.81	1.77

heat exchangers make them operate at much lower air frontal velocities, and thus the frontal area requirements are increased. For example, the spine fin design operates at 0.96-m/s frontal velocity, as compared to 1.42 m/s for the OSF plate-and-fin design. Because air-side frontal area is inversely proportional to frontal velocity, the OSF exchanger requires 31% less frontal area than does the spine fin design.

The convex louver fin (Figure 6.2d) provides the highest performance (lowest weight) of the round-tube exchangers—a 22% weight reduction relative to the spine fin. The weight reductions of the Figure 6.23 OSF and louver plate-and-fin designs (without splitter plates) are 53% and 48%, respectively.

The OSF and louver plate-and-fin exchangers are not constrained to an integer number of rows, as is required for the round tube design. The flat aluminum tubes may be extruded to any desired width using internal membranes to contain the internal pressure.

6.11   CONCLUSIONS

Enhanced surfaces are routinely used for application to gases. The dominant enhancement types are wavy or some form of interrupted strip fin. When used on circular tubes, a heat transfer enhancement of 80–100% is practically achieved.

Analytical or numerical models to predict the heat transfer performance of banks of high finned tubes do not exist. Power-law empirical correlations have been developed for plain fins. Although some work has been done to develop correlations for enhanced fins on circular tubes, they are not sufficiently general to account for the many geometric variables involved. The problem is complicated by the row effect of inline and staggered tube arrangements. Inline layouts provide significantly lower heat transfer performance than do the staggered tube layout. Few correlations are available for the inline layout. Low, integral fins used with liquids are not as susceptible to tube bank layout effects as are high fins. The row effect difference diminishes as e/d_o decreases.

Both the individually finned tube and plate fin-and-tube geometry are used. Enhanced surface geometries are available for both geometries. Preference of the geometry type is dependent on application and manufacturing cost considerations, rather than performance. High performance, per unit weight, can be obtained from both geometry types.

Higher performance can be obtained from oval or flat tubes. Oval tubes may not be practical for high tube-side design pressure. If aluminum is an acceptable material, extruded aluminum tubes having internal membranes offer potential for significant performance improvement, relative to round tube designs. High tube-side design pressures can be met. The resulting "brazed aluminum" heat exchangers offer many advanced technology possibilities. High-performance fin geometries are quite adaptable to this heat exchanger concept.

The advantages offered by the extruded aluminum tube having internal membranes can be achieved using different methods of manufacture. For example, a flat tube can be made by roll forming. By brazing a corrugated strip inside the tube, high design pressure can be realized. Virtually no work has been done on this.

Gas-side fouling may limit the permissible enhanced fin geometry and the fin spacing. Gas-side fouling is discussed in Chapter 10.

6.12 REFERENCES

Abbott, R. W., Norris, R. H., and Spofford, W. A., 1980. "Compact Heat Exchangers in General Electric Products—Sixty Years of Advances in Design and in Manufacturing Technologies," in *Compact Heat Exchangers—History, Technology, Manufacturing Technologies*, R. K. Shah, C. F. McDonald, and C. P. Howard, Eds., ASME Symposium, Vol. HTD-Vol. 10, ASME, New York, pp. 37–56.

Achaichia, A., and Cowell, T. A., 1988. "Heat Transfer and Pressure Drop Characteristics of Flat Tube and Louvered Plate Fin Surfaces," *Experimental Thermal and Fluid Science*, Vol. 1, pp. 147–157.

Beecher, D. T., and Fagan, T. J. 1987. "Effects of Fin Pattern on the Air-Side Heat Transfer Coefficient in Plate Finned-Tube Heat Exchangers," *ASHRAE Transactions*, Vol. 93, Part 2, pp. 1961–1984.

Benforado, D. M., and Palmer, J., 1964. "Wire Loop Finned Surface—A New Application (Heat Sink for Silicon Rectifiers)," *Chemical Engineering Progress Symposium Series*, No. 57, Vol. 61, pp. 315–321.

Brauer, H., 1964. "Compact Heat Exchangers," *Chemical Progress Engineering, London*, Vol. 45, No. 8, pp. 451–460.

Breber, G., 1991. "Heat Transfer and Pressure Drop of Stud Finned Tubes," *Chemical Engineering Progress Symposium Series,* No. 283, Vol. 87, pp. 383–390.

Briggs, D. E., and Young, E. H., 1963. "Convection Heat Transfer and Pressure Drop of Air Flowing Across Triangular Pitch Banks of Finned Tubes," *Chemical Engineering Progress Symposium Series*, No. 41, Vol. 59, pp. 1–10.

Carranza, R. G., and Holtzapple, M. T., 1991. "A Generalized Correlation for Pressure Drop Across Spined Pipe in Cross-Flow, Part I," *ASHRAE Transactions*, Vol. 97, Part 2, pp. 122–129.

Cox, B., 1973. "Heat Transfer and Pumping Power Performance in Tube Banks—Finned and Bare," ASME paper 73-HT-27.

Cox, B., and Jallouk, P. A., 1973. "Methods for Evaluating the Performance of Compact Heat Exchanger Surfaces," *Journal of Heat Transfer*, Vol. 95, pp. 464–469.

Davenport, C. J. 1984. "Correlations for Heat Transfer and Flow Friction Characteristics of Louvered Fin, *Heat Transfer—Seattle 1983*," AIChE Symposium Series, No. 225, Vol. 79, pp. 19–27.

Eckels, P. W., and Rabas, T. J., 1985. "Heat Transfer and Pressure Drop Performance of Finned Tube Bundles," *Journal of Heat Transfer*, Vol. 107, pp. 205–213.

ESDU, 1985. "Low-Fin Staggered Tube Banks: Heat Transfer and Pressure Loss for Turbulent Single-Phase Crossflow," Engineering Sciences Data Unit, ESDU Item No. 84016.

Fujii, M., Seshimo, Y., and Yoshida, T., 1991. "Heat Transfer and Pressure Drop of Tube-Fin Heat Exchanger with Trapezoidal Perforated Fins," in *Proceedings of the 1991 ASME-JSME Joint Thermal Engineering Conference*, Vol. 4, J. R. Lloyd and Y. Kurosake, Eds., ASME, New York, pp. 355–360.

Gianolio, E., and Cuti, F., 1981. "Heat Transfer Coefficients and Pressure Drops for Air Coolers Under Induced and Forced Draft," *Heat Transfer Engineering*, Vol. 3, No. 1, pp. 38–48.

Gray, D. L. and Webb, R. L., 1986. "Heat Transfer and Friction Correlations for Plate Fin-and-Tube Heat Exchangers Having Plain Fins," *Proceedings of the 9th International Heat Transfer Conference*, San Francisco.

Groehn, H. G., 1977. "Flow and Heat Transfer Studies of a Staggered Tube Bank Heat Exchanger with Low Fins at High Reynolds Numbers," Central Library of the Jülich Nuclear Research Center GmbH, Julich, Germany.

Haberski, R. J., and Raco, R. J., 1976. "Engineering Analysis and Development of an Advanced Technology, Low Cost, Dry Cooling Tower Heat Transfer Surface," Cardias–Wright Corporation Report C00–2774–1.

Hatada, T., and Senshu, T., 1984. "Experimental Study on Heat Transfer Characteristics of Convex Louver Fins for Air Conditioning Heat Exchangers," ASME paper ASME 84-H-74.

Hatada, D., Ueda, U., Oouchi, T., and Shimizu, T., 1989. "Improved Heat Transfer Performance of Air Coolers by Strip Fins Controlling Air Flow Distribution," *ASHRAE Transactions*, Vol. 95, Part 1, pp. 166–170.

Hitachi Cable, Ltd., 1984. "Hitachi High-Performance Heat Transfer Tubes," Cat. No. EA-500, Hitachi Cable, Ltd., Tokyo, Japan.

Holtzapple, M. T., and Carranza, R. G. 1990. "Heat Transfer and Pressure Drop of Spined Pipe in Cross Flow—Part 1: Pressure Drop Studies," *ASHRAE Transactions*, Vol. 96, Part 2, pp. 122–129.

Holtzapple, M. T., Allen, A. L., and Lin, K. 1990. "Heat Transfer and Pressure Drop of Spined Pipe in Cross Flow—Part 2: Heat Transfer Studies," *ASHRAE Transactions*, Vol. 96, Part 2, pp. 130–135.

Incropera, F. P. and DeWitt, D. P., 1990. *Fundamentals of Heat and Mass Transfer*, 3rd edition, John Wiley & Sons, New York, p. 426.

Jones, T. V., and Russell, C. M. B. 1980. "Heat Transfer Distribution on Annular Fins," ASME paper 78-H-30.

Kandlikar, S. G., 1987. "A General Correlation For Saturated Two-Phase Flow Boiling Heat

Transfer Inside Horizontal and Vertical Tubes," Boiling and Condensation in Heat Transfer Equipment, Vol. HTD-Vol. 85, pp. 9–19.

Kays, W. M., and London, A. L., 1984. *Compact Heat Exchangers*, McGraw–Hill, New York.

Krückels, S. W., and Kottke, V. 1970. Investigation of the Distribution of Heat Transfer on Fins and Finned Tube Models, *Chemical Engineering Technology*, Vol. 42, pp. 355–362.

LaPorte, G. E., Osterkorn, C. L., and Marino, S. M., 1979. "Heat Transfer Fin Structure," U. S. Patent 4,143,710.

McQuiston, F. C., 1978. "Correlation for Heat, Mass and Momentum Transport Coefficients for Plate-Fin-Tube Heat Transfer Surfaces with Staggered Tube," *ASHRAE Transactions*, Vol. 84, Part 1, pp. 294–309.

Mori, Y., and Nakayama, W., 1980. "Recent Advances in Compact Heat Exchangers in Japan," in *Compact Heat Exchangers—History, Technology, Manufacturing Technologies*, R. K. Shah, C. F. McDonald, and C. P. Howard, Eds., ASME Symposium, Vol. HTD-Vol. 10, ASME, New York, pp. 5–16.

Nakayama W., and Xu, L. P., 1983. "Enhanced Fins for Air-Cooled Heat Exchangers—Heat Transfer and Friction Factor Correlations," *Proceedings of the 1983 ASME-JSME Thermal Engineering Conference*, Vol. 1, pp. 495–502.

Neal, S. B. H. C., and Hitchcock, J. A. 1966. "A Study of the Heat Transfer Processes in Banks of Finned Tubes in Cross Flow, Using a Large Scale Model Technique," *Proceedings of the Third International Heat Transfer Conference*, Vol. 3, Chicago, IL, pp. 290–298.

O'Connor, J. M., and Pasternak, S. F., 1976. "Method of Making a Heat Exchanger," U.S. Patent 3,947,941.

Rabas, T. J., and Huber, F. V., 1989. "Row Number Effects on the Heat Transfer Performance of Inline Finned Tube Banks," *Heat Transfer Engeering*, Vol. 10, No. 4, pp. 19–29.

Rabas, T. J., and Taborek, J., 1987. "Survey of Turbulent Forced-Convection Heat Transfer and Pressure Drop Characteristics of Low-Finned Tube Banks in Cross Flow," *Heat Transfer Engeering*, Vol. 8, No. 2, pp. 49–62.

Rabas, T. J., Eckels, P. W., and Sabatino, R. A., 1981. "The Effect of Fin Density on the Heat Transfer and Pressure Drop Performance of Low Finned Tube Banks," *Chemical Engeering Communications*, Vol. 10, No. 1, pp. 127–147.

Rabas, T. J., Myers, G. A. and Eckels, P. W., 1986. "Comparison of the Thermal Performance of Serrated High-Finned Tubes Used in Heat-Recovery Systems," in *Heat Transfer in Waste Heat Recovery and Heat Rejection Systems*, J. P. Chiou and S. Sengupta, Eds., ASME Symposium, Volume HTD-Vol. 59, ASME, New York, pp. 33–40.

Rich, D. G., 1973. The Effect of Fin Spacing on the Heat Transfer and Friction Performance of Multi-row, Plate Fin-and-Tube Heat Exchangers," *ASHRAE Transactions*, Vol. 79, Part 2, pp. 137–145.

Rich, D. G., 1975. The Effect of the Number of Tube Rows on Heat Transfer Performance of Smooth Plate Fin-and-Tube Heat Exchangers," *ASHRAE Transactions*, Vol. 81, Part 1, pp. 307–319.

Robinson, K. K., and Briggs, D. E., 1966. "Pressure Drop of Air Flowing Across Triangular Pitch Banks of Finned Tubes," *Chemical Engineering Progress Symposium Series*, No. 64, Vol. 62, pp. 177–184.

Rozenman, T., 1976a. "Heat Transfer and Pressure Drop Characteristics of Dry Cooling Tower Extended Surfaces, Part I: Heat Transfer and Pressure Drop Data," report BNWL-PFR 7–100, Battelle Pacific Northwest Laboratories, Richland, WA, March 1.

Rozenman, T., 1976b. "Heat Transfer and Pressure Drop Characteristics of Dry Cooling Tower Extended Surfaces, Part II: Data Analysis and Correlation," report BNWL-PFR 7-102, Battelle Pacific Northwest Laboratories, Richland, WA.

Saboya, F. E. M., and Sparrow, E. M. 1974. "Local and Average Heat Transfer Coefficients for One-Row Plate Fin and Tube Heat Exchanger Configurations," *Journal of Heat Transfer*, Vol. 96, pp. 265–272.

Saboya, F. E. M., and Sparrow, E. M., 1976a. "Transfer Characteristics of Two-Row Plate Fin and Tube Heat Exchanger Configurations," *International Journal of Heat Mass Transfer*, Vol. 19, pp. 41–49.

Saboya, F. E. M., and Sparrow, E. M. 1976b. "Experiments on a Three-Row Fin and Tube Heat Exchanger," *Journal of Heat Transfer*, Vol. 98, pp. 26–34.

Schmidt, E., 1963. "Heat Transfer at Finned Tubes and Computations of Tube Bank Heat Exchangers," *Kaltetechnik*, No. 4, Vol. 15, p. 98 and No. 12, Vol. 15, p. 370.

Seshimo, Y., and Fujii, M., 1991. "An Experimental Study on the Performance of Plate and Tube Heat Exchangers at Low Reynolds Numbers," *Proceedings of the 1991 ASME-JSME Thermal Engineering Conference*, Vol. 4, pp. 449–454.

Shah, R. K., and Webb, R. L., 1982. "Compact and Enhanced Heat Exchangers," in *Heat Exchangers: Theory and Practice*, J. Taborek, G. F. Hewitt, and N. Afgan, Eds., Hemisphere Publishing Corp., Washington, D.C., pp. 425–468.

Webb, R. L., 1980. "Air-Side Heat Transfer in Finned Tube Heat Exchangers," *Heat Transfer Engeering*, Vol. 1, No. 3, pp. 33–49.

Webb, R. L., 1983. "Enhancement for Extended Surface Geometries Used in Air-Cooled Heat Exchangers," in *Low Reynolds Number Flow Heat Exchangers*, Hemisphere Publishing Corp., Washington, D.C., pp. 721–734.

Webb, R. L., 1987. "Enhancement of Single-Phase Heat Transfer," Chapter 17 in *Handbook of Single-Phase Heat Transfer*, S. Kakaç, R. K. Shah, and W. Aung, Eds., John Wiley & Sons, New York, pp. 17.1–17.62.

Webb, R. L., 1990. "Air-Side Heat Transfer Correlations for Flat and Wavy Plate Fin-and-Tube Geometries," *ASHRAE Transactions*, Vol. 96, Part 2, pp. 445–449.

Webb, R. L., and N. Gupte, 1990. "Design of Light Weight Heat Exchangers for Air-to-Two Phase Service," *Compact Heat Exchangers: A Festschrift for A. L. London*, R. K. Shah, A. Kraus, and D. E. Metzger, Eds., Hemisphere Publishing Corp., Washington, D.C., pp. 311–334.

Weierman, C., 1976. "Correlations Ease the Selection of Finned Tubes," *Oil and Gas Journal*, Vol. 74, pp. 94–100.

Weierman, C., Taborek, J., and Marner, W. J., 1978. "Comparison of of Inline and Staggered Banks of Tubes with Segmented Fins," *AIChE Symposium Series*, Vol. 74, No. 174, pp. 39–46.

Wieting, A. R., 1975. "Empirical Correlations for Heat Transfer and Flow Friction Characteristics of Rectangular Offset Fin Heat Exchangers," *Journal of Heat Transfer*, Vol. 97, pp. 488–490.

Zukauskas, A., 1972. "Heat Transfer form Tubes in Cross Flow," in *Advances in Heat Transfer*, Vol. 8, J. P. Hartnett and T. Irvine, Jr., Eds., Academic Press, New York, pp. 93–160.

6.13 NOMENCLATURE

A	Total heat transfer surface area (both primary and secondary, if any) on one side of a direct-transfer-type exchanger; total heat transfer surface area of a regenerator; m² or ft²
A_c	Flow cross-sectional area in minimum flow area, m² or ft²
A_f	Fin or extended surface area on one side of the exchanger, m² or ft²
A_{fr}	Heat exchanger frontal area, m² or ft²
A_t	Tube outside surface area considering it as if there were no fins, m² or ft²
AMTD	Arithmetic mean temperature difference, K or °F
c_p	Specific heat of fluid at constant pressure, J/kg-K or Btu/lbm-°F
D_{ab}	Diffusion coefficient for component a through component b, m²/s
D_h	Hydraulic diameter of flow passages, $4LA_c/A$; m or ft
D_v	Volumetric hydraulic diameter, m or ft
d_e	Fin tip diameter for a finned tube, m or ft
d_i	Tube inside diameter, or diameter to the base of internal fins or roughness; m or ft
d_o	Tube outside diameter, fin root diameter for a finned tube, minor diameter of rectangular cross-section tube; m or ft
$d_{o,max}$	Major diameter for rectangular tube cross section, m or ft
$d_{o,min}$	Minor diameter for rectangular tube cross section, m or ft
e	Fin height, m or ft
e_w	Wave height of wavy fin, m or ft
f	Fanning friction factor, $\Delta p \rho D_h/2LG^2$; dimensionless
f_f	Friction factor of fins in Equations 6.5 and 6.6 ($= 2\Delta p_f \rho A_c/A_f G^2$), dimensionless
f_t	Friction factor of fins tubes in Equations 6.5 and 6.6 ($= 2\Delta p_t \rho A_c/A_t G^2$), where $A_t = A - A_f$, dimensionless
f_{tb}	Tube bank friction factor ($= \Delta P \rho/2NG^2$), dimensionless
G	Mass velocity based on the minimum flow area, kg/m²-s or lbm/ft²-s
G_{ref}	Refrigerant-side mass velocity, kg/m²-s or lbm/ft²-s
Gz	Graetz number ($= \mathrm{RePr}D_h/L$), dimensionless
H	Louver fin height ($= P_t - d_{o,min}$), m or ft
h_a	h based on the arithmetic mean temperature difference, W/m²-K
j	Colburn factor ($= \mathrm{StPr}^{2/3}$), dimensionless
K	Thermal conductance $= hA$, W/K or Btu/hr-°F
K_m	Mass transfer coefficient, kg/m²-s
k	Thermal conductivity of fluid, W/m-K or Btu/hr-ft-°F
L	Fluid flow (core) length on one side of the exchanger, m or ft
L_p	Strip flow length of OSF or louver pitch of louver fin, m or ft
LMTD	Logarithmic mean temperature difference, K or °F
L_p	Strip flow length of OSF or louver pitch of louver fin, m or ft
N	Number of tube rows in the flow direction, dimensionless
NTU	Number of heat transfer units [$= UA/(Wc_p)_{min}$], dimensionless
Nu_a	Nusselt number based on $h_a D_v/k$
Nu_d	Nusselt number ($= hd_o/k$), dimensionless

Nu_{Dh}	Nusselt number ($= hD_h/k$), dimensionless
n_L	Number of louvers in air-flow depth, dimensionless
Pr	Prandtl number ($= c_p\mu/k$), dimensionless
p_f	Fin pitch, center-to-center spacing; m or ft
p_w	Wave pitch of wavy fin, m or ft
Δp	Air-side pressure drop, Pa or lbf/ft^2
Δp_f	Pressure drop assignable to fin area in finned tube exchanger, Pa
Δp_t	Pressure drop assignable to tubes in finned tube exchanger, Pa
Q	Heat transfer rate in the exchanger W or Btu/hr
Re_d	Reynolds number based on the tube diameter ($= Gd/\mu$, $d = d_i$ for flow inside tube and $d = d_o$ for flow outside tube), dimensionless
Re_{Dh}	Reynolds number based on the hydraulic diameter ($= GD_h/\mu$), dimensionless
Re_{Dv}	Reynolds number based on D_v, dimensionless
Re_L	Reynolds number based on the interruption length ($= GL_p/\mu$, Gx/μ), dimensionless
Re_{Sl}	Reynolds number based on S_l, dimensionless
S_d	Diagonal pitch $= [S_t^2 + S_l^2)/2]$, m or ft
S_f	Flow frontal area of heat exchanger, m^2 or ft^2
Sh	Sherwood number for mass transfer ($= K_m D_h/D_{ab}$)
S_l	Longitudinal tube pitch, m or ft
S_t	Transverse tube pitch, m or ft
St	Stanton number ($= h/Gc_p$), dimensionless
s	Spacing between two fins ($= p_f - t$), m or ft
ΔT	Temperature difference between hot and cold fluids, K or °F
t	Thickness of fin, m or ft
U	Overall heat transfer coefficient, W/m^2-K or Btu/hr-ft^2-°F
u	Fluid velocity, u_{fr}, frontal velocity; u_c, in minimum flow area; m/s
W	Mass flow rate, kg/s or lbm/s
w_{fin}	Weight of fins in heat exchangers, kg
w_s	Width of segmented fin (Figure 6.13), m or ft
w_{tot}	Weight of tubes and fins in heat exchanger, kg
w_{tub}	Weight of tubes in heat exchanger, kg
x	Coordinate direction from leading edge of fin, m or ft
X_{Dv}^+	RePrD$_v$/L, dimensionless

Greek Letters

η_f	Fin efficiency or temperature effectiveness of the fin, dimensionless
η_o	Surface efficiency of finned surface [$= 1 - (1 - \eta_f)A_f/A$], dimensionless
θ	Louver angle for louver fin, radians
μ	Fluid dynamic viscosity coefficient, Pa-s or lbm/s-ft
ν	Kinematic viscosity, m^2/s or ft^2/s
ρ	Fluid density kg/m^3 or lbm/ft^3

Subscripts

i	Tube inside surface
il	Inline tube arrangement
o	Outside (air-side) surface
p	Plain tube or surface
st	Staggered tube arrangement
w	Evaluated at wall temperature

7

INSERT DEVICES FOR SINGLE-PHASE FLOW

7.1 INTRODUCTION

Chapters 7, 8, and 9 address techniques to enhance the heat transfer coefficient inside a tube for single-phase flow. Hence, these three chapters discuss three different approaches to enhancement of tube-side convective heat transfer. Insert devices involve various geometric forms that are inserted in a smooth, circular tube. These devices are in competition with internal fins (Chapter 8) and integral roughness (Chapter 9). Which of the three methods is preferred depends on two factors: performance and initial cost. Integral internal fins and roughness require deformation of the material on the inside surface of a long tube. Cost-effective manufacturing technology to deform the inner surface of a tube has only recently been developed. Insert devices represent an early approach to tube-side enhancement, which allowed use of a plain tube. Anticipating the material to be discussed in Chapters 7, 8, and 9, we will show that insert devices are generally not competitive with the current performance and cost of internal roughness for turbulent flow. Only in laminar flow are insert devices an effective solution. However, insert devices may be used to upgrade the performance of an existing heat exchanger.

Work to conceive insert devices has been largely empirical. Figures 7.1 through 7.5 show the various insert devices. The dominant literature on insert devices involves the following five concepts:

1. Twisted-tape inserts (Figure 7.1a), which cause the flow to spiral along the tube length. The tape inserts generally do not have good thermal contact with the tube wall, so the tape does not act as a fin.
2. Extended surface inserts (Figure 7.2a), which are an extruded shape inserted in the tube. The tube is then drawn to provide good thermal contact between

166

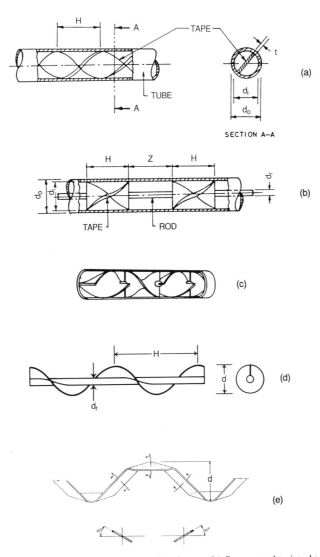

Figure 7.1 Tape inserts. (a) Continuous twisted tape. (b) Segmented twisted tape. (c) Kinex mixer. (d) Helical tape. (e) Bent strips.

the wall and the insert. The insert reduces the hydraulic diameter and acts as an extended surface.

3. Wire coil inserts (Figure 7.3a), which consist of a helical coiled spring that functions as a nonintegral roughness.

4. Mesh or brush inserts as shown in Figure 7.4.

5. An insert device that is displaced from the tube wall and causes periodic mixing of the gross flow. Figures 7.5a and 7.5b are representative devices.

(a)

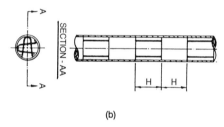

(b)

Figure 7.2 Extended surface inserts. (a) Extruded inserrt. (Courtesy of Wieland-Werke AG.) (b) Interrupted sheet metal. (From Jayaraj et al. [1989].)

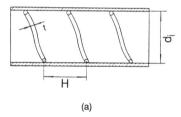

(a)

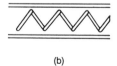

(b)

Figure 7.3 Wire coil insets. (a) Wires touching tube wall. (b) Wires displaced from tube wall.

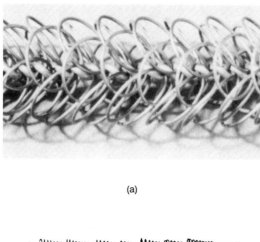

(a)

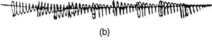

(b)

(c)

Figure 7.4 Mesh or brush inserts. (a) Mesh insert. (b) Helical coil inset. (From Colburn and King [1931].) (c) Brush insert.

For turbulent flow, it is more effective to mix the flow in the viscous boundary layer at the wall (e.g., Figure 7.3a) than to mix the gross flow using the devices shown in Figure 7.4 or 7.5. This is because the dominant thermal resistance is very close to the wall. Integral internal roughness (Chapter 9) can generally provide a given $E_i = h/h_p$ with a higher efficiency index (η) than is provided by the wire coil insert shown in Figure 7.3a.

For laminar flow, the dominant thermal resistance is not limited to a thin boundary layer adjacent to the flow. Thus, devices that mix the gross flow are more effective in laminar flow than in turbulent flow.

Although the extended surface insert of Figure 7.2a is an effective enhancement concept for both laminar and turbulent flow, it suffers from cost concerns. The twisted tape (Figure 7.1a) and mesh insert (Figure 7.4a) are of primary interest for

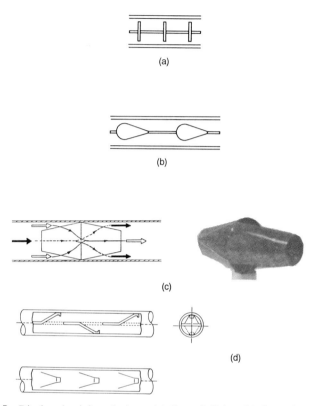

Figure 7.5 Displaced mixing devices. (a) Spaced disks. (b) Spaced streamline shapes. (c) Flow eversion device. (From Maezawa and Lock [1978].) (d) Louvered strip.

laminar flow, but their potential performance is diminished because the tape is not in good thermal contact with the wall.

We will discuss the relative merits of the various devices and will provide information that may be used to predict their heat transfer and friction characteristics.

7.2 TWISTED-TAPE INSERT

The continuous twisted-tape insert shown in Figure 7.1a has been extensively investigated for both laminar and turbulent flow. Variants of the twisted tape that have been evaluated include short sections of twisted tape at the tube inlet, or periodically spaced along the tube length (Figures 7.1b and 7.1c). Bergles and Joshi [1983] present a survey of the performance of the different types of swirl flow devices for laminar flow.

The insert shown in Figure 7.1a consists of a thin twisted strip that is slid into the tube. The axial distance for a 180-degree twist is dimension H. Authors describe their tapes by the "twist ratio," $y = H/d_i$. Alternate to the y parameter, one may use the helix angle of the tape, given by

$$\tan \alpha = \frac{\pi d_i}{2H} = \frac{\pi}{2y} \tag{7.1}$$

In order to allow easy insertion of the tape, there is usually a small clearance between the tape width and the tube inside diameter. This clearance results in poor thermal contact between the tape and the tube wall, so the heat transfer from the tape may be quite small. The blockage caused by the finite tape thickness increases the average velocity. Heat transfer enhancement may occur for three reasons:

1. The tape reduces the hydraulic diameter (D_h); this causes an increased heat transfer coefficient, even for zero tape twist.
2. The twist of the tape causes a tangential velocity component. Hence, the speed of the flow is increased, particularly near the wall. The heat transfer enhancement is a result of the increased shear stress at the wall and mixing by secondary flow.
3. Heat is transferred from the tape, if good thermal contact with the wall exists. However, little heat transfer is expected from a loosely fitting tape.

Thorsen and Landis [1968] show that centrifugal forces, caused by the tangential velocity component, provides enhancement by mixing fluid from the core region with fluid near the wall. However, this will occur only when the flowing fluid is being heated. The colder, high-density core region fluid is forced outward to mix with the warm, low-density fluid near the wall. If the fluid is being cooled, the centrifugal force acts to maintain thermal stratification of the fluid.

Although much work has been done on twisted tapes, recent papers by Manglik and Bergles [1992a, 1992b] present an updated understanding of the heat transfer and friction characteristics for both laminar and turbulent flow. The papers provide correlations valid for deep laminar flow through the turbulent regime, including the transition regime for constant wall temperature boundary condition. In the laminar flow regime, Manglik and Bergles [1992a] show that the laminar flow Nusselt number is influenced by entrance length and buoyancy forces (natural convection); these effects are controlled by the magnitude of the Graetz number ($Gz_d = \pi d_i Re_d$-Pr/4L) and the Rayleigh number (Ra = GrPr), respectively. They show that Nu is a function of Sw, Gz, Ra, Pr, and μ/μ_w. The friction factor for fully developed flow depends on Sw and μ/μ_w. The "swirl number," is defined as Sw $= Re_{Sw} y^{-1/2}$, where $y = H/d_i$. The Manglik and Bergles laminar and turbulent flow correlations are presented in Sections 7.2.2 and 7.2.3, respectively.

Note that there is no common agreement on the characteristic dimension used in the Reynolds number definition for twisted tapes. Manglik and Bergles [1992a, 1992b] use Re_d (based on d_i), whereas some others use Re_{Dh} (based on D_h).

7.2.1 Laminar Flow Data

Laminar flow is a relatively complex subject, because the flow is influenced by the
following conditions: (1) the thermal boundary condition, (2) entrance region effects, (3) natural convection at low Reynolds number, (4) fluid property variation
across the boundary layer, and (5) the duct cross-sectional shape. Furthermore, the
local Nusselt will be different for simultaneously developing velocity and temperature profiles, as compared to a fully developed velocity profile. To better understand
the twisted-tape data, we have provided Figure 7.6, which shows theoretical solutions for laminar, entrance region flow in a circular tube. These solutions, which are
taken from Shah and Bhatti [1987], assume fully developed velocity profile entering
the heating region, constant properties in the tube, and no natural convection effects. Figure 7.6 shows different results for the q = constant and T_w = constant
boundary conditions. Because Figure 7.6 shows the entrance region, the Nusselt
number is plotted versus the entrance region parameter, Nu_x versus x_d^* [= $x/(d_i$-
$Re_d Pr)$]. The figure shows the local Nusselt number (Nu_x) and the Nusselt number
averaged over the length x/d_i (Nu_m) for constant heat flux (subscript H) and constant
wall temperature (subscript T). Figure 7.6 shows that the fully developed condition
is not obtained until $x_d^* \cong 0.01$.

Laminar flow data are taken with either a constant heat flux (q = constant) or an
approximate constant wall temperature (T_w = constant) thermal boundary condition, usually with a developed velocity profile entering the heated test section. For
the q = constant boundary condition, the local heat transfer coefficient (h_x) is

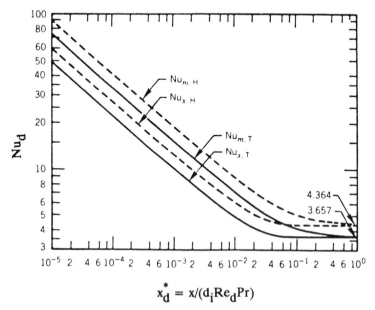

$$x_d^* = x/(d_i Re_d Pr)$$

Figure 7.6 Analytical solutions for developing laminar flow in circular tube with fully
developed velocity profile. (From Shah and Bhatti [1987].)

obtained using (a) the known heat flux and (b) the wall temperature measured by thermocouples along the tube length. It is usually possible to obtain the local Nusselt number for fully developed flow ($\text{Nu}_{\text{fd}} = h_{\text{fd}}d_i/k$) from wall thermocouples sufficiently far from the heated inlet. Using the T_w = constant boundary condition, it is nearly impossible to obtain data for the Nu_{fd} using a moderate test section length. In the T_w = constant test, the local heat flux is not known. Thus, experimenters measure the UA value over the test section length, and they subtract the shell-side resistance to obtain the average tube-side heat transfer coefficient over the test section length. This measurement corresponds to the $\text{Nu}_{x,T}$ dashed line on Figure 7.6. Even if Nu_{fd} could be obtained in the T_w = constant tests, Nu_{fd} would be different for q = constant and T_w = constant data, as shown by Figure 7.6.

There are two additional complicating factors in obtaining laminar flow data. First is the effect of fluid property variation across the boundary layer. Laminar flow is much more susceptible to the effect of fluid property variation than is turbulent flow. As shown in Tables 2.4 and 2.5, the effect of property variation across the boundary layer is accounted for by the term $(\mu/\mu_w)^{0.14}$. For heating with $\mu/\mu_w =$ 1.5, h_x would be increased 5.5%. For cooling with $\mu/\mu_w = 1/1.5$, h_x would be reduced 5.5%. A second complication is the effect of buoyancy force (natural convection), which is determined by the Rayleigh number, Ra = GrPr. Whether natural convection effects are important for a twisted-tape insert depends on the ratio Gr/Sw^2, which is the ratio of the buoyancy and centrifugal forces. Hence, one must be careful in comparing laminar flow data, since it is affected by entrance effects, thermal boundary condition, fluid property variation across the boundary layer, and buoyancy force.

Figure 7.7, from Bergles and Joshi [1983], shows Nusselt number data for the two boundary conditions. Figure 7.7a shows the average $\text{Nu}_{d,m}$ for T_w = constant, and Figure 7.7b shows the local Nu_x for q = constant. The data are presented as a function of entrance region parameters, because not all of the data points are for the fully developed condition. The Bergles and Joshi [1983] survey uses the independent variable x_d^* for the q = constant data and $\text{Gz} = \pi/(4x_d^*)$ for T_w = constant data.

Figure 7.7a shows the T_w = constant ethylene glycol data (20 < Pr < 100) for α = 17 degrees ($y = 5.1$) of Marner and Bergles [1985] which provides 300% enhancement over the smooth tube value. Note that the data are not fully developed, and the cooling data fall below the heating data at the lower Gz values. Figure 7.8a shows the friction factor data for the Figure 7.7a data. The friction data of Marner and Bergles [1978] are shown by curve 3. Figure 7.8b shows the numerically predicted fully developed friction factor of Date [1974]. As indicated on Figure 7.8b, the $\alpha = 0$ curve is for a tape of zero thickness and no twist. Increasing the twist causes $f\text{Re}_d$ to depart from the 42.2 value.

Marner and Bergles [1978] also tested an internally finned tube in laminar flow. Figure 7.9 compares h/h_p for the twisted tape and the internally finned tube for the PEC case FG-2b of Chapter 3. Figure 7.9 shows that the performance of the twisted tape is better in heating than in cooling. It also shows that the performance of the internally finned tube is better than that of the twisted tape.

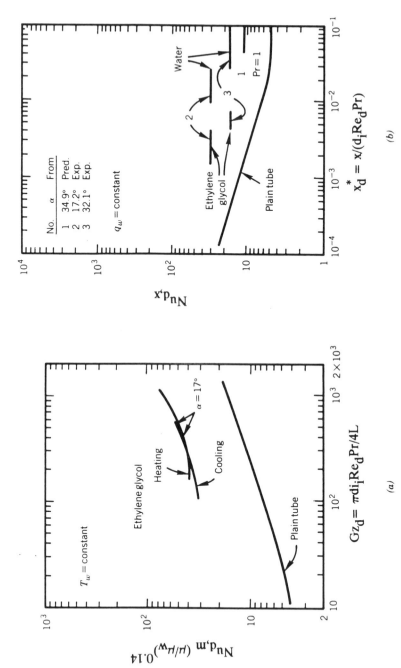

Figure 7.7 Nusselt number for laminar flow in tubes with twisted-tape insert. (a) Constant wall temperature. (b) Constant heat flux. (From Bergles and Joshi [1983].)

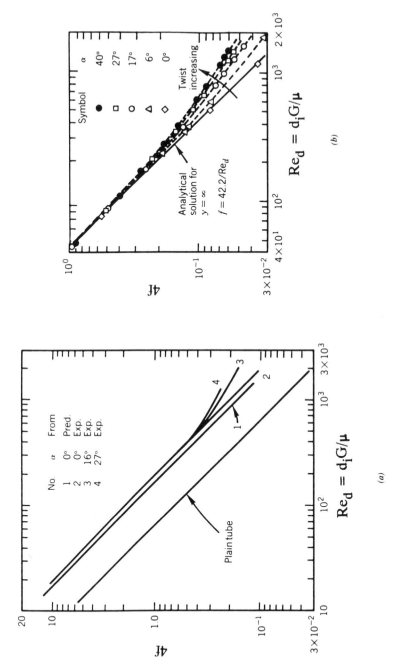

Figure 7.8 Friction factor in tubes with twisted-tape insert for laminar flow. (a) Experimental and predicted. (b) Predicted by Date [1974] for zero tape thickness. (From Bergles and Joshi [1983].)

175

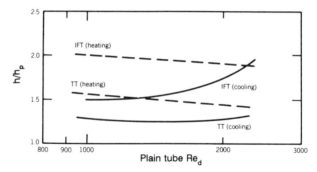

Figure 7.9 Performance comparison of internally finned tube for laminar flow of ethylene glycol (IFT, 16 fins, $e/D_i = 0.084$, $\alpha = 27$ degrees) and twisted-tape insert (TT, $\alpha = 30$ degrees, $y = 5.4$, $t/D_i = 0.053$) for laminar flow using case FG-2a. (Adapted from Marner and Bergles [1985]).

7.2.2 Predictive Methods for Laminar Flow

Date [1974] has performed numerical predictions for the twisted tape in fully developed laminar flow for q = constant with constant properties. His analysis (corrected) for a loosely fitting tape with $\alpha = 34.9$ degrees and Pr = 1 is shown by curve 1 on Figure 7.7b. Date's analysis also shows that the Nusselt number increases with increasing Prandtl and Reynolds numbers, contrary to laminar flow in plain tubes, which is independent of Re and Pr. Note that Date's [1974] reported Nu values are 50% low, due to a computational error, which is discussed by Hong and Bergles [1976]. Date and Singham [1972] developed empirical power law correlations to fit their numerical predictions of the friction factor.

The recommended friction factor correlation is the recent work of Manglik and Bergles [1992a]. Figure 7.10 shows the basis of the correlation and shows the ratio of $(f\mathrm{Re}_d)_{sw}/(f\mathrm{Re}_d)_{sw,\ y=\infty}$, where the $(f\mathrm{Re}_d)_{sw,\ y=\infty}$ is for a tape having no twist. The figure shows that the ratio approaches 1.0 as Sw → 0, and the $f\mathrm{Re}_d$ ratio approaches an asymptotic value ($0.10\mathrm{Sw}^{0.425}$) for Sw $\equiv \mathrm{Re}_{sw}/y^{1/2} \gg 1$. An asymptotic method is used to define a correlation, which approaches each of the two limiting asymptotes. The final correlation is given by

$$\frac{(f\mathrm{Re}_d)_{sw}}{(f\mathrm{Re}_d)_{sw,y=\infty}} = (1 + 10^{-6}\mathrm{Sw}^{2.55})^{1/6} \qquad (7.2)$$

where

$$(f\mathrm{Re}_d)_{sw,y=\infty} = 15.767 \left(\frac{\pi + 2 - 2t/d_i}{\pi - 4t/d_i} \right)^2 \qquad (7.3)$$

If Sw = $\mathrm{Re}_{sw}/y^{1/2} \to 0$, the friction factor approaches the value for fully developed flow in a tape having zero twist (Equation 7.3). The term in parentheses on the

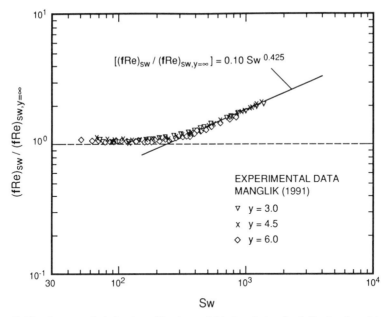

Figure 7.10 Asymptotic behavior of isothermal friction factor for fully developed laminar flow with twisted-tape insert. (from Manglik and Bergles [1992a].)

right-hand side of Equation 7.3 accounts for the effect of the finite tape thickness on the velocity. The friction factor, f_{sw} is defined as

$$f_{sw} = \frac{\Delta p d_i}{2\rho u_{sw}^2 L_s} = f \frac{L}{L_s} \left(\frac{u_c}{u_{sw}} \right)^2 \qquad (7.4)$$

Thus, f_{sw} is defined in terms of the swirl velocity (u_{sw}) and the length of the helically twisting streamline at the tube wall ($L_s = L/\cos \alpha$). The second form of Equation 7.4 shows how f_{sw} is related to the conventional friction definition, based on the axial velocity (u_c and pipe length L).

7.2.2.1 Constant Heat Flux. Hong and Bergles [1976] have developed an empirical correlation based on their q = constant, twisted-tape data for water and ethylene glycol. Two twisted tapes were used which had helix angles of $\alpha = 17$ and 32 degrees. The measured local heat transfer coefficient corresponds to fully developed condition, Nu_{fd}. Unfortunately, they did not use the measured wall temperature to determine the $(\mu/\mu_w)^{0.14}$ correction. Their heat transfer correlation is given by Equation 7.5, in which $y = H/d_i = \pi/(2 \tan \alpha)$.

$$Nu_d = 5.172 \left[1 + 0.005484 \left(\frac{Re_d}{y} \right)^{1.25} Pr^{0.7} \right]^{0.5} \qquad (7.5)$$

Xie et al. [1992] provided data for an oil ($40 \leq Pr \leq 80$) in an electrically heated tube for continuous tape with seven different twist ratios (y) between 1.67 and 5.0. The data were taken for $4000 \leq Re_d \leq 30,000$. Xie et al. provided the following empirical correlations for their data.

$$Nu_d = 0.0149Re_{d,f}^{0.833}Pr^{2/3}y^{-0.493}(\mu/\mu_w)^{0.14} \qquad (7.6)$$

$$f = (3.61 + 8.92y - 1.52y^2)Re_{d,f}^{-(0.028 + 0.14y-0.021y^2)}(\mu/\mu_w)^{1/3} \qquad (7.7)$$

where $Re_{d,f}$ is the Reynolds number evaluated at the film temperature. Using PEC FG-2a of Table 3.1, they show that the highest performance (h/h_p) is provided by $y = 2.5$ at $Re_d = 6000$ and by $y = 3$ for $Re = 20,000$.

7.2.2.2 Constant Wall Temperature.

Manglik and Bergles [1992a] developed a correlation for the T_w = constant case using (a) the data of Marner and Bergles [1978, 1985] for ethylene glycol and polybutene ($1000 < Pr < 7000$) and (b) the water and ethylene glycol data of Manglik [1991]. They show that $Nu_{d,m} = Nu_{d,m}(Sw, Gz, Ra, Pr, \mu/\mu_w)$. The Sw parameter accounts for swirl, Gz accounts for the entrance length, and Ra accounts for natural convection. This correlation is also based on limiting asymptotes. This correlation is built as follows. For fully developed flow in a tube containing a tape with $y = \infty$, $Nu_{d,m} = 4.612$. Accounting for entrance effects (Gz) and fluid property variation, the theoretical equation for flow in a tube containing a $y = \infty$ tape is

$$Nu_{d,m} = 4.612(1 + 0.0951Gz^{0.894})^{0.5}(\mu/\mu_w)^{0.14} \qquad (7.8)$$

For fully developed flow ($Gz \ll Sw$), the data are correlated by

$$Nu_{d,m} = 0.106Sw^{0.767}Pr^{0.3}(\mu/\mu_w)^{0.14} \qquad (7.9)$$

An asymptotic correlation is obtained for the limit $Sw \rightarrow \infty$, $Gz \rightarrow 0$ by combining the fully developed asymptote ($Nu_{d,m} = 4.162$) with Equation 7.9:

$$Nu_{d,m} = 4.612[1 + 6.413 \times 10^{-9}(SwPr^{0.391})^{3.385}]^{0.2}(\mu/\mu_w)^{0.14} \qquad (7.10)$$

Using the asymptotic correlation method, they included thermal entrance effects (Gz) by combining Equations 7.9 and 7.10 to obtain

$$Nu_{d,m} = 4.612[(1 + 0.0951Gz^{0.894})^{2.5} + 6.413 \times 10^{-9}(SwPr^{0.391})^{3.385}]^{0.2}(\mu/\mu_w)^{0.14} \qquad (7.11)$$

Natural convection effects will dominate, when $Gr > Sw^2$. The natural convection influence was determined by correlating the data versus $Re_d Ra$ to obtain

$$\frac{Nu_{d,m}}{4.612} = 4.294 \times 10^{-2} (Re_d Ra)^{0.223} \qquad (7.12)$$

The final correlation is obtained by using the following two asymptotes: 1) The swirl flow asymptote for Ra → 0 given by Equation 7.11, and 2) The buoyancy effects asymptote given by Equation 7.12. Using the asymptotic correlation method, the final form of the correlation is obtained:

$$Nu_{d,m} = 4.612\{[(1 + 0.0951Gz^{0.894})^{2.5} + 6.413 \times 10^{-9}(SwPr^{0.391})^{3.385}]^2 + 2.132 \times 10^{-14}(Re_d Ra)^{2.23}\}^{0.1}(\mu/\mu_w)^{0.14} \tag{7.13}$$

Figure 7.11 shows predicted values using Equation 7.13 and includes some of the Manglik and Bergles [1992a] data. Figure 7.11a shows the effect of natural convection and swirl flow. Figure 7.11b shows the effect of entrance effects and swirl flow. Note that Equation 7.13 is limited to constant wall temperature and is not applicable to a constant heat flux boundary condition.

7.2.3 Turbulent Flow

Smithberg and Landis [1964] developed a semiempirical model for the friction factor, which is based on adding the momentum losses for helical flow and tape-induced fluid mixing. They showed that the model reasonably predicted their data for tapes having $3.62 \le y \le \infty$. Zhuo et al. [1992] found that the Smithberg and Landis friction model underpredicted their friction data ($y = 3, 4.25, 5.25$) by approximately 15%. Based on his evaluation of published data, Date [1973] proposed an empirical correlation for the friction factor for the range $5000 < Re_{Dh} < 70,000$ and $0 < y < 1.5$, or $0 < \alpha < 46$ degrees. The correlation is

$$\frac{f}{f_p} = \left(\frac{y}{y - 1} \right)^m \tag{7.14}$$

where

$$m = 1.15 + \frac{1.25 \,(70,000 - Re_{Dh})}{65,000} \tag{7.15}$$

and f_p is the friction factor in a plain tube, given by

$$f_p = 0.046Re_d^{-0.2} \tag{7.16}$$

The recommended friction correlation is from Manglik and Bergles [1992b].

$$f = \frac{0.079}{Re_d^{0.25}} \left(\frac{\pi}{\pi - 4t/d_i} \right)^{1.75} \left(\frac{\pi + 2 - 2t/d_i}{\pi - 4t/d_i} \right)^{1.25} \left(1 + \frac{2.752}{y^{1.29}} \right) \tag{7.17}$$

Equation 7.17 accounts for tape thickness. The authors show that this is in good agreement with data and correlations published by others.

Smithberg and Landis [1964] developed the first semianalytical model for turbu-

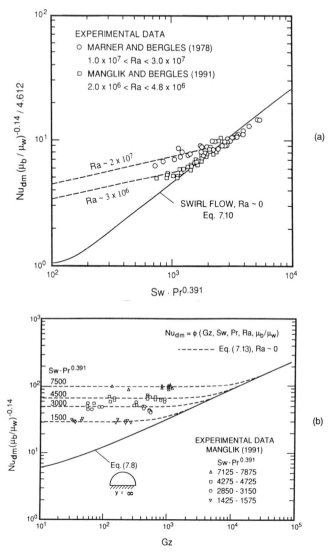

Figure 7.11 Comparison of predictions using Equation 7.11 with data illustrating the effect of (a) natural convection (Ra) and (b) entrance length (Gz). (From Manglik and Bergles [1992a].)

lent heat transfer with a twisted tape. The heat transfer model was outdated by Thorsen and Landis [1968], who suggested that buoyancy effects arising from density variations in the centrifugal field have an effect on heat transfer. They showed that the swirl flow-induced buoyancy effect should depend on the dimensionless group Gr/Re^2, which may be written as

$$\frac{\text{Gr}}{\text{Re}_{\text{Dh}}^2} = \frac{2D_h\beta_T\Delta T \tan \alpha}{d_i} \tag{7.18}$$

Thorsen and Landis measured the heat transfer coefficient for heating and cooling of water in tubes having tapes with three different helix angles, $\alpha = 11.1$, 16.9, and 26.5 degrees. The heating data were correlated by

$$\text{Nu} = 0.021F \left(1 + 0.25 \sqrt{\frac{\text{Gr}}{\text{Re}_{\text{Dh}}}} \right) \text{Re}_{\text{Dh}}^{0.8} \text{Pr}^{0.4} \left(\frac{T_w}{T_h} \right)^{-0.32} \tag{7.19}$$

and the cooling data were correlated by

$$\text{Nu} = 0.023F \left(1 - 0.25 \sqrt{\frac{\text{Gr}}{\text{Re}_{\text{Dh}}}} \right) \text{Re}_{\text{Dh}}^{0.8} \text{Pr}^{0.3} \left(\frac{T_w}{T_h} \right)^{-0.1} \tag{7.20}$$

where

$$F = 1 + 0.004872 \frac{\tan^2\alpha}{d_i(1 + \tan^2\alpha)} \tag{7.21}$$

and d_i is in meters.

Lopina and Bergles [1969] attempted to account for the increased speed of the flow (caused by the spiral flow) and the centrifugal buoyancy effect using a superposition model. The model also accounts for the possibility that the tape may act as an extended surface. The model may be expressed as

$$q = q_{\text{sc}} + q_{\text{cc}} + q_f \tag{7.22}$$

where q_{sc} is swirl convection, q_{cc} is centrifugal convection, and q_f is from the fin.

The swirl convection (q_{sc}) is predicted using an appropriate equation for turbulent flow in plain tubes with the Reynolds number calculated in terms of D_h and a modified velocity (u_{sw}) to account for the speed of the swirl flow at the wall. The heat transfer coefficient for the swirl convection term (q_{sc}) is given by

$$\text{Nu}_{\text{sc}} = \frac{h_{\text{sc}}D_h}{k} = 0.023 \, \text{Re}_{\text{Dh}}^{0.8} \text{Pr}^{0.4} \tag{7.23}$$

where

$$\text{Re}_{\text{Dh}} = \frac{u_{\text{sw}}D_h}{\nu}, \quad \text{where } u_{\text{sw}} - u_c \sqrt{1 + \tan^2\alpha} \tag{7.24}$$

The term q_{cc} term in Equation 7.22 represents the centrifugal convection effect identified by Thorsen and Landis. Lopina and Bergles [1969] use an equation for turbulent natural convection from a horizontal plate and replace the gravity force (g) by a radial acceleration (g_r), which is given by

$$g_r = \frac{2u_\theta^2}{d_i} = \frac{2}{d_i}\left(\frac{u_c\pi}{2_y}\right)^2 = \frac{4.94}{d_i}\left(\frac{u_c}{y}\right)^2 \tag{7.25}$$

The heat transfer coefficient for the centrifugal convection term is calculated by

$$\mathrm{Nu}_{cc} = \frac{h_{cc}D_h}{k} = 0.12(\mathrm{Gr}_d\mathrm{Pr})^{1/3} \tag{7.26}$$

where

$$\mathrm{Gr}_{Dh} = \frac{4.94\beta_T\Delta T_{fs}D_h\mathrm{Re}_{Dh}^2}{d_iy^2} \tag{7.27}$$

The term q_f in Equation 7.22 accounts for heat transfer from the tape as an extended surface. Evaluation of this term requires knowledge of the contact resistance between the tube wall and the tape and is usually unknown. For the poor thermal contact situation, $q_f = 0$. Refer to Lopina and Bergles [1969] for details on calculation of q_f, should this be of interest.

The most recent heat transfer correlation is by Manglik and Bergles [1992b]. Their correlation is based on the asymptotic method, and is valid for T_w = constant and q = constant with $\mathrm{Re}_d > 10,000$. The correlation is

$$\frac{\mathrm{Nu}_d}{\mathrm{Nu}_{d,y=\infty}} = 1 + \frac{0.769}{y} \tag{7.28}$$

where $\mathrm{Nu}_{d,\ y=\infty}$ is for a tape having no twist, and is given by

$$\mathrm{Nu}_{d,y=\infty} = 0.023\mathrm{Re}_d^{0.8}\mathrm{Pr}^{0.4}\left(\frac{\pi}{\pi - 4t/d_i}\right)^{0.8}\left(\frac{\pi + 2 - 2t/d_i}{\pi - 4t/d_i}\right)^{0.2}\phi \tag{7.29}$$

The term ϕ accounts for fluid property variation. For liquids, $\phi = (\mu/\mu_w)^n$, where $n = 0.18$ for heating and $n = 0.30$ for cooling. For gases, $\phi = (T/T_w)^m$, where $m = 0.45$ for heating and $m = 0.15$ for cooling with the temperature T in degrees K (or Rankine).

Manglik and Bergles [1992b] also propose a friction factor correlation for the laminar–turbulent transition region. They show that the transition is smooth, and their laminar (Equation 7.2) and turbulent correlations [Equation 7.17] may be combined to obtain a correlation for the transition region. Their recommended correlation is

$$f = (f_l^{10} + f_t^{10})^{0.1} \tag{7.30}$$

where f_l and f_t are given by Equations 7.2 and 7.17, respectively.

7.2.4   PEC Example 7.1

A heat exchanger to preheat 0.542 kg/s of combustion air from 26.7°C to 165.6°C using combustion products at 271°C is required. The capacity rate ratio is $C_a/C_g = 0.9$, where the subscript a is air and the subscript g is combustion gas. The heat exchanger has 100 steel tubes ($k = 41.5$ W/m-K) with 25.4-mm inside diameter and 1.5-mm wall thickness. The hot gas flows normal to the tubes, and air flows inside 100 tubes in parallel.

You have completed a plain tube design which has a pass length of $L_p = 2.51$ m and requires $UA = 863$ W/K with $h_a = 55.9$ and $h_g = 170$ W/m²-K. Because $h_a < h_g$, you plan to use a twisted tape ($H/d_i = 3.14$) inside the tube. Use PEC FN-2 of Table 3.1 ($Q/Q_p = W/W_w = 1$) to determine the tube length saving. For simplicity, assume zero tape thickness and neglect the tube wall resistance and any fouling resistances. Use the analytical model of Lopina and Bergles [1969] for the heat transfer coefficient.

Solution. This problem is straightforward, since the Reynolds number in the tube is directly calculable. The Nu for the tube is obtained using Equation 7.22, which requires calculating the Nu_{sc} (swirl convection) and Nu_{cc} (centrifugal convection) terms. We assume zero heat transfer from the tape ($q_f = 0$).

The hydraulic diameter is $\pi d_i/(\pi + 2) = 15.5$ mm, and the Reynolds number ($Re_{Dh} = GD_h/\mu$) is 8594. The Nu_{sc} is directly calculated using Equation 7.23, giving $Nu_{sc} = 27.9$. The Nu_{cc} is given by Equation 7.26 and requires an iterative calculation, since ΔT_{fs} (wall-to-fluid temperature difference) is not known. Substituting the known terms in Equation 7.26, we obtain $Nu_{cc} = 0.12(10,464\Delta T_{fs})^{1/3}$. We will guess $\Delta T_{fs} = 0.6\Delta T_{lm} = 0.6 \times 49.6$ K $= 29.77$ K and thus obtain $Nu_{cc} = 15.73$. Hence, $Nu = Nu_{sc} + Nu_{cc} = 27.9 + 15.73 = 43.63$, and $h_i = Nuk/D_h = 88.46$ W/m²-K. Solving for L/UA gives

$$\frac{L}{UA} = \frac{L}{h_i A_i} + \frac{L}{h_o A_o}$$

$$= \frac{1}{88.46 \times 0.0254\pi} + \frac{1}{170 \times 0.0284\pi} = 0.208 \qquad (7.31)$$

Using $UA/h_iA_i = \Delta T_{fs}/\Delta T_{lm}$, obtain $\Delta T_{fs} = 59.0$ K as compared to the guessed value of 29.8 K. Thus, Nu_{cc} is recalculated using $\Delta T_{fs} = 59.0$ K, which yields $Nu_{cc} = 44.33$ and results in $h_i = 89.9$ W/m²-K. Solving Equation 7.26 again gives $L/UA = 0.205$ and $\Delta T_{fs} = 58.6$ K, which is close enough to the second assumed $\Delta T_{fs} = 59.0$ K. The required pass length is $L = (UA/100) \times (L/UA) = 8.63 \times 0.205 = 1.77$ m, which compares to 2.51 m for the plain tube design.

Because the tube-side flow is turbulent, we may use Equation 7.14 and obtain $f/f_p = 1.635$. Then, the pressure drop ratio is

$$\frac{\Delta p}{\Delta p_p} = \frac{f}{f_p}\frac{L}{L_p} = 1.635 \times \frac{5.81}{8.25} = 1.152 \qquad (7.32)$$

7.2.5 Twisted Tapes in Annuli

Gupte and Date [1989] report data for three tape geometries in an annulus. Their data were taken for air with heat applied to the outer wall of the annulus for two radius ratios (r_i/r_o = 0.41 and 0.61). The three tapes tested had y = 2.66, 5.30 (r_i/r_o = 0.41), and 5.04 (r_i/r_o = 0.61). The Reynolds number for the heat transfer data spanned 10,000 < Re_{Dh} < 50,000.

They also developed a semianalytical model to predict their data. The model reasonably predicted the friction data for the three tapes. The heat transfer model worked well for the $y \cong 5$ data but underpredicted the y = 2.66 data.

Because only the outer wall was heated, centrifugal forces will act to stratify the flow. In practical applications, one would expect heat exchange to occur at only the inner wall. If the inner wall is heated, centrifugal forces will act to mix the flow. Hence, it is likely that the performance for a heated inner wall will be higher than the data of Gupte and Date [1989].

7.3 SEGMENTED TWISTED-TAPE INSERT

Saha et al. [1989] have conducted an extensive experimental program of the Figure 7.1b segmented twisted tape in laminar flow of water with constant heat flux. Their tape geometries had twist ratios (H/d_i) of 3.18, 5, 7.5, 10, and ∞, or helix angles (α) of 26.3, 17.4, and 8.82, and 0 degrees. Elements of twisted tape were spaced a distance $z = Z/d_i$. For each twist ratio y, they tested dimensionless spacings of z = 2.5, 5, 7.5, and 10. Figure 7.12 shows their results for y = 5. The Nusselt number is the average value over the 1.8-mm-long, 11-mm-inner-diameter test section. The line labeled z = 0 in Figure 7.12a is the Hong and Bergles [1976] correlation, Equation 7.5. The figure shows that the smallest z value gives the highest Nu, and Nu decreases for increasing z. At z = 2.5 and Re_d = 1000, the Nu is increased 15%, relative to the continuous twisted tape. The friction factor data are shown in Figure 7.12b. The solid line is for a tape with $y = \infty$ and has the value $f = 46.45/Re_d$. The dashed line is from a correlation for continuous tapes with y = 5. The highest friction factor also occurs for the smallest z. The segmented tapes yield friction factors higher than the continuous tape (z = 0) for $y \leq 7.5$. However, for z = 10, the friction factor is smaller than the continuous tape. Properly spaced tape segments provide higher enhancement than does a continuous tape, because the wake mixing region between the tape segments dissipates the thermal and velocity boundary layers. This mixing process soon dies out, and the re-initiation of the swirl flow is required. For y = 3.46, Saha et al. show that z values of 7.5 and 10 give smaller Nu values than does a continuous tape.

Table 7.1 compares the enhancement provided by the segmented tapes, relative to a continuous tape using PEC FG-2b in Chapter 3. The Nu_{st} is for the segmented tape, and Nu_{ct} is for a continuous tape. The enhancement ratios are constant for 679 $\leq Re_{d,c} \leq$ 1918, where $Re_{d,c}$ is the Reynolds number for the continuous tape.

The Reynolds number of the segmented tape is smaller than that of the continu-

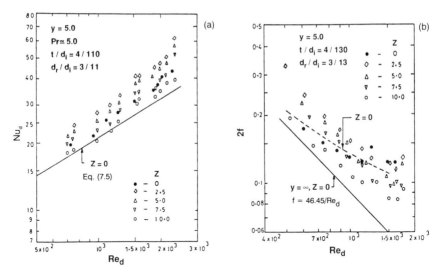

Figure 7.12 Developing laminar flow of water with q = constant in a tube having a segmental twisted tape of $y = 5$ and $0 < z < 10$. (a) Nusselt number. (b) Friction factor. (From Saha et al. [1989].)

ous tape, in order to meet the constant pumping power criterion. The Reynolds numbers are related by the equation

$$\frac{Re_{st}}{Re_{ct}} = \left(\frac{f_{st}}{f_{ct}} \frac{A_{c,ct}}{A_{c,st}} \right)^{1/3} \tag{7.33}$$

The cross-sectional flow area and hydraulic diameters used in Equation 7.33 are based on the volumetric average values.

Saha et al. [1989] give a series of rather complex empirical correlations to predict the Nu and f as a function of Re, y and z. The Nu correlation is an extension of Equation 7.5, and it converges to the Equation 7.5 value if $z = 0$.

Xie et al. [1992] provide data on the Figure 7.1c static mixer for an oil at Pr = 41 in an electrically heated tube with $2000 \leq Re_d \leq 20,000$. Data are provided for $y = 3, 3.5,$ and 4.0. No correlation was given.

TABLE 7.1   Case FG-1 PEC for Segmented Twisted Tapes

	Nu_{st}/Nu_{ct} for y values of			
z	3.18	5.0	7.5	10
2.5	1.44	1.47	1.28	1.08
5.0	1.34	1.31	1.13	1.05
7.5	1.17	1.16	1.12	1.04
10	1.09	1.11	1.09	1.00

7.4   DISPLACED ENHANCEMENT DEVICES

7.4.1   Turbulent Flow

These are some of the very earliest devices investigated and are probably the least effective for turbulent flow. Figures 7.5a and 7.5b illustrate two types of displaced insert devices tested by Koch [1958] and by Evans and Churchill [1963] in laminar and turbulent flow. Colburn and King [1931] tested the Figure 7.4b device and a variant of the Figure 7.5a device. These devices periodically mix the gross flow structure and accelerate the local velocity near the wall. Koch found that the Figure 7.5a and 7.5b devices have substantially higher pressure drop than does the twisted-tape insert (Figure 7.1a) or the Figure 7.3a wire coil insert. Theoretical reasoning argues that for turbulent flow the fluid should be mixed in the viscous dominated region near the wall, where the thermal resistance is large. The Figure 7.4 and 7.5 devices mix the flow in the core region and experience quite high profile drag forces, which substantially increase the pressure drop. Other displaced insert devices that have been tested in turbulent flow include (a) mesh or brush inserts of Figure 7.4, (b) bristle brushes, (c) static mixer devices (Figure 7.1c), and (d) flow-driven propellers, all of which promote mixing across the total flow cross section. Mergerlin et al. [1974] found that bristle brushes (Figure 7.4c) provide h/h_p as high as 8.5, but the pressure drop was increased a factor of 2800. Bergles [1985] provides a PEC evaluation of Koch's [1958] data for the Figure 7.5a and 7.5b displaced insert devices using PEC FG-2a of Table 3.1. PEC FG-2a calculates h/h_p for constant pumping power, and it also calculates tube diameter ($d_o = 50$ mm) and tube length. The devices provide very poor performance relative to other competing devices and are not recommended for operation in the turbulent regime. The velocity in the enhanced tube must be reduced so low that little benefit, if any, occurs.

The displaced wire coil insert device shown in Figure 7.3b is somewhat different from the other displaced insert devices. It is similar to the wall-attached wire coil insert shown in Figure 7.3a, except it causes mixing in a narrow region close to the tube wall. Thomas [1967] tested displaced wires for turbulent flow of water in an annulus. He concluded the most favorable St/f performance was obtained when the wires were axially spaced in pairs separated approximately nine wire diameters, followed by a second pair separated approximately 75 wire diameters from the first pair. The enhancement was provided by an increased velocity gradient at the surface and by interaction of the cylinder wake with the fluid in the boundary layer. Tests of this displaced wire coil in a circular tube have not been reported. It would be relatively difficult to install such a displaced wire coil in a tube.

Xie et al. [1992] provide data on the Figure 7.5d louvered insert for an oil (Pr = 41) in an electrically heated tube with $2000 \leq \text{Re}_d \leq 30,000$. This consists of a flat strip which is slit and bent at an approximately 45-degree angle. The strips, spaced at distance p, are bent in alternate directions, so that when it is slipped in the tube, the flat strip is centered in the tube. Their data were taken for an interrupted element pitch of $y = p/d_i = 1, 2, 4$, and 6.

$$Nu_d = 0.0202Re_{d,f}^{0.767}Pr^{2/3}y^{-0.282}(\mu/\mu_w)^{0.14} \qquad (7.34)$$

$$f = (5.68 - 2.07y + 0.43y^2)Re_{d,f}^{-(0.34-1.41y-0.021y^2)}(\mu/\mu_w)^{1/3} \qquad (7.35)$$

7.4.2 Laminar Flow

Because the thermal resistance is not confined to a thin boundary layer region near the wall in laminar flow, the devices shown in Figures 7.4 and 7.5 offer greater potential than for turbulent flow. Such devices may be helpful for cooling of viscous fluids, such as an oil. When oil is cooled, the higher viscosity at the wall causes reduced velocity and temperature gradients at the tube wall. If the insert promotes bulk fluid mixing, and/or increases the temperature and velocity gradient at the wall, enhancement will result. The principal questions are: (1) Which devices give the highest heat transfer enhancement? and (2) What is the pressure drop penalty for the enhancement?

Bergles and Joshi [1983] and Bergles [1985] compare the performance of such displaced insert devices and twisted tapes for laminar flow. The comparisons are somewhat inconclusive, since they are not made for the same fluids (e.g., air versus ethylene glycol).

The commercially available "HEATEX" mesh insert device shown in Figure 7.4a provides significantly higher enhancement than does a twisted-tape insert for cooling oil in laminar flow. We will compare the HEATEX insert (subscript H) with a twisted-tape insert (subscript T). The geometries compared are HEATEX Geometry D in Oliver and Aldington [1986] with a twisted tape ($y = 4.8$) that operates at Re $= 500$ using PEC VG-1 of Table 3.1. PEC VG-1 seeks A_H/A_T at $P_H/P_T = Q_H/Q_T = 1$, assuming all of the thermal resistance is on the tube side. The analysis shows $G_H/G_T = 0.89$, $A_H/A_T = 0.38$, and $L_H/L_T = 0.34$. Thus, to obtain equal hA and friction power, the HEATEX device requires only 38% as much surface area. The tube length is reduced 66%, and the number of tubes is increased 12%.

7.4.3 PEC Example 7.2

Compare the performance of the three insert devices tested by Xie et al. [1992] for oil flow (Pr $= 41$) at $Re_{d,p} = 5000$. The three insert devices are twisted tapes (Figure 7.1a), the static mixer insert (Figure 7.1c), and the louvered strip insert device (Figure 7.5d). The performance is compared for PEC FG-2a of Table 3.1. This PEC compares h/h_p for fixed flow rate, tube length, and pumping power. The flow rate in the enhanced tube may be reduced to compensate for the increased pressure drop in the enhanced tube. The specific geometries compared are listed in Table 7.2. Figure 7.13 shows that the Figure 7.5d louver strip insert with $y = 1.0$ (Code C-1) provides the best performance, followed by the Figure 7.1c static mixer with $y = 3.0$ (Code B-2). The lowest performance is provided by the twisted tape. At $Re_{d,p} = 5000$, the Code C-1 louvered strip provides $h/h_p = 1.65$, as compared to only 20% for the $y = 1.67$ twisted tape. At $Re_{d,p} = 40,000$, only the C-1 device is better than a plain tube.

TABLE 7.2   Case FG-2a PEC Comparison for Insert Devices Tested by Xie et al. [1992]

Geometry	Figure	Code	y
Twisted tape	7.1a	A-1	1.67
		A-2	2.5
		A-3	3.5
Static mixer	7.1c	B-1	3.0
		B-2	3.5
		B-3	4.0
Louver strip	7.5d	C-1	3.0
		C-2	2.0
		C-3	4.0

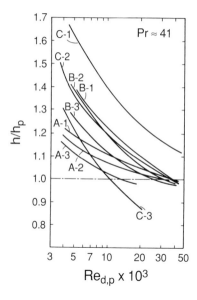

Figure 7.13   FG-2a comparison of twisted tapes (A-1, A-2, A-3), static mixers (B-1, B-2, B-3), and louvered strip inserts (C-1, C-2, C-3). The insert geometry dimensions are defined in Table 7.2.

7.5   WIRE COIL INSERTS

The Figure 7.3a wire coil insert is made by tightly wrapping a coil of spring wire (wire diameter e) on a circular rod. The coil outside diameter, d_c, is made slightly larger than the tube inside diameter, d_i. When the coil spring is pulled through the tube, the wires form a helical roughness of height e at helix angle $\sin^{-1}(d_i/d_c)$ and spacing $p = \pi d_c \cos \alpha$. It is necessary that the coil spring forces the wire tightly against the tube wall to hold the wire in place and prevent tube wall erosion. This requires a helix angle of 25 degrees or more. To obtain $\alpha = 25$ degrees in a 17.6-mm-diameter tube using wire of diameter $e = 1.0$ mm would require a wire coil diameter of 19.36 mm. The resulting coil insert would have $e/d_i = 0.057$, $p/d_i = 3.47$, and $p/e = 61$. The dimensionless geometric parameters that influence the heat transfer and friction characteristics are α, e/d_i, and p/e (or p/d_i). This example is given to show that practical considerations limit the dimensions that influence the performance of the wire coil insert.

The wire coil insert provides enhancement by flow separation at the wire, causing fluid mixing in the downstream boundary layer. Because the boundary layer mixing will dissipate downstream from the wire, the local enhancement quickly dissipates. Hence, rational selection of the wire diameter and spacing requires knowledge of how the local heat transfer coefficient varies with dimensionless distance (p/e) downstream from the trip wire. Two measurements of the local heat transfer coefficient downstream from a wall attached wire have been reported. Edwards and Sheriff [1961] used a boundary layer on a flat plate, and Emerson (1961) worked with pipe flow. In both air flow studies, the thermal and velocity boundary layers were large compared to the rib height. Figure 7.14 shows h_{max}/h_p versus x/e at $Re_x = u_\infty x/v = 530,000$, where h_p is for the undisturbed boundary layer flow. The maximum local enhancement ratio of 1.45 is insensitive to Re_x and is attained 10 rib heights downstream from the rib, which corresponds to the boundary layer reattachment point. Figure 7.15 shows the local enhancement ratio for $0.4 \leq e \leq 6.35$ mm plotted versus e/δ_L, where δ_L is the laminar sublayer thickness of the undisturbed boundary layer at the rib and is given by $\delta_L u^*/v = 5$. Figure 7.15 shows a linear increase of h_{max}/h_p with increasing e/δ_L for $2 \leq e/\delta_L \leq 8$, and only a small increase of h_{max}/h_p for $e/\delta_L > 8$. The dimensionless thickness of the "viscous dominated boundary layer" is typically assumed to be defined by $yu^*/v \simeq 30$, or $6\delta_L$. Hence, the maximum of h_{max}/h_p occurs when the wire diameter is 1.3 times that of the viscous dominated boundary layer thickness. Larger wires diameters protrude into the turbulence-dominated boundary layer and produce very little additional benefit. However, the larger wires will significantly increase the pressure drop, because of profile drag. These studies suggest an approximate rule for picking the maximum effective wire size.

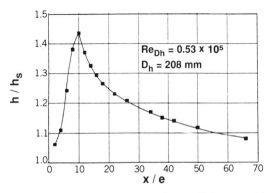

Figure 7.14 Enhancement of the local heat transfer coefficient provided by a 1.58-mm-diameter transverse wire for boundary layer flow of air over a flat plate at $Re_L = 530,000$.

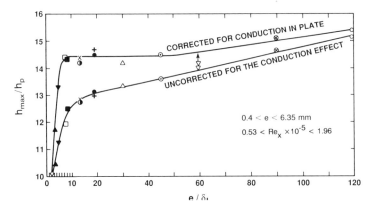

Figure 7.15   Effect of Re_L and wire diameter on h_{max}/h_p for boundary layer flow of air over a flat plate for wire sizes between 0.4 and 6.35 mm.

The wire coil insert amounts to "wall-attached roughness" of spaced helical ribs. Roughness is discussed in detail in Chapter 9. A wire coil insert having a helix angle large enough to fix the wire coil in the tube will have a p/e ratio of 40 or higher. Figure 7.14 shows that h/h_p drops to 1.18 at $p/e = 30$. So, little enhancement will exist in the range of p/e > 30, unless a large diameter wire is used. Generally, integral roughness geometries discussed in Chapter 9 provide better heat transfer and pressure drop performance than can be obtained from wire coil inserts, which typically have a p/e ratio of 50 or higher.

7.5.1 Laminar Flow

Uttawar and Raja Rao [1985] tested seven different wire coil insert geometries in laminar flow ($30 < Re_d < 675$) for heating of an oil ($300 < Pr < 675$). The range of insert geometries tested were $0.08 < e/d_i < 0.13$ and $32 < \alpha < 61$ degrees. Because the heated tube length was only 60 diameters, the flow was not fully developed. The measured enhancement levels were $1.5 \leq Nu_d/Nu_{d,p} \leq 4.0$. The friction increases were considerably less than the Nusselt number increase. The heat transfer data were correlated by

$$Nu_{Dv} = 1.65 \tan \alpha \ Re_{Dv}^{m} \ Pr^{0.35} \left(\frac{\mu}{\mu_w} \right)^{0.14} \tag{7.36}$$

where $m = 0.25(\tan \alpha)^{-0.38}$. Nu, f, and Re are based on the volumetric hydraulic diameter, D_v. A friction factor correlation was not developed. However, when defined in terms of D_v, the friction factor was only 5–8% higher than the smooth tube value for $Re_{Dv} < 180$.

7.5.2 Turbulent Flow

The correlations for turbulent flow are discussed in Section 9.3.3. This is because the wire coil insert essentially acts as a "wall-attached roughness." Its enhancement mechanism is the same as that of integral roughness, which is covered in detail in Section 9. The correlation of Sethumadhavan and Raja Rao [1983] given in Section 9.3.3 is recommended for turbulent flow of liquids and gases. The lowest Re tested by Sethumadhavan and Raja Rao [1983] for their turbulent flow correlation is 4000. The largest Re tested by Uttawar and Raja Rao [1985] for their laminar flow correlation is 675. What correlation should be used in the range $675 \leq \text{Re} \leq 4000$? We cannot make a firm recommendation. However, it is probable that the turbulent flow correlation should be applicable to Re values lower than 4000—possibly as low as 1500.

7.6 EXTENDED SURFACE INSERT

Figure 7.2a shows this device. The insert device is formed as an aluminum extrusion. After inserting the extrusion in the tube, a tube drawing process is employed to obtain a tight mechanical joint between the tube and the insert. The aluminum extrusion is normally formed with five legs, although the number of legs is a design choice and may be between four and eight. By twisting the extrusion before its insertion in the tube, one may also promote a swirling flow. However, the helix angles would be 15 degrees or less.

This insert device is not often used for liquids and gases. Apparently, this is because the pressure drop and cost are unfavorable compared to other enhancement devices (e.g., roughness).

Hilding and Coogan [1964] provide j and f versus Re data for a six-legged straight extrusion in turbulent flow. For turbulent flow, one may predict the turbulent flow j and f characteristics using an appropriate turbulent flow equation for smooth tubes with the tube diameter replaced by the hydraulic diameter (D_h).

Trupp and Lau [1984] have predicted the Nu and f for laminar flow in a tube having full height fins of infinite thermal conductivity. The included angle between the fin legs was varied from 8 to 180 degrees.

Analytical predictions for the Nusselt number of the device must account for the fin efficiency and consider the possibility that a thermal contact resistance may exist between the aluminum and tube contact surfaces. The contact resistance may be negligible for flow of gases but may be appreciable for liquids.

7.7 TANGENTIAL INJECTION DEVICES

This concept involves tangentially injecting part of the flow at locations around the tube circumference, at the tube inlet end. This provides a swirl flow along the tube length. One method of fluid injection is shown in Figure 7.16. Hay and West [1975]

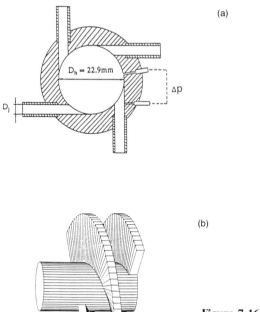

Figure 7.16 Swirl flow injector designs. (a) Design used by Dhir et al. [1989]. (b) An alternate injector design used by Dhir and Chang [1992]. (From Dhir and Chang [1992].)

found that the enhancement depends on the ratio of the injected momentum (M_t) to the total momentum of the axial flow (M_T). Although high enhancement levels are obtained at the tube inlet, they decay along the tube length. Razgaitis and Holman [1976] provide an early survey of heat transfer to swirling flows. Dhir et al. [1989] and Dhir and Chang [1992] provide recent information on such swirl flows. Figure 7.17 shows the enhancement ratio obtained by Dhir et al. [1989] for air flow in a 22.9-mm-inside-diameter tube at Re = 25,000, for different values of M_t/M_T. They used the injector design shown in Figure 7.16a. The tangential-to-total momentum ratio is given by

$$\frac{M_t}{M_T} = \frac{W_t^2}{W^2}\frac{A_c}{A_{c,j}} \tag{7.37}$$

Data by Dhir et al. [1989] show that the enhancement ratio is not significantly affected by either Reynolds number or Prandtl number. For the total flow rate entering four injectors ($W_t/W = 1$) and $M_t/M_T = 5.08$, the enhancement ratio drops from 5.8 at the tube inlet to approach 1.0 near 80 pipe diameters. The enhancement is caused by the high swirl velocity in the boundary layer at the wall. Because this velocity decreases along the pipe length, the enhancement decreases. Figure 7.17 shows that the enhancement ratio drops to 1.5 at $x/d_i = 50$. Consider a 3-m-long, 18-mm-inside-diameter heat exchanger tube operated at Re = 16,400. Hence, the

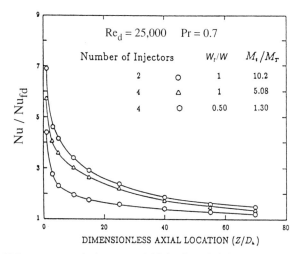

Figure 7.17   Enhancement ratio for tangential injection of air in a 22.9-mm-inside-diameter tube at Re = 25,000. (From Dhir et al. [1989].)

1.5 enhancement factor would be obtained at $x = 0.018 \times 50 = 0.9$ m from the inlet. A significant disadvantage of this enhancement concept is that it does not provide a uniform enhancement level along the tube length, and that it dies out after a relatively short length. Furthermore, a very high enhancement level at the tube inlet may not be of much value, if the controlling resistance is on the shell side at this location. Although it is conceivably possible to inject fluid at points along the tube length, this may not be practical.

Dhir et al. [1989] and Dhir and Chang [1992] correlate the enhancement level by

$$\frac{Nu}{Nu_{fd}} = 1 + 1.93 \left(\frac{M_t}{M_T} \right)^{0.6} Pr^{-1/7} \exp[-m(x/d_i)^{0.6}] \tag{7.38}$$

where

$$m = 0.89 \left(\frac{M_t}{M_T} \right)^{0.2} Re^{-0.18} Pr^{-0.083} \tag{7.39}$$

The correlation is based on data for water and air and correlated 400 data points within ±15%.

7.8   CONCLUSIONS

This chapter has discussed insert devices for single-phase flow in tubes. In the turbulent regime, insert devices are not competitive with internally finned tubes or roughness. However, they may offer advantages for laminar flow of very viscous liquids. In laminar flow, the dominant thermal resistance is not limited to a very thin thermal boundary layer at the wall. The commercially used integral roughness is generally not a good candidate for laminar flow, because the roughness is too small to enhance a laminar flow.

Insert devices may be used to upgrade the performance of an existing heat exchanger having a plain inner tube surface. In this case, the performance improvement would be calculated using the FG-1 or FG-2 PEC of Table 3.1. A consequence of using insert devices to upgrade an existing heat exchanger is that the pressure drop will be significantly increased (for fixed flow rate), or the flow rate must be decreased to compensate for the increased pressure drop.

Cost-effective manufacturing technology to form tube-side roughness or internal fins is a fairly recent development. Hence, insert devices represent an early approach to tube-side enhancement, which allowed use of a plain tube. A variety of insert devices are available. The major types discussed in this chapter are twisted-tape inserts, wire coil inserts, extended surface inserts, mesh or brush inserts, and insert devices displaced from the tube wall.

Design correlations exist for use of twisted-tape inserts in laminar, transition, and turbulent flow regimes. Although design equations are well established for the twisted-tape insert, it may not necessarily be the best insert device. The potential performance of the twisted-tape insert is diminished, because the tape is not in good thermal contact with the wall.

Tangential swirl injection is probably not competitive with integral roughness for turbulent flow. A significant disadvantage of swirl injection is that the heat transfer coefficient varies with axial location. This enhancement has not been sufficiently investigated for laminar flow. Practical solutions of how to practically inject a tangential flow are yet to be identified.

Application of insert devices to laminar flow is a complex issue, because this flow is sensitive to entrance length, thermal boundary conditions, and natural convection effects. Significant performance differences may exist for a given insert device in heating and cooling. The performance characteristics of the twisted-tape insert has been well-defined for these various characteristics. However, this is not the case for the other insert devices.

Potentially interesting insert devices that require additional research to define their laminar flow characteristics are extended surface devices (Figure 7.2a). The wire mesh insert (Figure 7.4a) appears to offer significant enhancement for laminar flow of viscous liquids. Comparisons of several insert devices are given. However, a complete comparison of all insert devices applied to laminar flow of very viscous liquids for both heating and cooling is yet to be completed.

Virtually no work has been done to establish fouling characteristics of insert devices. It is probable that insert devices, which introduce internal obstructions to the flow (e.g., a mesh insert), would cause fouling problems. The twisted-tape insert or extended surface insert probably will not experience fouling problems.

7.9 REFERENCES

Bergles, A. E., 1985. "Techniques to Augment Heat Transfer," Chapter 1 in *Handbook of Heat Transfer Applications*, 2nd edition, W. M. Rohsenow, J. P. Hartnett, and E. N. Ganic, Eds., McGraw–Hill, New York.

Bergles, A. E., and Joshi, S. D., 1983. "Augmentation Techniques for Low Reynolds Number In-Tube Flow, in *Low Reynolds Number Flow Heat Exchangers*, Hemisphere Publishing Corp., Washington, D.C., pp. 694–720.

Colburn, A. P., and King, W. J., 1931. "Relationship Between Heat Transfer and Pressure Drop," *Industrial and Engineering Chemistry*, Vol. 23, No. 8, pp. 918–923.

Date, A. W., 1973. "Flow in Tubes Containing Twisted Tapes," *Heating and Ventilating Engineering*, Vol. 47, pp. 240–249.

Date, A. W., 1974. "Prediction of Fully-Developed Flow in a Tube Containing a Twisted Tape," *International Journal of Heat Mass Transfer*, Vol. 17, pp. 845–859.

Date, A. W., and Singham, J. R., 1972. "Numerical Prediction of Friction and Heat Transfer Characteristics of Fully Developed Laminar Flow in Tubes Containing Twisted Tapes," ASME paper 72-HT-17.

Dhir, V. K., and Chang, F., 1992. "Heat Transfer Enhancement Using Tangential Injection," *ASHRAE Transactions*, Vol. 98, Part 2, pp. 383–390.

Dhir, V. K., Tune, V. X., Chang, F., and Yu, J., 1989. "Enhancement of Forced Convection Heat Transfer Using Single and Multi-Stage Tangential Injection," in *Heat Transfer in High Energy Heat Flux Applications*, R. J. Goldstein, L. C. Chow, and E. E. Anderson, Eds., ASME Symposium, Vol. HTD-Vol. 119, ASME, New York, pp. 61–68.

Edwards, F. J., and Sheriff, N., 1961. "The Heat Transfer and Friction Characteristics for Forced Convection Air Flow over a Particular Type of Rough Surface," *International Developments in Heat Transfer*, ASME, New York, pp. 415–425.

Emerson, W. H., 1961. "Heat Transfer in a Duct in Regions of Separated Flow," *Proceedings of the Third International Heat Transfer Conference*, Vol. 1, pp. 267–275.

Evans, L. B., and Churchill, S. W., 1963. "The Effect of Axial Promoters on Heat Transfer and Pressure Drop inside a Tube," *Chemical Engineering Progress Symposium Series 59*, Vol. 41, pp. 36–46.

Gupte, N., and Date, A. W., 1989. "Friction and Heat Transfer Characteristics of Helical Turbulent Air Flow in Annuli, *Journal of Heat Transfer*, Vol. 111, pp. 337–344.

Hay, N., and West, P. D., 1975. "Heat Transfer in Free Swirl Flow in a Pipe," *Journal of Heat Transfer*, Vol. 97, pp. 411–416.

Hilding, W. E., and Coogan, C. H., Jr., 1964. "Heat Transfer and Pressure Drop in Internally Finned Tubes," in *ASME Symposium on Air Cooled Heat Exchangers*, ASME, New York, pp. 57–84.

Hong, S. W., and Bergles, A. E., 1976. "Augmentation of Laminar Flow Heat Transfer in Tubes by Means of Twisted Tape Inserts," *Journal of Heat Transfer*, Vol. 98, pp. 251–256.

Jayaraj, D., J. G., Masilamani, J. G., and K. N. Seetharamu, 1989. "Heat Transfer Augmentation by Tube Inserts in Heat Exchangers," SAE Technical Paper 891983.

Koch, R., 1958. "Druckverlust und Wäermeueubergang bei Verwirbeiter Stroemung," *Verein Deutscher Ingenieure-Forschungsheft*, Series B, Vol. 24, No. 469 pp. 1–44.

Lopina, R. F., and Bergles, A. E., 1969. "Heat Transfer and Pressure Drop in Tape-Generated Swirl Flow of Single-Phase Water," *Journal of Heat Transfer*, Vol. 91, pp. 434–442.

Maezawa, S., and Lock, G. S. H., 1978. "Heat Transfer inside a Tube with a Novel Promoter," *Proceedings of the 6th International Heat Transfer Conference*, Vol. 2, pp. 596–600.

Manglik, R. M., 1991. "Heat Transfer Enhancement of In-tube Flows in Process Heat

Exchangers by Means of Twisted-Tape Inserts," Ph.D. Thesis, Department of Mechanical Engineering, Rensselaer Polytechnic Institute, Troy, NY.

Manglik, R. M., and Bergles, A. E., 1992a. "Heat Transfer and Pressure Drop Correlations for Twisted-Tape Inserts in Isothermal Tubes: Part I—Laminar Flows," in *Enhanced Heat Transfer*, M. B. Pate and M. K. Jensen, Eds., ASME Symposium, Vol. HTD-Vol. 202, ASME, New York, pp. 89–98.

Manglik, R. M., and Bergles, A. E., 1992b. Heat Transfer and Pressure Drop Correlations for Twisted-Tape Inserts in Isothermal Tubes: Part II—Transition and Turbulent Flows, in *Enhanced Heat Transfer*, M. B. Pate and M. K. Jensen, Eds., ASME Symposium, Vol. HTD-Vol. 202, ASME, New York, pp. 99–106.

Marner, W. J., and Bergles, A. E., 1978. "Augmentation of Tube-Side Laminar Flow Heat Transfer by Means of Twisted Tape Inserts, Static Mixer Inserts and Internally Finned Tubes," in *Heat Transfer 1978, Proc. 6th International Heat Transfer Conference*, Vol. 2, Hemisphere Publishing Corp., Washington, D.C., pp. 583–588.

Marner, W. J., and Bergles, A. E., 1985. "Augmentation of Highly Viscous Laminar Tube-side Heat Transfer by Means of a Twisted-Tape Insert and an Internally Finned Tube," in *Advances in Enhanced Heat Transfer—1985*, S. M. Shenkman, J. E. O'Brien, I. S. Habib, and J. A. Kohler, Eds., ASME Symposium, Vol. HTD-Vol. 43, ASME, New York, pp. 19–28.

Mergerlin, F. E., Murphy, R. W., and Bergles, A. E., 1974. "Augmentation of Heat Transfer in Tubes by Means of Mesh and Brush Inserts," *Journal of Heat Transfer*, Vol. 96, pp. 145–151.

Oliver, D. R., and Aldington, R. W. J., 1986. "Enhancement of Laminar Flow Heat Transfer Using Wire Matrix Turbulators," *Heat Transfer—1986, Proceedings of the Eighth International Heat Transfer Conference*, Vol. 6, pp. 2897–2902.

Razgaitis, R., and Holman, J. P., 1976. A Survey of Heat Transfer in Confined Swirl Flows. *Heat and Mass Transfer Processes*, Vol. 2, pp. 831–866.

Saha, S. K., Gaitonde, U. N., and Date, A. W., 1989. "Heat Transfer and Pressure Drop Characteristics of Laminar Flow in a Circular Tube Fitted with Regularly Spaced Twisted-Tape Elements," *Experimental Thermal and Fluid Science*, Vol. 2, pp. 310–322.

Sethumadhavan, R., and Raja Rao, M., 1983. "Turbulent Flow Heat Transfer and Fluid Friction in Helical Wire Coil Inserted Tubes," *International Journal of Heat Mass Transfer*, Vol. 26, pp. 1833–1845.

Shah, R. K. and Bhatti, M. S., 1987. "Laminar Convective Heat Transfer in Ducts," Chapter 3 in *Handbook of Single-Phase Heat Transfer*, Sadik Kakaç, R. K. Shah, and W. Aung, Eds., John Wiley & Sons, New York, p. 3.20.

Smithberg, E., and Landis, F., 1964. "Friction and Forced Convection Heat Transfer Characteristics in Tubes with Twisted Tape Swirl Generators," *Journal of Heat Transfer*, Vol. 87, pp. 39–49.

Thomas, D. G., 1967. "Enhancement of Forced Convection mass Transfer Coefficient Using Detached Turbulence Promoters," *Industrial Engineering and Chemical Process Design Developments*, Vol. 6, pp. 385–390.

Thorsen, R., and Landis, F., 1968. "Friction and Heat Transfer Characteristics in Turbulent Swirl Flow Subjected to Large Transverse Temperature Gradients," *Journal of Heat Transfer*, Vol. 90, pp. 87–89.

Trupp, A. C., and Lau, A. C. Y., 1984. "Fully Developed Laminar Heat Transfer in Circular Sector Ducts with Isothermal Walls," *Journal of Heat Transfer*, Vol. 106, pp. 467–469.

Uttawar, S. B., and Raja Rao, M., 1985. "Augmentation of Laminar Flow Heat Transfer in Tubes by Means of Wire Coil Inserts," *Journal of Heat Transfer*, Vol. 105, pp. 930–935.

Xie, L., Gu., R., and Zhang, X., 1992. "A Study of the Optimum Inserts for Enhancing Convective Heat Transfer of High Viscosity Fluid in a Tube," in *Multiphase Flow and Heat Transfer, Second International Symposium*, Vol. 1, X-J. Chen, T. N. Veziroğlu, and C. L. Tien, Eds., Hemisphere Publishing Corp., New York, pp. 649–656.

Zhuo, N., Ma, Q. L., Zhang, Z. Y., Sun, J. Q., and He, J., 1992. "Friction and Heat Transfer Characteristics in a Tube with a Loose Fitting Twisted-Tape Insert," in *Multiphase Flow and Heat Transfer, Second International Symposium*, Vol. 1, X-J. Chen, T. N. Veziroğlu, and C. L. Tien, Eds., Hemisphere Publishing Corp., New York, pp. 657–661.

7.10 NOMENCLATURE

A	Heat transfer surface area, m² or ft²
A_c	Flow cross-sectional area in minimum flow area, m² or ft²
$A_{c,j}$	Cross-sectional area for tangential injection, m² or ft²
C	Capacity rate ($= \rho c_p$), kJ/kg or Btu/lbm-°F
c_p	Specific heat of fluid at constant pressure, J/kg-K or Btu/lbm-°F
D_h	Hydraulic diameter of flow passages, $4LA_c/A$, m or ft
D_v	Volumetric hydraulic diameter, 4 × void volume/total surface area, m or ft
d_c	Outside diameter of wire coil used to make Figure 7.3a insert, m or ft
d_i	Tube inside diameter, or diameter to the base of internal fins or roughness, m or ft
d_o	Tube outside diameter, or fin root diameter for a finned tube, m or ft
e	Wire diameter for wire coil insert, m or ft
f	Fanning friction factor, $\Delta p_f d_i/2LG^2$, dimensionless
f_{Dh}	Fanning friction factor based on D_h, $\Delta p_f D_h/2LG^2$, dimensionless
f_{sw}	Fanning friction factor for swirl flow, $\Delta p_f d_i/2L(G_{sw})^2$, dimensionless
G	Mass velocity based on the minimum flow area, kg/m²-s or lbm/ft²-s
G_{sw}	Mass velocity based on u_{sw} and the minimum flow area, kg/m²-s or lbm/ft²-s
Gr_d	Grashof number (= $g_r\beta\Delta T d_i^3/\nu^2$), dimensionless
Gr_{Dh}	Grashof number (= $g_r\beta\Delta T D_h^3/\nu^2$), dimensionless
Gz	Graetz number (= $\pi d_i RePr/4L = \pi/4L^*$), dimensionless
g	Acceleration due to gravity, 9.806 m/s² or 32.17 ft/s²
g_r	Radial acceleration defined by Equation 17.25, m/s² or ft/s²
H	Length for 180 degree revolution of twisted tape, m
h	Heat transfer coefficient based on A, h_x (local value) W/m²-K or Btu/hr-ft²-°F
j	Colburn factor (= $StPr^{2/3}$), dimensionless
k	Thermal conductivity of fluid, W/m-K or Btu/hr-ft-°F
L	Fluid flow length, m or ft
L_s	Swirl length of twisted tape (= $L/\cos\alpha$), m or ft

L^*	$L/d_i\mathrm{Re}_d\mathrm{Pr}$, dimensionless
LMTD	Logarithmic mean temperature difference, K or °F
M_T	Axial momentum $(= W^2/\rho A_c)$, kg-m/s^2 or lbm-ft/s^2
M_t	Tangential injected momentum $(= W_t^2/\rho A_{c,j})$, kg-m/s^2
NTU	Number of heat transfer units $[= UA/(Wc_p)_{\mathrm{min}}]$, dimensionless
$\mathrm{Nu}_{\mathrm{Dh}}$	Nusselt number based on D_h $(= hD_h/k)$, dimensionless
$\mathrm{Nu}_{\mathrm{Dv}}$	Nusselt number based on D_v $(= hD_v/k)$, dimensionless
Nu_d	Nusselt number based on d_i $(= hd_i/k)$, dimensionless
P	Fluid pumping power, W or hp
Pr	Prandtl number $(= c_p\mu/k)$, dimensionless
p	Axial pitch of wire elements, m or ft
p_t	$2H$, m or ft
Δp	Fluid static pressure drop on one side of a heat exchanger core, Pa or lbf/ft^2
q	Heat transfer rate in the exchanger, W or Btu/hr
Ra	Rayleigh number $(= \mathrm{GrPr})$, dimensionless
R_{fi}	Tube-side fouling resistance, m^2-K/W or ft^2-hr-°F/Btu
$\mathrm{Re}_{\mathrm{Dh}}$	Reynolds number based on the hydraulic diameter $(= GD_h/\mu)$, dimensionless
$\mathrm{Re}_{\mathrm{Dv}}$	Reynolds number based on D_v, dimensionless
Re_d	Reynolds number based on the tube diameter $(= Gd_i/\mu)$, dimensionless
$\mathrm{Re}_{\mathrm{sw}}$	Swirl Reynolds number $(= d_iu_{\mathrm{sw}}\rho/\mu)$, dimensionless
Re_x	Reynolds number based on axial distance $(= u_\infty x/\nu)$, dimensionless
r	Tube or annulus radius, m or ft
S_l	Longitudinal tube or element pitch, m or ft
St	Stanton number $(= h/Gc_p)$, dimensionless
Sw	Swirl number $(\mathrm{Re}_{\mathrm{sw}}y^{-1/2})$, dimensionless
ΔT_{fs}	Temperature difference between fluid and surface, K or °F
ΔT_i	Temperature difference between hot and cold inlet fluids, K or °F
ΔT_{lm}	Logarithmic mean temperature difference between fluid and surface, K or °F
t	Thickness of fin or twisted tape, m or ft
U	Overall heat transfer coefficient, W/m^2-K or Btu/hr-ft^2-°F
u_c	Fluid mean axial velocity at the minimum free flow area, m/s or ft/s
u_∞	Free stream velocity over flat plate, m/s or ft/s
u_{sw}	$u_c(1 + \tan^2\alpha)^{1/2} = u_c[1 + (\pi/2y)^2]^{1/2}$, m/s or ft/s
u_θ	Tangential velocity, m/s or ft/s
u^*	Friction velocity $[= (\tau_w/\rho)^{1/2}]$, m/s or ft/s, dimensionless
W	Fluid mass flow rate, kg/s or lbm/s
W_t	Tangentially injected fluid mass flow rate, kg/s or lbm/s
x	Cartesian coordinate along the flow direction, m or ft
x_d^*	$x/(d_i\mathrm{Re}_d\mathrm{Pr})$, dimensionless
y	Twist ratio $= H/d_i = \pi/(2 \tan \alpha)$, dimensionless
Z	Spacing between tape segments (Figure 7.1b)
z	Dimensionless pitch of segmented inserts (Z/d_i) on Figure 7.1b

Greek Letters

α Helix angle relative to tube axis $(= \pi d_i/p_t)$, radians (or degrees)
β_t Volume coefficient of thermal expansion, 1/K or 1/R
δ_L Thickness of laminar sublayer on flat plate, m or ft
η_f Fin efficiency or temperature effectiveness of the fin, dimensionless
μ Fluid dynamic viscosity coefficient, Pa-s or lbm/s-ft
ν Kinematic viscosity, m/s² or ft/s²
ρ Fluid density, kg/m³ or lbm/ft³
τ_w Wall shear stress, Pa or lbf/ft²

Subscripts

ct	continuous tape
fd	Fully developed flow
m	Average value over tube cross section or flow length
p	Plain tube or surface
H	Constant heat flux thermal boundary condition
st	segmented tape
T	Constant wall temperature thermal boundary condition
w	Evaluated at wall temperature
x	Local value

8

INTERNALLY FINNED TUBES
AND ANNULI

8.1 INTRODUCTION

Internally finned tubes are primarily used for liquids, although they may also be used for pressurized gases. An example for in-tube gas flow is an air compressor intercooler. Internally finned tubes used for condensation and vaporization inside tubes are discussed in Chapters 12 and 14, respectively. Because liquids have higher heat transfer coefficients than gases, fin efficiency considerations require shorter fins with liquids than with gases. Figures 8.1a and 8.1b show internally finned having integral internal fins. These tubes are made with either axial or helical fins. The thermal conductance per unit tube length is $\eta_o hA/L$. Typical finned surfaces provide an A/L in the range of 1.5–3 times that of a bare tube. Fin efficiency concern will limit the practical fin height for use with liquids. For plane fins, the fin efficiency is given by

$$\eta_f = \frac{\tanh(me)}{me} \tag{8.1}$$

where

$$m = \left(\frac{2h}{k_f t} \right)^{1/2} \tag{8.2}$$

Low material thermal conductivity (k_f) and/or high fin height (e) will result in small fin efficiency. Use of water, which has a high heat transfer coefficient, will require low fin height (e.g., less than 2.0 mm). Tube materials other than copper

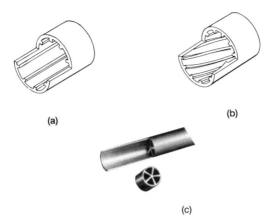

(a)   (b)

(c)

Figure 8.1   Illustration of integral-fin tubes for liquids. (a) Axial internal fins. (b) Helical internal fins. (c) Extruded aluminum insert device.

and aluminum may have relatively low thermal conductivity, which will limit the practical fin height. High fins would be applicable only for liquids having low heat transfer coefficients. The insert device shown in Figure 8.1c could be used for gases or liquids having low heat transfer coefficient. The insert device must have good thermal contact with the tube wall.

8.2   INTERNALLY FINNED TUBES

The flow in internally finned tubes (Figure 8.1c) may be either laminar or turbulent. When used to cool viscous fluids, such as oil, it is possible that the flow will be laminar. Figure 8.2 defines the key dimensions of the internally finned tube. In addition to the dimensions defined in Figure 8.2, the fins may be at a helix angle. The helix angle of the fins, relative to the tube axis, is given the symbol α.

After reviewing the enhancement geometries described in Chapters 8 and 9, the reader will observe that some of the roughness geometries discussed in Chapter 9 appear quite similar to those of an internally finned tube. This is the case for the internally roughened Turbo-C tube shown in Figure 9.13a. In order to fall within the

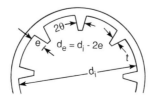

Figure 8.2   Definition of dimensions of internally finned tube.

internal fin classification, we require that no flow separations exist on the internal fin. As the helix angle is increased, we expect flow separation will occur at some critical helix angle. It is doubtful that this condition is attained for the geometries discussed in this chapter. This may not be the case for the Turbo-C tube, which has 33-degree helix angle. Flow separation is a key feature of the enhancement mechanism of rough tubes. However, the theoretical analysis and correlations for internally finned tubes presume that flow separation does not exist. An internally finned tube should provide a significant surface area increase (e.g., 50% or more).

8.2.1   Laminar Flow

Watkinson et al. [1975a] report Nu and f data ($50 < \text{Re}_d < 3000$) for steam heating of oil ($180 < \text{Pr} < 350$) in 18 different internally finned tubes. Marner and Bergles [1985] report Nu and f for a tube with 16 axial fins ($e/d_i = 0.026$) for both heating and cooling conditions ($24 < \text{Pr} < 85$, and $380 < \text{Re}_d < 3470$) taken with $T_w \simeq$ constant. The data are not fully developed, and the reported heat transfer coefficients and friction factors are the average value for over the $L/d_i = 96$ test section length. Figure 7.9 shows h/h_p of the Marner and Bergles [1985] data for PEC FG-2a (Table 3.1). This figure also contains results for a twisted-tape insert ($\alpha = 30$ degrees) illustrated by Figure 7.1. Figure 7.9 shows the following: (1) The internal fin performance is better in heating than in cooling, and (2) the internal fin geometry is superior to that of the twisted tape. Marner and Bergles [1989] provide further interpretation of the Marner and Bergles [1985] data for polybutene, which has $1260 \leq \text{Pr} \leq 8130$. They provide power law factors to correct the heating and cooling data for fluid property variation across the boundary layer.

Rustum and Soliman [1988b] report water data on four internally finned tubes and a smooth tube. They report isothermal friction data, as well as local and fully developed heat transfer coefficients for water flow in four internally finned tubes using electrically heated tubes.

The problem of laminar flow in internally finned channels having axial fins is ideally suited to numerical solution. A number of studies are reported in the literature. Early studies by Hu and Chang [1973], Nandakumar and Masliyah [1975], and Soliman and Feingold [1977] addressed fully developed flow in tubes having zero-thickness fins. Patankar and Chai [1991] analyze laminar flow in finned horizontal annuli. Soliman and Feingold [1977] and Soliman et al. [1980] account for finite fin thickness and thermal conductivity for fully developed flow. Prakash and Patankar [1981] and Rustum and Soliman [1988b] account for the effects of natural convection in vertical and horizontal tubes, respectively. Entrance region solutions are provided by Prakash and Liu [1985], Choudhury and Patankar [1985], and Rustum and Soliman [1988a].

Soliman and co-workers numerically solved the energy equation for fully developed laminar flow in internally finned tubes and predicted the Nusselt number for constant wall temperature (Soliman et al. [1980]) and constant heat flux (Soliman [1979]) boundary conditions. Their heat flux boundary condition has uniform heat input axially, with uniform circumferential wall (and fin) temperature. They assume constant fluid properties and no free convection effects. The range of fin geometries analyzed are $0.1 < e/d_i < 0.4$, $4 < n_f < 32$, and $1.5 < \theta < 3$ degees, where n_f is the number of fins and 2θ is the fin included angle shown in Figure 8.2. The analysis includes the effect of fin thermal conductivity as defined by the parameter $\theta^* \equiv k_f/k$, where k and k_f are the thermal conductivities of the fluid and fin material respectively. Soliman and Feingold [1977] have analytically solved the momentum equation for the friction factor. Table 8.1 shows the calculated Nu_d and f_d, relative to the plain tube value (subscript p). The friction factor and Nu_d refer to the diameter and surface area of a plain tube. Hence, the ratio Nu_d/Nu_p is the ratio of the conductance (hA/L) of the finned tube to that of the plain tube. Similarly, the pressure drop increase at fixed velocity is given by f/f_p. Table 8.1 also shows the area ratio (A/A_p). The table includes Nu_d/Nu_p for $\theta k_f/k$ of 5 and ∞. Several important conclusions may be drawn from Table 8.1:

1. For $e/d_i < 0.3$, the maximum Nu_d/Nu_p occurs with $n_f = 8$. However, f_d/f_p continues to increase for $n_f > 8$. Hence the use of $n_f > 8$ is of no value in providing increased hA/L.
2. $Nu_d/Nu_p < A/A_p$, except for the highest fins at $4 < n_f < 16$. However, the friction factor increase is greater than the area increase for all geometries.

TABLE 8.1 Enhancement Ratios Provided by Internally Finned Tubes for Fully Developed Laminar Flow, as Predicted by Soliman [1979] and Soliman et al. [1980]

				$\theta k_m/k$			
				$(Nu_d/Nu_p)_T$		$(Nu_d/Nu_p)_{H1}$	
n_f	e/d_i	A/A_p	f/f_p	5	∞	5	∞
4	0.1	1.26	1.24	1.04	1.04	1.05	1.07
	0.2	1.51	1.91	1.28	1.30	1.38	1.45
	0.3	1.76	3.28	2.25	2.44	2.47	2.84
	0.4	2.02	4.80	3.73	4.40	3.61	4.52
8	0.1	1.51	1.57	1.06	1.06	1.08	1.10
	0.2	2.02	3.53	1.29	1.31	1.50	1.56
	0.3	2.58	8.67	2.41	2.50	3.79	4.84
	0.4	3.07	14.5	8.07	9.64	7.81	10.45
16	0.1	2.02	2.02	1.03	1.03	1.06	1.06
	0.2	3.04	5.93	1.09	1.10	1.18	1.21
	0.3	4.06	22.2	1.45	1.47	2.07	2.18
	0.4	5.07	60.8	8.59	8.66	17.4	24.4
24	0.1	2.53	2.26	1.01	1.01	1.02	1.02
	0.2	4.06	6.99	1.02	1.03	1.05	1.06
	0.3	5.58	31.7	1.13	1.13	1.29	1.32
	0.4	7.11	172.0	3.29	3.30	9.31	10.5
32	0.1	3.04	2.36	1.00	1.00	1.00	1.01
	0.2	5.08	6.99	1.00	1.00	1.01	1.02
	0.3	7.11	36.3	1.03	1.03	1.08	1.09
	0.4	9.15	355.0	1.66	1.66	2.81	2.91

Subscript p refers to plain tube.

3. Nu_d/Nu_p for $\theta k_f/k = 5$ is within 10% of that for $\theta k_f/k = \infty$. The $\theta k_f/k = \infty$ case corresponds to 100% fin efficiency.

Table 8.2 shows values of $\theta k_m/k$ of interest. Examination of this table shows that the values listed in Table 8.1 for $\theta k_f/k = \infty$ are of primary interest. For $\theta k_f/k = \infty$, one may apply the conventionally used formulas for fin efficiency (η_f) to the Nu for $\theta k_f/k = \infty$ and obtain acceptable accuracy. Soliman et al. [1977, 1979, 1980] have not attempted to compare their predicted Nu and f values for constant properties, fully developed flow with the available data of Watkinson et al. [1975a] and Marner and Bergles [1978, 1985, 1989]. However, Rustum and Soliman [1988b] compare their experimental friction results with the numerical results of Nandakumar and Masilyah [1975] and Soliman and Feingold [1977]. The numerical results generally agree within $+10/-25\%$ of the experimental friction factor. It is likely that experimental values for heating may exceed the Nusselt numbers given in Table 8.1.

Rustum and Soliman [1988a] extend the analysis of Soliman et al. [1979, 1980] to predict the local entrance region Nusselt number for hydrodynamically fully developed flow. Although they studied the same n_f and e/d_i range of Soliman et al. [1979, 1980], they used zero fin thickness. The analysis was performed for constant heat flux and constant wall temperature boundary conditions. Table 8.3 lists the thermal entrance lengths for the two thermal boundary conditions. This is the value of X^+ $[x/(d_i Re_d Pr)]$ at which the fully developed Nusselt number is attained. For $e/d_i \leq 0.2$, the entrance length L_t^+ for the heat flux boundary condition is not very sensitive to the number of fins (n_f), and L_t^+ is close to that for plain tubes (0.0442). Smaller L_t^+ values exist for the wall temperature boundary condition, and they are also smaller than for plain tubes (0.0357). Figure 8.3 shows their results for 24 internal fins plotted in the form of Nu_d versus X^+. Note the rapid drop of Nu_d for the constant temperature boundary condition. The entrance length analysis of Prakash and Liu [1985] is also for zero-thickness fins, but is for simultaneously developing temperature and velocity profiles. The thermal entrance length for $e/d_i = 0.15$ found by Prakash and Liu was approximately 50% smaller for the predictions of Rustum and Soliman [1988a]. Prakash and Liu [1985] also solved for the entrance region friction factor. Table 8.4 gives the hydrodynamic entrance lengths (L_h^+) and the incremental pressure drop, $K(\infty)$, in the entrance region calculated by Prakash and Liu. The hydrodynamic entrance length is defined as the length for the local friction factor to fall to a value 1.05 times the fully developed value. The $K(\infty)$ is the ratio of the entrance region drop to that for fully developed value in the same flow length.

The experimental data are typically for a certain tube length, which may include

TABLE 8.2   Values of $\theta k_m/k$ **for Different Fluid–Material Combinations at 25°C**

Fluid	Pr	Copper	Aluminum	Steel
Air	0.7	16,000	8,000	2000
Water	6.0	670	330	80
Oil	1,200	2,700	1,300	400

TABLE 8.3 **Thermal Entrance Lengths for Internally Finned Tubes, L_t^+**

e/d_i	n_f				
	0	4	8	16	24
Heat Flux Boundary Condition					
0.0	0.0442				
0.1		0.0451	0.0462	0.0475	0.0466
0.2		0.0458	0.0524	0.0544	0.0518
0.3		0.0287	0.0412	0.0596	0.0630
0.4		0.0109	0.0589	0.0028	0.0102
Constant Wall Temperature Boundary Condition					
0.0	0.0357				
0.1		0.0259	0.0209	0.0157	0.0131
0.2		0.0255	0.0197	0.0128	0.0107
0.3		0.0274	0.0234	0.0093	0.0064
0.4		0.0088	0.0055	0.0296	0.0093

an entrance region and free convection effects. If the fluid is heated, such experimental Nu values may be substantially greater than the values predicted by Soliman. However, cooling tends to reduce the Nusselt number, relative to the enhancement provided by free convection and entrance region effects; this is seen on Figure 7.9. The effects of natural convection are influenced by the magnitude of the Raleigh number (Ra), defined by

$$Ra = GrPr = \left(\frac{G\beta q d_i^4}{\nu^2 k} \right)\left(\frac{c_p \mu}{k} \right) \tag{8.3}$$

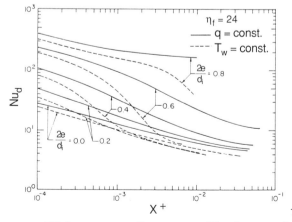

Figure 8.3 Nu_d versus X^+ for constant wall temperature (T) and constant heat flux boundary conditions. (From Rustum and Soliman [1988a].)

TABLE 8.4   Hydrodynamic Entrance Length Parameters for Internally Finned Tubes, L_h^+

e/d_i	n_f			
	0	8	16	24
Hydrodynamic Entrance Length				
0.00	0.0415			
0.15		0.0433	0.0438	0.0417
0.30		0.0320	0.0540	0.0622
0.50		0.0052	0.0024	0.0014
Incremental Pressure Drop, $K(\infty)$				
0.00	1.25			
0.15		2.44	4.11	5.40
0.30		2.85	10.70	23.50
0.50		1.58	1.79	1.93

Figure 8.4 shows the influence of Ra on water flow in the entrance region of a 13.9-mm-inside-diameter tube having 10 fins 1.5 mm high. The solid line shows the predictions of Rustum and Soliman [1988a] for Ra = 0 (pure forced convection). Figure 8.4 shows that increasing Ra significantly increases Nu_d over the Ra = 0 value. As Ra approaches zero, the data approach the predicted value for Ra = 0. Also note that the entrance length decreases with natural convection effects present.

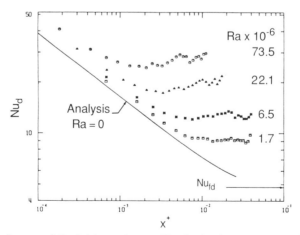

Figure 8.4   Influence of Rayleigh number on Nu_d for laminar entrance region flow with electric heat input (d_i = 13.9 mm, n_f = 10, e/d_i = 0.11). (From Rustum and Soliman [1988a].)

For the four internal fin geometries tested, Rustum and Soliman [1988b] show that the effect of Rayleigh number may be correlated by the empirical expression

$$\frac{\text{Nu}_{fd}}{\text{Nu}_p} = 1 + \left(\frac{\text{Ra}_m}{C_1} \right)^{C_2} \qquad (8.4)$$

where Nu_p is for fully developed forced convection (Ra = 0), Nu_{fd} is the fully developed Nusselt number (Ra > 0), and Ra_m is evaluated at the mean bulk temperature. The constants are geometry-dependent. For the Figure 8.4 geometry, $C_1 = 1.98E - 6$, and $C_2 = 0.44$. Their data also show that Nu_d is insensitive to Re_d for constant Ra.

Zhang and Ebadian [1992] numerically investigated the influence of free convection (mixed flow) on laminar flow horizontal internally finned tubes with constant heat flux and constant wall temperature. They analyzed tubes having 0, 3, 7, and 11 fins for $0.15 \leq e/d_i \leq 0.5$. Their analysis shows that buoyancy force enhances the Nusselt number for plain and finned tubes, although the enhancement effects are smaller for a finned tube. Figure 8.5 shows the effect of Rayleigh and fin height on a tube having 11 fins. The figure plots $\text{Nu}_{\text{Dh}}/\text{Nu}_p$ versus Ra for $e/d_i = 0.15, 0.3$, and 0.5. The figure shows that $\text{Nu}_{\text{Dh}}/\text{Nu}_p$ increases with increasing Ra, and that the enhancement is greatest for the smaller fin heights. Also note that the friction factor ratio decreases with increasing Ra.

Numerical solutions for the entrance region have been performed by Prakash and Liu [1985], Choudhury and Patankar [1985], and Rustum and Soliman [1988a]. Choudhury and Patankar [1985] numerically solve the momentum and energy equations for zero fin thickness in the entrance region flow in internally finned tubes at Pr = 0.7 and 5.0. Their analysis shows that the entrance length (based on tube diameter) is not much different for internally finned and plain tubes. The analysis is performed for 6, 12, and 24 fins with $0.05 \leq e/d_i \leq 0.3$. Similar analysis is reported by Prakash and Liu [1985] for zero fin thickness.

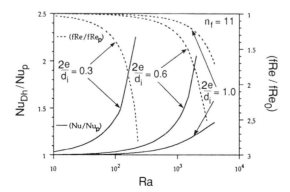

Figure 8.5   Effect of natural convection on $f_{\text{Dh}}\text{Re}_{\text{Dh}}$ and Nu_{Dh} for an internally finned tube having 11 fins, for constant heat flux. (From Zhang and Ebadian [1992].)

Kelkar and Patankar [1990] numerically analyze internally finned tubes, whose fins are segmented along their length. The fin segment is of length H in the flow direction, separated by an equal distance H before the next fin. Figure 8.6 illustrates the staggered and inline segmented fins analyzed. Table 8.5 shows their calculated results for $e/d_i = 0.15$, $n_f = 12$ fins with $L/(d_i\text{Re}_d) = 0.001$ with Pr = 0.7. The tabled values are for the Nu averaged over the finned and unfinned axial increments.

Table 8.5 shows that the inline segmented fins give only 6% higher Nusselt than continuous fins, and the staggered arrangement is 6% below that of continuous fins. However, the inline arrangement gives 22% lower friction than continuous fins. Note that staggered arrangement uses the same fin surface area as continuous fins. However, the inline arrangement has half the fin surface area, because of the axial spaces between fins.

Bergles and Joshi [1983] provide additional comparisons of theoretical and experimental performance of internally finned tubes in laminar flow. Watkinson et al. [1975a] also present data on helical internally finned tubes. It appears that the fin helix angle adds to the enhancement.

8.2.2 Turbulent Flow

Among the earliest data reported on internally finned tubes are that of Hilding and Coogan [1964], who worked with air flow in 13.6-mm-inside-diameter tubes. Their tubes contained four or eight fractional- or full-height fins. They also investigated use of axial interruptions of the internal fins. This improved performance in laminar flow, but not in the turbulent regime.

Carnavos [1980] developed empirical correlations of Nu and f versus Re_{Dh} for correlations for turbulent flow in internally finned tubes. His correlation was based

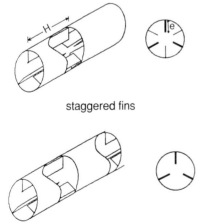

staggered fins

inline fins

Figure 8.6 Illustration of inline and staggered segmented internal fins analyzed by Kelkar and Patankar [1990]. (From Kelkar and Patankar [1990].)

TABLE 8.5   Results for Segmented and Continuous Internal Fins for e/d$_i$ = 0.15

Geometry	$f_d \text{Re}_d/\text{fRe}p$	Nu_d/Nu_p
Continuous fins	2.17	1.18
Inline segmented fins	1.71	1.25
Staggered segmented fins	2.19	1.12

on Carnavos [1979, 1980] data on 21 surface geometries, including helix angles up to 30 degrees. The data are for heating of fluids (6 < Pr < 30) and span 10,000 < Re$_{\text{Dh}}$ < 60,000. Carnavos attempted to correlate the data using standard heat transfer and friction factor correlations (based on the hydraulic diameter) for turbulent flow in plain tubes. He found that this method underpredicted the Nu and f for axial fins. He then developed geometry-dependent correction factors to correlate the data. Webb and Scott [1980] restated the correlations in terms of the fundamental dimensional geometry parameters. The Carnavos correlations for straight and helical fins as stated by Webb and Scott [1980] are

$$\frac{\text{Nu}_{\text{Dh}}}{\text{Nu}_p} = \frac{hD_h/k}{h_p d_i/k} = \left[\frac{d_i}{d_{\text{im}}} \left(1 - \frac{2e}{d_i} \right) \right]^{-0.2} \left(\frac{d_i D_h}{d_{\text{im}}^2} \right)^{0.5} \sec^3\alpha \qquad (8.5)$$

$$\frac{f_{\text{Dh}}}{f_p} = \frac{d_{\text{im}}}{d_i} \sec^{0.75}\alpha \qquad (8.6)$$

The term d_{im} is the tube inside diameter that would exist if the fins were melted and the material returned to the tube wall. The friction factor and Nusselt number in the above equations are defined in terms of the total surface area and hydraulic diameter, and Re$_{\text{Dh}}$ = $D_h G/\mu$. Carnavos used the Dittus–Boelter and the Blasius equations to calculate the plain tube Nusselt number (Nu$_p$) and friction factor (f_p), respectively. These equations are

$$\text{Nu}_p = \frac{h_p d_i}{k} = 0.023\text{Re}_p^{0.8}\text{Pr}^{0.4} \qquad (8.7)$$

$$f_p = 0.046\text{Re}_p^{-0.2} \qquad (8.8)$$

Equations 8.4 and 8.5 correlated the Carnavos data on 21 tube geometries within +8%. The range of dimensionless geometric parameters covered by the correlated data are 0.03 < e/d_i < 0.24, 1.4 < $\pi d_i/n_f e$ < 7.3, 0.1 < t/e < 0.3, and 0 < α < 30 degrees. If one desires to calculate the enhancement ratio in terms of h based on $A_i/L = \pi d_i L$, one writes Nu$_d$/Nu$_p$ as

$$\frac{\text{Nu}_d}{\text{Nu}_p} = \frac{\text{Nu}_{\text{Dh}}}{\text{Nu}_p} \cdot \frac{d_i}{D_h}\left(1 + \frac{2n_f e}{\pi d_i} \right) \qquad (8.9)$$

Similarly, the friction factor ratio, f_d/f_p, is given by

$$\frac{f_d}{f_p} = \frac{f_{Dh}}{f_p} \cdot \frac{d_i}{D_h} \tag{8.10}$$

Webb and Scott [1980] have applied cases VG-1, -2, and -3 of the Table 3.1 PEC to turbulent flow in internally finned tubes and determined their performance relative to plain tubes. A key purpose of this analysis was to determine preferred internal fin geometries. This analysis assumes all of the thermal resistance is on the tube side. The results for the VG-1 analysis are presented in PEC Example 8.1.

Trupp and Haine [1989] measured turbulent flow heat transfer and friction data in five internally finned tubes having $10 \leq n_f \leq 16$. These are tube numbers 9, 10, 13, 14, and 20 also tested by Carnavos [1979, 1980]. Electric heat input was provided by heating wire wrapped on the tube circumference, and local heat transfer coefficients were determined using wall thermocouples. The data spanned $500 \leq Re_{Dh} \leq 10,000$, and they allow comparison with both the laminar friction results of Watkinson et al. [1975a] and the turbulent flow data and correlation of Carnavos [1979, 1980]. The turbulent friction data may also be compared with that of Watkinson et al. [1973, 1975b]. The geometrical dimensions of the tubes are defined in Table 8.6. Figure 8.7 shows the friction data. The data show reasonably good agreement with the extrapolated Carnavos [1980] correlation, Equation 8.5. Note that the lowest Reynolds number (Re_{Dh}) data on which the Carnavos friction correlation was developed is 15,000. The Watkinson et al. [1973] data are higher than the Trupp and Haine [1989] data and the Carnavos [1980] correlation. Figure 8.8 shows the Trupp and Haine fully developed heat transfer data compared with the Carnavos [1980] correlation, Equation 8.5. The figure shows excellent agreement with Equation 8.4. Figure 8.8 also shows the transition Reynolds number for the internally finned tubes. Transition occurs at Reynolds numbers (based on hydraulic diameter) between 500 and 1000. Trupp and Haine also investigated the effect of natural convection by varying the Rayleigh number. Their data showed a very weak influence of Rayleigh number for $0.9 \leq Ra \times 10^{-6} \leq 1.8$.

Trupp et al. [1981] performed measurements of the flow structure in a scaled-up

TABLE 8.6 Tube Geometries Tested by Trupp and Haine [1989]

Parameter	9	10	13	14	20
d_o (mm)	12.7	9.53	9.53	15.9	12.7
d_i (mm)	10.3	8.00	7.04	13.9	10.4
e_i (mm)	1.28	1.27	2.29	1.50	1.47
n_f	10	16	10	10	16
α (degrees)	0	0	0	0	2.5

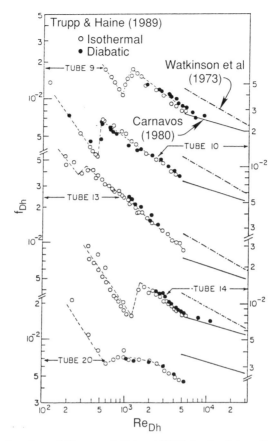

Figure 8.7 Friction data of Trupp and Haine [1989] for the internally finned tubes described in Table 8.6. (From Trupp and Haine [1989].)

internally finned tube 114.3-mm inside diameter for $50,000 < \text{Re}_{\text{Dh}} < 71,000$. The tube had six axial internal fins with $e/d_i = 0.33$ and $t/e = 0.13$. The fin height is considerably higher than would be expected for practical applications. They measured the velocity distribution in the inter-fin region, secondary velocities, Reynolds stresses, and the wall shear stress. Figure 8.9a shows the measured velocity profile, and Figure 8.9b shows the surface shear stress distribution. Trupp et al. [1981] also observed two counter-rotating corkscrew vortices in the inter-fin region. Said and Trupp [1984] developed a two-equation turbulence model to predict the heat transfer coefficient and friction factor for air flow in axial internally finned tubes. Their analysis was performed for $0.1 < e/d_i < 0.4$, $6 \leq n_f \leq 14$, $\theta = 3$ degrees, and $25,000 \leq \text{Re}_{\text{Dh}} \leq 150,000$. They developed the following empirical correlations to predict their numerical results:

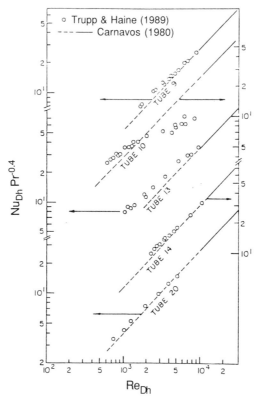

Figure 8.8   Fully developed Nusselt number data of Trupp and Haine [1989] for the internally finned tubes described in Table 8.6. (From Trupp and Haine [1989].)

$$f = 0.0525 \text{Re}_{\text{Dh}}^{-0.216} \left(\frac{p}{D_h} \right)^{0.351} \left(\frac{d_i D_h}{d_{\text{im}}^2} \right)^{0.148} \tag{8.11}$$

$$\text{Nu} = 0.027 \text{Re}_{\text{Dh}}^{0.774} \left(\frac{p}{D_h} \right)^{0.397} \left(\frac{d_i D_h}{d_{\text{im}}^2} \right)^{0.168} \tag{8.12}$$

Patankar et al. [1979] have also numerically solved the momentum and energy equations for turbulent flow in internally finned tubes having zero fin thickness. They used one adjustable constant from the data of Carnavos [1979]. Although they do not provide a correlation of their results, they show good ability to predict the Carnavos [1979] data. Ivanović et al. [1990] analyzed the same geometries of Patankar et al. [1979] and extended the work using numerical methods, which can reveal secondary flow features. Their analysis used the low Reynolds number version of the k–ϵ turbulence model.

Kim and Webb [1993] developed an analytical model to predict the turbulent heat transfer coefficient and friction factor in internally finned tubes having axial fins.

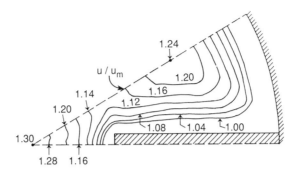

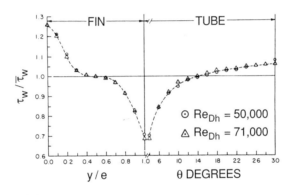

Local Wall Shear Stress Distributions Along
Tube Wall and Fin Surface

Figure 8.9 Experimental measurements of Trupp et al. [1981]. $e/d_i = 0.33$, $t/e = 0.13$, $Re_d = 71,000$. (a) Measured velocity profile for (b) surface shear stress distribution. (From Trupp et al. [1981].)

This model applies the Law of the Wall to the flow in the interfin region. The Law of the Wall states that the velocity and temperature profile is independent of the channel shape, and is described by Hinze [1975]. The "universal velocity profile" for $y^+ > 26$ is given by

$$\frac{u}{u^*} = 2.5 \ln \frac{y u^*}{\nu} + 5.5 \tag{8.13}$$

The friction factor is predicted by integrating Equation 8.13 over the interfin region and over the core region between the fin tips. Similarly, the dimensionless temperature profile is obtained by integrating the corresponding "universal temperature profile" over the same flow regions. Kim and Webb used the universal temperature profile ($y^+ > 0$) proposed by Gowen and Smith [1967], which is given by

$$\frac{T}{T^*} = 2.5 \ln \frac{yu^*}{\nu} + 5 \ln \left(\frac{5\text{Pr} + 1}{30} \right) + 5\text{Pr} + 8.55 \qquad (8.14)$$

Carnavos [1979] friction data for 11 internally finned tubes were predicted within $\pm 10\%$. The Carnavos [1979, 1980] air- and water-flow Nusselt numbers were predicted within $\pm 15\%$.

8.2.3 PEC Example 8.1

Assume that a plain tube heat exchanger has been designed to provide heat duty q with specified flow rates and inlet temperatures. The tubes have 19.2-mm outer diameter and 0.70 mm wall thickness. The plain tube design operates at $\text{Re}_{d,p}$ with N tubes per pass and tube length L_p per pass. One desires to use internally finned tubes to reduce the total length of tubing (NL). The design constraints are those of case VG-1 of Table 3.1. Assuming that the total thermal resistance is on the tube side, the equations in Section 3.3.4 apply. Using Equations 8.7 and 8.8, one solves Equation 3.7 for G/G_p. For constant inside diameter, $G/G_p = (\text{Re}_{\text{Dh}}/\text{Re}_p)(d_i/D_h)$.

The geometric parameters of the finned tube are e, t, n_f, and α. For a given tube geometry, one iteratively solves Equation 3.7 for G/G_p and calculates $\text{Re}_{\text{Dh}}/\text{Re}_p$. Constraints on Equations 3.1 and 3.2 are $hA/h_pA_p = P/P_p = 1$. With the known Re_{Dh}, one calculates S_t and f of the finned tube and then solves for A/A_p using Equation 3.2 or 3.4. The $A/A_p = NL/N_pL_p$. The ratio of tube material in the finned and plain tube exchangers is given by

$$\frac{V_m}{V_{m,p}} = \frac{N}{N_p} \frac{L}{L_p} \frac{M}{M_p} \qquad (8.15)$$

where M is the tube weight per unit length. Figure 8.10a shows the effect of varying the geometry parameters e/d_i and n_f with $e/t = 3.5$ and $\alpha = 0$. This figure shows that the smallest $V_m/V_{m,p}$ occurs with $e = 1$ to 1.5 mm and decreases with increasing n_f when $e = 1$ to 1.5. Figure 8.10b shows the effect of helix angle for $e/d_i = 0.084$ and $e/t = 3.5$. The $V_m/V_{m,p}$ may be reduced nearly 50% with $\alpha = 30$ degrees. The curve for $\alpha = 30$ degrees shows that n_f has little effect on $V_m/V_{m,p}$. However, as n_f increases, G_p/G must increase to meet the $P/P_p = 1$ constraint. Since heat exchanger cost is quite sensitive to shell diameter, selection of $12 \leq n_f \leq 16$ appears to be a good choice. Because $A_c/A_{c,p} = G_p/G$, selection of the smaller G_p/G will reduce the shell diameter, and hence the shell cost. The preferred tube geometry ($e/d_i = 0.084$, $n_f \simeq 12$, and $\alpha = 30$ degrees) provides a 58% reduction of tubing length and a 48% reduction of tubing material (weight) relative to a plain tube design.

This example assumed no thermal resistance on the shell side. Webb [1981] describes how the analysis may be extended to account for shell-side, fouling, and

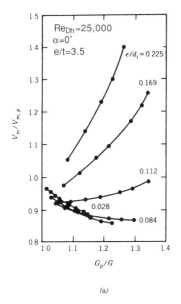

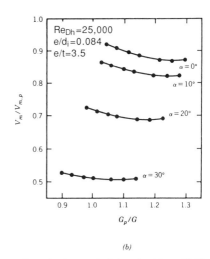

(a)   (b)

Figure 8.10   Performance comparison of internally finned tubes and plain tube ($d_i = 17.78$ mm) for case VG-2 of Table 3.1. (a) Effect of e and n_f for $e/t = 3.5$ and $\theta = 0$. (b) Effect of n_f and θ for $e/d_i = 0.084$ and $e/t = 3.5$ mm. The points on each curve define n_f (from the left, n_f = 5, 8, 12, 16, 25, 32, and 40 fins). (From Webb and Scott [1980].)

tube wall resistances. The value of $V_m/V_{m.p}$ will decrease when these additional resistances are present.

8.3   SPIRALLY FLUTED TUBES

Figure 8.11 shows two variants of spirally fluted tubes. These tubes provide extended surface by deforming the tube wall to form spiral flutes. Note that the tubes have spiral flutes on both the inner and outer surfaces. The tubes shown in Figure 8.11a and 8.11b were developed for General Atomics Corp. by Yampolsky [1983, 1984]. This tube is formed by corrugating strip material (Figure 8.11a) and then rolling it in a circular form, such that the corrugations are at a helix angle. It may also be made by extruding aluminum sheet with the fluted pattern and then twisting the extrusion to form the helix angle as shown in Figure 8.11b. Versions of the tube have been made with helix angles (α) of 30 and 40 degrees.

The tube shown in Figure 8.11c, which we will described as "spirally indented," is manufactured by several companies and is made in a variety of diameters. The tube is pulled through a die to deform the wall into the spirally indented shape. The tubes are made in several diameters with different flute heights, helix angles, and number of starts.

Because these tubes provide enhancement on both sides, it is necessary to mea-

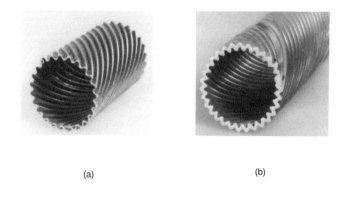

(a)   (b)

(c)

Figure 8.11 Spirally fluted tubes. (a) Stainless steel tube developed by Yampolsky [1983]. (b) Yampolsky tube made of aluminum. (c) The spirally indented tube.

sure the overall heat transfer coefficient and to independently determine the thermal resistance on one side. This is typically done using the modified Wilson plot method, as described by Shah [1990]. Frequently investigators have measured the overall heat transfer coefficient for condensation on the outer tube surface. Accurate use of the Wilson plot method to derive the tube-side heat transfer coefficient should involve two key elements in the test procedure:

1. Maintainence of constant condensing saturation temperature and heat flux.
2. The tube-side Reynolds number exponent should be a constant over the range of water velocities used for determination of the shell-side heat transfer coefficient.

Careful evaluation method of the test methods employed sometimes reveals errors or inaccuracies in using the Wilson plot method. Hence readers should be careful in accepting the tube-side coefficients derived for the spirally fluted tubes. An example is data by Marto et al. [1979] for the tubes shown in Figures 8.11a and 8.11c. These data appear to show higher tube-side coefficients than have been measured by others.

8.3.1 The General Atomics Spirally Fluted Tube

Yampolsky [1983, 1984] has developed and tested an aluminum, spirally fluted tube shown in Figure 8.11b. Test results are also reported by Marto et al. [1979], Panchal and France [1986], Ravigururajan and Bergles [1986], and Obot et al. [1991]. Figure 8.12 shows the friction and heat transfer characteristics reported by Obot et al. [1991], which include test results of other investigators. Tubes 1 and 2 are made of stainless steel with 0.5 mm wall thickness (Figure 8.11a) and Tube 3 is aluminum with 1.5 mm wall thickness (Figure 8.11b). The dimensions are given in Table 8.7. The flutes of Tube 3 (Table 8.7) provide a surface area enhancement of 1.50, relative to a plain tube of the same d_i. The Nu_d is defined as $h'd_i/k$, where h' is based on $A/L = \pi d_i$ and the f and Re_d are based on the maximum internal diameter (d_i). The heat transfer data are plotted as $Nu_d/Pr^{0.4}$ versus Re_d using an assumed Prandtl number dependency. Figure 8.12b shows this reasonably correlates the heat transfer data of Obot et al. [1991] and Panchal and France [1986] for air (Pr = 0.71)

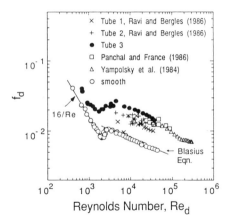

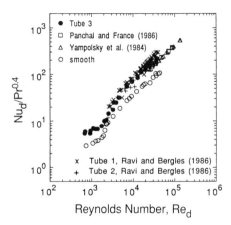

Figure 8.12 Test results of the aluminum spirally fluted tube shown in Figure 8.11b. Tube 3 has d_i = 28.5 mm, with n_f = 31, α = 30 degrees, e/d_i = 0.056, p/e = 3.57, t = 1.5 mm, and D_h = 16.7 mm. (From Obot et al. [1991].) a) Friction factor, b) Nu/$Pr^{0.4}$.

TABLE 8.7   Geometry of Figure 8.10b Tubes Reported by Obot et al. [1991][a]

Tube	d_i	e/d_i	p/e	n_f	α	Material
1	21.54	0.044	3.64	20	40.6	Stainless steel
2	23.96	0.056	4.11	25	30.2	Stainless steel
3	28.49	0.056	3.57	30	30.0	Aluminum

[a]Dimensions are in millimeters.

and the data of Yampolsky [1984] and Ravigururajan and Bergles [1986] for water (Pr $\simeq$ 5.50). Figure 8.12b shows that the correlated data of all investigators for tube 3 are in close agreement. The tube provides $Nu_d/Nu_p = 1.9$ and $f_d/f_p = 3.1$ at $Re_d = 30,000$. Compared to a plain tube of $d_o = d_e$, operated at the same mass velocity, the fluted tube has a thermal conductance (hA/L) 1.5 times higher than the plain tube and 2.0 times the pressure drop. However, the fluted tube operates at 10% lower flow rate, because of its smaller flow area (A_c).

The heat transfer data of Yampolsky [1984], taken with condensing steam on the outer surface, show the same Nusselt number for both heating and cooling. Earlier data reported by Yampolsky [1983] shows higher heat transfer coefficients for heating than for cooling. A possible explanation of this is the very high condensing heat fluxes used in the heating tests. A potential application problem with this tube is how to make a plain tube end for insertion in a tube sheet.

The data of Obot et al. [1991] show significant enhancement in the laminar flow regime. Barba et al. [1983] numerically study laminar flow in the spirally fluted tube. Their analysis is done for Pr = 0.71, 5.0, and 93 with $Re_d < 2500$. The analysis shows increasing values of h/h_p with increasing Pr.

8.3.2   Spirally Indented Tube

Ravigururajan and Bergles [1986] provide a literature survey on various types of corrugated tubes, including the spirally indented tube. Blumenkrantz and Taborek [1971] show that the spirally indented tube provides up to 200% enhancement for heating a viscous fluid in laminar flow. However, no enhancement was provided for cooling the same fluid. This is apparently because centrifugal force tends to prevent mixing of the boundary layer, as discussed for twisted tapes in Chapter 7.

Richards et al. [1987] tested 12 different spirally indented tubes in turbulent water flow using condensing steam on the outer surface. They assumed a 0.8 exponent on the Reynolds number for the tube-side flow. The validity of the 0.8 exponent has not been confirmed. Table 8.8 lists the geometric parameters of the 12 tubes tested. Table 8.9 gives the curve fit coefficients (C_i and B) and exponents (n) for the heat transfer and friction equations given below:

$$\frac{hd_e}{k} = C_i \left(\frac{d_e G}{\mu} \right)^{0.8} Pr^{1/3} \left(\frac{\mu}{\mu_w} \right)^{0.14} \tag{8.16}$$

$$f = B \left(\frac{d_e G}{\mu} \right)^n \tag{8.17}$$

TABLE 8.8 Dimensionless Geometric Parameters for Tubes Tested by Richards et al. [1987]

Tube	d_e/d_c	e/d_c	e/p	e/d_c	p/d_c	A_o/A_e
1	1.56	0.179	0.238	0.278	1.168	0.89
2	1.59	0.186	0.479	0.296	0.618	1.41
3	1.77	0.217	0.416	0.385	0.925	1.48
4	1.23	0.093	0.179	0.114	0.637	0.92
5	1.38	0.139	0.272	0.192	0.706	1.02
6	1.49	0.165	0.515	0.247	0.408	1.11
7	1.93	0.241	0.349	0.465	1.332	1.60
8	1.90	0.237	0.704	0.449	0.638	2.04
9	1.71	0.208	0.275	0.356	1.295	1.22
10	1.83	0.226	0.221	0.414	1.873	1.28
11	2.07	0.258	0.225	0.534	2.373	1.39
12	1.68	0.202	0.500	0.388	0.776	1.36

The mass velocity and the characteristic dimension in the Nusselt number and Reynolds numbers are based on the envelope tube diameter (d_e). Table 8.9 also lists the tube-side enhancement level (j/j_p) and the efficiency index [$\eta = (j/j_p)/(f/f_p)$] for $Re_d = 10{,}000$. Examination of Table 8.9 shows $1.84 \leq j/j_p \leq 3.14$ and $0.13 \leq \eta \leq 0.5$. The efficiency index (η) values are quite low compared to the performance of the fluted tubes shown in Figures 8.11a and 8.11b. In fact, the η values of the Table 8.9 tubes are typically lower than the other tube-side enhancements discussed in Chapters 7–9.

TABLE 8.9 Curve Fit and Performance Parameters for the Doubly-Fluted Tubes Described in Table 8.8 (j/j_p and η at Re = 10,000)

Tube	C_i	B	n	j/j_p	η
1	0.0442	4.07	0.297	1.84	0.21
2	0.0681	3.79	0.253	2.96	0.24
3	0.0440	6.33	0.276	2.15	0.13
4	0.0455	0.45	0.125	2.45	0.52
5	0.0496	1.37	0.208	2.30	0.34
6	0.0632	0.76	0.117	3.14	0.37
7	0.0596	8.73	0.305	3.02	0.20
8	0.0501	6.81	0.235	2.56	0.12
9	0.0487	4.24	0.244	2.61	0.17
10	0.0480	5.55	0.260	2.25	0.14
11	0.0526	14.73	0.365	2.63	0.30
12	0.0495	3.51	0.220	2.56	0.25

8.4 ADVANCED INTERNAL FIN GEOMETRIES

One class of tubes, which is difficult to classify as an internal-fin, or as a roughness has been developed. Examples are the internal enhancements shown in Figures 1.13b and 1.13c. They provide a relatively large internal area increase (e.g., 50%), but have a high helix angle (25–45 degrees). These tubes have some of the characteristics of roughness and internal fins. The tube is similar to an internally finned tube, because it provides significant internal surface area increase. Yet it has a higher helix angle than a helical internal finned tube. It is likely that the flow separates at the fin tip, which probably does not occur in an internally finned tube having a helix angle less than 25 degrees. The Figure 1.13 geometries are discussed in Chapter 9. The low fins (e.g., 0.5 mm) are typically much more closely spaced than an internally finned tube.

A quite different internally finned tube, discussed in Chapters 13 and 14, is the "microfin tube," which is shown in Figure 13.7d. This tube has 0.2- to-0.25-mm-high triangular-shaped fins spaced at approximately 2.0 mm. The fins are formed with a helix angle of 15–30 degrees. These *microfins* may be formed in a copper tube at high speed by drawing the tube over a grooved slug. This tube has found major application for convective vaporization and condensation of refrigerants. Khanpara, et al. [1987] measured the enhancement for subcooled R-22 and R-113 liquid flow in the microfin tube shown in Figure 14.5. The particular tube tested had 8.83-mm inside diameter, 60 fins, 0.22-mm height and 17-degree helix angle. Figure 8.13 shows their Nu_d versus Re_d data, for which the heat transfer coefficient is based on $A_i/L = \pi d_i$. For $Re_d > 10,000$, $h/h_p \approx 2.0$. The smooth tube data are

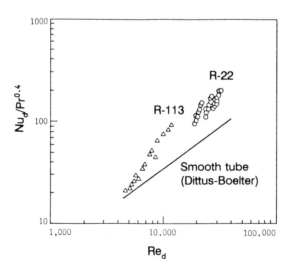

Figure 8.13 Nu_d versus Re_d for single-phase flow in the microfin tube shown in Figure 13.7d. ($d_i = 8.83$ mm, 60 fins, 0.22-mm fin height, and 17-degree helix angle). (From Khanpara et al. [1987].)

predicted by the Dittus–Boelter equation for cooling, which uses $Pr^{0.4}$. The enhancement is somewhat higher than the 54% internal surface area increase. Note that the enhancement level significantly decreases for $Re_d < 10,000$.

It is expected that variants of the tubes shown in Figures 1.13b and 13.7d will be developed and investigated for application to single-phase flow of liquids.

8.5   FINNED ANNULI

Figure 8.14a shows axial fins, which are used for a double-pipe heat exchanger, or for axial flow on the outer surface of a tube bundle. When used in a double-pipe heat exchanger, the fins may span the full width of the annulus, as shown by Figure 8.14b. Multitube designs as illustrated in Figure 8.14c are also used. A common use of double-pipe exchangers is for gases or high-viscosity liquids (e.g., oil) for which the flow may be in the laminar or transition regime. Then some form of enhance-

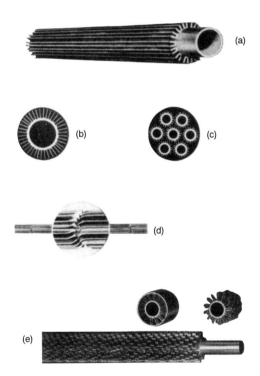

Figure 8.14   (a) Axial fins on external surface. (b) Axial fins used in double-pipe heat exchanger. (c) Axial fins with multitubes. (d) Cut-and-twist axial fins. (e) Offset strip fins. (Parts a–d are from Brown Fin-Tube brochure. Part e is reproduced courtesy of Wieland-Werke AG.)

ment is of interest. Various enhanced fin geometries may be employed. Figure 8.14d shows the "cut-and-twist" geometry. This provides intermittent mixing along the length. The length between cuts is in the range of 0.3-to-1.0 m. Figure 8.14e shows the offset strip fin applied to double-pipe geometry. The fin material is wrapped around the inner tube and soldered. Guy [1983] discusses design and application of this geometry.

The geometry of the annulus flow passage varies from trapezoidal (for small d_i/d_o and a small number of fins) to rectangular (for $d_i/d_o \simeq 1.0$ with a large number of fins). In the limiting condition of rectangular channels with small clearance between the fin tip and the tube wall, one may reasonably predict the performance using appropriate equations for flow in rectangular channels. Note that the outer annulus tube is not a heat transfer surface.

DeLorenzo and Anderson [1945] presents finned annulus data for 24, 28, and 36 fins on a 48.3-mm-diameter tube for heating and cooling of oils ($11 < Pr < 1600$). Gunter and Shaw [1942] present data for the cut-and-twist geometry shown in Figure 8.14d. The correlations presented in Chapter 5 are applicable to the offset strip fin geometry shown in Figure 8.14e.

Taborek [1993] provides friction correlations for laminar, transition, and turbulent flow in finned annuli, including the cut-and-twist geometry. The correlations are based on the data of Gunter and Shaw [1942], DeLorenzo and Anderson [1945], and unpublished work of Gardner.

8.6 CONCLUSIONS

Internally finned tubes and annuli are among the earliest tube-side enhancement geometries. The early versions used extruded aluminum insert devices, which provide full-height fins. The first integral, internal fin tubes were made of copper using a cold swaging process. Performance can be improved by forming the fins at a helix angle (e.g., up to 25 degrees). Although the copper tubes provide high performance, they are quite expensive. External, axially finned tubes for annuli are made by welding steel fins on a steel tube. The same axial fin geometries may be made at lower cost in aluminum, by a hot extrusion process. Helical fins may be provided by twisting the aluminum tube. The spirally indented tubes shown in Figure 8.11 are variants of internally finned tubes, which provide enhancement on both the inside and outside surfaces. Making a corrugated strip, followed by rolling into a circular shape and seam welding, offers a low cost possibility. However, it is difficult to make a plain tube end.

Advanced internally roughened tubes, as illustrated in Figures 1.13b and 1.13c, provide a relatively large internal area increase, approaching that of internally finned tubes. The Figure 1.13 geometries are discussed in Chapter 9. They are similar to an internally finned tube, because they provide significant internal surface area increase. Yet they have a higher helix angle than a helical internal finned tube. It is likely that the flow separates at the fin tip, which does not occur in an internally finned tube. The low fins (e.g., 0.5 mm) are typically much more closely spaced than an internally finned tube.

The "microfin" tube shown in Figure 13.7d has closely spaced fins 0.20 to 0.25 mm high, and is used for convective vaporization and condensation of refrigerants. The only test data for single-phase flow is the R-113 data of Khanpara et al. [1987]. It is expected that variants of the tubes shown in Figures 1.13b and 13.7d will be developed and investigated for application to single-phase flow of liquids.

Many numerical solutions have been provided for laminar flow in internally finned tubes. These solutions cover both fully developed and entrance region conditions, and constant heat flux and constant wall temperature boundary conditions. Further work is needed to account for fluid property variation across the boundary layer.

The Carnavos [1980] empirical correlation is applicable for prediction of turbulent flow heat transfer and friction. Taborek [1993] provides correlations for laminar and turbulent flow in finned annuli. Kim and Webb [1993] have shown that the Law of the Wall may be used to model turbulent flow in internally finned tubes.

As discussed in Chapter 10, the conventional internally finned tube provides good fouling characteristics.

8.7 REFERENCES

Barba, A., Bergles, G., Gosman, A. D., and Launder, B. E., 1983. "The Prediction of Convective Heat Transfer in Viscous Flow Through Spirally Fluted Tubes," ASME paper 83-WA/HT-37.

Bergles, A. E., and Joshi, S. D., 1983. " Augmentation Techniques for Low Reynolds Number In-tube Flow," in *Low Reynolds Number Flow in Heat Exchangers*, S. Kakaç, R. K. Shah, and A. E. Bergles, Eds., Hemisphere Publishing Corp., Washington, D.C., pp. 694–720.

Blumenkrantz, A., and Taborek, J., 1971. "Heat Transfer and Pressure Drop Characteristics of Turbotec Spirally Deep Grooved Tubes in the Laminar and Transition Regime," Report 2439–300–8, April 1971, Heat Transfer Research, Inc.

Carnavos, T. C., 1979. "Cooling Air in Turbulent Flow with Internally Finned Tubes," *Heat Transfer Engineering*, Vol. 1, No. 2, pp. 41–46.

Carnavos, T. C., 1980. "Heat Transfer Performance of Internally Finned Tubes in Turbulent Flow," *Heat Transfer Engineering*, Vol. 4, No. 1, pp. 32–37.

Choudhury, D., and Patankar, S. V., 1985. "Analysis of Developing Laminar Flow and Heat Transfer in Tubes with Radial Fins," in *Advances in Enhanced Heat Transfer—1985*, ASME Symposium, Vol. HTD-Vol. 43, S. M. Shenkman, J. E. O'Brien, I. S. Habib, and J. A. Kohler, Eds., ASME, New York, pp. 57–63.

DeLorenzo, B., and Anderson, E. D., 1945. "Heat Transfer and Pressure Drop of Liquids in Double Pipe Fintube Exchangers," *ASME Transactions*, Vol. 67, pp. 697–702.

Gowen, R. A., and Smith, J. W., 1967. "Heat Transfer Performance of Internally Finned Tubes in Turbulent Flow," *Chemical Engineering Science*, Vol. 22, pp. 1701–1711.

Gunter, A. Y., and Shaw, W. A., 1942. "Heat Transfer, Pressure Drop, and Fouling Rates of

Liquids for Continuous and Noncontinuous Longitudinal Fins," *ASME Transactions*, Vol. 64, pp. 795–802.

Guy, A. R., 1983. "Double-Pipe Heat Exchangers," Section 3.2. in *Heat Exchanger Design Handbook*, Vol. 3, E. U. Schlünder, Ed., Hemisphere Publishing Corp., Washington, D.C.

Hilding, W. E., and Coogan, C. H. , 1964. "Heat Transfer and Pressure Drop Measurements of Internally Finned Tubes," in *Symposium on Air-Cooled Heat Exchangers*, ASME, New York, pp. 57–85.

Hinze, J. O., 1975. *Turbulence*, 2nd edition, McGraw–Hill, New York.

Hu, M. H., and Chang, Y. P., 1973. "Optimization of Finned Tubes for Heat Transfer in Laminar Flow," *Journal of Heat Transfer*, Vol. 95, pp. 332–338.

Ivanović, M., Selimović, R., and Bajramović, R., 1990. "Mathematical Modeling of Heat Transfer in Internally Finned Tubes," in *Mathematical Modeling and Computer Simulation of Processes in Energy Systems*, H. Hanjalić, Ed., Hemisphere Publishing Corp., Washington, D.C., pp. 147–153.

Kelkar, K. M., and Patankar, S. V., 1990. "Numerical Prediction of Fluid Flow and Heat Transfer in a Circular Tube with Longitudinal Fins Interrupted in the Streamwise Direction," *Journal of Heat Transfer*, Vol. 112, pp. 342–348.

Khanpara, J. C., Pate, M. B., and Bergles, A. E., 1987. "Local Evaporation Heat Transfer in a Smooth Tube and a Micro-fin Tube Using Refrigerants 22 and 113," in *Boiling and Condensation in Heat Transfer Equipment*, E. G. Ragi, Ed., ASME Symposium, Vol. HTD-Vol. 85, ASME, New York, pp. 31–39.

Kim, N-H., and Webb, R. L., 1993. "Analytic Prediction of the Friction and Heat Transfer for Turbulent Flow in Axial Internal Fin Tubes," *Journal of Heat Transfer*, Vol. 115, pp. 553–559.

Marner, W. J., and Bergles, A. E., 1978. "Augmentation of Tube-Side Laminar Flow Heat Transfer by Means of Twisted Tape Inserts, Static Mixer Inserts and Internally Finned Tubes," in *Heat Transfer 1978, Proceedings 6th International Heat Transfer Conference*, Vol. 2, Hemisphere Publishing Corp., Washington, D.C., pp. 583–588.

Marner, W. J., and Bergles, A. E., 1985. "Augmentation of Highly Viscous Laminar Tube-side Heat Transfer by Means of a Twisted-Tape Insert and an Internally Finned Tube," in *Advances in Enhanced Heat Transfer—1985*, S. M. Shenkman, J. E. O'Brien, I. S. Habib, and J. A. Kohler, Eds., ASME Symposium, Vol. HTD-Vol. 43, ASME, New York, pp. 19–28.

Marner, W. J., and Bergles, A. E., 1989. "Augmentation of Highly Viscous Laminar Heat Transfer Inside Tubes with Constant Wall Temperature," *Experimental and Thermal Fluid Science*, Vol. 2, No. 3, pp. 252–267.

Marto, P. J., Reilly, D. J., and Fenner, J. H., 1979. "An Experimental Comparison of Enhanced Heat Transfer Condenser Tubing," in *Advances in Enhanced Heat Transfer*, J. M. Chenoweth, J. Kaellis, J. Michel, and S. M. Shenkman, Eds., ASME, New York, pp. 1–10.

Nandakumar, K., and Masliyah, H. H., 1975. "Fully Developed Viscous Flow in Internally Finned Tubes," *Chemical Engineering Journal*, Vol. 10, pp. 113–120.

Obot, N. T., Esen, E. B., Snell, K. H., and Rabas, T. J., 1991. "Pressure Drop and Heat Transfer for Spirally Fluted Tubes Including Validation of the Role of Transition," in *Fouling and Enhancement Interactions*, T. J. Rabas and J. M. Chenoweth, Eds., ASME Symposium, Vol. HTD-Vol. 164, ASME, New York, pp. 85–92.

Panchal, C. B., and France, D. M., 1986. "Performance Tests of the Spirally Fluted Tube Heat Exchanger for Industrial Cogeneration Applications," Argonne National Laboratory Report ANL/CNSV-59.

Patankar, S. V., and Chai, J. C., 1991. "Laminar Natural Convection in Internally Finned Horizontal Annuli," ASME paper 91-HT-12.

Patankar, S. V., Inanović, M., and Sparrow, E. M., 1979. "Analysis of Turbulent Flow and Heat Transfer in Internally Finned Tubes and Annuli," *Journal of Heat Transfer*, Vol. 101, pp. 29–37.

Prakash, C., and Liu, Y-D., 1985. "Analysis of Laminar Flow and Heat Transfer in the Entrance Region of an Internally Finned Circular Duct," *Journal of Heat Transfer*, Vol. 107, pp. 84–91.

Prakash, C., and Patankar, S. V., 1981. "Combined Free and Forced Convection in Internally Finned Tubes with Radial Fins," *Journal of Heat Transfer*, Vol. 103, pp. 566–572.

Ravigururajan, T. S., and Bergles, A. E., 1986. "Study of Water-Side Enhancement for Ocean Thermal Conversion Heat Exchangers," HTL-44/ERI Project 1718, Iowa State University.

Richards, D. E., Grant, M. M., and Christensen, R. N., 1987. "Turbulent Flow and Heat Transfer Inside Doubly-Fluted Tubes," *ASHRAE Transactions*, Vol. 93, Part 2, pp. 2011–2026.

Rustum, I. M., and Soliman, H. M., 1988a. "Numerical Analysis of Laminar Forced Convection in the Entrance Region of Tubes with Longitudinal Internal Fins," *Journal of Heat Transfer*, Vol. 110, pp. 310–313.

Rustum, I. M., and Soliman, H. M., 1988b. "Experimental Investigation of Laminar Mixed Convection in Tubes with Longitudinal Internal Fins," *Journal of Heat Transfer*, Vol. 110, pp. 366–372.

Said, N. M. A., and Trupp, A. C., 1984, "Predictions of Turbulent Flow and Heat Transfer in Internally Finned Tubes," *Chemical Engineering Communications*, Vol. 31, pp. 65–99.

Shah, R. K., 1990. "Assessment of Modified Wilson Plot Techniques Used for Obtaining Heat Exchanger Design Data," *Proceedings of the 1990 International Heat Transfer Conference*, Vol. 5, Hemisphere Publishing Corp., Washington, D.C., pp. 51–56.

Soliman, H. M., 1979. "The Effect of Fin Material on Laminar Heat Transfer Characteristics of Internally Finned Tubes," in *Advances in Enhanced Heat Transfer*, J. M. Chenoweth, J. Kaellis, J. W. Michel, and S. Shenkman, Eds., ASME, New York, pp. 95–102.

Soliman, H. M., and Feingold, A., 1977. "Analysis of Fully Developed Laminar Flow in Longitudinally Internally Finned Tubes," *Chemical Engineering Journal*, Vol. 14, pp. 119–128.

Soliman, H. M., Chau, T. S., and Trupp, A. C., 1980. "Analysis of Laminar Heat Transfer in Internally Finned Tubes with Uniform Outside Wall Temperature," *Journal of Heat Transfer*, Vol. 102, pp. 598–604.

Taborek, J., 1993. "Double Pipe Heat Exchanger Design," in *Heat Transfer Engineering*, Vol. 14, No. 3, Hemisphere Publishing Co., Washington, D.C.

Trupp, A. C., and Haine, H., 1989. "Experimental Investigation of Turbulent Mixed Convection in Horizontal Tubes with Longitudinal Internal Fins," in *Heat Transfer in Convective Flows*, R. K. Shah, Ed., ASME Symposium, Vol. HTD-Vol. 107, ASME, New York, pp. 17–25.

Trupp, A. C., Lau, A. C. Y., Said, N. N. A., and Soliman, H. M., 1981. "Turbulent Flow

Characteristics in an Internally Finned Tube," in *Advances in Enhanced Heat Transfer—1981*, R. L. Webb, T. C. Carnavos, E. L. Park, Jr., and K. M. Hostetler, Eds., ASME Symposium, Vol. HTD-Vol. 18, ASME, New York.

Watkinson, A. P., Miletti, P. L., and Tarassoff, P., 1973. "Turbulent Heat Transfer and Pressure Drop in Internally Finned Tubes," *AIChE Symposium Series*, Vol. 69, No. 131, pp. 94–103.

Watkinson, A. P., Miletti, P. L., and Kubanek, G. R., 1975a. "Heat Transfer and Pressure Drop of Internally Finned Tubes in Laminar Oil Flow," ASME paper 75-HT-41.

Watkinson, A. P., Miletti, P. L., and Kubanek, G. R., 1975b. "Heat Transfer and Pressure Drop of Internally Finned Tubes in Turbulent Air Flow," *ASHRAE Transactions*, Vol. 81, Part 1, pp. 330–349.

Webb, R. L., 1981. "Performance Evaluation Criteria for Use of Enhanced Heat Transfer Surfaces in Heat Exchanger Design," *International Journal of Heat and Mass Transfer*, Vol. 24, pp. 715–726.

Webb, R. L., and Scott, M. J., 1980. "A Parametric Analysis of the Performance of Internally Finned Tubes for Heat Exchanger Application," *Journal of Heat Transfer*, Vol. 102, pp. 38–43.

Yampolsky, J. S., 1983. "Sprially Fluted Tubing for Enhanced Heat Transfer," in *Heat Exchangers—Theory and Practice*, J. Taborek, G. F. Hewitt, and N. Afgan, Eds., Hemisphere Publishing Corp., Washington, D.C., pp. 945–952.

Yampolsky, J. S., Libby, P. A., Launder, B. E., and LaRue, J. C., 1984. "Fluid Mechanics and Heat Transfer Spirally Fluted Tubing," GA Technologies Report GA-A17833.

Zhang, H. Y., and Ebadian, M. A., 1992. "The Influence of Internal Fins on Mixed Convection Inside a Semicircular Duct," in *Enhanced Heat Transfer*, M. B. Pate and M. K. Jensen, Eds., ASME Symposium, Vol. HTD-Vol. 202, ASME, New York, pp. 17–24.

8.8   NOMENCLATURE

A	Total heat transfer surface area (both primary and secondary), m^2
A_c	Cross-sectional flow area, m^2 or ft^2
A_f	Fin or extended surface area, m^2 or ft^2
A_{nom}	Heat transfer surface area based on $A/L = \pi d$, inside tube ($d = d_i$), outside tube ($d = d_o$), m^2 or ft^2
c_p	Specific heat of fluid at constant pressure, J/kg-K or Btu/lbm-°F
D_h	Hydraulic diameter of flow passages, $4LA_c/A$, m or ft
d_c	Internal core flow diameter between fin tips for internally finned tube (Figure 8.2), m or ft
d_e	External diameter at tip of fins (or flutes) for a finned (or fluted) tube, m or ft
d_i	Plain tube inside diameter, or diameter to the base of internal fins or roughness, m or ft
d_{im}	Internal diameter if enhancement material is uniformly returned to tube wall thickness, m or ft
d_o	Tube outside diameter, fin root diameter for a finned tube, m or ft
e	Fin height or roughness height, m or ft

f_{Dh}	Fanning friction factor based on D_h $(= \Delta p_f D_h / 2LG^2)$, dimensionless
f_d	Fanning friction factor based on d_i $(= \Delta p_f d_i / 2LG^2)$, dimensionless
G	Mass velocity based on the minimum flow area, kg/m²-s or lbm/ft²-s
Gr	Grashof number $(= g_r \beta \Delta T D_h^3 / \nu^2)$, dimensionless
Gz	Graetz number $(= \pi d_i \text{RePr}/4L)$, dimensionless
h	Heat transfer coefficient, W/m²-K or Btu/hr-ft²-°F
h′	Heat transfer coefficient based on A/L = πd_i, W/m²-K or Btu/h-ft²-°F
k	Thermal conductivity of fluid, W/m-K or Btu/hr-ft-°F
k_f	Thermal conductivity of fin material, W/m-K or Btu/hr-ft-°F
L	Fluid flow (core) length on one side of the exchanger, m or ft
L_h	Hydraulic entrance length, m or ft
L_t	Thermal entrance length, m or ft
L^+	Hydraulic entrance length, $L_h^+/D_h \text{Re}$, dimensionless
L_t^+	Thermal entrance length, $L_h^+/D_h \text{RePr}$, dimensionless
l	Strip flow length of OSF, m or ft
M	Mass of tube material, kg or lbm
N	Number of tube rows in the flow direction, or number of tubes in heat exchanger, dimensionless
Nu_{Dh}	Nusselt number $(= hD_h/k)$, dimensionless
Nu_d	Nusselt number $(= hd/k)$, inside tube $(d = d_i)$, outside tube $(d = d_o)$; dimensionless
n_f	Number of fins in internally finned tube, dimensionless
P	Fluid pumping power, W or HP
Pr	Prandtl number $(= c_p \mu/k)$, dimensionless
p	Aaxial spacing between roughness elements, m or ft
p_f	Fin pitch, center-to-center spacing; m or ft
Δp	Fluid static pressure drop, Pa or lbf/ft²
Q	Heat transfer rate in the exchanger W or Btu/h
Ra	Rayleigh number, GrPr; dimensionless
Re_{Dh}	Reynolds number based on the hydraulic diameter $(= GD_h/\mu)$, dimensionless
Re_d	Reynolds number based on the tube diameter $(= Gd/\mu, d = d_i$ for flow inside tube and $d = d_o$ for flow outside tube), dimensionless
R_{fi}	Tube-side fouling resistance, m²-K/W or ft²-°F-hr/Btu
St	Stanton number $(= h/Gc_p)$, dimensionless
s	Spacing between two fins $(= P_f - t)$, m or ft
ΔT_i	Temperature difference between hot and cold inlet fluids, K or °F
ΔT_{lm}	Logarithmic mean temperature difference, K or °F
t	Thickness of tube wall, fin or twisted tape, m or ft
U	Overall heat transfer coefficient, W/m²-K or Btu/hr-ft²-°F
u^*	Friction velocity $[= (\tau_w/\rho)^{1/2}]$, m/s or ft/s, dimensionless
u_m	Fluid mean axial velocity at the minimum free flow area, m/s or ft/s
V	Heat exchanger total volume, m³ or ft³
V_m	Heat exchanger tube material volume, m³ or ft³
W	Fluid mass flow rate, kg/s or lbm/s

X^+ $x/D_h\text{RePr}$, dimensionless
x Cartesian coordinate along the flow direction, m or ft

Greek Letters

α Helix angle relative to tube axis $= \pi d_i/p_{tv}$, radians or degrees
γ Reciprocal of fin pitch, m or ft
η_f Fin efficiency or temperature effectiveness of the fin, dimensionless
η_o Surface efficiency of finned surface $= 1 - (1 - \eta_f)A_f/A$, dimensionless
θ^* $\theta k_m/k$, dimensionless
2θ Included angle of fin cross section normal to flow, radians
μ Fluid dynamic viscosity coefficients, Pa-s or lbm/hr-ft
ν Kinematic viscosity, m²/s or ft²/s
ρ Fluid density kg/m3 or lbm/ft³
τ_w Wall shear stress, Pa or lbf/ft²

Subscripts

fd   Fully developed flow
H1   Heat flux boundary condition
m Average value over flow length
p Plain tube or surface
T Wall temperature boundary condition
w Evaluated at wall temperature
x Local value

9

INTEGRAL ROUGHNESS

9.1 INTRODUCTION

Considerable data exist for single-phase forced convection flow over rough surfaces. Data exist for six different flow geometries:

1. Flat plates
2. Circular tubes
3. Noncircular channels in gas turbine blades
4. Longitudinal flow in rod bundles
5. Annuli having roughness on the outer surface of the inner tube
6. Flow normal to circular tubes

Internally roughened tubes are becoming quite important in commercial applications. The refrigeration industry routinely uses roughness on the water side of evaporators and condensers in large refrigeration equipment, as described by Webb and Robertson [1988] and Webb [1991]. Water-side roughness also offers economic benefits for use in electric utility steam condensers, as described by Webb et al. [1984] and Jaber et al. [1991].

Figure 9.1 shows a modern gas turbine blade, containing roughened channels. Air from the compressor flows in the rough channels. The work of Han [1984, 1988] and Han et al. [1991] and Metzger et al. [1983, 1987] have significantly advanced this technology.

Many papers have been published on work related to the use of roughened fuel

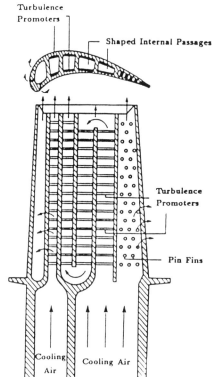

Figure 9.1 Cooling concepts of a modern multipass turbine blade. (From Han et al. [1988].)

rods in gas-cooled nuclear reactors (e.g., Dalle-Donne and Meyer [1977]).

There are many possible roughness geometries. Figure 9.2 is the author's attempt to catalog the possible geometries. This figure shows three basic roughness families. For any basic type, the key dimensionless variables are the dimensionless roughness height (e/d), the dimensionless roughness spacing (p/e), the dimensionless rib width (w/e), and the shape of the roughness element. The ridge-and-groove-type roughness may also be applied at a helix angle, as shown in Figure 9.2. For a specific roughness type, a family of geometrically similar roughnesses is possible simply by changing e/d while maintaining constant p/e and w/e. Thus, the designer is faced with choosing among virtually thousands of possible specific roughness geometries and sizes. Figure 9.3 shows two of the early commercially used roughness geometries. Their geometric parameters are e (rib height), p (rib spacing), α (helix angle), and the rib shape. The tube shown in Figure 9.3a has internal helical ribs and is normally made with low integral fins on the external surface. The internal ribs are made by cold deformation of the metal into a grooved internal mandrel to form the helical ridges. The helically corrugated tube shown in Figure 9.3b is made by rolling a sharp edged wheel on the outer surface of the tube.

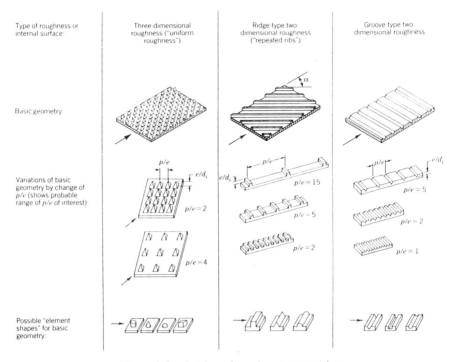

Figure 9.2   Catalog of roughness geometries.

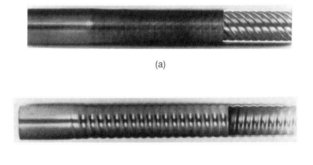

Figure 9.3   Illustration of commercially used enhanced tubes. (a) Helical-rib Turbo-Chil tube. (b) Corrugated Korodense tube. (Courtesy of Wolverine Tube Division, Decatur, AL.)

9.2   HEAT-MOMENTUM TRANSFER ANALOGY CORRELATION

Correlations are needed to predict the friction and heat transfer characteristics of rough surfaces. Rationally based correlations for friction and heat transfer have been developed for geometrically similar roughness. Consider a two-dimensional roughness, such as illustrated by Figure 9.3a. A geometrically similar family of roughness geometries will exist if p/e, w/e, α, and the rib shape are held constant. Such a family of roughened tubes would differ in their e/d_i values. The friction and heat transfer correlations described in this section are believed to apply to any arbitrary family of geometrically similar roughness.

9.2.1   Friction Similarity Law

A "friction similarity law" was developed by Nikuradse [1933] and described by Schlichting [1979]. Nikuradse showed the similarity law to be valid for closely packed sand-grain roughness. The velocity profile, valid for $y/e \geq 1$, is given by

$$\frac{u}{u^*} = 2.51 \ln \frac{y}{e} + B(e^+) \tag{9.1}$$

The velocity profile is based on the "Law of the Wall" velocity distribution for rough surfaces and is discussed by Hinze [1975] and Schlicting [1979]. The Law of the Wall for rough surfaces states that the velocity profile for $y/e \geq 1$ is independent of the duct shape and depends on the "roughness Reynolds number," e^+. The e^+ may be written in the following two equivalent forms:

$$e^+ = \frac{eu^*}{\nu} = \frac{e}{d_i} \, \mathrm{Re}_d \, \sqrt{\frac{f}{2}} \tag{9.2}$$

where u^* is the "friction velocity, $(\tau_o/\rho)^{1/2}$. $B(e^+)$ is the dimensionless velocity at the tip of the roughness elements ($y/e = 1$), and τ_o is the average apparent wall shear stress. For flow in a tube, τ_o is given by the force balance

$$\tau_o = -\frac{d_i}{4\rho} \frac{dp}{dx} \tag{9.3}$$

The friction factor is obtained by integrating the velocity profile over the flow area. For a circular pipe, the integration gives

$$\frac{\bar{u}}{u^*} = \left(\frac{2}{f}\right)^{1/2} = -2.5 \ln \left(\frac{2e}{d_i}\right) - 3.75 + B(e^+) \tag{9.4}$$

Nikuradse's [1933] data for six different e/d_i values (Figure 9.4a) were correlated using Equation 9.4 as shown in Figure 9.4b. Figure 9.4a shows that the friction factor approaches (and equals) the smooth tube value in the laminar regime. In the turbulent regime, the smaller e/d_i values remain on the smooth tube curve to a higher Reynolds number and then drift above the smooth tube line. The friction factors all attain an asymptotic, constant value, which is termed the "fully rough"

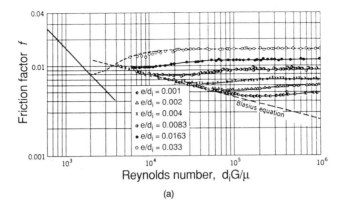

(a)

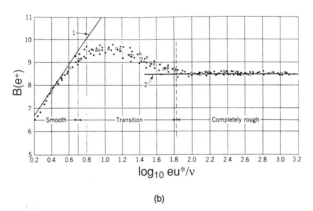

(b)

Figure 9.4 (a) Friction factors for artificially roughened tubes, as measured by J. Nikuradse. (b) Roughness parameter B for Nikuradse's sand roughness. Curve 1, hydraulically smooth; curve 2, completely rough. (From Schlichting [1960].)

condition. As shown by Figure 9.4b, $B(e^+) = 8.48$ for the fully rough condition, which is attained at $\log_{10}e^+ = 1.82$, or $e^+ = 70$.

Equations 9.1 and 9.4 should hold for any family of geometrically similar roughness. But, the function $B(e^+)$ will be different for different basic roughness types.

9.2.2 PEC Example 9.1

Water flows at 2.4 m/sec in a 15.9-mm-diameter tube at 27°C having closely packed sand-grain roughness. Determine the roughness size (e) required to attain the fully rough condition, and then calculate f/f_s.

Figure 9.4b shows that the fully rough condition is attained at $\log_{10}e^+ = 1.8$, or $e^+ = 70$. At the specified flow condition ($\nu = 0.857E - 6$ m²/sec), we obtain Re_d

$= 44{,}530$. Use Equation 9.2 with $(f/2)^{1/2}$ from Equation 9.4 to solve for e/d_i. Thus, one iteratively solves for $e/d_i = 0.020$ using

$$70 = \frac{e}{d_i} \, \mathrm{Re}_d \, \sqrt{\frac{f}{2}} = \frac{(e/d_i) \, 44{,}530}{8.48 - 2.5 \, \ln \, (2e/d_i) - 3.75} \tag{9.5}$$

With e/d_i known, $e = 0.020 \times 15.9 = 0.31$ mm. Use Equation 9.2 to calculate the friction factor, $f = 0.01227$. Using the Blasius friction factor for a smooth tube, obtain $f_s = 0.00544$. Hence, $f/f_s = 2.25$.

9.2.3   Heat Transfer Similarity Law

Dipprey and Sabersky [1963] developed a rational correlation based on the heat-momentum transfer analogy for rough surfaces. The model is applicable to any type of geometrically similar surface roughness. The correlating function is obtained by combining the momentum and energy equations for turbulent flow and integrating across the boundary layer thickness. We will briefly describe the basis of the model applied to boundary layer flow over a plate for both smooth and rough surfaces. Including the analogy for smooth surfaces allows a better understanding of the concepts applied for rough surfaces.

9.2.3.1   Smooth Surfaces   The analogy equation for smooth surfaces is developed in detail by Webb [1971]. Written in dimensionless form, these equations for flow over smooth or rough surfaces are

$$\frac{\tau}{\tau_o} = \frac{\nu_e}{\nu} \frac{du^+}{dy^+} \tag{9.6}$$

$$\frac{q}{q_o} = \frac{\nu_e}{\nu} \frac{1}{\mathrm{Pr}_e} \frac{dT^+}{dy^+} \tag{9.7}$$

Eliminating the effective viscosity (ν_e) from Equations 9.6 and 9.7 to combine the equations, along with integrating over the boundary layer thickness (δ^+), gives

$$T_w^+ - T_\infty^+ = \int_0^{\delta^+} \mathrm{Pr}_e \frac{du^+}{dy^+} \tag{9.8}$$

The assumptions listed in Table 9.1 are made for the integration of Equation 9.7. For convenience, the following two identities are used

$$T_w^+ - T_\infty = \frac{(f/2)^{1/2}}{\mathrm{St}}, \qquad u_\delta^+ = \frac{u_\delta}{u^*} = \left(\frac{2}{f} \right)^{1/2} \tag{9.9}$$

Separate integrals are written for the viscous and turbulence dominated regions, and the term du^+/dy^+ integrated over $0 \le y^+ \le y_b^+$ is added and subtracted to make

TABLE 9.1   Assumptions for Integration of Equation 9.7

Assumption	Smooth	Rough
$\epsilon_m/\nu \gg 1,\ \epsilon_h/\alpha \gg 1$	$y > y_b$	Same
$\mathrm{Pr}_e = 1$	$y > y_b$	Same
$(q/q_w)(\tau_o/\tau) = 1$	$y > 0$	$y < e$

use of the second identity of Equation 9.9. The integration for a smooth surface gives

$$\frac{(f_s/2)}{\mathrm{St}_s} = \int_0^{y_b^+} (\mathrm{Pr}_e - 1) \frac{du^+}{dy^+}\, dy^+ + \int_0^{\delta^+} \frac{du^+}{dy^+}\, dy^+ \tag{9.10}$$

Using the second identity of Equation 9.9 for the second integral allows writing Equation 9.10 as

$$\frac{f_s/(2\mathrm{St}_s) - 1}{(f_s/2)^{1/2}} = \int_0^{y_b^+} (\mathrm{Pr}_e - 1) \frac{du^+}{dy^+}\, dy^+ \tag{9.11}$$

The integral of Equation 9.11 is a function only of the molecular Prandtl number, which may be written as $F(\mathrm{Pr})$. $F(\mathrm{Pr})$ is obtained by plotting data in the form of the left-hand side of Equation 9.11 versus the Prandtl number. Petukhov [1970] has shown that the Prandtl number function is given by

$$F(\mathrm{Pr}) = 12.7\ (\mathrm{Pr}^{2/3} - 1) \tag{9.12}$$

Combining Equations 9.11 and 9.12 gives the heat–momentum analogy relation for smooth surfaces.

$$\frac{f_s/(2\mathrm{St}_s) - 1}{(f_s/2)^{1/2}} = 12.7(\mathrm{Pr}^{2/3} - 1) \tag{9.13}$$

Solving Equation 9.13 for the smooth surface Stanton number gives

$$\mathrm{St}_s = \frac{f_s/2}{1.0 + 12.7(f_s/2)^{1/2}(\mathrm{Pr}^{2/3} - 1)} \tag{9.14}$$

For flow in tubes, the first term in the denominator (1.0) should be replaced by 1.07. Webb [1971] compares the ability of Equation 9.14 to predict heat transfer data in smooth tubes. He shows that Equation 9.14 (using the 1.07 term) predicts smooth tube, constant property data of four investigators with $0.7 \leq \mathrm{Pr} \leq 75$ within

$\pm 8\%$. Petukhov [1970] recommends that the smooth tube friction factor be predicted by

$$f_s = (1.58 \ln \mathrm{Re}_d - 3.28)^{-2} \qquad (9.15)$$

9.2.3.2 Rough Surfaces Development of the Dipprey and Sabersky [1963] analogy model for rough surfaces parallels that of the smooth surface model, and it uses the approximations listed in Table 9.1. For $y_b^+ \le e^+$, the integration of Equation 9.10 results in

$$\frac{f/(2\mathrm{St}) - 1}{(f/2)^{1/2}} + B(e^+) = \int_0^{e^+} (\mathrm{Pr}_e - 1) \frac{du^+}{dy^+} \, dy^+ \qquad (9.16)$$

As compared to the smooth surface analogy (Equation 9.11), the integral is a function of both e^+ and Pr. The functional dependence is written as $g(e^+, \mathrm{Pr})$, and it is assumed that this function can be written as the product of $\bar{g}(e^+)\mathrm{Pr}^n$. Hence the heat transfer correlation is given by

$$\bar{g}(e^+)\mathrm{Pr}^n = \frac{f/(2\mathrm{St}) - 1}{\sqrt{f/2}} + B(e^+) \qquad (9.17)$$

Figure 9.5 shows Dipprey and Sabersky's [1963] correlation of sand-grain data ($0.0024 \le e/d_i \le 0.049$ and $1.2 \le \mathrm{Pr} \le 5.94$). The $B(e^+)$ term is obtained from Figure 9.4b. The Prandtl number dependence, $\mathrm{Pr}^{-0.44}$, is obtained by cross-plotting the $g(e^+, \mathrm{Pr})$ versus e^+ data for each of the three Prandtl numbers. Solving Equation 9.17 for St gives

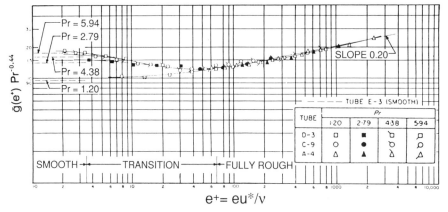

Figure 9.5 Heat transfer correlation of Dipprey and Sabersky [1963] sand-grain roughness data. (From Dipprey and Sabersky [1963].)

$$\text{St} = \frac{f/2}{1 + \sqrt{f/2}\,[\bar{g}(e^+)\text{Pr}^n - B(e^+)]} \tag{9.18}$$

where $\text{Pr}^n = \text{Pr}^{0.44}$. One calculates the friction factor and Stanton number as follows:

1. Select e^+ and obtain $B(e^+)$ from Figure 9.4b and g(e^+) from Figure 9.5.
2. Calculate the friction factor using Equation 9.4.
3. Calculate the Stanton number using Equation 9.18.

9.3   TWO-DIMENSIONAL ROUGHNESS

Substantial work has been done on two-dimensional rib roughness. Figure 9.6 shows five variants of two-dimensional roughness. Figures 9.6a and 9.6b show "transverse ribs," which are normal to the flow. Figures 9.6c, 9.6d, and 9.6f show "helical-rib" roughness, which have helix angles less than 90 degrees. The roughness elements may be integral to the base surface (Figure 9.6c and 9.6d), or they may be in the form of wire coil inserts (Figure 9.6e). A wire coil insert is an example of attached helical-rib roughness. The integral helical-rib roughness may be made as single- or multi-start elements, whereas a wire coil insert is a single-start roughness. The element axial pitch (p) may be made with a smaller dimensionless axial spacing (p/e) than is possible with wire coil inserts. The relation between the

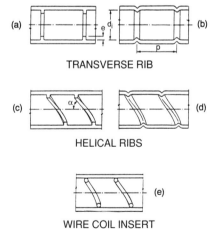

TRANSVERSE RIB

HELICAL RIBS

WIRE COIL INSERT

RIB PROFILE SHAPES

Figure 9.6 Illustrations of different methods of making two-dimensional roughness in a tube. (a) Integral transverse rib. (b) Corrugated transverse rib. (c) Integral helical rib. (d) Helically corrugated. (e) Wire coil insert. (f) Different possible profile shapes of roughness. (From Ravigururajan and Bergles [1985].)

axial pitch of the roughness elements (p), the number of starts (n_s) and the helix angle (α) measured from the tube axis is

$$p = \frac{\pi d_i}{n_s \tan \alpha} \qquad (9.19)$$

The shape of the roughness element in a corrugated tube is also different from that of a helical integral roughness or a wire coil insert. The geometric factors that influence the j and f factors are e/d_i, p/e, α, or (e/d_i, n_s, α) and the shape of the roughness element. We will discuss each of the variants of repeated-rib roughness in separate sections.

9.3.1 Transverse-Rib Roughness

Figure 9.7 shows the flow patterns cataloged by Webb et al. [1971] for transverse-rib roughness ($\alpha = 90$ degrees) as a function of the dimensionless rib spacing (p/e). The flow separates at the rib and reattaches six-to-eight rib heights downstream from the rib. Measurements of Edwards and Sheriff [1961] for a two-dimensional rib on a flat plate show that the heat transfer coefficient attains its maximum value near the reattachment point. The Edwards and Sheriff [1961] data are shown in Figure 7.14. Figure 9.7 shows that reattachment does not occur for $p/e < 8$. The highest average coefficient occurs for $10 \leq p/e \leq 15$.

The heat transfer correlation of Section 9.2.3.2 should apply to any family of geometrically similar roughness. Equations 9.4 and 9.17 may be used to correlate the data for any family of geometrically similar roughnesses. However, the $B(e^+)$ and $g(e^+)$ functions may be different for different roughness families. For any geometrically similar roughness type, the data for different e/d_i will fall on the same curve when plotted in the form $\overline{g}(e^+)Pr^n$ versus e^+.

Webb et al. [1971] applied the model to geometrically similar, $p/e = 10$, transverse-rib roughness ($0.01 \leq e/d_i \leq 0.04$ and $0.7 \leq Pr \leq 38$). Figure 9.8 shows the $\overline{g}(e^+)$ function determined by Webb et al. [1971], and Figure 9.9 shows the $B(e^+)$ correlation. The figure also contains data for $e/d_i = 0.02$ with $p/e = 20$ and 40, which are not geometrically similar to the $p/e = 10$ data. Figure 9.8 shows that Equation 9.17 does an excellent job of correlating the three e/d_i values (0.01, 0.02, and 0.04) for $p/e = 10$. The figure also shows that the nonsimilar $p/e = 20$ and 40 data also fall on the same correlating line as the $p/e = 10$ data. There is no rational reason to expect this, so it is regarded as a fortuitous occurrence. The data are correlated by a Prandtl number exponent of n = 0.57 (cf. Equation 9.18) as compared to $n = 0.44$ found by Dipprey and Sabersky [1963] for sand-grain roughness. Further commentary on the Prandtl number dependency will be given later. The $B(e^+)$ function for the transverse-rib roughness of Webb et al. [1971] is shown in Figure 9.9.

Webb et al. [1972] investigated the effect of the p/e and rib cross-sectional shape on the correlation of transverse-rib data taken by other investigators. They show that

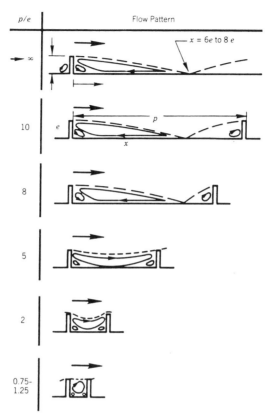

Figure 9.7 Catalog of flow patterns over transverse-rib roughness as a function of rib spacing. (From Webb et al. [1971].)

the $g(e^+)$ function for different p/e and rib shape reasonably agrees with that of Figure 9.8. However, the $B(e^+)$ function is sensitive to p/e and rib shape.

Almeida and Souza-Mendes [1992] used the naphthalene sublimation technique to measure (a) the length of the thermal entrance region and (b) the local distribution of the mass transfer coefficient between ribs. They showed that the mass transfer coefficients are fully developed in three tube diameters for $e/d_i = 0.01$.

Hijikata et al. [1987] investigated the effect of the rib shape on performance. Their work was done for transverse corrugations in the wall of a parallel plate channel having $e/D_h = 0.01$ and $p/e = 15$. They investigated arc, sine, and square-rib shapes. They concluded that the arc shape provides the same heat transfer coefficient as the other shapes, but has a lower pressure drop. The base width of the arc-shaped element was four times its height.

Tanasawa et al. [1983] tested the novel forms of transverse-rib roughness illustrated in Figures 9.10b and 9.10c. The tests were performed with air flow in a

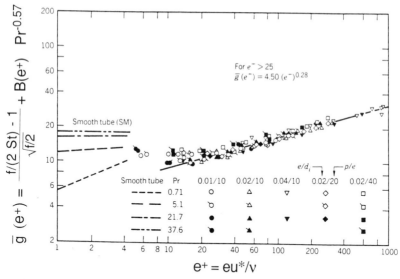

Figure 9.8 Heat transfer correlation for transverse-rib roughness (α = 90 degrees). (From Webb et al. [1971].)

rectangular cross-section channel. The best performance was provided by the Figure 9.10c geometry, which had (1) 3.0-mm-diameter perforations in the lower half of the 9-mm-high rib ($e/D_h = 0.045$) and (2) 10% open area in the rib. This geometry provided only 3% higher St than did the Figure 9.10a rib. However, the friction factor was only 80% that of the unperforated rib of the same height. Nearly the same performance was obtained with 1.0-mm-diameter perforations and 10% open area. The friction is smaller because of reduced profile drag.

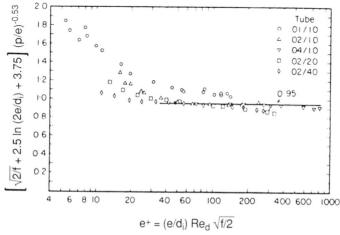

Figure 9.9 Final friction correlation for repeated-rib tubes. (From Webb et al. [1971].)

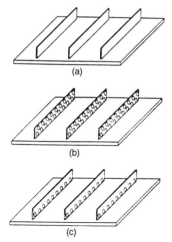

Figure 9.10 Transverse-rib promoters investigated by Tanasawa et al. [1983]. (From Tanasawa et al. [1983].)

9.3.2 Integral Helical-Rib Roughness

How does the helix angle (α) affect the St and f characteristics of repeated rib roughness? Figure 9.11 shows the St and f data of Gee and Webb [1980] for $p/e = 15$ and $e/d_i = 0.01$ as a function of helix angle. As α increases, the friction factor drops faster than does the Stanton number. Gee and Webb found that the maximum St/f occurs at $\alpha = 45$ degrees. They give correlations for $g(e^+)$ and $B(e^+)$ versus e^+. Nakayama et al. [1983] provide additional data for helical ribs, including $0 \leq \alpha$

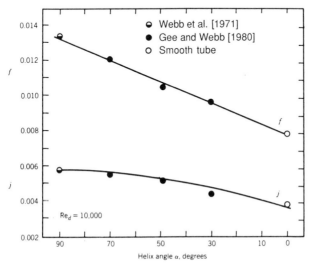

Figure 9.11 Effect of helix angle on f and St for $e/d_i = 0.01$, $p/e = 15$. (From Gee and Webb [1980].)

≤ 80 degrees. Han et al. [1978] provide transverse-rib data, along with $g(e^+)$ and $B(e^+)$ correlations for 17 transverse-rib geometries ($5 \leq p/e \leq 20$) and two helical-rib ($\alpha = 40$ and 55 degrees) geometries. An error in their 1978 data reduction is corrected by Han et al. [1979]. The Figure 9.3a helical-rib tube is commercially available and is known as the Turbo-Chil™ tube. The Turbo-Chil™ tube has 10 rib starts, $\alpha = 47$ degrees, $p/e = 11.1$, and $e/d_i = 0.0264$. The $\bar{g}(e^+)$ and $B(e^+)$ functions for this tube are given by Equations 9.20 and 9.21 (which were developed by the author), respectively, and are valid for $e^+ > 25$.

$$\bar{g}(e^+) = 7.68(e^+)^{0.136} \tag{9.20}$$

$$B(e^+) = 1.7 + 2.06 \ln e^+ \tag{9.21}$$

Equations 9.20 and 9.21 are used in Equation 9.18 (with $n = 0.57$) to predict the Stanton number of the Turbo-Chil™ tube. The friction factor is calculated using Equation 9.21 in Equation 9.4. This correlation may be used to calculate St and f for any value of e/d_i, provided that geometric similarity is maintained ($p/e = 11.1$, $\alpha = 47$ degrees, and the same rib shape). Withers [1980b] provides additional information on helical-rib roughness for different values of p/e and α.

9.3.3   Wire Coil Inserts

Although wire coil inserts were introduced in Section 7.5, we have elected to discuss their turbulent flow performance here, because they may also be classified as a "wall-attached roughness." Their enhancement mechanism is the same as that of helical-rib roughness. A key difference between integral helical-rib roughness and wire coil inserts is that the wire coil insert is a "single-start" roughness and typically has a larger wire pitch (and p/e) than the "multi-start" helical-rib roughness.

The correlation of Sethumadhavan and Raja Rao [1983] is recommended for turbulent flow of liquids, because the data included a wide Prandtl number range ($5.2 \leq \text{Pr} \leq 32$) and also because they provide both heat transfer and friction correlations. The heat transfer correlation is probably valid for air (Pr = 0.7). The researchers also measured St and f for a smooth tube, and found good agreement with the accepted correlations. They tested inserts having helix angles of 30, 45, 60, and 75 degrees for $e/d_i = 0.08$ and 0.12 for $4000 \leq \text{Re}_d \leq 100{,}000$. For the same Pr, they found that the St for $e/d_i = 0.08$ was the same as for the $e/d_i = 0.12$ inserts. However, the larger wire size gives significantly higher friction factor for the same Re_d. The friction correlation, based on the friction similarity law (Equation 9.4), is

$$B(e^+) = 7.0(\tan \alpha)^{-0.18}(e^+)^{0.13} \tag{9.22}$$

The heat transfer data were correlated using the heat–momentum transfer analogy model (Equation 9.17). The correlated results are given by

$$g(e^+) = 8.6(\tan \alpha)^{-0.18}\text{Pr}^{-0.55}(e^+)^{0.13} \tag{9.23}$$

Equations 9.22 and 9.23 correlated data very well. They showed that Equation 9.23 closely predicts the water data of Kumar and Judd [1970]. Although the correlation does not explicitly include e/d_i, it should be applicable to smaller e/d_i values. In this author's opinion, the heat transfer correlation should also be valid for air (Pr $\simeq$ 0.7). Note that the Prandtl number exponent (0.55) agrees very closely with the 0.57 value determined by Webb et al. [1971] for transverse-rib roughness.

Sethumadhavan and Raja Rao [1983] provide a performance evaluation of the wire coil inserts using PEC FN-1 of Chapter 3. They evaluated the effect of helix angle for α = 30, 45, 60 and 75 degrees with e/d_i = 0.08. The highest (and equal) performance is provided by the α = 60 and 75 degree helix angles.

Two other correlations have also been developed for turbulent flow of water. The two correlations span significantly different ranges of p/d_i. Kumar and Judd [1970] generated electric heat in the wall of a stainless steel tube. The wire used for the inserts was coated with a plastic electrical insulator. The empirical heat transfer correlation was based on their test data for: $0.108 < e/d_i < 0.15$; $1.12 < p/d_i < 5.5$, and $6000 \leq \mathrm{Re}_d \leq 100{,}000$. The relationship

$$\mathrm{Nu}_d = 0.175 \left(\frac{p}{d_i} \right)^{-0.35} \mathrm{Re}_d^{0.7} \mathrm{Pr}^{1/3} \qquad (9.24)$$

correlated their data with 7.5% root mean square (rms) deviation. The correlation does not contain the parameter e/d_i. Apparently, they operated in the "fully rough" regime where the heat transfer coefficient is not a function of e/d_i. A dependence on e/d_i should become evident at some unknown lower e/d_i. Hence, care should be used in applying the correlation at e/d_i values less than the smallest they tested (0.108). A friction correlation was not developed.

Prasad and Brown [1988] subtracted the annulus side resistance to obtain a correlation for their water data: $0.058 \leq e/d_i \leq 0.11$; $0.2 \leq p/d_i \leq 0.6$; and 40,000 $\leq \mathrm{Re}_d \leq 97{,}000$. The p/d_i of the wire coils used was so small that a thin, axial wire was soldered to the coil to maintain the wire pitch. Refer to their paper for the heat transfer and friction correlations.

Zhang et al. [1991] measured the wall temperature in an air-to-air heat exchanger for their tests of coil inserts using round and rectangular wire shapes. Their circular wire data are for: $0.037 \leq e/d_i \leq 0.09$; $0.35 \leq p/d_i \leq 2.48$; and $6000 \leq \mathrm{Re}_d \leq 80{,}000$. The empirical correlations for the circular wire data are

$$\mathrm{Nu}_d = 0.253\mathrm{Re}_d^{0.716}\left(\frac{e}{d_i} \right)^{0.372} \left(\frac{p}{d_i} \right)^{0.171} \qquad (9.25)$$

$$f = 62.36 \ln \mathrm{Re}_d^{-2.78}\left(\frac{e}{d_i} \right)^{0.816} \left(\frac{p}{d_i} \right)^{-0.689} \qquad \text{where } 6 \leq \mathrm{Re}_d \times 10^3 < 15$$

$$(9.26)$$

$$f = 5.153(\ln \mathrm{Re}_d)^{-1.08} \left(\frac{e}{d_i} \right)^{0.796} \left(\frac{p}{d_i} \right)^{-0.707} \qquad \text{where } 15 \leq \mathrm{Re}_d \times 10^3 < 100$$

$$(9.27)$$

They also give correlations for the rectangular wire coil inserts, whose performance is very close to that of the circular wire inserts.

9.3.4 Corrugated Tube Roughness

Substantial data have also been reported on single- and multi-start corrugated tubes of the type illustrated in Figure 9.3b. The reported data are summarized in Table 9.2. Additional data by Newson and Hodgson [1973] are discussed in Section 12.3.5. The Newson and Hodgson work involves steam condensation on vertical, corrugated tubes with tube-side cooling water flow.

All authors except Sethumadhavan and Raja Rao [1986] and Li et al. [1982] used only single-start roughness. The Sethumadhavan and Raja Rao [1986] tubes were either 1, 2, 3, or 4 starts with α = 65 degrees. Four of the Li et al. geometries have 2, 3, or 4 starts. Li et al. provide data for helix angles as small as 45 degrees using multi-start roughness. Only Sethumadhavan and Raja Rao [1986] significantly altered the Prandtl number from that of water for Newtonian fluids ($5.2 \leq \mathrm{Pr} \leq 32$). Raja Rao [1988] used water and power law non- Newtonian fluids to obtain $5.1 \leq \mathrm{Pr} \leq 82$, whereas Mehta and Raja Rao [1988] used only water.

The dimensionless geometric factors that affect the heat transfer coefficient and friction factor are p/e, e/d_i, and α. For a given pitch, $\tan \alpha \propto 1/n_s$, where n_s is the number of starts. All authors used the friction similarity law (Equation 9.4) and the analogy model (Equation 9.17) to correlate their data. Table 9.3 gives the correlations for the $\bar{g}(e^+)$ and $B(e^+)$ of several investigators. The variable e/d_i does not appear in the correlations, because the model inherently includes the effect of e/d_i.

The term n' in the Table 9.3 Raja Rao [1988] correlation is a "flow behavior power law index," which is 1.0 for water. The Prandtl number dependence of Raja Rao et al. ($n = 0.55$) agrees closely with that found by Webb et al. [1971] for transverse-rib roughness. Comparison of the Withers [1980a] and Mehta and Raja Rao [1988] correlations shows that the two $\bar{g}$ correlations have significantly different exponents for the effect of the helix angle (α). The heat transfer coefficient increases with increasing helix angle. Note that the Sethumadhavan and Raja Rao [1986] correlation does not contain the (tan α) term, since the data were for α = 65°. Table 9.4 compares the $\bar{g}$ function predicted by the four correlations.

TABLE 9.2 Data Range for Corrugated Tubes

Author	Fluid	No.	e/d_i	p/e	α (degrees)
Withers [1980a]	Water	14	0.016–0.043	11–24	79–85
Mehta and Raja Rao [1988]	Water	11	0.008–0.097	4.5–49	69–84
Sethumad havan and Raja Rao [1986]	Water Glycerine	5	0.012–0.030	13–45	65
Raja Rao [1988]	Water Power law	12	0.020–0.060	4.6–30	65–82
Li et al. [1982]	Water	20	0.008–0.069	7.7–29	41–85

TABLE 9.3 Correlations for $\bar{g}(e^+)$ and $B(e^+)$

Author	$\bar{g}(e^+)$	$B(e^+)$
Withers [1980a]	$4.95(\tan\alpha)^{0.33}(e^+)^{0.127}Pr^{0.5}$	Does not use $B(e^+)$
Mehta and Raja Rao [1988]	$7.92(\tan\alpha)^{0.15}(e^+)^{0.11}Pr^{0.55}$	$0.465(p/e)^{0.53}(\ln e^+ + 0.25)$
Raja Rao [1988]	$6.06(\tan\alpha)^{0.15}(e^+)^{0.13}Pr^{0.55}$	$0.465(p/e)^{0.53}(\ln e^+ + 0.25)(n')^{2.5}$
Sethumadhavan and Raja Rao [1986]	$8.6(e^+)^{0.13}Pr^{0.55}$	$0.40(e^2/pd_i)^{1/3}(e^+)^{0.64}$

The Withers [1980a] and Mehta and Raja Rao [1988] correlations give approximately equal values for $\alpha = 85°$. For $\alpha = 70°$, the larger exponent on the $(\tan\alpha)$ term results in a smaller $\bar{g}$ value than given by Mehta and Raja Rao [1988]. The Raja Rao [1988] $\bar{g}$ correlation is approximately 16% below that of Mehta and Raja Rao [1988]. All of the Table 9.3 correlations contain statistical correlating errors. Withers did not choose to use the $B(e^+)$ correlation for the friction factor, because he felt the error was too large. He based his friction correlation on an equation previously developed by Churchill [1973] to correlate the friction factor in commercially rough tubes. The Churchill equation for commercial pipe roughness is

$$\sqrt{\frac{2}{f}} = -2.46 \ln \left[\frac{e}{d_i} + \left(\frac{7}{Re_d} \right)^{0.9} \right] \qquad (9.28)$$

Withers replaced the e/d_i and 0.9 terms with empirical constants. Thus, he fitted his friction data for each tube to determine the curve fit constants r and m. The resulting equation predicted the data for each tube with a standard deviation of 3%. The resulting equation is

$$\sqrt{\frac{2}{f}} = -2.46 \ln \left[r + \left(\frac{7}{Re_d} \right)^{m} \right] \qquad (9.29)$$

TABLE 9.4 Predicted g Function for $\alpha = 70°$ and $90°$

Correlation	$\bar{g}(e^+)(\tan\alpha)^n$ for $\alpha = 70°$	$\alpha = 85°$	Comments
Withers [1980a]	10.38	16.64	Water
Mehta and Raja Rao [1988]	13.11	16.25	Water
Raja Rao [1988]	10.72	13.28	Water and power law
Sethumadhavan and Raja Rao [1986]	13.07	NA[a]	Water and water/glycol

[a]NA, not available.

Table 9.5 lists the geometries tested by Withers [1980a], in order of increasing e/d_i, and gives the parameters m and r needed to calculate the friction factor using Equation 9.29. Wolverine provides commercial versions of helically corrugated tubes. These are called the Wolverine Korodense™ LPD and MHT tubes described in the Wolverine data book, (Wolverine [1984]), which are also listed in Table 9.5. The heat transfer coefficient of the LPD tube is approximately 25% less than that of the MHT geometry. The friction factor of the LPD version is about half that of the MHT tube.

Ravigururajan and Bergles [1985] empirically correlated the r and m terms as functions of p/e, p/d_i ($= \pi/\tan \alpha$), and α (degrees). It was not necessary to include both p/d_i and α, since $p/d_i = \pi/\tan \alpha$. The correlations are

$$r = 0.17 \left(\frac{e}{d_i} \right)^{-1/3} \left(\frac{p}{d_i} \right)^{0.03} \left(\frac{\alpha}{90} \right)^{-0.29}$$

$$m = 0.0086 + 0.033 \left(\frac{e}{d_i} \right) + 0.005 \left(\frac{p}{d_i} \right) + 0.0085 \left(\frac{\alpha}{90} \right)$$

(9.30)

The previously discussed correlations were developed by each investigator using only their own test data. Rabas et al. [1988] sought to develop an empirical correlation for single-start corrugated roughness using data from eight sources. The data bank included 60 roughness geometries which spanned $17.3 \leq d_i \leq 28.4$ mm, $0.13 \leq e \leq 1.42$ mm, $5 \leq p \leq 30$. The reader is referred to the published paper for this rather complex empirical correlation. The rms error was 10.9% and 9.5% for the Nusselt number and friction factor, respectively. They also compare the ability of

TABLE 9.5 Tubes Tested by Withers [1980a]

Tube No.	d_i (inches)	e/d_i	p/e	α (degrees)	m	r
2300	0.892	0.016	20.10	83.03	0.71	0.0001
2100	0.869	0.016	18.76	84.58	0.68	0.00094
2200	0.862	0.020	14.45	84.73	0.564	0.0
2	0.805	0.024	19.46	81.70	0.649	0.0
1	0.805	0.024	19.46	81.67	0.622	0.0
LPD	All	0.025	20.00	81.00	0.61	0.00088
9	1.175	0.030	11.31	83.88	0.50	0.0035
1100	0.547	0.031	14.70	81.72	0.65	0.0023
7	0.922	0.033	12.14	82.73	0.47	0.0015
20	0.855	0.036	12.42	81.85	0.445	0.0
6	0.921	0.038	11.00	81.53	0.45	0.0073
15	1.152	0.040	10.98	82.06	0.48	0.00995
MHT	All	0.040	12.	81.	0.44	0.00595
14	0.866	0.043	14.20	78.98	0.52	0.0006
33	0.863	0.047	12.28	79.52	0.47	0.0081
5	0.682	0.052	17.45	73.83	0.51	0.0027

several other correlations discussed here to predict their data set. It should be noted that they adjusted the original heat transfer data of Mehta and Raja Rao [1979]. This is discussed in their paper.

Tan and Xaio [1989] measured the Prandtl number dependence of a single-start corrugated roughness ($e/d_i = 0.053$, $p/e = 10.8$) for five Prandtl numbers between 0.70 and 20.2. They found $\bar{g}(e^+, \text{Pr}) \propto \text{Pr}^{0.55}$, which agrees well with Webb et al. [1971].

9.3.5 PEC Example 9.2

Water flows at 2.4 m/sec in a 15.9-mm-diameter tube having $p/e = 10$ transverse-rib roughness at 27°C ($\nu = 0.857\text{E}-6$ m²/sec). Figure 9.9 shows that the fully rough condition is attained at $e^+ = 30$, for which $B = 3.5$.

a. Calculate the roughness size and spacing (e and p) for the incipient fully rough condition. Substituting Equation 9.4 in Equation 9.2 and solving for e/d_i gives

$$\frac{e}{d_i} = \frac{e^+}{\text{Re}_d}\left(\frac{2}{f}\right)^{1/2} = \frac{e^+}{\text{Re}_d}[2.5 \ln (2e/d_i) - 3.75 + B(e^+)] \quad (9.31)$$

Substituting $e^+ = 30$ and $B(e^+) = 35$ iteratively, solve for $e/d_i = 0.007$. Then, $e = 0.007 \times 15.9 = 0.111$ mm and $p = 10\ e = 1.11$ mm.

b. Calculate f/f_s and compare the e and f/f_s with that calculated for PEC Example 9.10. Equation 9.2 gives $f = 0.0183$. Using $f_s = 0.00542$, obtain $f/f_s = 3.37$.

c. What is the "equivalent sand-grain" roughness size? The equivalent sand-grain roughness may be obtained from

$$\left(\frac{2}{f}\right)^{1/2} = B_{\text{tr}}(e^+) - 2.5 \ln \left(\frac{2e}{d_i}\right) - 3.75$$

$$= B_{\text{sg}}(e^+) - 2.5 \ln \left(\frac{2e}{d_i}\right) - 3.75 \quad (9.32)$$

where $B_{\text{tr}}(e^+) = 3.5$ and $B_{\text{sg}}(e^+) = 8.48$ (Figure 9.4b). Solve Equation 9.32 to obtain $e_{\text{sg}}/e = 7.38$. This shows that sand-grain roughness size must be 7.38 times the two-dimensional rib height to obtain the same friction factor!

9.4 THREE-DIMENSIONAL ROUGHNESS

Cope [1945] was the first to test three-dimensional roughness. He used a knurling process to roughen the inner tube surface and provided water-flow data on three roughness geometries. Internal knurling is a relatively expensive process. The sand-grain roughness studied by Nikuradse [1933] and Dipprey and Sabersky [1963] did not involve a practical method of making a commercial three-dimensional roughness. Gowen and Smith [1968] embossed a three-dimensional roughness pattern on a thin sheet, which was rolled into tubular form, and then the axial seam was soldered. They tested two roughness geometries and provide data for $0.7 \leq Pr \leq 15.5$. Again, the Gowen and Smith fabrication method is not commercially viable.

Significant progress has been made in manufacturing practical three-dimensional roughness configurations, which are shown in Figure 9.12. The Figure 9.12a Tred-19D™ tube has 19 fins/in. on the outer surface and dimples on the inner surface.

The Figure 9.12b steel tube and the Figure 9.12c copper tubes are "cross-rifled" using internal mandrels. Takahashi et al. [1988] provide a comprehensive experimental study using 11 variants of the Figure 9.12c geometry with water flow. The best performing geometries (A6 and A8) are listed in Table 9.6, which compares the 3-D roughness with a corresponding 2-D helical-rib roughness. The table gives the roughness dimensions and compares the tube performance using PEC VG-1 ($Q/Q_s = P/P_s = W/W_s = 1$). The PEC analysis is for a refrigerant evaporator having shell-side boiling ($h_o = 12,800$ W/m²-K) and 10°C water flowing in the tubes. The smooth-inner-diameter tube operates at $Re_s = 30,000$. A tube-side fouling resistance of $R_{fi} = 8.6E-5$ m²-K/W is assumed. The thermal resistance of the smooth-inner-diameter tube is 1.9 times the shell-side resistance. Tube B3 is a high-performance 2-D helical rib roughness. The 3-D roughness tube A6 has the same

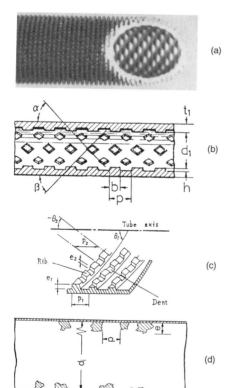

(a)

(b)

(c)

(d)

Figure 9.12 Three-dimensional roughness geometries. (a) Sumitomo Tred-19D™ tube of Sumitomo [1983]. (Courtesy of Sumitomo Light Metal Industries.) (b) "Cross-rifled" steel boiler tube of Nakamura and Tanaka [1973]. (c) Three-dimensional ribs of Cuangya et al. [1991]. (d) Attached particle roughness of Fenner and Ragi [1979].

TABLE 9.6 Case VG-1 PEC for Takahashi et al. [1988] Tests (Copper Tubes, $d_i = 13.9$ mm)

Geometry	Type	e/d_i	α (degrees)	p/e	A/A_s	G/G_s
B3	2-D	0.021	80	13.1	0.79	0.71
A6	3-D	0.022	30	12.6	0.78	0.80
A8	3-D	0.036	30	7.60	0.73	0.72

e/d_i and p/e as tube B3. Tubes B3 and A6 have approximately equal heat transfer performance, since they provide the same A/A_s. However, tube A6 has smaller friction, which allows operation at higher G/G_s. The larger e/d_i of tube A8 provides higher heat transfer performance than tube A6 (lower A/A_s) and operates at approximately equal G/G_s as tube B3. Thus, at the same Reynolds number, the 3-D Tube A8 has slightly lower friction than the 2-D B3 tube, but tube A8 provides significantly higher heat transfer coefficient.

Cuangya et al. [1991] provide data for low Reynolds number air flow ($4000 \leq Re_d \leq 9000$ for three of the Figure 9.12c geometries. At $Re_d = 5000$, they obtained a 2.8 enhancement factor over a plain tube.

The Figure 9.12d tube is formed by sintering a single layer of metal particles to the interior tube surface. The preferred geometry has 50% area coverage. The Fenner and Ragi [1979] patent presents heat transfer and friction data for water (Pr = 10) in a 17.2-mm inner-diameter tube for different particle diameters (e). The best performance was provided by $e/d_i = 0.0092$, which provides 105% heat transfer enhancement and 96% friction factor increase. McLain [1975] describes a method to emboss a three-dimensional roughness pattern on a flat strip, which is then rolled into a circular strip and axially welded. Modern high-speed tube welding techniques make this a very practical concept for commercial production. Some of the micro-fin tubes discussed in Chapter 13 are made by this method.

9.5 PRACTICAL ROUGHNESS APPLICATIONS

9.5.1 Tubes with Inside Roughness

Considerable success has been achieved in the commercial development of tubes having inside roughness. This section discusses the commercially used tubes, which have tube-side roughness. Figure 9.3 shows the 19-mm outer-diameter Wolverine, Turbo-Chil™, and Korodense™ tubes. The Wolverine [1984] engineering data book provides design correlations. Yorkshire Imperial Metals [1982] also provides corrugated tubes, similar to the Korodense™ tube. The referenced Yorkshire publication also provides design correlations. The Turbo-Chil™ tube has 10 internal helical ribs and is made with 26, 35, or 40-fin/in. integral fins on the outer diameter.

Figure 9.13 shows currently used tube-side enhancements in large water-cooled evaporators and condensers. Commercial tube manufacturers have developed a new

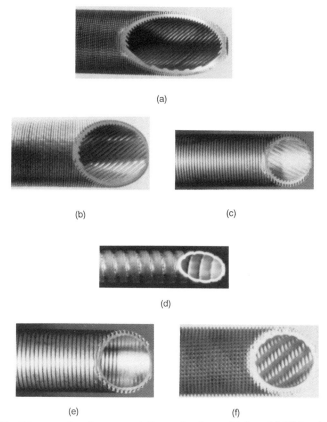

(a)

(b)   (c)

(d)

(e)   (f)

Figure 9.13 Photographs of commercially used enhanced tubes. (a) Wolverine Turbo-C™. (Courtesy of Wolverine Tube Division.) (b) Carrier Super-B™ (Courtesy of Carrier Corp.) (c) Wieland GEWA-SC™. (Courtesy of Wieland-Werke AG.). (d) Hitachi Thermoexcel-CC™. (Courtesy of Hitachi Cable, Ltd.). (e) Wieland GEWA-TW™ (Courtesy of Wieland-Werke AG.). (f) Sumitomo Tred-26D™. (Courtesy of Sumitomo Light Metal Industries.)

class of helical-rib roughened tubes, which are shown in Figures 9.13a, 9.13b, and 9.13c. These tube-side enhancements are all similar, and they differ only in the number of starts, fin height, and helix angle. The geometries are neither a classic "roughness" (little surface area increase) nor an internally finned tube (significant surface area increase, but no flow separation). They have some of the characteristics of both enhancement types. They are similar to an internally finned tube, because they provide significant internal surface area increase (compared to Figure 9.3b). Yet they have a higher helix angle than the helical internal finned tubes of Chapter 8. It is likely that the flow separates at the fin tip, which does not occur in an internally finned tube. The Figure 9.13a Wolverine Turbo-C™ condenser tube has 0.5-mm fin height and 30 internal helical ribs, as compared to the Figure 9.13b Carrier Super-

B™ and the Figure 9.13c Wieland GEWA-SC™ boiling tubes, which have 18 and 25 internal ribs, respectively.

The tube-side roughness in the Figures 9.13d through 9.13f tubes is formed by applying pressure to the external tube surface. Figure 9.13d shows the Hitachi Thermoexcel-CC condenser tube. The commercial Hitachi Thermoexcel-CC™ corrugated tube (Hitachi [1984]) has $e/d_i = 0.025$ and $p/e = 47$ and an enhanced outer surface. Ito et al. [1983] provide Nu and f data for five variants of the Hitachi Thermoexcel-CC™ tube, which they call the Thermoexcel-CPC™ tube.

Figure 9.13e shows the Wieland 19-fin/in. GEWA-TW™ (used for boiling and condensing). The GEWA-TW™ tube has a wavy inside surface, whose wave pitch is equal to the outside fin pitch. Figures 9.12a and 9.13f show the Sumitomo Tred-19D™ (19 external fins/in.) and Tred-26D™ (26 external fins/in.) condenser tubes, respectively, having a three-dimensional roughness.

We have previously given equations for prediction of the performance of the Turbo-Chil™ and Korodense™ tubes. The Wolverine Turbo-B™ performance (cooling) is given by the following equations, which are taken from Wolverine [1984].

$$\frac{hd_i}{k} = 0.06\mathrm{Re}_d^{0.8}\mathrm{Pr}^{0.4} \tag{9.33}$$

$$f = 0.198\mathrm{Re}_d^{-0.267} \tag{9.34}$$

where the Reynolds number is $\mathrm{Re}_d = d_i u_m / \nu$, with d_i being to the base of the enhancement. The heat transfer coefficient is based on the nominal surface area $(A_i/L = \pi d_i)$.

The author has calculated the enhancement ratios, h/h_p and f/f_p, for most of the commercially available enhanced tubes. Table 9.7 shows the enhancement ratios for water (Pr = 10.4) at $\mathrm{Re}_d = 25{,}000$ and lists the internal enhancement dimensions. The tubes are listed in order of increasing h/h_p. The greatest heat transfer enhancement is provided by the three-dimensional roughness Tred-19D™ tube. According

TABLE 9.7 Performance Comparison of Commercial Enhanced Tubes
($d_o = 19$ mm, Re = 25,000, Pr = 10.4)

Tube	d_i	e/d_i	n_s	p/e	α	h/h_p	f/f_p	η
GEWA-TW™	15.3	0.016	1	5.3	89	1.40	1.40	1.00
Thermoexcel-CC™	14.97	0.025	1	46.7	73	1.59	1.90	0.84
GEWA-SC™	15.02	0.035	25	2.67	30	1.87	1.65	1.13
Korodense™ (LPD)	17.63	0.025	1	20.3	81	1.89	2.26	0.84
Tubro-Chil™	14.60	0.026	10	11.1	47	1.98	1.83	1.08
Korodense™ (MHT)	17.63	0.04	1	12.0	81	2.50	4.63	0.54
Tred- 26D™	14.45	0.024	10	7.63	45	2.24	1.88	1.19
Turbo-B™	16.05	0.028	30	1.94	35	2.34	2.14	1.09
Tred-19D™	14.45	0.024	10	7.63	57	2.55	1.76	1.45
B3 (Table 9.7)	13.5	0.021	2	7.6	30	3.75	3.35	1.11

to Sumitomo [1983] test data the Tred-19D™ provides $h/h_p = 2.55$ with $f/f_p = 1.76$. The pressure drop of the Tred-19D™ tube appears to be lower than is expected. Although the three-dimensional roughness tube A8 of Table 9.7 is not a commercial geometry, it is included in Table 9.6 for comparison. It provides even higher performance than the Tred-19D™ tube. The B3 tube provides 3.7 times higher heat transfer coefficient and 3.3 times higher friction than a plain tube. The B3 tube provides the highest heat transfer performance, along with the highest η this author has seen reported.

The method used to make the Tred-19D™ and Tred-26D™ integral-fin tubes represents an innovative and important manufacturing method. The Figure 9.12a Tred-19D™ tube has a symmetrical roughness pattern. However, the Figure 9.13f Tred-26D™ tube has an intermittent helical-rib roughness. Rabas et al. [1993] discuss the method and show how it can be used to make a variety of internal 3-D roughness patterns. The dimples on the inner surface are formed by spaced protrusions on the finning disks, which are rolled on the outer tube surface. These protrusions apply pressure to the outer tube surface in the root region of the external fins. The Figure 9.3b Korodense™ and the Figure 9.3a helical ribs are formed by applying constant external pressure on the outer tube surface. Using the dimpled protrusion method, it is possible to make intermittent helical ribs. The technique is described in the U.S. Patent of Kuwahara et al. [1989] and illustrated in Figure 9.14a. The roughness dimensions of the Figure 9.13d Thermoexcel-CC™ helical-rib roughness are the axial rib pitch (p), the roughness height (e), and the helix angle. The Figure 9.14a tube has the same dimensional variables, plus the dimple pitch (z) and the dimple width (w) along the rib. Figure 9.14b shows different dimple pitches (z), which result in vortices generated at the dimples. Kuwahara et al. [1989] present heat transfer and friction data that show the effect of varying e, p, and w on performance. Figure 9.14c compares the performance of a 15.8-mm-inside-diameter tube having continuous and intermittent helical roughness spaced at $p = 7$ mm. The $e = 0.45$-mm intermittent rib provides the same Nu as the $e = 0.5$-mm continuous helical rib. However, the friction factor of the intermittent helical rib is 48% less than that of the continuous helical rib. The solid lines show the smooth tube performance.

9.5.2 Rod Bundles and Annuli

Considerable data have been reported for transverse- and helical-rib roughness for flow in annuli and rod bundles. The annulus data are for roughness on the outer surface of the inner tube. Flow is axial to the tubes in a rod bundle. Much of this work was done between 1965 and 1978 in support of gas-cooled nuclear reactors. The ribs may be integral with the tube wall (Sheriff and Gumley [1966], Wilkie [1966], Wilkie et al. [1967]) or may be in the form of a helical wire wrap (White and Wilkie [1970], Williams et al. [1970]). White and Wilkie [1970] show that a 45-degree helix angle is preferred. Webb et al. [1972] show that their correlation for transverse-rib roughness inside tubes predicts the transverse-rib roughness data for axial flow in rod bundles.

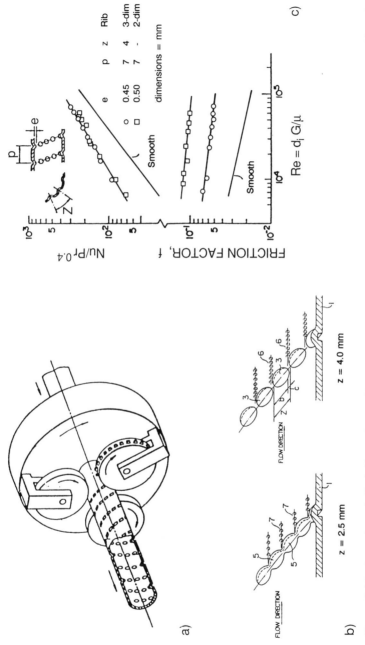

Figure 9.14 Intermittent helical-rib roughness of Kuwahara et al. [1989]. (a) Method of making the interior dimples. (b) Effect of transverse dimple pitch on local vortices. (c) Performance of intermittent helical ribs.

253

Much of the data supporting the rod bundle geometry have been taken in annuli, in which the inner wall is rough and the outer surface is smooth. Other investigators have used plane channels, having one rough and one smooth wall. It is necessary to "transform" the data, so that they are applicable to the rod bundle geometry, for which all bounding surfaces are rough. Hall [1962] proposes a special definition of equivalent diameter that should be used to convert the measured annulus data to apply to the rod bundle problem. Lewis [1974] provides an excellent assessment of the fundamental principles involved in the "Hall transformation." See the references by Sheriff and Gumley [1966], Wilkie [1966], Wilkie et al. [1967], Maubach [1972], Dalle-Donne and Meyer [1977], Meyer [1980], and Hudina [1979] for data and specific details of the method they used to transform their data. Several of these papers give velocity profile data for rough surfaces.

9.5.3   Rectangular Channels

A gas turbine blade with internal roughness is shown in Figure 9.1. Figure 9.1 shows that the hollow interior of the turbine blade is separated into channels by material membranes, which join the top and bottom walls of the blade. These membranes are required to provide the structural integrity. The top and bottom walls are rough and the side walls are smooth. Four references on use of roughness to cool the blades were given in Section 9.1. The channel aspect ratio A_r is defined as $A_r = L_s/L_r$, where L_s is the height of the smooth wall and L_r is the width of the rough surface. The composite friction factor $(\bar{f})$ and heat and transfer coefficient $(\bar{h})$ are made up of the rough and smooth wall contributions. Han [1988] shows that the composite friction factor of a channel having rough L_r walls and smooth L_s walls may be calculated by

$$\bar{f} = \frac{L_s f_s + L_r f}{L_s + L_r} \tag{9.35}$$

where f applies to a channel, whose bounding surfaces are rough, and f_s is for a smooth wall channel.

Han [1984] showed that his data for transverse-rib roughness in rectangular channel of $A_r = 1.0$ was well predicted by the correlation of Webb et al. [1971] given in Figure 9.8 with $B(e^+)$ from Figure 9.9. Hence, the friction factor data for a channel bounded by rough walls may be used with Equation 9.35 to predict the composite friction factor for a rough/smooth wall channel of any aspect ratio. Calculation of the Stanton number for a rectangular channel having a combination of rough and smooth surfaces may be similarly predicted, as described by Han [1984]. Han [1988] measured friction and St data for transverse-rib roughened channels having $A_r = 1/4, 1/2, 1, 2,$ and 4. The friction factor for the rough surface wall was extracted using Equation 9.35 and was correlated using Equation 9.4. A similar treatment was applied to the heat transfer data.

Much of the recent work on roughness for gas turbine blades has focused on different roughness geometries or roughness patterns. These data were taken in rectangular channels having roughness on two opposite walls and a smooth surface on the other two opposite walls. Taslim and Spring [1988] tested channels having ribs at α = 45 degrees, and Han et al. [1989] investigated angles of 30, 45, 60, and 90 degrees. Han et al. [1991] tested nine rib configurations in a square channel with p/e = 10. The nine rib configurations tested are shown in Figure 9.15a. Figure 9.15b shows the enhancement ratio, Nu/Nu_s, for the ribbed and smooth sides plotted versus f/f_s. For fixed f/f_s, this figure shows that the highest enhancement is provided by the V-rib configuration. Hishida and Takase [1987] measured the local Nusselt number for air flow in a 0.106-aspect-ratio channel having rectangular ribs (one heated rib) on one wall. Their data spanned $2.5 \leq p/e \leq 60$. They provide empirical correlations for Nu_x versus Re_x downstream from the ribs.

9.5.4 Outside Roughness for Cross Flow

Enhancement may be desired on the outside tube surface for shell-and-tube heat exchangers or tube banks. Several studies of knurled roughness on the outside of tubes have been performed using either air or transformer oil as flowing fluid. Groehn and Scholz [1976] tested knurled roughness (e/d_o = 0.017 and 0.03) with cross flow of air at 40 bar pressure in an inline tube bank. For $Re_d < 25,000$ the roughness provided no heat transfer enhancement. At higher Re_d the coefficient was increased above that for the smooth tube bank. For $Re_d \geq 100,000$, both tubes provided approximately 50% enhancement. The friction factor was slightly below that of the smooth tubes. Achenbach [1977] performed local measurement on a single knurled cylinder in cross flow using air at 40 bar. He found that roughness decreases the critical Reynolds number, at which the boundary layer transitions to a turbulent boundary layer. Žukauskas and Ulinskas [1983] tested knurled roughness on staggered tube banks with air and transformer oil. They tested four roughness sizes, $0.008 \leq e/d_o \leq 0.04$. Heat transfer enhancement was obtained for the oil at $Re_d \geq 2500$. But enhancement did not occur for air flow until $Re_d \geq 10,000$–70,000. Figure 9.16 shows Nu/Nu_s versus Re_d for their tests with air and oil. The authors provide curves of Nu_d and f versus Re_d for the banks tested, and also give design correlations. A later publication by Žukauskas and Ulinskas [1988] provides (a) curves of Nu_d and f versus Re_d for the banks tested, (b) design correlations, and (c) recommendations for selection of the roughness size. For $1000 \leq Re_d \leq 100,000$ the Nusselt number for knurled roughness is given by

$$Nu_d = 0.5 Re_d^{0.65} Pr^{0.36} \left(\frac{S_t}{S_l} \right)^{0.2} \left(\frac{e}{d_o} \right)^{0.1} \left(\frac{Pr}{Pr_w} \right)^{0.25} \qquad (9.36)$$

Knurled roughness is not commonly used as shell-side enhancement in tube banks. The favored shell-side enhancement is low, integral-fin tubes, which are discussed in Chapter 6.

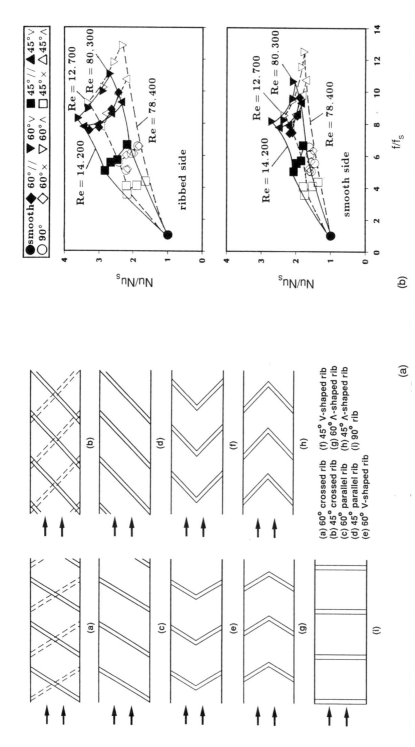

Figure 9.15 (a) The nine rib configurations tested by Han et al. [1991] for a channel having two opposite rough and two opposite smooth walls in a square channel. (b) Enhancement ratio versus f/f_p. (From Han et al. [1991].)

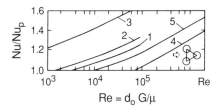

Figure 9.16 Nu/Nu_p versus Re for knurled roughness on tube bank in cross flow. Curves 1, 2, and 3 are for $Pr_f = 84$ with $e/d_o = 6.67E - 3$ (1), $15.0E - 3$ (2), and $15.0E - 3$ (3), and curves 4 and 5 are $Pr_f = 0.7$ with $e/d_o = 1.0E - 3$ (4), $8.0E - 3$ (5).

9.6 GENERAL PERFORMANCE CHARACTERISTICS

9.6.1 St and f Versus Reynolds Number

The distinct advantage of the rough surface heat–momentum analogy correlation (Equation 9.16) is that data for one e/d_i are applicable to those of other e/d_i, provided that geometric similarity of the roughness geometry is maintained. However, the correlations do not allow easy interpretation, because the correlated data are plotted as $g(e^+)$ and $B(e^+)$ versus the roughness Reynolds number e^+. Engineers are accustomed to viewing St and f data plotted versus the pipe Reynolds number. However, one may use the correlations to generate St and f data versus Re data for constant e/d_i, which may then be plotted in the familiar format and compared against that of a smooth surface. This will provide a better physical understanding of the Reynolds and Prandtl number characteristics of rough surfaces.

Figure 9.17 shows curves of St, f, and 2St/f versus Re_d for the sand-grain roughness tested by Dipprey and Sabersky [1963] for four values of e/d_i at $Pr = 5.1$. These curves were generated from curve fits of the correlated data shown in Figures 9.4b and 9.5. Figure 9.17 also shows the Stanton number and friction curves for a smooth tube at $Pr = 5.1$. The smooth-tube values are calculated from the Petukhov equation (Equation 9.14). The salient features shown in Figure 9.17 are as follows:

1. The St and f values for each roughness size asymptotically approach the smooth surface St and f at sufficiently low Re_d. Above this Re_d, the roughness elements begin to project into the turbulent region, causing a deviation from the smooth tube turbulent St and f curves. When the Re_d is sufficiently high, the friction factor attains a constant value and the St attains a maximum value. The dashed line corresponding to $e^+ = 42$ is the condition at which the friction factor attains 95% of its "fully rough" value. Here we define the fully rough condition (FR) as the e^+ or Re_d at which the friction factor attains $0.95f_{FR}$. The St_{max} occurs at approximately e^+_{FR}.

2. The term 2St/f describes heat transfer per unit friction. We have previously defined $(St/f)/(St_s/f_s)$ as the efficiency index (η). The 2St/f attains a maximum value if $Pr > 1$. The maximum 2St/f may exceed that of a smooth surface. This maximum occurs at a value of Re_d less than the fully rough condition.

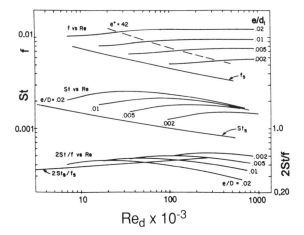

Figure 9.17   Heat transfer and friction characteristics of sand-grain roughness calculated from the Dippery and Sabersky [1963] correlation for Pr = 5.1. (From Webb [1979].)

3. The Re_d at which St and 2St/f attain maxima, and f attains its fully rough value, decreases with increasing e/d_i. But, the e^+ values at which the maxima occur are independent of e/d_i.

4. In the fully rough region, 2St/f monotonically decreases, yielding very small values for the efficiency index (η). As e/d_i is increased, f/f_s increases much faster than St/St_s. Thus, at a fixed Re_d in the fully rough region, larger e/d_i will yield smaller values of 2St/f.

Webb [1979] proposes that all geometrically similar roughnesses of interest will exhibit the above four qualitative characteristics. The minimum Pr at which a maximum of 2St/f occurs apparently depends on the roughness type. To illustrate the generality of these basic roughness characteristics, Figure 9.18 is presented for $p/e = 10$ transverse-rib roughness, based on the study of Webb et al. [1971]. The parameters of Figures 9.17 and 9.18 are the same. Both roughness types exhibit the same qualitative behavior. However, major differences exist. The transverse-rib roughness gives a higher f and St than the sand-grain type for the same e/d_i (cf. Example 9.2). For example, with transverse-rib roughness the maximum St for e/d_i = 0.005 is about the same as for e/d_i = 0.02 sand-grain roughness. The "equivalent sand-grain roughness" of the transverse-rib is $e_s/e = 4$, based on equivalent heat transfer performance. The transverse rib does not attain a maximum of 2St/f at Pr = 5.1; a maximum does not occur until Pr $\geq$ 10. Thus, the sand-grain roughness yields a higher efficiency index. Dashed lines at $e^+ = 14$ and 35 are drawn in Figure 9.18. Inspection of the figure shows that St and 2St/f are within 5% of their maximum values within the range $14 \leq e^+ \leq 35$. Hence, this e^+ range is suggested for this roughness geometry.

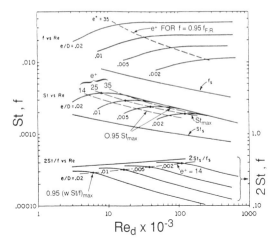

Figure 9.18   Heat transfer and friction characteristics of $p/e = 10$ transverse-rib roughness calculated from Webb et al. [1971] correlation for Pr $= 5.1$. (From Webb [1979].)

9.6.2   Other Correlating Methods

Extending our understanding to other roughness types may not be as complex as suggested previously. In their work on transverse-rib roughness, Webb et al. [1971] also tested $p/e = 20$ and 40 and found that the heat transfer data were correlated with the $p/e = 10$ data, as shown in Figure 9.8. Webb and Eckert [1972] also evaluated the effect of rib geometry (e.g., circular and hemispherical shaped ribs) by testing the correlation with different rib shapes. They found that these data were correlated on the same $\bar{g}(e^+)$ versus e^+ line, thus accounting for the different value of the friction similarity parameter. We emphasize that there is no rational reason to expect that nongeometrically similar roughness will fall on the same correlating line as geometrically similar roughness. Clearly, the $\bar{g}(e^+)$ functions of sand-grain roughness (Figure 9.5) and transverse-rib roughness (Figure 9.8) are not equal. At $e^+ = 50$, the $\bar{g}(e^+)$ function is 11.34 for sand-grain roughness and 6.85 for transverse-rib roughness.

Webb et al. [1971] found that the friction similarity function $B(e^+)$ was different for each p/e ratio. They were able to *empirically* correlate the p/e effect. Is there any possibility for correlating heat transfer data for all roughness types on the same line? Burck [1970] attempted to do this. He determined the "equivalent sand-grain roughness" for each roughness geometry (e_{sg}), and calculated the roughness Reynolds number as $(e_{sg})^+ = e_{sg}u^*/\nu$. The equivalent sand-grain roughness size is determined at the fully rough condition by solving the following equation for e_{sg}:

$$2.5 \ln(e_{sg}/e) = 8.48 - B(e^+) \tag{9.37}$$

Edwards [1966] did this for his repeated-rib air flow data. Webb and Eckert [1972] show the transverse-rib data for $10 \leq p/e \leq 40$ of Webb et al. [1971] plotted in this format. The correlation was not as good as the original correlation in terms of the actual e^+.

Burck chose to plot the efficiency index (η) versus $(e_{sg})^+$ for the data of 14 investigators. Although the correlation was not totally satisfactory, it reasonably grouped the data. It would seem that a better way to use the equivalent sand-grain roughness is to maintain the $\bar{g}$ versus e^+ format and plot $\bar{g}(e_{sg})$ versus $(e_{sg})^+$. Although different roughness types may be "normalized" to equivalent sand-grain roughness, the correlation will not be as good as that obtained by maintaining the actual e^+ format.

There are assumptions in the rough surface heat–momentum analogy that can affect the accuracy of the correlation. Use of the law-of-the-wall velocity distribution (Equation 9.1) assumes that the fully developed velocity profile u/u^* is independent of pipe Reynolds number. This is a good approximation, but is not precise for either rough or smooth surfaces. Ye et al. [1987] measured the velocity profile in a tube having corrugated roughness. Their velocity profiles were measured for $12{,}000 \leq \mathrm{Re}_d \leq 20{,}000$, which is not a large enough range to determine the Reynolds number effect.

Some investigators (e.g., Sheriff and Gumley [1966]) have attempted to correlate $\mathrm{St}/\mathrm{St}_s$ versus e^+. This also provides an approximate correlation. Finally, Ravigururajan and Bergles [1985] attempted to correlate transverse-rib roughness data from many investigators using a linear multiple regression method. A poor correlation was obtained, since the data were correlated within $\pm 50\%$. In the author's opinion, multiple regression methods should be employed *only* as a last resort—that is, when no other rationally based method is available. The heat–momentum analogy method presented in this chapter has been shown to be far superior to the empirical multiple regression method. Use of the equivalent sand-grain roughness method with the heat–momentum analogy correlation should yield a significantly better correlation of the data set used by Ravigururajan and Bergles [1985].

9.6.3   Prandtl Number Dependence

Rough surfaces show a significantly different Prandtl number dependency than do smooth surfaces. The Pr dependency of rough tubes, relative to smooth tubes, may be observed by examining the behavior of $\mathrm{St}/\mathrm{St}_s$ versus e^+ for constant e/d_i. We have generated curves of St versus e^+ for $p/e = 10$ transverse-rib roughness for different Prandtl numbers using curve-fits of the $\bar{g}(e^+)$ and $B(e^+)$ functions shown in Figures 9.8 and 9.9, respectively. Figure 9.19 shows the results plotted as $\mathrm{St}/\mathrm{St}_s$ versus e^+ for $0.71 \leq \mathrm{Pr} \leq 200$ with $e/d_i = 0.01$. In this comparison, St and St_s are calculated at the same Re_d. If the rough and smooth surfaces have the same Pr dependence, $\mathrm{St}/\mathrm{St}_s$ will be the same for all Pr. These curves show that $\mathrm{St}/\mathrm{St}_p$ increases with Pr. Figure 9.19 also shows that a maximum value of $\mathrm{St}/\mathrm{St}_p$ is attained at $e^+ = 20$ for $\mathrm{Pr} > 5.1$. However, at large e^+, the curves tend to asymptotically approach a common value. The interpretation of these curves is as follows: (1) At high e^+ (high Re_d), the rough and smooth tubes have the same Prandtl number

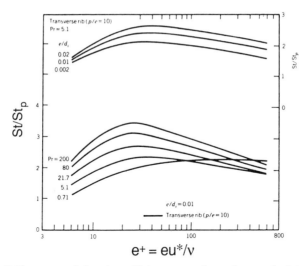

Figure 9.19 St/St_s versus e^+ for $p/e = 10$ transverse-rib roughness calculated from Webb et al. [1971] correlation for different Prandtl numbers. (From Webb [1979].)

dependency. (2) At lower e^+, the rough surface has a stronger Prandtl number effect than a plain surface. Figure 9.19 shows that the rough surface will be of greater benefit for high-Prandtl-number fluids than for low-Prandtl-number fluids. This means that the heat transfer coefficient increase is greater than the friction increase for higher-Prandtl-number fluids. The curves in the upper part of Figure 9.19 are for different e/d_i at Pr $= 5.1$.

The basic form of the rough surface heat–momentum analogy correlation (Equation 9.18) is

$$St = \frac{f/2}{1 + \sqrt{f/2}\,[\bar{g}(e^+)\mathrm{Pr}^n - B(e^+)]} \tag{9.38}$$

At high e^+ and/or Pr, $\bar{g}\,\mathrm{Pr}^n \gg B$, so the asymptotic value of Equation 9.38 may be written

$$\frac{f/2\mathrm{St} - 1}{\sqrt{f/2}} \simeq \bar{g}(e^+)\mathrm{Pr}^n \tag{9.39}$$

This algebraic form for the Prandtl number dependence is unfamiliar to those who use power-law-type equations for smooth surfaces, such as the Sieder–Tate equation, which shows St $\propto$ Pr$^{-2/3}$. However, the Petukhov [1970] equation for smooth tubes (Equation 9.13) has an algebraic form similar to that of Equation 9.39. For Pr $\gg 1$, the asymptotic value of Equation 9.13 is

$$\frac{f_s/2\mathrm{St}_s - 1}{\sqrt{f_s/2}} \cong 12.7\mathrm{Pr}^{2/3} \tag{9.40}$$

TABLE 9.8   Prandtl Number Exponents for Rough Surfaces

Reference	n	Pr (or Sc)	Roughness Type
Dipprey and Sabersky [1963]	0.44	1.2–5.94	Sand-grain
Webb et al. [1971]	0.57	0.7–37.6	Ribs ($10 < p/e < 40$)
Sethumadhavan and Raja Rao [1986]	0.55	5.2–32	Corrugated
Sethumadhavan and Raja Rao [1983]	0.55	5.2–32	Wire coil inserts
Dawson and Trass [1972]	0.58	320–4500	Ribs ($p/e \simeq 3.7$)
Withers [1980a]	0.50	2.5–10	Corrugated
Withers [1980b]	0.50	4.7–9.3	Helical rib

If the asymptotic Pr dependence of rough tubes is the same as that of smooth tubes, the exponent n in Equation 9.39 would have the value 2/3. However, several studies have shown $n \neq 2/3$ for rough tubes. It is not yet fully established how the roughness type may affect the exponent of Pr.

Table 9.8 summarizes the results of work to determine the Prandtl number dependence (Pr^n) for rough surfaces. The Dawson and Trass data are for mass transfer tests, so Sc^n corresponds to Pr^n. Equations 9.38 and 9.39 show that higher St values are obtained if the exponent n of Pr is small.

Table 9.8 shows that rough tubes do not have the same $\mathrm{Pr}^{2/3}$ exponent as smooth tubes. The Sethumadhavan and Raja Rao [1983, 1986] and Dawson and Trass [1972] data give 0.55 and 0.58 exponents, respectively, which agree closely with the 0.57 value of Webb et al. [1971]. Thus, the value $n \simeq 0.57$ appears to be applicable to a wide range of repeated-rib roughness geometries. Although Withers [1980a, 1980b] shows $n = 0.5$, the vast majority of his data points are for $3 \leq \mathrm{Pr} \leq 7$. Hence, his Prandtl number test range may be too small to determine the Prandtl number exponent with good precision. The Webb [1971], Sethumadhavan and Raja Rao [1983, 1986], and Dawson and Trass [1972] data cover a much wider Prandtl number range.

The Pr exponent for sand-grain roughness (sg) is different from that for repeated-rib roughness (tr), 0.44 versus 0.57. Again, the Prandtl number range tested by Dipprey and Sabersky [1963] was quite small, $1.2 \leq \mathrm{Pr} \leq 5.9$. Because both roughness types have $n < 2/3$, the ratio $\mathrm{St}/\mathrm{St}_s$ will increase with increasing Pr. The smaller exponent for sand-grain roughness (sg) means that the ratio $\mathrm{St}_{sg}/\mathrm{St}_{tr}$ will increase with increasing Pr. It would be desirable to test the sand-grain roughness over a larger Prandtl number range to verify the 0.44 exponent.

9.7   HEAT TRANSFER DESIGN METHODS

Because there are a large number of different basic roughness types and a range of possible e/d_i, p/e, and α for each type, the designer is faced with a confusing array of choices. In order to approach the question, one must first define the design objective and establish design constraints. Possible design objectives are listed in

Table 3.1. Assume that tube-side roughness will be used to reduce the size of a heat exchanger for case VG-1 of Table 3.1. Thus, one is interested in calculating A/A_p for constant heat duty, flow rate, and pumping power. One may place some limits on the roughness choices by defining a hierarchy of questions:

1. What is the A/A_p for a given roughness type and e/d_i?
2. Will another e/d_i (for the same basic roughness type) give a smaller A/A_p?
3. Will another basic roughness type (and e/d_i) provide further reduction of A/A_p?

First, Question 1 will be addressed for the VG-1 criterion, which is described in Section 3.3.4. Assume, for example, that a heat exchanger has been designed that has a plain surface on the tube side. The tube-side mass flow rate, the pressure drop, and the UA value of the plain tube heat exchanger are known. Will a tube-side roughness be of benefit? If the dominant thermal resistance is on the tube side, a tube-side roughness is of potential benefit. Hence, the designer wants to calculate A/A_p for $P/P_p = UA/U_pA_p = 1$.

For simplicity, assume that the total thermal resistance is on the tube side. Then the constraints are satisfied by Equation 3.7. The required velocity in the rough tube that will satisfy the pumping power constraint (Equation 3.6) is unknown. Hence, one must use an iterative calculation procedure to solve Equations 3.6 and 3.7 for A/A_p and G/G_p (or $\text{Re}_d/\text{Re}_{d,p}$). The solution procedure is complicated by the fact that the equations for the rough tube St and f are written in terms of the roughness Reynolds number (e^+) rather than the pipe Reynolds number (Re_d). Two solution procedures are possible and will be described.

9.7.1 Design Method 1

Assume that empirical curve fits of the St and f versus Re_d (or G) characteristics for the selected e/d_i have been developed. Using these equations, one solves Equation 3.7 for G/G_p. With G/G_p known, the Re_d of the rough tube is calculated and St and f are readily calculated. Next, one calculates the U value for the heat exchanger, and the area ratio (A/A_p) is given by U_p/U. Note that the pumping power constraint is inherently satisfied by Equation 3.7.

9.7.2 Design Method 2

The second method avoids the need for the St and f versus Re_d curve fits and uses Equation 9.18. Using Equation 9.2, one may write

$$\frac{\text{Re}_d}{\text{Re}_{d,p}} = \frac{e^+}{\text{Re}_{d,p}} \frac{d_i}{e} \sqrt{\frac{2}{f}} \qquad (9.41)$$

If the rough and smooth tubes are of the same inside diameter, $\text{Re}_d/\text{Re}_{d,p} =$

G/G_p. Using Equations 9.18 with $n = 0.57$ and Equation 9.41 in Equation 3.7, $\mathrm{Re}_d/\mathrm{Re}_{d,p} = [(f/f_p)/(\mathrm{St}/\mathrm{St}_p)]^{1/2}$ gives

$$\frac{e^+}{\mathrm{Re}_p} \frac{d_i}{e} \sqrt{\frac{2}{f}} = \left(\frac{f_p/(2\mathrm{St}_p)}{1 + \sqrt{f/2}\,[\bar{g}(e^+)\mathrm{Pr}^{0.57} - B(e^+)]} \right)^{1/2} \qquad (9.42)$$

The values of $\mathrm{Re}_{d,p}$, St_p and f_p for the smooth tube and the e/d_i are known. Equation 9.42 is iteratively solved for e^+ using Equation 9.4 to eliminate $2/f$. One uses the $g(e^+)$ and $B(e^+)$ equations for the roughness family of interest. The iterative solution gives the value of e^+ that satisfies Equation 9.42. With e^+ known, one easily computes St, f, and Re_d of the rough tube and proceeds to calculate A/A_p as for Method 1.

Using either Method 1 or Method 2, one obtains the predicted A/A_p and G/G_p. One desires $G/G_p \simeq 1$. If the calculated $G/G_p \leq 1$, it will be necessary to increase the number of tubes in the heat exchanger. For $\mathrm{Pr} \geq 20$, one may find $G/G_p > 1$. Is it possible that a different e/d_i will provide a lower A/A_p or a more desirable G/G_p? This is Question 2. One may repeat the analysis for different e/d_i to answer this question. Webb and Eckert [1972] address this question for $p/e = 10$, $\alpha = 90$ degree rib roughness using PEC VG-1, VG-2, and VG-3 of Table 3.1. For this roughness type, they conclude that the criterion $14 \leq e^+ \leq 20$ will provide the best value for the objective function. If one can specify the desired e^+, Equation 9.42 may be iteratively solved for the required e/d_i. Such evaluations have not been done for different roughness types (e.g., $p/e = 15$, $\alpha = 45$-degree helical-rib roughness). Hence, one cannot state the preferred e^+ for different helix angles (α).

9.8 PREFERRED ROUGHNESS TYPE AND SIZE

9.8.1 Roughness Type

What is the best roughness type (Question 3)? There is no direct way to answer this question without performing a parametric study of candidate roughness types. Webb [1982] uses a case study method to compare the performance improvements of different roughness types. This analysis was performed for case VG-1 of Table 3.1. The design conditions of the reference plain tube heat exchanger are $\mathrm{Re}_{d,p} = 46{,}200$, $h_o = 27{,}260$ W/m-K, $h_i = 6814$ W/m-K, and $R_{fi} = 0.000044$ m²-K/W (fouling resistance). Five basic types of tube-side enhancements were evaluated, including internal fins. The enhanced tubes are listed in order of decreasing material savings $(V_m/V_{m,p})$ where V_m is the tube material volume in the heat exchanger. The ratio of the tube lengths in the enhanced and plain tube designs is L/L_p, and the relative number of tubes in the exchangers is N/N_p, which is equal to G_p/G for the roughness geometries. A summary of the key results of the study is shown in Table 9.9. Table 9.9 provides several conclusions:

1. The greatest material savings are provided by the helical internal fins. The savings for the helical fins is 12% greater than that for the axial internal fins.

TABLE 9.9 Performance Comparison of Tube-Side Enhancement Geometries (Case VG-1 of Table 3.1)

Geometry	α (degrees)	e/d_i	N/N_p	L/L_p	$1 - V_m/V_{m,p}$
Helical fins ($n_f = 32$)	30	0.0420	1.01	0.37	0.42
Transverse ribs ($p/e = 10$)	90	0.0060	1.16	0.46	0.53
Axial fins ($nf = 32$)	0	0.0420	1.04	0.49	0.54
Corrugated ($p/e = 19.5$)	82	0.0090	1.02	0.56	0.57
Helical ribs ($p/e = 15$)	45	0.0069	0.96	0.60	0.58

The helical fins can be operated at approximately the same velocity as that of the plain tube design.

2. The smallest material savings (42%) is provided by the helical ribs. However, the helical ribs can be operated at a velocity only 4% lower than that of the plain tube design. The material savings for the corrugated tube is approximately the same as that for the helical ribs. However, the corrugated tube must be operated at a lower velocity.

3. The transverse ribs provide 5% greater material savings than do the helical ribs; however, the velocity must be reduced to 84% of the plain tube value, which requires a larger heat exchanger shell diameter.

4. The heat exchanger length using the helical internal fins is only 37% as long as that of the plain tube design.

Although the helical internal fin geometry appears to be superior to the roughness geometries, one must consider if it can be made in the material of interest, one must also consider the relative cost of the enhanced and plain tubes. Presently, it appears that the helical-rib roughness is of significantly lower cost than an internally finned tube. Second, one must consider whether the design fouling factor (R_{fi}) can be maintained. This is discussed in Chapter 10.

9.8.2 PEC Example 9.3

Your company plans to manufacture tubes having $p/e = 10$, transverse-rib roughness [$\alpha = 90$ degrees] applicable to $5,000 \leq Re_d \leq 80,000$ and $1 \leq Pr \leq 100$. The tubes will be produced in 19.05-mm outside diameter with 0.70-mm wall thickness. Select the number of different roughness sizes needed to support the desired operating range, and select the specific roughness dimensions, p and e. Use the information given in Figures 9.8, 9.9, 9.18, and 9.19 to support your recommendations.

Based on the discussion of Figure 9.18 in Section 9.6.1, we set $14 \leq e^+ \leq 35$ as the design range for each tube. This will provide St and $2St/f$ within 0.95 of the maximum values. Using Figure 9.9 with $p/e = 10$, find the values of $B(e^+)$ at $e^+ = 14$ and 35. The $B(e^+)$ values are 4.5 and 3.2, respectively. Use Equation 9.2 with $(f/2)^{1/2}$ defined by Equation 9.4 to solve for the required e/d_i, giving

$$\frac{e}{d_i} = \frac{e^+}{Re_d} \left[2.5 \ln \left(\frac{2e}{d_i} \right) - 3.75 + B(e^+) \right] \qquad (9.43)$$

Using $B(35) = 3.2$, iteratively solve for the $e/d_i = 0.0048$ at $Re_d = 80,000$. Then repeat the procedure using $B(14) = 4.5$ to solve for the minimum Re_d at $e^+ = 14$ with $e/d_i = 0.0048$. This gives $Re_d = 36,060$. Thus a tube having $e/d_i = 0.0048$ is applicable to $36,060 \leq Re_d \leq 80,000$. Then choose the next required e/d_i following the same procedure. For the second tube, solve for the required e/d_i using $e^+ = 35$ at $Re_d = 36,060$ and obtain $e/d_i = 0.00916$. Following the same procedure, the minimum Reynolds number for tube 2 is 16,430. Continuing the procedure for tube 3, obtain $e/d_i = 0.0169$ at $e^+ = 35$ and $Re_d = 16,430$. Tube 3 will have $Re_d = 7,635$ at $e^+ = 14$, which approximately satisfies the minimum e^+. Hence, three e/d_i values (0.0169, 0.00916, and 0.0048) are required to satisfy the operating Re_d. A similar procedure can be applied for any type of enhancement geometry.

One may now use Equation 9.18 to generate curves of St and f versus Re and Pr for each of the selected rough tubes. This example shows the following: (1) If an enhanced tube is used below its design e^+ range, small St/St_s will result, and (2) if the tube is used above its e^+ range, a small efficiency index will result.

9.9   NUMERICAL SOLUTIONS

Significant advances have been made in numerically solving the momentum and energy equations for complex flow geometries. As discussed in Chapter 8, the heat transfer and friction characteristics of internally finned tubes (axial fins) have been numerically modeled. Another example of numerical simulation is the offset-strip fin geometry discussed in Chapter 6. However, roughness is a more difficult problem, because local flow separations are involved for two-dimensional roughness. Three-dimensional roughness geometries involve further complications. The simplest problem to attack is laminar flow over spaced two-dimensional ribs. This is similar to the offset-strip fin problem analyzed by Patankar [1990]. The analysis assumes a periodically, fully developed flow. Hence, the numerical simulation zone is limited to the geometrically repeating pattern. The problem becomes more difficult when the flow is turbulent.

Numerical predictions have been made for laminar and turbulent two-dimensional transverse ribs. However, no progress has been made on helical ribs or three-dimensional roughness. Numerical predictions for two-dimensional flow obstructions include Benodekar et al. [1985], Fujita et al. [1986], Hijikata et al. [1987], Hung et al. [1987], Durst et al. [1988], Fodemski and Collins [1988], and Becker and Rivir [1989].

9.9.1 Predictions for Transverse-Rib Roughness

Recent work by Arman and Rabas [1991, 1992] and Rabas and Arman [1992] is described to show the potential for prediction of practical two-dimensional roughness geometries. Figure 9.20 shows the near-wall flow field for turbulent flow over a two-dimensional rib. The near-wall flow consists of stable recirculation and boundary layer development zones. Above this zone is a free shear region and large eddies. These authors used the two-layer turbulence model of Chen and Patel [1988]. This model divides the flow into (1) a near-wall region and (2) the fully developed core region. The standard $k–\epsilon$ model was used in the core region, whereas a one-equation model was used in the near-wall region. The standard $k–\epsilon$ two-equation model failed to adequately predict the reattachment length. The problem occurs because the model requires the friction velocity (u^*), for which the local-wall shear stress varies with axial position and changes sign at the reattachment point.

Figure 9.21a shows the ability of Arman and Rabas [1991] to predict the experimental friction factors for five repeated-rib geometries of Webb et al. [1971]. The geometry code used in Figure 9.21 is as follows: The first two digits are the e/d_i, and the second two digits are the p/e of the roughness. Excellent agreement between predictions and experiment were obtained for the smaller e/d_i and the larger p/e. Figure 9.21b shows the predicted axial-wall shear stress distribution for $e/d_i = 0.02$ and $p/e = 20$. The figure shows that the shear stress changes sign at $x/e \simeq 7$, which reasonably corresponds to the reattachment point shown in Figure 9.20. The small recirculation zones just upstream and downstream from the rib are responsible for the local variation of the shear stress at these locations.

Arman and Rabas [1992] showed good ability to predict the Nusselt number and friction factor of the two-dimensional roughness data of Webb et al. [1971]. Figure 9.22 shows a comparison between the predicted and experimental Nusselt number data of Webb et al. [1971] for $e/d_i = 0.02$ and $p/e = 20$ for Pr = 0.71, 5.1, and 21.7. Their paper also presents predictions for other transverse-rib geometries tested by Webb et al. [1971]. The average error of all of the predictions was 15%.

The discussion in Section 9.6.4 showed that the Prandtl number dependency of transverse-rib roughness is significantly different from that for a smooth tube. Rabas

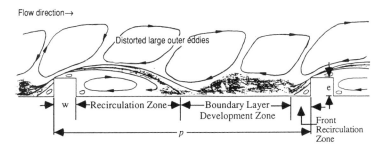

Figure 9.20 Structure of turbulent flow over a two-dimensional rib. (From Arman and Rabas [1971].)

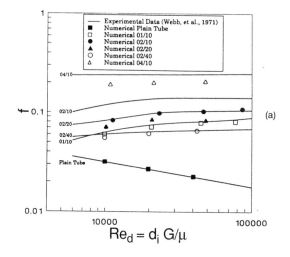

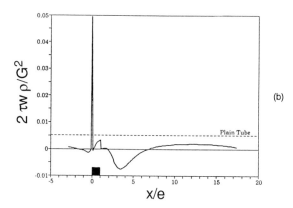

Figure 9.21 (a) Predicted friction factors compared with data of Webb et al. [1971]b). (b) Predicted axial variation of wall shear stress for $e/d_i = 0.01$ and $p/e = 20$ at Re $= 47,000$. (From Arman and Rabas [1991].)

and Arman [1992] sought to explain the physical mechanism for this difference. They predicted the axial distribution of the local Nusselt number. Their interpretation of these curves for different Pr provides valuable insights on the mechanism of the Prandtl number influence.

Fodemski and Collins [1988] predict heat transfer and friction in a rectangular channel containing square ribs spaced at $p/e = 6$. They solve the conjugate problem, which includes conduction in the tube wall. The scope of this work parallels that of Arman and Rabas [1991, 1992].

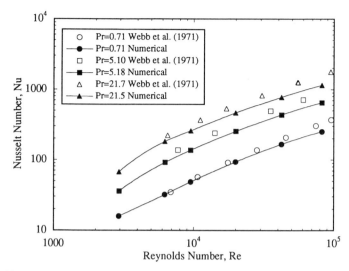

Figure 9.22 Comparison of predicted and Webb et al. [1971] experimental data for transverse-rib roughness [$e/d_i = 0.02$, $p/e = 20$] for $0.71 \leq Pr \leq 21.7$. (From Rabas and Arman [1992].

9.9.2 Effect of Rib Shape

Arman and Rabas [1992] predicted the effect of the rib shape on the Nusselt number and friction factor. The numerical model was validated by predicting the Nu and f of two-dimensional roughness geometries tested by Hijikata and Mori [1987] and Nunner [1956]. The Nu and f of the Hijikata and Mori arc-shaped geometries ($e/d_i = 0.02$, $p/e = 44.4$; and $e/d_i = 0.04$, $p/e = 22.2$) were overpredicted by 17% and 26%, respectively. The Nunner semicircular rib geometry ($e/d_i = 0.04$, $p/e = 20$) Nu and f were overpredicted by 5% and 0%, respectively.

They next predicted the effect of the four different rib shapes shown in Figure 9.23 for $e/d_i = 0.02$ and $p/e = 20$ and 40 at $Re_d = 9400$ and 39,000. Table 9.10

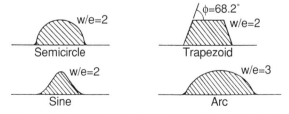

Figure 9.23 Different rib shapes analyzed by Arman and Rabas [1992]. (From Arman and Rabas [1992].

TABLE 9.10   Comparison of Mean Nusselt Numbers and Efficiency Index
($e/d_i = 0.02$)

Shape	p/e	Re = 9400		Re = 39,000	
		Nu/Nu$_p$	η/η_{trap}	Nu/Nu$_r$	η/η_{trap}
Sine	10	2.17	0.95	2.22	0.96
Sine	20	1.84	0.97	1.88	0.99
Semicircle	10	2.13	0.93	2.21	0.96
Semicircle	20	1.79	0.95	1.83	0.96
Arc	10	1.98	0.86	2.16	0.91
Arc	20	1.71	0.91	1.78	0.94
Trapezoid	10	2.29	1.00	2.32	1.00
Trapezoid	20	1.89	1.00	1.91	1.00

shows the enhancement ratios Nu/Nu$_p$ and the efficiency index (η) referred to a plain tube. The table shows that the trapezoidal shape gives the highest E_h and η. The arc-shaped rib gives 29% lower E_h and 9% lower η than does the trapezoid-shaped rib. They also provide plots of the local shear stress and pressure distributions, along with a table of the reattachment lengths.

9.9.3   The Discrete-Element Predictive Model

Taylor et al. [1988] and Taylor and Hodge [1992] developed a seminumerical "discrete-element" method to predict the j and f factors for flow in tubes having a three-dimensional roughness. In this method, one numerically integrates the velocity and temperature profiles across the pipe radius, and accounts for the flow blockage caused by the roughness elements. The drag coefficient on the discrete roughness elements is included using empirical expressions of the form $C_D = a\text{Re}^n$. The solution to the momentum equation provides the velocity profile and the friction factor. The velocity profile is used in a similar approach to solving the energy equation. Again, the heat transfer from the discrete roughness elements is accounted for by empirical expressions of the form Nu $= b\text{Re}^m$.

The discrete-element roughness model was originally developed by Taylor et al. [1984]. The major physical phenomena which distinguish flow over a rough surface are blockage of the flow, form drag on the roughness elements, and skin friction on the base surface. The heat transfer formulation accounts for the local heat transfer from the roughness elements and from the base surface between the elements. The momentum and energy equations are developed by writing momentum and energy balances on the control volume of thickness δy and length δx shown in Figure 9.24. The momentum equation is

$$0 = (pA_x)|x - (pA_x)|_{x+dx} - F_D + (\tau A_y)|x - (\tau A_y)|_{y+dy} \qquad (9.44)$$

The areas A_x and A_y are given by $A_x = 2\pi r \delta y \beta_x$, and $A_y = 2\pi r \delta x \beta_y$, where β_x and β_y are the fraction of the control surface that is not obstructed by the roughness

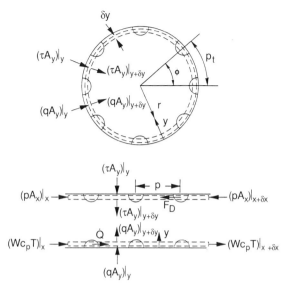

Figure 9.24 Control volume for discrete-element method of Taylor and Hodge [1992]. (From Taylor and Hodge [1992].)

elements. Taylor et al. [1984, 1988] have shown that $\beta_x = \beta_y = 1 - \pi d_e^2/[4p(1 - 2y/d_i)]$ for a uniform array of elements spaced at p, where d_e is the local element diameter at level y. The drag force is given by

$$F_D = 0.5\rho C_D A_p N_{el} \tag{9.45}$$

where the drag coefficient on the roughness element (C_D) is obtained from empirical data. N_{el} is the number of roughness elements contained in the control volume, and A_p is the projected area of the roughness elements to the flow.

The energy equation is given by

$$0 = (W c_p T)|_{x+\delta x} - (W c_p T)|_x + (q A_y)|_{y+dy} - (q A_y)|_y \\ - h_{el} \delta A_{el} N_{el} (T_w - T) \tag{9.46}$$

The Nusselt number for heat transfer from the element is given by an empirical correlation. Letting $d_e(y)$ be the local diameter of the element at distance y from the surface, $h_{el}\, \delta A_{el}$ is given by

$$h_{el} \delta A_{el} = \frac{\mathrm{Nu}_{el} k}{d_e(y)} \pi d_e(y) \delta y \tag{9.47}$$

The solution of Equations 9.45–9.47 requires a turbulence model for ϵ_m and ϵ_h and a roughness model for C_D and Nu_{el}. The turbulence model is not modified to include roughness effects, because the physical effects of the roughness are included explicitly in the differential equations. The Prandtl mixing-length model with Van Driest damping and constant turbulent Prandtl number $Pr_t = 0.9$ is used. Following Kays and Crawford [1980] the pressure gradient effects on the turbulence are accounted for using a wall parameter.

The boundary conditions are applied at the smooth base wall ($y = 0$ at the smooth base surface) and the free stream. At $y = 0$, $u = v = 0$ and $T = T_w$. At $y = d_i/2$, $du/dy = dT/dy = 0$.

Empirical correlations are used for C_D and Nu_{el}. Taylor et al. [1984, 1988] give

$$\log_{10}C_D = -0.125\log_{10}Re_{de} + 3.75, \quad \text{where } Re_{de} < 6000$$
$$C_D = 0.6, \quad \text{where } Re_{de} > 6000 \quad (9.48)$$

and Hosni et al. [1989, 1991] proposed

$$Nu_{de} = 1.7Re_{de}^{0.49}Pr^{0.4}, \quad \text{where } Re_{de} < 13,800$$
$$Nu_{de} = 0.0605Re_{de}^{0.84}Pr^{0.4}, \quad \text{where } Re_{de} > 13,800 \quad (9.49)$$

An iterative solution method is required to provide closure for integration of the momentum and energy equations. After solution for the velocity and temperature profiles, the friction factor and Stanton numbers are obtained from Equations 9.50 and 9.51:

$$f = \frac{\beta_{y,w}\mu \left.\frac{du}{dy}\right|_w + \frac{1}{2p^2}\int_0^e \rho d_e C_D u^2 \, dy}{0.5\rho u_m^2} \quad (9.50)$$

where $\beta_{y,w} = 1 - \pi d_e^2/4p$ is the total blockage factor for the elements.

$$St = \frac{\beta_{y,w}kc_p \left.\frac{dT}{dy}\right|_w + \frac{\pi}{p^2}\int_0^e kNu_{de}(T_w - T) \, dy}{\rho u_m(T_w - T_m)} \quad (9.51)$$

The Equation 9.48 correlation is based on the friction factor correlation of Žukauskas [1972] for tube banks, reinterpreted as a drag coefficient. The Equation 9.49 correlation for Nu_{de} is based on the Žukauskas [1972] tube bank heat transfer correlation, and modified by Taylor et al. [1984] and Hosni et al. [1989, 1991] to obtain the best prediction of several data sets. The weak point of the analysis is in the accuracy of Equations 9.48 and 9.49. Thus, can these empirical correlations be applied to different element shapes? Although the concept makes sense, the methodology to define C_D and Nu_{de} is relatively weak. A more rationally based method of defining these important parameters would strengthen the method.

Taylor and Hodge [1992] used the discrete-element model to predict the j and f factors for the three-dimensional roughness data of Dipprey and Sabersky [1963],

Takahashi et al. [1988], Gowen and Smith [1968], and Cope [1945]. They show quite good ability to predict the j and f versus Re_d data. The effect of Prandtl number is also well-predicted. They note that some inconsistencies may be due to errors in some of the experimental data. Figure 9.25 shows the predictions for the closely packed sand-grain roughened tube A-4 ($e/d_i = 0.0488$) of Dipprey and Sabersky [1963]. Figure 9.25 shows very good agreement with the data.

Hosni et al. [1991] and Taylor and Hodge [1992] use the method to predict heat transfer and friction on rough plates in boundary layer flow. Chakroun and Taylor [1992] treat accelerating boundary layer flow over rough plates.

It is noted that a successful application of the method to two-dimensional rough-

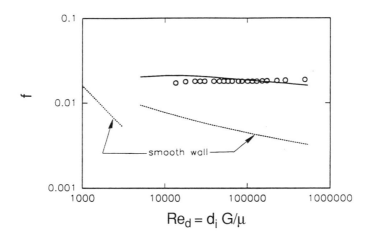

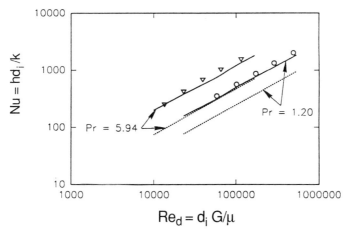

Figure 9.25 Prediction of Dipprey and Sabersky [1963] data on tube A-4 ($e/d_i = 0.0488$) using the discrete-element method of Taylor and Hodge [1992]. (From Taylor and Hodge [1992].)

ness has not yet been reported. Correlations different from those given in Equations 9.48 and 9.49 are required to predict the Nu and f of the roughness elements.

9.10   CONCLUSIONS

Significant progress has been made in the design and manufacture of roughened surfaces. Innovative manufacturing methods have resulted in the development of tubes having internal roughness (with external enhancement), which provide capital cost savings relative to plain tube designs. Rough surfaces are extensively used for water-side enhancement in air-conditioning condensers and evaporators and in large steam condensers. Roughness is also used to cool the interior surfaces of gas-turbine blades. Table 9.7 shows that the commercially used roughness geometries provide enhancement of 225% or more. Roughness appears to offer higher performance than other enhancement techniques, with the possible exception of internal fins. However, internal fins are difficult to make in many practical materials of interest. The low fin, helical-rib geometries of Figure 9.13a–c approach those of helical internal fins and can be made in copper. Engineers are encouraged to seek new engineering applications of tube-side roughness.

Roughness has a higher Prandtl number dependency than exists for plain tubes. This means that the heat transfer enhancement ratio (h/h_p) increases with increasing Prandtl number. For Prandtl numbers above, say, 20, a properly designed rough tube may have an efficiency index greater than 1.0.

The roughness size must be selected for the Reynolds number range of interest. If the tube is operated above this range, the efficiency index will drop to small values. If the tube is operated below this range, little enhancement will occur.

There are many possible roughness geometries, as suggested by Figure 9.2. Presently, the highest-performance roughness types are helical ribs and three-dimensional roughness. The three-dimensional roughness appears to offer the highest enhancement level, at a high efficiency index. Recent advances have been made in the manufacture of three-dimensional roughness, as shown in Figure 9.12.

Techniques to correlate the performance of roughness have moved well past the empirical correlation method. A semiempirical Reynolds analogy for rough surfaces has been found to be a powerful tool to correlate roughness data. Using data on one roughness geometry, the model may be used to predict the performance of geometrically similar roughness.

There have been recent advances in hydraulic and thermal modeling of roughness. This includes (a) the "discrete-element" method of Taylor and Hodge [1992] and (b) numerical solutions, as evidenced by Arman and Rabas [1992] and others. The numerical solution methods are presently limited to two-dimensional roughness geometries. The discrete-element method has been found to work well for three-dimensional roughness. Future developments are expected in predictive methods.

Heat transfer surface fouling is a concern for rough surfaces. This subject is discussed in Chapter 10.

9.11   REFERENCES

Achenbach, E., 1977. "The Effect of Surface Roughness on the Heat Transfer from a Circular Cylinder to the Cross Flow of Air," *International Journal Heat Mass Transfer*, Vol. 20, pp. 359–369.

Almeida, J. A., and Souza-Mendes, P. R., 1992. "Local and Average Transport Coefficients for the Turbulent Flow in Internally Ribbed Tubes," *Experimental and Thermal Fluid Science*, Vol. 5, pp. 513–523.

Arman, B., and Rabas, T. J., 1991. "Prediction of the Pressure Drop in Transverse, Repeated-Rib Tubes with Numerical Modeling," in *Fouling and Enhancement Interactions*, T. J. Rabas and J. M. Chenoweth, Eds., ASME HTD-Vol. 164, ASME, New York, pp. 93–99.

Arman, B., and Rabas, T. J., 1992. "Disruption Shape Effects on the Performance of Enhanced Tubes with the Separation and Reattachment Mechanism," in *Enhanced Heat Transfer*, M. B. Pate and M. K. Jensen, Eds., ASME Symposium, Vol. HTD-Vol. 202, ASME, New York, pp. 67–76.

Becker, B. R., and Rivir, R. B., 1989. "Computation of the Flow Field and Heat Transfer in a Rectangular Passage with a Turbulator," ASME paper 89-GT-30.

Benodekar, R. W., Goddard, A. J. H., Gosman, A. D., and Issa, R. I., 1985. "Numerical Prediction of Turbulent Flow over Surface-Mounted Ribs" *AIAA Journal*, Vol. 23, pp. 359–366.

Burck, E., 1970. "The Influence of Prandtl Number on Heat Transfer and Pressure Drop of Artificially Roughened Channels," in *Augmentation of Convective Heat and Mass Transfer*, A. E. Bergles and R. L. Webb, Eds., ASME Symposium Volume, ASME, New York, pp. 27–35.

Chakroun, W., and Taylor, R. P., 1992. "The Effects of Modestly Strong Acceleration on Heat Transfer in the Turbulent Rough-Wall Boundary Layer," in *Enhanced Heat Transfer*, M. B. Pate and M. K. Jensen, Eds., ASME Symposium Vol. HTD-Vol. 202, ASME, New York, pp. 57–66.

Chen, H. C., and Patel, V. C., 1988. "Near-Wall Turbulence Models for Complex Flows Including Separation," *AIAA Journal*, Vol. 26, pp. 641–648.

Churchill, S. W., 1973. "Empirical Expressions for the Shear Stress in Turbulent Flow in Commercial Pipe," *AIChE Journal*, Vol. 19, pp. 375–376.

Cope, W. G., 1945. "The Friction and Heat Transmission Coefficients of Rough Pipes," *Proceedings of the Institute of Mechanical Engineers*, Vol. 145, pp. 99–105.

Cuangya, L., Chuanyun, G., Chaosu, W., Jinshu, H., and Cun, J., 1991. "Experimental Investigation of Transitional Flow Heat Transfer of Three-Dimensional Internally Finned Tubes," in *Advances in Heat Transfer Augmentation and Mixed Convection*, M. A. Ebadian, D. W. Pepper, and T. Diller, Eds., ASME Symposium, Vol. HTD-Vol. 169, ASME, New York, pp. 45–48.

Dalle-Donne, M., and Meyer, L., 1977. "Turbulent Convection Heat Transfer from Rough Surfaces with Two-Dimensional Rectangular Ribs," *International Journal of Heat Mass Transfer*, Vol. 20, pp. 583–620.

Dawson, D. A., and Trass, O., 1972. "Mass Transfer at Rough Surfaces," *International Journal of Heat Mass Transfer*, Vol. 15, pp. 1317–1336.

Dipprey, D. F., and Sabersky, R. H., 1963. "Heat and Momentum Transfer in Smooth and Rough Tubes at Various Prandtl Numbers," *International Journal of Heat Mass Transfer*, Vol. 6, pp. 329–353.

Durst, F., Fonti, M., and Obi, S., 1988. "Experimental and Computational Investigation of the Two-Dimensional Channel Flow Over Two Fences in Tandem," *Journal of Fluids Engineering, ASME* Vol. 110, pp. 48–54.

Edwards, F. J., 1966. "The Correlation of Forced Convection Heat Transfer Data from Rough Surfaces in Ducts Having Different Shapes of Flow Cross Section," *Proceedings of the Third International Heat Transfer Conference*, Vol. 1, pp. 32–44.

Edwards, F. J., and Sheriff, N., 1961. "The Heat Transfer and Friction Characteristics of Forced Convection Air Flow Over a Particular Type of Rough Surface," *International Developments in Heat Transfer*, ASME, New York, pp. 415–426.

Fenner, G. W., and Ragi, E., 1979. "Enhanced Tube Inner Surface Heat Transfer Device and Method," U.S. Patent 4,154,291, May 15, 1979.

Fodemski, T. R., and Collins, M. W., 1988. "Flow and Heat Transfer Simulations for Two- and Three-Dimensional Smooth and Ribbed Channels," *Proceedings of the 2nd U.K National Conference on Heat Transfer*, C138/88, University of Stratchlyde, Glasgow, U.K.

Fujita, H., Hajime, Y., and Nagata, C., 1986. "The Numerical Prediction of Fully Developed Turbulent Flow and Heat Transfer in a Square Duct with Two Roughened Facing Walls," *Proceedings of the 8th International Heat Transfer Conference*, Vol. 3, pp. 919–924.

Gee, D. L., and Webb, R. L., 1980. "Forced Convection Heat Transfer in Helically Rib-Roughened Tubes," *International Journal of Heat Mass Transfer*, Vol. 23 pp. 1127–1136.

Gowen, R. A., and Smith, J. W., 1968. "Turbulent Heat Transfer from Smooth and Rough Surfaces," *International Journal of Heat Mass Transfer*, Vol. 11, pp. 1657–1673.

Groehn, H. G., and Scholz, F., 1976. "Heat Transfer and Pressure Drop of In-Line Tube Banks with Artificial Roughness," *Heat and Mass Transfer Sourcebook: Fifth All-Union Conf.*, Minsk, Scripta Publishing Co., Washington, D.C., pp. 21–24.

Hall, W. B., 1962. "Heat Transfer in Channels Having Rough and Smooth Surfaces," *Journal of Mechanical Engineering Science*, Vol. 4, pp. 287–291.

Han, J. C., 1984. "Heat Transfer and Friction in Channels with Two Opposite Rib-Roughened Walls," *Journal of Heat Transfer*, Vol. 106, pp. 774–781

Han, J. C., 1988. "Heat Transfer and Friction Characteristics in Rectangular Channels With Rib Turbulators," in *Journal of Heat Transfer*, Vol. 110, pp. 321–328.

Han, J. C., Glicksman, L. R., and Rohsenow, W. M., 1978. "An Investigation of Heat Transfer and Friction for Rib-Roughened Surfaces," *International Journal of Heat Mass Transfer*, Vol. 21, pp. 1143–1156.

Han, J. C., Glicksman, L. R., and Rohsenow, W. M., 1979. "An Investigation of Heat Transfer and Friction for Rib-Roughened Surfaces," *International Journal of Heat Mass Transfer*, Vol. 22, pp. 1587.

Han, J. C., Chandra, P. R., and Lau, S. C., 1988. "Local Heat/Mass Transfer Distributions around Sharp 180 deg Turns in Two-Pass Smooth and Rib-Roughened Channels," *Journal of Heat Transfer*, Vol. 110, pp. 91–98.

Han, J. C., Ou, S., Park, J. S., and Lei, C. K., 1989. "Augmented Heat Transfer in Rectangular Channels of Narrow Aspect Ratios with Rib Turbulators," *International Journal Heat Mass Transfer*, Vol. 32, No. 9, pp. 1619–1630.

Han, J. C., Zhang, Y. M., and Lee, C. P., 1991. "Augmented Heat Transfer in Square Channels with Parallel, Crossed, and V-Shaped Angled Ribs," in *Journal of Heat Transfer*, Vol. 113, pp. 590–596.

Hijikata, K., and Mori, Y., 1987. "Fundamental Study of Heat Transfer Augmentation of Tube Inside Surface by Cascade Smooth Turbulence Promoters and Its Application to Energy Conversion," *Wärme-und-Stoffübertragung*, Vol. 21, pp. 115–124.

Hijikata, K., Ishiguro, H., and Mori Y., 1987. "Heat Transfer Augmentation in a Pipe Flow with Smooth Cascade Turbulence Promoters and Its Application to Energy Conversion," *Heat Transfer in High Technology and Power Engineering*, W. J. Yang, and Y. Mori, Eds. Hemisphere Publishing Corp., Washington, D.C., pp. 368–397.

Hinze, J. O., 1975. *Turbulence*, 2nd Edition, McGraw–Hill, New York.

Hishida, M., and Takase, K., 1987. "Heat Transfer Coefficient of the Ribbed Surface," *Proceedings of the Third ASME/JSME Joint Thermal Engineering Conference*, Vol. 3, ASME, New York, pp.103–110.

Hitachi, 1984. "Hitachi High-Performance Heat-Transfer Tubes," Catalog EA-500, Hitachi Cable, Ltd, Tokyo, Japan.

Hosni, M. H., Coleman, H. W., and Taylor, R. P., 1989. "Measurement and Calculation of Surface Roughness Effects on Turbulent Flow and Heat Transfer, Report TFD-89–1, Mechanical and Nuclear Engineering Department, Mississippi State University.

Hosni, M. H., Coleman, H. W., and Taylor, R. P., 1991. "Measurements and Calculations of Rough-Wall Heat Transfer in the Turbulent Boundary Layer," *International Journal of Heat Mass Transfer*, Vol. 34, pp. 1067–1081.

Hudina, M., 1979. "Evaluation of Heat Transfer Performances of Rough Surfaces from Experimental Investigation in Annular Channels," *International Journal of Heat Mass Transfer*, Vol. 22, pp. 1381–1392.

Hung, Y. H., Liou, T. M., and Syang, Y. C., 1987. "Heat Transfer Enhancement of Turbulent Flow in Pipes with an External Circular Rib," in *Advances in Enhanced Heat Transfer-1987*, M. K. Jensen and V. P. Carey, Eds., ASME Symposium Vol. HTD-Vol. 68, ASME, New York, pp. 55–64.

Ito, Y., Hozumi, H., Noguchi, K., Oizumi, K., and Seki, K., 1983. "Characteristics of Newly Developed THERMOEXCEL with Pin-Shaped Fins for Water Cooled Coaxial Condensers," *Hitachi Cable Review*, No. 2 (August), pp. 45–49.

Jaber, M. H., Webb, R. L., and Stryker, P., 1991. "An Experimental Investigation of Enhanced Tubes for Steam Condensers," ASME paper 91-HT-5.

Kays, W. M., and Crawford, M. E., 1980. *Convective Heat Transfer*, McGraw–Hill, New York, pp. 174 and 188.

Kumar, R., and Judd, R. L., 1970. "Heat Transfer with Coiled Wire Turbulence Promoters," *Canadian Journal of Chemical Engineering*, Vol. 48, pp. 378–383.

Kuwahara, H., Takahashi, K., Yanagida, T., Nakayama, W., Hzgimoto, S., and Oizumi, K., 1989. "Method of Producing a Heat Transfer Tube for Single-Phase Flow," U.S. Patent 4,794,775, Jan. 3, 1989.

Lewis, M. J., 1974. "Roughness Functions, the Thermohydraulic Performance of Rough Surfaces and the Hall Transformation—An Overview," *International Journal of Heat Mass Transfer*, Vol. 17, pp. 809–814.

Li, H. M., Ye, K. S., Tan, Y. K., and Den, S. J., 1982. "Investigation of Tube-Side Flow Visualization, Friction Factors and Heat Transfer Characteristics of Helical-Ridging

Tubes," *Proceedings of the 7th International Heat Transfer Conference*, Vol. 3, Hemisphere Publishing Corp., Washington, D.C., pp. 75–80.

Maubach, K., 1972. "Rough Annulus Pressure Drop—Interpretation of Experiments and Recalculation for Square Ribs," *International Journal of Heat Mass Transfer*, Vol. 15, pp. 2489–2498.

McLain, C. D., 1975. "Process for Preparing Heat Exchanger Tube," U.S. Patent 1,906,605 issued to Olin Corp.

Mehta, M. H., and Raja Rao, M., 1979. "Heat Transfer and Friction Characteristics of Spirally Enhanced Tubes for Horizontal Condensers," in *Advances in Enhanced Heat Transfer*, J. M. Chenoweth et al. [Eds.], ASME Symposium Volume, ASME, New York, pp. 11–22.

Mehta, M. H., and Raja Rao, M., 1988. "Analysis and Correlation of Turbulent Flow Heat Transfer and Friction Coefficients in Spirally Corrugated Tubes for Steam Condenser Application," *Proceedings of the 1988 National Heat Transfer Conference*, HTD-Vol. 96, Vol. 3, ASME, New York, pp. 307–312.

Metzger, D. E., Fan, C. Z., and Pennington, J. W., 1983. "Heat Transfer and Flow Friction Characteristics of Very Rough Transverse Ribbed Surfaces with and without Pin Fins," *Proceedings of the 1983 ASME-JSME Thermal Engineering Conference*, Vol. 1, pp. 429–436.

Metzger, D. E., Vedula, R. P., and Breen, D. D., 1987. "The Effect of Rib Angle and Length on Convection Heat Transfer in Rib-Roughened Triangular Ducts," *Proceedings of the 1987 ASME-JSME Thermal Engineering Conference*, Vol. 3, pp. 327–333.

Meyer, L., 1980. "Turbulent Flow in a Plane Channel Having One or Two Rough Walls," *International Journal of Heat Mass Transfer*, Vol. 23, pp. 591–608.

Nakamura, H., and Tanaka, M., 1973. "Cross-Rifled Vapor Generating Tube," U.S. Patent 3,734,140, May 22, 1973.

Nakayama, W., Takahashi, K. and Daikoku, T., 1983. "Spiral Ribbing to Enhance Single-Phase Heat Transfer Inside Tubes," *Proceedings of the ASME-JSME Thermal Engineering Joint Conference*, Honolulu, HI, Vol. 1, ASME, New York, pp. 503–510.

Newson, I. H., and Hodgson, T. D., 1973. "The Development of Enhanced Heat Transfer Condenser Tubing," *Desalination*, Vol. 14, pp. 291–323.

Nikuradse, J., 1933. "Laws of Flow in Rough Pipes," *VDI Forschungsheft*, pp. 361. English translation, NACA TM-1292 (1965).

Nunner, W., 1956. "Heat Transfer and Pressure Drop in Rough Pipes," *VDI-Forschungsheft*, 455, Ser. B, Vol. 22, pp. 5–39, English translation, AERE Lib./Trans./786 (1958).

Patankar, S.V., 1990. "Numerical Prediction of Flow and Heat Transfer in Compact Heat Exchanger Passages," in *Compact Heat Exchangers*, R. K. Shah, A. D. Kraus and D. Metzger, Eds., Hemisphere Publishing Corp., Washington, D.C., pp. 191–204.

Petukhov, B. S., 1970. "Heat Transfer in Turbulent Pipe Flow with Variable Physical Properties," in *Advances in Heat Transfer*, Vol. 6, T. F. Irvine and J. P. Hartnett, Eds., Academic Press, New York, pp. 504–564.

Prasad, R. C., and Brown, M. J., 1988. "Effectiveness of Wire-Coil Inserts in Augmentation of Convective Heat Transfer," in *Experimental Heat Transfer, Fluid Mechanics and Thermodynamics*, R. K. Shah, E. N. Ganic and K. T. Yang, Eds., Elsevier, New York, pp. 502–509.

Rabas, T. J., and Arman, B., 1992. "The Influence of the Prandtl Number of the Thermal

Performance of Tubes with the Separation and Reattachment Enhancement Mechanism," in *Enhanced Heat Transfer*, M. B. Pate and M. K. Jensen, Eds., ASME Symposium, Vol. HTD-Vol. 202, ASME, New York, pp. 77–88.

Rabas, T. J., Bergles, A. E., and Moen, D. L., 1988. "Heat Transfer and Pressure Drop Correlations for Spirally Grooved (Rope) Tubes Used in Surface Condensers and Multistage Flash Evaporators," *Augmentation of Heat Transfer in Energy Systems*, ASME Symposium, Vol. HTD-Vol. 52, ASME, New York, pp. 693–704.

Rabas, T. J., Thors, P., Webb, R. L., and Kim, N-H, 1993. "Influence of Roughness Shape and Spacing on the Performance of Three-Dimensional Helically Dimpled Tubes," *Journal of Enhanced Heat Transfer*. Vol. 1, pp. 53–64.

Raja Rao, M., 1988. "Heat Transfer and Friction Correlations for Turbulent Flow of Water and Viscous Non-Newtonian Fluids in Single-Start Spirally Corrugated Tubes," *Proceedings of the 1988 National Heat Transfer Conference*, HTD-96, Vol. 1, ASME, New York, pp. 677–683.

Ravigururajan, T. S., and Bergles, A. E., 1985. "General Correlations for Pressure Drop and Heat Transfer for Single-Phase Turbulent Flow in Internally Ribbed Tubes," in *Augmentation of Heat Transfer in Energy Systems*, ASME Symposium, Vol. HTD-Vol. 52, ASME, New York, pp. 9–20.

Schlicting, H., 1979. *Boundary-Layer Theory*, 7th edition, McGraw–Hill, New York, pp. 600–620.

Sethumadhavan, R., and Raja Rao, M., 1983. "Turbulent Flow Heat Transfer and Fluid Friction in Helical Wire Coil Inserted Tubes," *International Journal of Heat Mass Transfer*, Vol. 26, pp. 1833–1845.

Sethumadhavan, R., and Raja Rao, M., 1986. "Turbulent Flow Friction and Heat Transfer Characteristics of Single- and Multi-Start Spirally Enhanced Tubes," *Journal of Heat Transfer*, Vol. 108, pp. 55–61.

Sheriff, N., and Gumley, P., 1966. "Heat Transfer and Friction Properties of Surfaces with Discrete Roughness," *International Journal of Heat and Mass Transfer*, Vol. 9, pp. 1297–1320.

Sumitomo, 1983. "Technical Data of Tred-Fin," Sumitomo Light Metal Industries, Aichi, Japan.

Takahashi, K., Nakayama, W., and Kuwahara, H., 1988. "Enhancement of Forced Convective Heat Transfer in Tubes Having Three-Dimensional Spiral Ribs," *Heat Transfer—Japanese Research*, Vol. 17, No. 4, pp. 12–28.

Tan, Y. K., and Xaio, J. W., 1989. "Influence of Prandtl Number to Fluid Heat Transfer Characteristics of Spirally Corrugated Tubes," *The Fourth Asian Congress of Fluid Mechanics*, Hong Kong.

Tanasawa, I., Nishio, S., Takano, K., and Tado, M., 1983. "Enhancement of Forced Convection Heat Transfer in Rectangular Channel Using Turbulence Promoters," *1983 ASME-JSME Thermal Engineering Joint Conference*, Vol. 1, Y. Mori and I. Tanasawa, Eds., pp. 395–402.

Taslim, M. E., and Spring, S. D., 1988. "An Experimental Investigation of Heat Transfer Coefficients and Friction Factors in Passages of Different Aspect Ratios Roughened with 45° Turbulators," in *Proceedings of the 1988 National Heat Transfer Conference*, HTD-96, Vol. 1, ASME, New York, pp. 661–668.

Taylor, R. P., and Hodge, B. K., 1992. "Fully-Developed Heat Transfer and Friction Factor

Predictions for Pipes with 3-Dimensional Roughness," in *Fundamentals of Forced Convection Heat Transfer*, M. S. Ebadian and P. H. Oosthuizen, Eds, ASME Symposium Vol. HTD-Vol. 210, ASME, New York, pp. 75–84.

Taylor, R. P., Coleman, H. W., and Hodge, B. K., 1984. "A Discrete Element Prediction Approach for Turbulent Flow Over Rough Surfaces," Report TFD-84–1, Department of Mechanical Engineering, Mississippi State University.

Taylor, R. P., Scaggs, and Coleman, H. W., 1988. "Measurement and Prediction of the Effects of Nonuniform Surface Roughness on Turbulent Flow Friction Coefficients," *Journal of Heat Transfer*, Vol. 110, pp. 380–384.

Webb, R. L., 1971. "A Critical Evaluation of Analytical and Reynolds Analogy Equations for Turbulent Heat and Mass Transfer in Smooth Tubes, *Wärme-und-Stoffübertragung*, Vol. 4, pp. 197–204.

Webb, R. L., 1979, "Toward a Common Understanding of the Performance and Selection of Roughness for Forced Convection," *Studies in Heat Transfer: A Festschrift for E. R. G. Eckert*, J. P. Hartnett et al., Eds., Hemisphere Publishing Corp., Washington, D.C., pp. 257–272.

Webb, R. L., 1982. "Performance Cost Effectiveness and Water-Side Fouling Considerations of Enhanced Tube Heat Exchangers for Boiling Service with Tube-Side Water Flow," *Heat Transfer Engineering*, Vol. 3, No. 3, pp. 84–98.

Webb, R. L., 1991. "Advances in Shell Side Boiling of Refrigerants," *Journal of the Institute of Refrigeration*, Vol. 87, pp. 75–86.

Webb, R. L., and Eckert, E. R. G., 1972. "Application of Rough Surfaces to Heat Exchanger Design," *International Journal of Heat Mass Transfer*, Vol. 15, pp. 1647–1658.

Webb, R. L., Eckert, E. R. G., and Goldstein, R. J., 1971. "Heat Transfer and Friction in Tubes with Repeated-Rib Roughness," *International Journal of Heat Mass Transfer*, Vol. 14, pp. 601–617.

Webb, R. L., and Robertson, G. F., 1988. "Shell-Side Evaporators and Condensers Used in the Refrigeration Industry, in *Heat Transfer Equipment Design*, R. K. Shah, E. C. Subbarao, and R. A. Mashelkar, Eds., Hemisphere Publishing Corp., Washington, D.C., pp. 559–570.

Webb, R. L., Eckert, E. R. G., and Goldstein, R. J., 1972. "Generalized Heat Transfer and Friction Correlations for Tubes with Repeated-Rib Roughness," *International Journal of Heat Mass Transfer*, Vol. 15, pp. 180–184.

Webb, R. L., Hui, T. S., and Haman, L., 1984. "Enhanced Tubes in Electric Utility Steam Condensers," *Heat Transfer in Heat Rejection Systems*, S. Sengupta and Y.F. Mussalli, Eds., Book No. G00265, HTD-Vol. 37, ASME, New York, pp. 17–26.

White, L., and Wilkie, D., 1970. "The Heat Transfer and Pressure Loss Characteristics of Some Multi-start Ribbed Surfaces," in *Augmentation of Convective Heat and Mass Transfer*, ASME, New York, pp. 55–62.

Wilkie, D., 1966. "Forced Convection Heat Transfer from Surfaces Roughened by Transverse Ribs," *Third International Heat Transfer Conference*, Vol. 1, pp. 1–19.

Wilkie, D., Cowan, M., Burnett, P., and Burgoyne, T., 1967. "Friction Factor Measurements in a Rectangular Channel with Walls of Identical and Non-identical Roughness," *International Journal of Heat Mass Transfer*, Vol. 10, pp. 611–621.

Williams, F., Pirie, M. A. M., and Warburton, C., 1970. "Heat Transfer from Surfaces Roughened by Ribs," in *Augmentation of Convective Heat and Mass Transfer*, ASME, New York, pp. 55–62.

Withers, J. G., 1980a. "Tube-Side Heat Transfer and Pressure Drop for Tubes Having Helical Internal Ridging with Turbulent/Transitional Flow of Single-Phase Fluid, Part 1, Single-Helix Ridging," *Heat Transfer Engineering*, Vol. 2, No. 1, pp. 48–58.

Withers, J. G., 1980b. "Tube-Side Heat Transfer and Pressure Drop for Tubes Having Helical Internal Ridging with Turbulent/Transitional Flow of Single-Phase Fluid, Part 2, Multiple-Helix Ridging," *Heat Transfer Engineering*, Vol. 2, No. 2, pp. 43–50.

Wolverine, 1984. *Engineering Data Book II*, K. J. Bell and A. C. Mueller, Eds., Wolverine Tube Corp., Decatur, AL.

Ye, Q. Y., Chen, W. L., and Tan, Y. K., 1987. "Prediction of Heat Transfer Characteristics in Spirally Fluted Tubes by Studying the Velocity Profiles and Turbulence Structure," in *1987 ASME-JSME Thermal Engineering Joint Conference*, Vol. 5, P. D. Marto and I. Tanasawa, Eds., pp. 189–194.

Yorkshire, 1982. "YIM Heat Exchanger Tubes: Design Data for Horizontal Rope Tubes in Steam Condensers," Technical Memorandum 3, Yorkshire Imperial Metals, Ltd., Leeds, England.

Zhang, Y. F., Li, F. Y., and Liang, Z. M., 1991. "Heat Transfer in Spiral-Coil-Inserted Tubes and its Application," in *Advances in Heat Transfer Augmentation*, M. A. Ebadian, D. W. Pepper, and T. Diller, Eds, ASME Symposium, Vol. HTD-Vol. 169, ASME, New York, pp. 31–36.

Žukauskas, A. A., 1972. "Heat Transfer from Tubes in Crossflow," in *Advances in Heat Transfer*, Vol. 8, J. P. Hartnett and T. F. Irvine, Jr., Eds., Academic Press, New York.

Žukauskas, A. A., and Ulinskas, R. V., 1983. "Surface Roughness as Means of Heat Transfer Augmentation for Banks of Tubes in Crossflow," in *Heat Exchangers: Theory and Practice*, J. Taborek, G. P. Hewitt and N. Afgan, Eds., Hemisphere Publishing Corp., Washington, D.C., pp. 311–321.

Žukauskas, A. A., Ulinskas, R. V., 1988. *Heat Transfer in Tube Banks in Crossflow*, Hemisphere Publishing Corp., New York, pp. 94–118.

9.12 NOMENCLATURE

A	Heat transfer surface area ($= \pi d_i L$ for tube), m² or ft²
A_c	Flow cross-sectional area ($= \pi d_i^2/4$ for tube), m² or ft²
A_p	Projected area of roughness element, m² or ft²
A_r	Aspect ratio of channel having two smooth walls opposite two rough walls ($= L_s/L_r$), dimensionless
A_x	Cross-sectional flow shown in Figure 9.24, m² or ft²
$B(e^+)$	Correlating function for rough tubes, Equation 9.4, dimensionless
b	Height of rectangular channel, m² or ft²
C_D	Drag coefficient, dimensionless
c_p	Specific heat of fluid at constant pressure, J/kg-K or Btu/lbm-°F
D_h	Hydraulic diameter of flow passages, $4LA_c/A$, m or ft
d_e	Roughness element base diameter, m or ft
d_i	Tube inside diameter (diameter to the base of roughness), m or ft
d_o	Tube outside diameter, m or ft
e	Fin height or roughness height, m or ft

e_{sg}	Equivalent sand-grain roughness, m or ft
e^+	Roughness Reynolds number ($= eu^*/\nu$), dimensionless
F_D	Drag force on roughness element, N or lbf
f	Fanning friction factor, $\Delta p_f d_i/2LG^2$; dimensionless
G	Mass velocity (W/A_c), kg/m²-s or lbm/ft²-s
g	$[f/(2St) - 1]/(f/2)^{1/2} - B(e^+)$, dimensionless
$\bar{g}$	$g\,Pr^{-n}$, dimensionless
g	Acceleration due to gravity, 9.806 m/s² or 32.17 ft/s²
h	Heat transfer coefficient based on A (h for all rough tubes is based on $A_i/L = \pi d_i$), W/m²-K or Btu/hr-ft²-°F
h_o	Shell-side heat transfer coefficient, W/m²-K or Btu/hr-ft²-°F
j	Colburn factor ($= StPr^{2/3}$), dimensionless
K	Thermal conductance ($= hA$), W/K or Btu/hr-°F
k	Thermal conductivity of fluid, W/m-K or Btu/hr-ft-°F
L	Fluid flow (core) length on one side of the exchanger, m or ft
L_s	Height of smooth wall in rectangular channel having one rough and one smooth wall, m or ft
L_r	Width of rough wall in rectangular channel having one rough and one smooth wall, m or ft
LMTD	Logarithmic mean temperature difference, K or °F
M	Mass of tube material, kg or lbm
Nu_d	Nusselt number ($= hd_i/k$), dimensionless
NTU	Number of heat transfer units [$= UA/(Wc_p)_{min}$], dimensionless
n_f	Number of fins in internally finned tube, dimensionless
n_s	Number of starts around the tube circumference for a helical-rib roughness, dimensionless
P	Fluid pumping power, W or hp
Pr	Prandtl number ($= c_p\mu/k$), dimensionless
Pr_e	Effective Prandtl number (ν_e/α_e), dimensionless
Pr_t	Turbulent Prandtl number (ϵ_m/ϵ_h), dimensionless
p	Axial spacing between roughness elements, m or ft
Δp	Fluid static pressure drop, Pa or lbf/ft²
Q	Heat transfer rate, W or Btu/hr
q	Heat flux, W/m² or Btu/hr-ft²-°F
q_o	Heat flux at wall, W/m² or Btu/hr-ft²-°F
Re_{Dh}	Reynolds number based on the hydraulic diameter ($= GD_h/\mu$), dimensionless
Re_d	Reynolds number based on the tube diameter ($= Gd/\mu$, $d = d_i$ for flow inside tube and $d = d_o$ for flow outside tube), dimensionless
Re_{de}	Reynolds number based on the tube diameter ($= Gd_e/\mu$), dimensionless
R_{fi}	Tube-side fouling resistance, m²-K/W or ft²-hr-°F/Btu
St	Stanton number $= (h/Gc_p)$, dimensionless
T	Temperature, T_w (wall), T_∞ (free stream), T_{av}(average), K or °F
T^+	Dimensionless temperature, $T/(Q/\rho c_p u^*)$, dimensionless

ΔT	Temperature difference between hot and cold fluids, K or °F
ΔT_i	Temperature difference between hot and cold inlet fluids, K or °F
U	Overall heat transfer coefficient, W/m²-K or Btu/hr-ft²-°F
u	Local fluid velocity, m/s or ft/s
u_m	Fluid mean axial velocity at the minimum free flow area, m/s or ft/s
u^*	Friction velocity [$= (\tau_w/\rho)^{1/2}$], m/s or ft/s
u^+	Dimensionless velocity (u/u^*)
V	Heat exchanger total volume, m³ or ft³
V_m	Heat exchanger tube material volume, m³ or ft³
v	Velocity component normal to wall, m/s or ft/s
W	Fluid mass flow rate ($= \rho u_m A_c$), kg/s or lbm/s
w	Width of roughness element at base, m or ft
x	Cartesian coordinate along the flow direction, m or ft
y	Coordinate distance normal to wall, m or ft
y_b	Thickness of viscous influenced fluid layer, m or ft
y^+	yu^*/v, dimensionless
y_b^v	$y_b u^*/v$, dimensionless
z	Dimple pitch in helix angle direction, m or ft

Greek Letters

α	Thermal diffusivity ($= k/\rho c_p$), dimensionless
α	Helix angle relative to tube axis ($= \pi d_i/p_{tv}$), radians or degrees
δ	Boundary layer thickness, m or ft
δ^+	Dimensionless boundary layer thickness ($\delta u^*/v$)
ϵ_m	Eddy diffusivity for momentum, m²/s or ft²/s
ϵ_h	Eddy diffusivity for heat, m²/s or ft²/s
η	Efficiency index [$= (h/h_s)/(f/f_s)$], dimensionless
μ	Fluid dynamic viscosity coefficient, Pa-s or lbm/s-ft
v	Kinematic viscosity, m²/s or ft/s²
v_e	Effective viscosity in turbulent region ($= \beta + \epsilon_m$), m/s²
ρ	Fluid density, kg/m³ or lbm/ft³
τ	Shear stress in fluid, Pa or lbf/ft²
τ_w	Wall shear stress, Pa or lbf/ft²
τ_o	Apparent wall shear stress, Pa or lbf/ft²

Subscripts

el	Refers to roughness elements of Figure 9.24
fd	Fully developed flow
FR	Fully rough condition
m	Average value over flow length
max	Maximum value
p	Plain tube or surface
s	Smooth tube or surface

sg	Sand-grain roughness
tr	Transverse-rib roughness
w	Evaluated at wall temperature
x	Local value
δ	Viscous boundary layer thickness

10

FOULING ON ENHANCED SURFACES

10.1 INTRODUCTION

The commercial viability of enhanced heat transfer surfaces is dependent on their long-term fouling characteristics. The problem of heat transfer surface fouling is of concern for both plain and enhanced surfaces. Although fouling is generally regarded as a more serious problem for liquids than for gases, gas-side fouling can be important in certain situations. Section 2.7 describes the importance of fouling in heat exchanger design and explains why fouling is an important concern for enhanced surfaces. A primary concern for enhanced surfaces is their fouling rate, relative to a plain surface, when operated at the same velocity.

Presently, the state of the art does not allow quantitative prediction of the fouling resistance that will occur on smooth or enhanced surfaces in actual field installations. However, recent and ongoing research is advancing our quantitative understanding of fouling phenomena. Readers interested in general information describing the state of the art on fouling are referred to books by Somerscales and Knudsen [1979], Garrett-Price et al. [1985], and Melo et al. [1987]. Watkinson [1990, 1991] provides review articles of fouling on enhanced heat transfer surfaces.

Fouling research has resulted in the definition of six different types of fouling that may occur with liquids or gases:

1. Precipitation fouling (scaling) occurs when salts precipitate on a heat transfer surface, if the temperature is sufficiently hot (or cold) to cause supersaturation of the salts at the surface temperature. The salts may have either increasing or decreasing solubility with increasing temperature. Evaporation of dissolved salts in a liquid beyond the solubility limit is one example. Precipitation of $CaCO_3$ from heated cooling water is an example of inverse solubility.

2. Particulate fouling takes place when suspended solids deposit on the surface. Examples are: dirt contained in cooling water; corrosion products formed elsewhere and deposited on the surface; and airborne particulates (such as dust in air, or soot in combustion products).

3. Corrosion fouling occurs when the heat transfer surface material reacts with the fluid to yield corrosion deposits.

4. Chemical reaction at the heat transfer surface (the surface material is not a reactant) may yield surface deposits. Examples are polymerization, cracking, and coking of hydrocarbons.

5. Biofouling deposits are formed when biological mechanisms attach and grow on the heat transfer surface. Untreated water-cooling systems are susceptible to biofouling.

6. Solidification fouling can occur with either liquids or gases. Water vapor may be condensed from air and frozen on a surface below the freezing point. Similarly, a component such as sodium sulfate may precipitate from a gas and solidify on the heat transfer surface. A layer of ice will form on the surface if the surface temperature is below the freezing point of water. Paraffin waxes in hydrocarbon solutions may deposit as solids on a cooled surface.

Measurement of the fouling resistance is typically performed by measurement of the total thermal resistance $(1/UA)$ for the clean and fouled conditions. The fouling resistance (R_f) is obtained by subtraction of the $1/UA$ values for the fouled and clean conditions, respectively, giving

$$\frac{R_f}{A} = \frac{1}{(UA)_f} - \frac{1}{(UA)_c} \qquad (10.1)$$

where A is the surface area, on which the R_f is based. For tube-side fouling, the UA values for the clean and fouled conditions are defined by Equations 10.2 and 10.3, respectively:

$$\frac{1}{(UA)_c} = \frac{1}{h_i A_i} + \frac{t_w}{k_w A_w} + \frac{1}{h_o A_o} \qquad (10.2)$$

$$\frac{1}{(UA)_f} = \left(\frac{1}{h_i} + R_f \right) \frac{1}{A_i} + \frac{t_w}{k_w A_w} + \frac{1}{h_o A_o} \qquad (10.3)$$

The inside (h_i) and outside (h_o) heat transfer coefficients must be equal in the dirty and clean tube conditions. Otherwise, the measured fouling resistance, R_f, will be erroneous.

10.2   FOULING FUNDAMENTALS

Fouling is a rate-dependent phenomena. The net fouling rate is the difference between the solids deposition rate and their removal rate:

$$\frac{dm_f}{dt} = \dot{m}_d - \dot{m}_r \qquad (10.4)$$

Depending on the magnitudes of the deposition and removal terms, several fouling rate characteristics are possible. Figure 10.1 illustrates the increase of the fouling factor (R_f) with time for different possible fouling situations. Crystallization fouling has an initial delay period (t_d), during which nucleation sites are established; then, the fouling deposit begins to accumulate. A linear growth of the fouling deposit occurs if the removal rate (m_r) is negligible or if $\dot{m}_d$ and $\dot{m}_r$ are constant with $\dot{m}_d > \dot{m}_r$. Crystallization, chemical reaction, corrosion, and freezing fouling will show behavior in the linear or falling rate categories. The fouling resistance will attain an asymptotic value if $\dot{m}_d$ is constant and $\dot{m}_r$ eventually attains a constant value. Particulate fouling typically shows asymptotic behavior. Which type of fouling characteristics occurs depends on the fouling mechanism.

Epstein [1983, 1988a, 1988b] provides detailed discussion of the different fouling mechanisms. He also describes models to predict the deposition and removal rates for the different fouling mechanisms. The deposition model is a function of the fouling mechanism. The removal rate model depends on the re-entrainment rate, which is proportional to the shear stress at the surface.

The heat exchanger operating parameters that may influence the deposition or removal rates are as follows:

1. The bulk fluid temperature will typically increase chemical reaction rates and crystallization deposition rates.
2. Increasing surface temperature will increase chemical reaction rates, or crystallization from inverse solubility salts. Reducing surface temperature will increase solidification fouling.
3. The combination of surface material and fluid will influence corrosion foul-

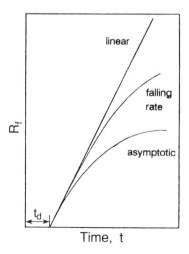

Time, t

Figure 10.1 Characteristic fouling curves.

ing. Copper surfaces act as a biocide to biological fouling. Rough surfaces may promote nucleation based phenomena.

4. The liquid or gas velocity may influence both the deposition and removal rates. Removal of soft deposits is affected by surface shear stress, which increases with velocity. Increased velocity will increase the mass transfer coefficient, and hence the deposition rate for diffusion controlled particulate fouling.

10.2.1   Particulate Fouling

To limit our discussion of fouling mechanisms, we will focus on particulate fouling, which is important for both gases and liquids. The particulate fouling deposition rate term is given by

$$\dot{m}_d = SK_m(C_b - C_w) \tag{10.5}$$

The deposition process is controlled by the mass transfer coefficient (K_m) and the concentration difference between that in the bulk fluid (C_b) and that in the liquid at the foulant surface (C_w). Typically, $C_w = 0$. The term S is called the "sticking probability." This is the probability that a particle transported to the wall will stick to the wall.

The foulant removal rate model was proposed by Taborek et al. [1972] and is assumed to be proportional to surface shear stress (τ_w) and foulant deposit thickness (x_f) and inversely proportional to the deposit bond strength factor (ξ). The removal rate is given by

$$\dot{m}_r = \frac{m_f \tau_w}{\zeta} = \frac{\rho_f x_f \tau_w}{\zeta} \tag{10.6}$$

Substitution of Equations 10.5 and 10.6 in Equation 10.2 using $R_f \equiv x_f / k_f$ and $R_f^* = dR_f / dt$ gives the differential equation to be solved:

$$\frac{dR_f}{dt} = SK_m C_b - BR_f \tag{10.7}$$

where

$$B = \frac{\rho_f \tau_w k_f}{\zeta} \tag{10.8}$$

The solution of Equation 10.7 is

$$R_f = R_f^*(1 - e^{-Bt}) \tag{10.9}$$

where R_f^* is the asymptotic fouling resistance defined by

$$R_f^* = \frac{SK_m C_b}{B} \tag{10.10}$$

Equations 10.7 and 10.8 show that the R_f and R_f^* can be predicted if K_m, τ_w, S, and ξ are known. One must also know the foulant density (ρ_f) and thermal conductivity (k_f). Equation 10.7 was first developed by Kern and Seaton [1959], who assumed that S and ξ are unity.

Papavergos and Hedley [1984] have shown that there are three regimes for particle deposition—the diffusion, inertia, and impaction regimes. Which regime controls the deposition process is determined by the dimensionless particle relaxation time defined by

$$t^+ = \frac{\rho_p d_p^2 (u^*)^2}{18\mu\nu} \tag{10.11}$$

Papavergos and Hedley [1984] show that the t^+ values associated with the diffusion, inertia, and impaction regimes are $t^+ < 0.10$ (diffusion), $0.10 \leq t^+ \leq 10$ (inertia), and $t^+ > 10$ (impaction). Fine-particle transport will likely be diffusion-controlled. For example, the deposition process will be diffusion controlled for $d_p \leq$ 10 μm for Re = 30,000 in a 15-mm-diameter tube. If the particle transport is diffusion-controlled, the mass transfer coefficient may be predicted from heat transfer data on the enhanced surface, using the heat–mass transfer analogy, as reported by Kim and Webb [1991] and Chamra and Webb [1993b] on rough surfaces. If the heat transfer coefficient for the enhanced surface (h) is known, the mass transfer coefficient (K_m) is given by

$$\frac{K_m}{u} Sc^{2/3} = \frac{h}{\rho u c_p} Pr^{2/3} \tag{10.12}$$

Sc (ν/D) and the diffusion coefficient (D) is defined in the nomenclature. Small particles result in high Schmidt numbers. Equation 10.12 shows that a high heat transfer coefficients will result in a high mass transfer coefficient. Hence, enhanced heat transfer surfaces should result in higher foulant deposition rates than occur with plain surfaces at the same operating velocity.

Equation 10.6 shows that the foulant removal rate is directly proportional to the surface shear stress. One may also expect higher surface shear stress for an enhanced surface. The net effect of deposition and removal is defined by the asymptotic fouling resistance given by Equation 10.10. Equation 10.10 shows that $R_f^* \propto$ $\dot{m}_d \tau_w \propto K_m/\tau_w$. If $K_m/\tau_w > 1$, the enhanced surface will have a higher asymptotic fouling resistance than a plain surface. More discussion of this is provided by Watkinson [1991] and in Section 10.8.

10.3   FOULING OF GASES ON FINNED SURFACES

Marner and Webb [1983] provide a survey and bibliography of gas-side fouling. The bibliography provides 206 references dating from 1970. As noted in Chapter 2, gas-side fouling may not significantly increase the gas-side thermal resistance in finned tube gas-to-liquid heat exchangers. This may occur if the gas-side surface area is much greater than the liquid-side surface area. However, the fouling deposit on the heat exchanger surface increases the gas pressure drop. The increased pressure drop may result in a lower gas-flow rate, because of the lower gas-flow rate at the balance point on the fan curve. Hence, the heat transfer rate is reduced, because of the lower gas-flow rate through the heat exchanger. An example of this is provided by the test data of Bott and Bemrose [1981]. They measured the effects of particulate fouling (3- to 30-μm calcium carbonate dust) on a finned heat exchanger as illustrated in Figure 6.1b. The four-row heat exchanger was composed of spiral wound fins at 2.4-mm pitch on a 25.4-mm-diameter tube with 15.9-mm-high fins arranged in a staggered layout. Dust concentrations of 0.65 to 1.5 g/m^3 were injected, which the authors state "is equivalent to desert storm concentration." The heat exchanger was sprayed with a bonding agent prior to starting the fouling tests. This was intended to simulate field conditions and stabilize the foulant layer. Their test #24, taken at 4.8-m/s air frontal velocity (Re = 2400) and 0.78-g/m^3 dust loading, resulted in a 67% friction factor increase over the 26-hr test period. However, the air-side heat transfer coefficient decreased only 12%. The air-flow rate reduced during the test period, because of the increased pressure drop. Although the flow rate reduction was not documented for test #24, they indicate that it may have been on the order of 10%. These data clearly show that the severe fouling caused a relatively small effect in the j factor, but a large increase in pressure drop. The associated increase of fan power is a costly penalty resulting from the fouling. Fouling factor curves were not given by the authors.

Gas-side fouling is generally not a problem of concern for air-conditioning and automotive heat exchangers, where heating of the air is involved. This is because ambient air is relatively clean. In an air-cooled refrigerant evaporator, freezing fouling (frost or ice) occurs if the surface temperature is below the freezing point. This results in a thermal resistance, along with plugging of the flow passage. The foulant deposit is removed by melting the frost layer. Heat exchangers used in agricultural equipment are subject to particulate fouling (e.g., dust and chaff of threshed grains). Industrial heat exchangers used in dusty environments are subject to particulate fouling by dust. Such freezing and particulate fouling situations generally cause limitations on the minimum fin pitch.

Zhang et al. [1990] measured accelerated particulate fouling for a plate fin-and-tube automotive heat exchanger containing oval tubes and louvered fins, similar to that illustrated in Figure 6.15. The heat exchanger contained eight rows of tubes (27.6/12-mm major/minor diameter, 8.3-mm fin height, and 2.11-mm fin pitch). The air contained 5- to 12-μm CaCO$_3$ particulates. A foulant "cake" formed on the front surface of the heat exchanger, and fouling rates were higher for the smaller particles and at higher air velocities. The foulant deposit caused a significant air pressure drop increase. These authors present fouling rate curves for different particle loadings.

Zhang et al. [1992] investigated the benefits of placing "spoilers" at the upstream face of a four-row bank of circular finned tubes. The spoilers consists of a shaped flow obstruction placed just upstream of each tube in the first row of tubes. They propose that the turbulence generated by the spoiler may reduce fouling. They measured the R_f^* for six different spoiler configurations. They asserted that the largest fouling deposit will occur on the front face of the finned tube heat exchanger. This may be the case for the plate fin-and-tube geometry studied by Zhang et al. [1990], although it may not be true for the circular fin geometry. In the author's opinion, this study was not conclusive in justifying the spoiler concept. Furthermore, the spoilers add to the gas pressure drop.

More severe fouling situations exist for combustion products of coal and fuel oil. Combustion products may also contain vapors, which precipitate upon cooling. Possible fouling mechanisms include particulate, corrosion, and precipitation fouling. Gases which contain SO_2 will form corrosive H_2SO_3 (sulfuric acid) if both SO_2 and water vapor are condensed. Combustion products from glass furnaces contain H_2SO_4, which forms a sticky deposit when combined with condensed water vapor. Two examples, described by Webb et al. [1983], are provided to illustrate fouling resulting from combustion products, for two types of extended surface heat exchangers. Burgmeier and Leung [1981] measured the fouling characteristics of a plate-and-fin geometry using a simulated glass furnace exhaust. The 677°C gases were composed of natural gas combustion products with dilution air, 1600 ppm of 11- to 75-μm particulates, and added contaminants primarily containing SO_2, with SO_3 and HCl. They tested a plain fin geometries (Figure 6.1b) having 7.4-mm fin height with 295 fins/m and an offset strip fin having 3.7-mm fin height with 197 fins/m. They used a traversing air lance for cleaning, which consisted of a series of small holes drilled along the axis of the lance tube. The lance provides a line array of high-velocity air jets which discharge directly onto the flow passages, at the upstream exchanger face. Figure 10.2 shows the fouling data for the two surface geometries with the air lance operating at 10-hr intervals. The air lance successfully cleaned the plain fin surface, but was ineffective for the offset strip fin.

Roberts and Kubasco [1979] measured fouling on a bank of helical-finned tubes (Figure 6.1b) in the exhaust of a gas turbine. The tubes had a 19.05-mm diameter, 275 fins/m, and approximately 9-mm fin height. The tubes were exposed to 450°C exhaust gases, and the fin temperatures were 150–315°C. The deposits were "fluffy" and easily removed by air lancing. The deposit thickness increased as the metal temperature was decreased. Analysis of the deposit suggested a two-stage fouling mechanism. Trace quantities of hydrocarbons condense on the surface, forming a thin coating of an adhesive nature, which trapped particulates. A self-cleaning method was evaluated, in which the tube-side coolant flow was stopped and the hot gases baked away the deposit.

Kindlman and Silvestrini [1979] measured fouling and corrosion on 32-mm-diameter, 118- and 197-fin/m finned tubes (Figure 6.1b) in the 200–260°C exhaust of a diesel engine, whose acid condensation temperature was 117°C. The test

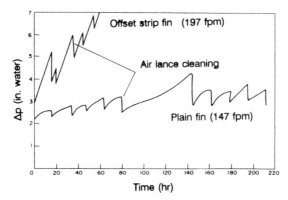

Figure 10.2   Fouling and cleaning characteristics of the plate-and-fin heat exchangers using 677°C/149°C gas inlet/outlet temperatures, with 1600-ppm particle loading, as reported by Burgmeier and Leung [1981].

compared fouling on uncooled tubes with that on cooled tubes (104°C wall temperature). No measurable soot accumulation was observed on the uncooled tubes. However, the cooled tubes exhibited severe fouling and plugging.

These three studies show a consistency in several respects: (1) The precipitation fouling is dependent on the wall temperature and the surface-to-gas temperature difference. (2) If vapors do not condense, an easily removed fluffy deposit occurs. (3) Above the acid and water vapor dewpoints, some vapors may condense. These yield a hard deposit on the tube surface that is difficult to remove by air lancing. However, it may be washed away with water. (4) Below the acid and water dewpoints, the deposit is sticky and cannot be removed by air lancing, although it may be water washed.

It is interesting to speculate on the relative fouling rates of the plate-and-fin and finned-tube exchanger types, and their cleanability. The plate-and-fin geometry has a constant area flow channel, and no stagnation or separated flow zones within the fin passage. One would expect a relatively uniform fouling deposit on the fins. Air or steam lancing will affect foulant removal as a result of high shear stress at the fin surface. This high velocity should be quite effective over the entire finned surface area. Conversely, finned tubes experience (a) stagnation/accelerating flow on the upstream half of the tube and (b) a decelerating flow with flow separation on the downstream half of the tube. Because the surface shear stress is greater on the front half of the tube, smaller fouling deposits are expected on the upstream face. This foulant deposit pattern was observed by Kindlman and Silvestrini [1979] for heat recovery from diesel exhaust. The fouling deposit thickness distribution at the fin tip for the upstream and downstream faces are shown in Figure 10.3. Using air or steam lancing, the smallest local surface shear stresses will occur on the downstream face—where the greatest foulant deposit exists. Thus, the desired cleaning shear stress distribution is opposite to the desired distribution.

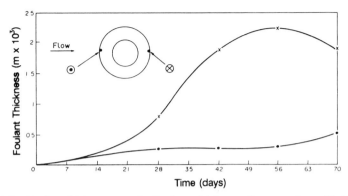

Figure 10.3   Fouling thickness near the fin tip on 31.8-mm-diameter tubes with 118 fins/m and 15.9-mm fin height, as reported by Kindlman and Silverstrini [1979].

10.4   SHELL-SIDE FOULING OF LIQUIDS

10.4.1   Low Radial Fins

Several papers have been published on fouling of low, integral-finned tubes (Figure 6.1c) in actual heat exchanger service. These long-term studies in field fouling situations typically show that the integral-fin tubes experience equal or less scaling fouling (R_f/A_o) than do plain tubes, and the scale is easier to remove than on plain tubes. Moore [1974] argues that the small axial thermal expansion between adjacent fins acts to break the scale or keep it looser than that on plain tubes. Moore's conclusions are based on actual examination of fouling deposits and cleaning ability using water jets. Webber [1960] conducted laboratory and field studies of precipitation (scaling) fouling on 19-mm-diameter integral-fin tubes (748 fins/m, 1.5 mm). Their single-tube laboratory studies indicated that the finned tubes would not foul as readily as plain tubes, and that once dirty, they would be easier to clean. Moore recommends integral-fin tubes in "any dirty service, as long as coking does not occur."

Webber reports several case studies, in which plain and finned tube bundles are compared in service involving scaling fouling. He concludes that the finned tube bundles may be operated for a longer time period, before heat transfer limitations make cleaning necessary.

Katz et al. [1954] compare the performance of plain and finned tube bundles (725 fins/m) having hot No. 5 fuel oil on the shell side and cooling water on the tube side. In the clean condition, the finned tube bundle provided a *UA* value 2.2 times that of the plain tube bundle. After 670 hours, the *UA* value of the finned tube bundle was 2.5 times that of the plain tube bundle. These data are also discussed by Watkinson [1990].

Bemrose and Bott [1984] tested finned tubes in one to four tube row configurations with calcium carbonate dust. They used a multiple regression technique to develop a correlation for the fouling resistance. No comparison with plain tube fouling was given.

10.4.2   Axial Fins and Ribs in Annulus

Sheikholeslami and Watkinson [1986] measured accelerated precipitation fouling from hard water on flow in an annulus. They tested a plain tube and an axially finned tube, as illustrated in Figure 8.13a. The 25°C hard water was heated by tube side steam at 129 kPa. The R_f was based on the nominal tube surface area ($A_o/L = \pi d_o$). Surprisingly, the finned tube (12 fins, 6.0 mm high on a 19-mm-diameter tube) showed a fouling resistance 50–60% less than that for a steel plain tube of the same diameter. The deposit on the finned tube was heaviest on the prime surface. The scale thickness decreased with distance from the fin base. This is expected, since high surface temperature promotes scaling, and the fin temperature decreases from the base.

Freeman et al. [1990] measured particulate fouling on the outer surface of several 12.7-mm-diameter finned and rough tubes in a double-pipe heat exchanger. Their tests were performed using 0.3-μm Al_2O_3 particles in a nonpolar fluid called X-2, which contained C_6 and C_7 hydrocarbons. They measured the asymptotic fouling resistance for particle concentrations between 700 and 2000 ppm and liquid Reynolds numbers between 4000 and 14,000. Figure 10.4 shows their asymptotic fouling results on three enhanced surfaces and two plain surfaces (one commercially rough and one polished), apparently taken at 650-ppm particulate concentration. The R_f^* of threaded and knurled surfaces is not much different from that of the plain

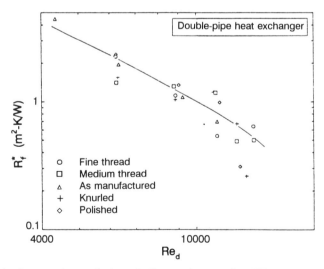

Figure 10.4 Asymptotic particulate fouling resistance for 650-ppm concentration of 0.3-μm Al_2O_3 particles in X-2 fluid tested in double-pipe heat exchangers by Freeman et al. [1990]. (From Freeman et al. [1990].)

tubes. The thread or knurl depth roughness of the three enhanced surfaces, based on British Standard BS 1134, are fine thread (0.01 mm), medium thread (0.018 mm), and knurled (0.087 mm). These compare to 0.0055 for the commercial smooth plain tube. The Figure 10.4 data shows that $R_f^* \propto \text{Re}^{-1.85}$, which compares with $R_f^* \propto \text{Re}^{-2.0}$ predicted by theory of Müller-Steinhagen et al. [1988] for adhesion controlled particulate fouling.

10.4.3 Ribs in Rod Bundle

Owen et al. [1987] investigated particulate fouling effects that may occur in gas-cooled nuclear reactors, which use fuel rods having transverse-rib roughness. They observed submicron deposits (0.1- to 0.2-μm-diameter) on the fuel rods. They developed a theoretical model to predict the deposition process. The model accounts for the effect of thermophoresis and surface roughness. The model shows that thermophoresis significantly affects the deposition of particles in the 0.1- to 0.7 μm-size range.

10.5 FOULING OF LIQUIDS IN INTERNALLY FINNED TUBES

Watkinson et al. [1974] and Watkinson [1975] conducted accelerated scaling experiments on the internally finned and spirally indented tubes listed in Table 10.1. Calcium carbonate was deposited from a solution of $NaHCO_3$, $CaCl_2$, and NaCl containing 400-ppm suspended solids and 3000-ppm dissolved solids. The 330 K hard water was heated by steam condensing at 380 K. The fouling factor was based on the total internal surface area of the tubes (A_{tot}). Figure 10.5 shows the asymptotic fouling resistance R_f^* versus water velocity for the internally finned tubes. Table 10.1 gives the ratio $R_{f,t}/R_{f,p}$, where $R_{f,t}$, and $R_{f,p}$ are based on the total and the nominal surface areas of the enhanced tubes, respectively. The fouling factor ratio is compared at 1.2 m/s and the same fouling operating conditions. Table 10.1 shows that the tubes have higher fouling resistance than the plain tube, at the same operating conditions. This is somewhat misleading, since the plain and enhanced tube surface areas (A_i/L) are different. As shown by Equation 10.1, the fouling resistance

TABLE 10.1 Ratio of Fouling Resistances for Enhanced and Plain Tubes

Geometry	n_f	d_i (mm)	e_i/d_i	A_i/L	A_{tot}/A_n	$R_{f,t}/R_{f,p}$
Plain	0	10.4	0.00	0.033	1.00	1.00
Internal fin 1	6	11.8	0.16	0.052	1.42	1.15
Internal fin 2	10	10.4	0.13	0.060	1.85	1.27
Internal fin 3	16	12.7	0.14	0.082	2.46	1.32
Spirally indented	4	17.4	0.23	0.054	1.0	1.15
Spirally indented	4	17.6	0.24	0.055	1.0	2.30

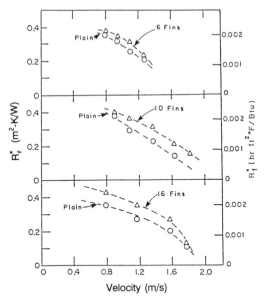

Figure 10.5   Asymptotic fouling resistances for the internally finned and plain tubes shown in Figure 10.1, based on the data of Watkinson et al. [1974]. (From Watkinson [1990].)

is R_f/A. The ratio of the R_f/A_i values for internal fin #3 and the plain tube is $1.32 \times 0.033/0.082 = 0.53$. Hence, the fouling penalty $(iR_f/A_i t)$ for the internally finned tube is actually less than that for the plain tube!

Somerscales et al. [1991] tested two internally finned tubes using 3-μm magnesium oxide foulant at 2500-ppm concentration in water. The tests were performed at two different water velocities, 0.9/1.0 and 1.5 m/s. Heat was supplied by electric heat generation in the tube wall (1.2-m long), and the heat transfer coefficient was measured using a wall thermocouple. The tube geometries tested were: Tube 1 (12.3-mm inside diameter with 10 axial fins 1.57-mm high), Tube 2 (16.2-mm inside diameter with 38 helical fins 0.64 mm high and 27-degree helix angle).

Data were also obtained using a smooth tube and three rough tubes. The rough-tube results are discussed in Section 10.6. Their fouling resistance was based on the internal envelope area ($A_i/L = \pi d_i$). None of the six tubes showed fouling at 1.5 m/s. After 12 hours, neither internal fin tube showed much fouling ($R_f \approx 0.3E - 5$ m^2-K/W). However, the helical finned tube had the smaller R_f. At 1.0 m/s, the helical finned tube showed initial build-up of high R_f, but it apparently broke away and the R_f settled down to a small value.

The fouling resistance based on total internal area may be converted to a resistance based on the nominal internal surface area ($A_n = \pi d_i/L$) using the relationship,

$$R_{f,n} = R_{f,\text{tot}} \left(\frac{A_n}{A_\text{tot}} \right) \qquad (10.13)$$

Definition of R_f based on Equation 10.3 allows the R_f to be directly compared with the plain tube R_f for enhanced and plain tubes having the same nominal inside diameter.

10.6 LIQUID FOULING IN ROUGH TUBES

Webb and Kim [1989] measured accelerated particulate fouling in water for three different internally ribbed tubes and a plain tube. The purpose of the tests was to compare the fouling rate of the enhanced tubes, relative to a plain tube, and to investigate the use of on-line brush cleaning of the tubes. The three 19.05-mm-external-diameter tube geometries installed in the apparatus are the GEWA-SC tube (Figure 9.13c), GEWA-SPIN tube (same as Figure 9.3a), and the GEWA-TWX tube (Figure 9.13e). Ferric oxide powder was used as the foulant material. Ferric oxide is the pigment used in red primer paint. Data were also obtained using 3-μm-diameter aluminum oxide foulant material. Similar results were obtained with both foulants.

Figure 10.6 shows the fouling resistance measured in a 100-hr test at 1.7-m/s water velocity with 2000-ppm foulant concentration. The maximum fouling resistance occurred in the SC tube. Very little fouling occurred in the SPIN, TWX, and plain tubes. These results are quite different from those at low velocities (1.2 m/s), where significant fouling occurred in the TWX and SPIN tubes. The 2000 ppm used is much higher than one would expect in actual heat exchange systems. The data cannot be used to predict long-term fouling rates using typical field quality water.

Flow-driven nylon bristle brushes (Figure 10.7) may be used to clean the tubes during service. This cleaning system is described by Leitner [1980]. Figure 10.8 shows the accelerated fouling test results of Webb and Kim [1989] for operation at 1.2 m/s with 1500-ppm ferric oxide foulant. The figure shows three brushing cycles. The R_f of the SC tube shows a moderately rapid increase, whereas the other tubes show negligible fouling resistance. As shown in Figure 10.8, each brushing

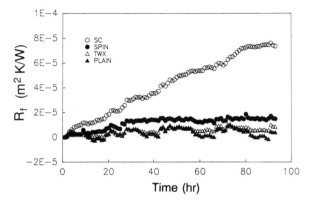

Figure 10.6 Fouling curves for Wieland GEWA tubes for 2000 ppm ferric oxide at 1.8-m/s water velocity. (From Webb and Kim [1989].)

Figure 10.7   Flow-driven nylon brush cleaning system. (Courtesy of Water Services of America.)

cycle returned the fouling resistances back to zero except the SC tube, where a slight residual R_f is observed. This occurred, because the brush in the SC tube was slightly undersized.

Leitner [1980] reports fouling tests of an on-line brush cleaning system used in the condenser tubes of a large refrigeration system. Leitner reports that the brush system maintained the U value of the Turbo-Chil™ tube (Figure 9.3a) within 8% of the clean tube value for cooling tower water.

Webb and Chamra [1991] performed particulate fouling tests of the Figure 12.9c Wieland GEWA-NW™ and the Figure 12.9a Wolverine Korodense™ tubes using 0.3-μm aluminum oxide particles. Figure 10.9 shows the test results for the GEWA-NW™ (18.0-mm inner diameter), the Korodense™ (19.7-mm inner diameter), and a plain tube (17.9-mm inner diameter) with 1.8-m/s water velocity and 1500-ppm foulant concentration. Both the GEWA-NW™ and Korodense™ tubes foul significantly faster than the plain tube.

Taprogge Corp. has developed an on-line recirculating sponge ball cleaning system that has been installed in many electric utility plants having plain tubes. This system is described by Keysselitz [1984] and Renfftlen [1991]. Webb and Chamra [1991] tested the ability of flow driven nylon brushes and sponge balls to clean their tubes. Both the nylon brushes and sponge-ball cleaning system was very effective in cleaning all of the enhanced tubes.

Somerscales et al. [1991] used 3-μm magnesium oxide foulant at 2500-ppm

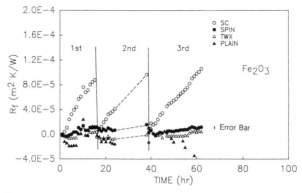

Figure 10.8   On-line brush cleaning tests of Wieland GEWA tubes using 1500-ppm ferric oxide with 1.2-m/s water velocity. (From Webb and Kim [1989].)

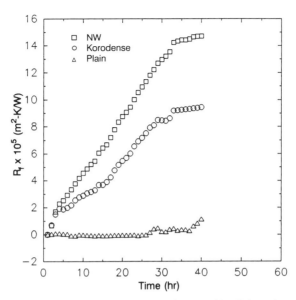

Figure 10.9 Fouling curves for 1500-ppm aluminum oxide (0.3 μm) at 1.8-m/s water velocity. (From Webb and Chamra [1991].)

concentration in water. They tested the Wolverine Turbo-Chil™ (Figure 9.3a), the Turbo-B™ (internal surface of Figure 9.13a), the Korodense™ (Figure 9.3b), and the two internally finned tubes discussed in Section 10.5. The tests were performed at two different water velocities, 1.0 and 1.5 m/s. Contrary to the tests of Webb and Chamra [1991], they conclude that the Korodense™ tube is less susceptible to fouling than a smooth tube. The Wolverine Turbo-Chil™ tube tested by Somerscales et al. [1991] has the same internal geometry as the Wieland GEWA-SPIN™ tube tested by Webb and Kim [1989]. Again, the Somerscales data are below those of Webb and Kim [1989]. One possible reason for the difference of results may be the tube length used. Somerscales et al. [1991] used a relatively short tube length (1.2-m), as compared to the 3.05-m length used by Webb and Kim [1989].

Boyd et al. [1983] describe operating experience of the Wolverine Korodense™ tube in steam condensers. Rabas et al. [1990] provide an update of operating experience with the Korodense™ tube since the publication of the Boyd et al. [1983] paper. Rabas et al. [1991] describes long-term operating experience of 12 electric utility steam condensers, nine of which were retubed with Korodense™ tubes. Although the results show that the Korodense™ tube fouls faster than plain tubes, a relatively small fouling resistance occurred within nine months after cleaning. Table 10.2 compares the fouling resistance of plain and Korodense™ tubes at the three plant locations 10 months after cleaning.

No cleaning was performed during the 10-month fouling period. Although Table 10.2 shows that the Korodense™ tube had a higher fouling rate than the plain tube, large fouling resistances were not attained. The thermal performance of the Ko-

TABLE 10.2 **Fouling Resistance of Steam Condensers 10 Months After Cleaning (hr-ft²-F/Btu)**

Plant	Plain	Korodense
Galatin	0.00035	0.00048
Shawnee	0.0001–0.0002	0.0003–0.0005

rodense™ tubes remained superior to that of the plain tubes for more than a year without cleaning. In addition, the thermal performance of both the enhanced and plain tubes was restored to the new, clean levels after mechanical brush cleaning. As previously noted, Renfftlen [1991] reports that enhanced tubes can be successfully cleaned using the sponge-ball cleaning system.

10.7 CORRELATIONS FOR FOULING IN ROUGH TUBES

Kim and Webb [1990] tested transverse-rib roughened tubes having $0.015 \leq e/d_i \leq 0.030$ and $10 \leq p/e \leq 20$, at $14,000 \leq \text{Re} \leq 26,000$ using 0.3-μm aluminum oxide particles at 1500-ppm concentration. The rib cross section was semicircular, with $w/e \simeq 3.0$. They developed an empirical correlation to predict the effect of the roughness parameters and the flow velocity on the asymptotic fouling rate (R_f^*). Their results are correlated by

$$R_f^* = (e/d_i)^{-0.3}(p/e)^{0.3}\text{Re}_d^{-3.9} \qquad (10.14)$$

Equation 10.14 shows that $R_f^* \propto u^{-3.9}$. Hence increasing velocity will significantly reduce the fouling rate. The velocity dependency is significantly higher than the value predicted by the Kern and Seaton [1959] model ($R_f^* \propto u^{-1.1}$) for plain tubes. Actually, the Kern and Seaton model underpredicts the velocity dependency of R_f. Chamra and Webb [1993b] found that $R_f^* \propto u^{-2.94}$ for plain tubes. Also note that smaller e/d_i gives higher fouling rate for the same p/e.

Chamra and Webb [1993a] determined the effect of particle size, particle distribution, and foulant concentration on the Figure 12.9c Wieland GEWA-NW™ and the Figure 9.3b Wolverine Korodense™ tubes. They used 2-, 4-, and 16-μm diameter with concentrations between 800 and 2000 ppm at 1.2- to 2.4-m/s water velocity. Tests were also performed with a distribution of the three particle sizes. The tests were intended to simulate particulate fouling in the Mississippi River, which contains a range of particle size distributions. Figure 10.10 shows the asymptotic fouling resistance (R_f^*) as a function of particle diameter at 1.8-m/s water velocity and 2000-ppm particle concentration. The figure shows that R_f^* increases as the particle size is reduced. This is because $K_m \propto \text{Sc}^{2/3}$, and $\text{Sc} \propto 1/d_p$. A mixture of containing 33% 2-μm particles and 67% 4-μm particles has a lower R_f than for 100% 2-μm particles, because the Schmidt number is reduced. Figure 10.11 shows the effect of particle concentration on R_f^* for 1.8 m/s water velocity.

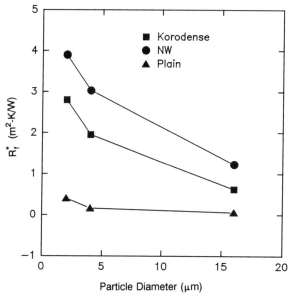

Figure 10.10 Asymptotic fouling resistance as a function of particle diameter for 1.8-m/s water velocity and 2000-ppm particle concentration. (From Chamra and Webb [1993a].)

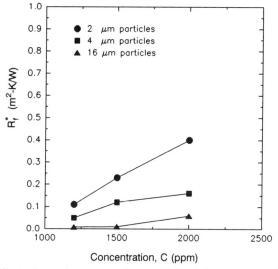

Figure 10.11 Effect of particle concentration on R_f^* for 1.8-m/s water velocity for 2-, 4-, and 16-μm diameter particles. (From Chamra and Webb [1993a].)

10.8 MODELING OF FOULING IN ENHANCED TUBES

Kim and Webb [1991] and Chamra and Webb [1993b] have worked to develop predictive models for particulate fouling in rough tubes. They used the heat–momentum analogy (Equation 10.12) to predict the mass transfer coefficient (K_m) from the heat transfer coefficient. This is valid, if the particle deposition occurs in the diffusion regime, as discussed by Kim and Webb [1991]. If the particle deposition occurs in the inertia regime, the mass transfer coefficient (K_m) must be corrected to account for particle inertia effects, as discussed by Chamra and Webb [1993b]. The corrected mass transfer coefficient is called the particle deposition coefficient, K_D, and is described by Chamra [1992]. The wall shear stress was predicted based on an analytical model proposed by Lewis [1975]. Because the apparent shear stress (τ_a) based on pressure drop includes contributions due to profile drag on the roughness elements, $\tau_w < \tau_a$. The fouling data were curve-fitted to the asymptotic form of Equation 10.9, which provided B (Equation 10.8) and R_f^* (Equation 10.10). With B and R_f^* known, they then calculated the sticking probability (S) and the deposit strength factor (ξ) using Equations 10.10 and 10.8, respectively. Chamra and Webb [1993b] show that the ξ and S functions for plain and enhanced tubes (GEWA-NW™ and Korodense™) are reasonably correlated by

$$\frac{S_{\text{enh}}}{S_{\text{ref}}} \propto \tau_w^{-0.721} d_p^{-0.319} C_b^{1.02} \left(\frac{e}{D} \right)^{-0.307} \tag{10.15}$$

$$\frac{\zeta_{\text{enh}}}{\zeta_{\text{ref}}} \propto \tau_w^{-0.435} d_p^{-0.0769} C_b^{0.421} \left(\frac{e}{d_i} \right)^{-0.396} \tag{10.16}$$

The S and ξ values are presented relative to those of the smooth tube to avoid the evaluation of the deposit density (ρ_f), and the deposit thermal conductivity (k_f). The reference values S_{ref} and ξ_{ref}, are for the smooth tube at Re = 24,000, d_p = 2 μm, and C_b = 2000-ppm. The sticking probability decreases as τ_w and d_p increase. The deposit bond strength factor decreases as the Reynolds number and particle diameter increase. The exponents on τ_w, d_p, and C_b agree closely for the three geometries tested. The Equation 10.15 and 10.16 results agree closely with those of Kim and Webb [1991] for repeated-rib tubes. Figures 10.12 and 10.13 show the measured S/S_{ref} and ξ/ξ_{ref} for the Korodense™ tube at 1500-ppm foulant concentration. Equations 10.15 and 10.16 may be used with Equations 10.6 and 10.10 to predict the fouling rate, relative to a plain tube.

R_f^*, S, and ξ are influenced by the Reynolds number (Re$_d$). Tables 10.3 and 10.4 compare the Reynolds dependency (exponent on Re$_d$) of R_f^*, S, and ξ for the referenced models with the test results for the plain and enhanced tubes. Table 10.3 shows that the experimental results are in good agreement with the results of Watkinson and Epstein [1970] for smooth tubes. In addition, Table 10.5 shows that the Kern and Seaton [1959] model underpredicts the experimental data. Table 10.4 shows that the experimental results closely match the Kim and Webb [1991] correlation for repeated-rib tubes. The difference between the exponents occurs because the current results are applicable to both the diffusion and inertia regimes.

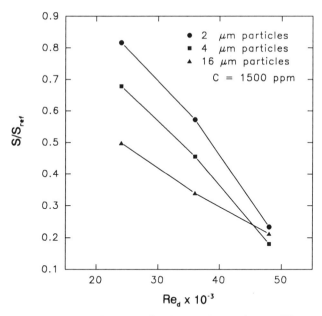

Figure 10.12 Sticking probability curves for the Korodense tube at 1500-ppm foulant concentration. (From Chamra and Webb [1993b].)

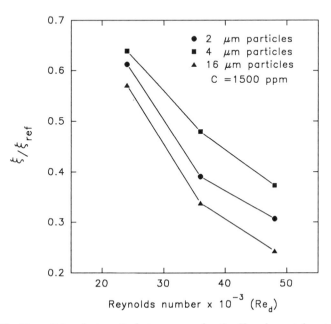

Figure 10.13 Deposit bond strength factor curves for the Korodense tube at 1500-ppm foulant concentration. From Chamra and Webb [1993b].)

TABLE 10.3 Comparison of Reynolds Number Exponents with Different Models for a Plain Tube

Model	R_f^*	S	ζ
Kern and Seaton [1959]	−1.10	—	—
Watkinson and Epstein [1970]	−3.00	−2.00	—
Chamra and Webb [1993b]	−2.94	−1.57	−0.95

TABLE 10.4 Comparison of Reynolds Number Exponents with Different Models for Plain Tube

Model	R_f^*	S	ζ
Kim [1989]	−3.90	−2.20	−0.400
Chamra and Webb [1993b]	−3.93	−1.59	−0.659

The detailed model of Chamra [1992] may be used to predict the R_f^* for different water velocities, foulant concentrations, particle diameters, or particle size distributions. This model was used to prepare Tables 10.5 and 10.6. Table 10.5 shows the ratio of the Korodense™-to-smooth R_f^* for $300 \leq C \leq 2000$ at 1.8 m/s (Re = 36,000) and $d_p = 2$ μm. Table 10.5 shows that the asymptotic fouling resistance for the GEWA-NW™ tube is slightly higher than that for the Korodense™ tube. However, The Korodense™ tube fouls faster than a plain tube. In addition, both of the asymptotic fouling resistance ratios increase as the concentration decreases.

Table 10.6 shows the predicted R_f^* ratio for different particle sizes and size distributions. The concentration and Reynolds number are held constant at 2000 ppm and 36,000, respectively. The Schmidt number for the mixture is calculated by taking the weighted average of the particle diameter. Table 10.6 shows the asymptotic fouling resistance ratios decrease as the particle diameter increases.

TABLE 10.5 Predicted Asymptotic Resistance Ratio for Different Concentrations ($u = 1.8$ m/s, $d_p = 2$ μm)

Concentration, C (ppm)	$(R_f^*)_{KD}/R_{f,s}^*$	$(R_f^*)_{NW}/(R_f^*)_{KD}$
2000	1.470	1.075
1500	1.640	1.114
1200	1.800	1.130
800	2.100	1.170
600	2.340	1.200
300	3.060	1.260

TABLE 10.6 Predicted Asymptotic Resistance Ratio for Different Particle Sizes (y = 1.8 m/s, c = 2000 ppm)

Particle Diameter, d_p (μm)	$(R_f^*)_{KD}/R_{f,s}^*$	$(R_f^*)_{NW}/(R_f^*)_{KD}$
2	2.870	1.670
4	1.940	1.394
16	1.750	1.230
67% 2 and 33% 4	2.210	1.460
33% 2 and 67% 4	2.040	1.230
50% 2 and 50% 4	2.160	1.306

10.8.1 Example Problem 10.1

Calculate the ratio of the asymptotic fouling resistance (R_f^*) of the Korodense tube, relative to a plain tube for particulate fouling of 35°C water in a 19.6-mm-inside-diameter tube at 1.8 m/s (Re_d = 36,000, Pr = 4.8) with C = 800 ppm using d_p = 2-μm particles. How will the results change if the particle size is increased to 4 μm?

The results are obtained from Table 10.5. This table shows that $(R_f^*)_{KD}/R_{f,s}^*$ = 2.10. If the particle size is increased to 4 μm, Table 10.6 shows that there is negligible change in the R_f^* ratio.

10.9 CONCLUSIONS

The fouling data on enhanced tubes appears to show a higher fouling rate than plain tubes. However, cleaning studies using on-line bristle brushes or sponge balls shows that the tubes can be effectively cleaned. Significant progress is being made in developing correlations to predict the particulate fouling rate of enhanced tubes, relative to plain tubes. The particulate fouling rate is sensitive to the tube geometry and particle size. Small particles result in a higher fouling rate than do large particles. Scaling fouling tests on internally finned tubes show that the fouling resistance (R_f/A_i) is less than that of a plain tube. This lower fouling resistance is a result of the surface area increase provided by the internally finned tube.

For a finned-tube, gas-to-liquid heat exchanger, gas-side fouling may be of less concern than liquid-side fouling. This is because the gas-side fouling thermal resistance (R_f/A_o) is apportioned over the total air-side surface area (A_o), as discussed in Section 2.8. However, if the foulant reduces the air-flow rate, heat exchanger thermal performance will suffer.

Much less work has been done on gas-side fouling than on liquid-side fouling. Particulate, solidification, and freezing fouling are the major modes of concern. All three fouling modes may cause plugging of the fin channels. Particulate fouling may not be of great concern for finned tube heat exchangers, unless the gases contain very high particulate loading. Gas-side fouling problems are most severe in cooling of combustion products, which may contain vapors which condense and solidify on the fin surface.

10.10   REFERENCES

Bemrose, C. R., and Bott, T. R., 1984. "Correlations for Gas-Side Fouling of Finned Tubes," *The Institution of Chemical Engineers, Symposium Series*, No. 86, presented at First U.K. National Conference on Heat Transfer, University of Leeds. pp. 357 –367.

Bott, T. R., and Bemrose, C. R., 1981. "Particulate Fouling on the Gas-Side of Finned Tube Heat Exchangers," in *Fouling in Heat Exchange Equipment*, J. M. Chenoweth and M. Impagliazzo, Eds., ASME Symposium, Vol. HTD-Vol. 17, ASME, New York, pp. 83–88.

Boyd, L. W., Hammon, J. C., Littrel, J. J., and Withers, J. G., 1983. "Efficiency Improvement at Gallatin Unit 1 with Corrugated Condenser Tubing," ASME paper, 83-JPGC-PWR-4.

Burgmeier, L., and Leung, S., 1981. "Heat Exchanger and Cleaning System Report," AiResearch Mfg. Co. report 81–17932 on DOE contract EC-77-C-03–11557, August 21.

Chamra, L. M., 1992. "A Theoretical and Experimental Study of Particulate Fouling in Enhanced Tubes," Ph.D. Thesis, The Pennsylvania State University, August 1992.

Chamra, L. M., and Webb, R. L., 1993a. "Effect of Particle Size and Size Distribution on Particulate Fouling in Enhanced Tubes," to be published in *Journal of Enhanced Heat Transfer*, Vol. 1, No. 2.

Chamra, L. M., and Webb, R. L, 1993. "Modeling Particulate Fouling in Enhanced Tubes having a Particle Size Distribution, to be published in *International Journal of Heat Mass Transfer*, 1993.

Epstein, N., 1983. "Fouling of Heat Exchangers," in *Heat Exchangers: Theory and Practice*, J. Taborek, G. F. Hewitt, and N. H. Afgan, Eds., Hemisphere Publishing Corp., New York, pp. 795–815.

Epstein, N., 1988a. "General Thermal Fouling Models," in *Fouling Science and Technology*, *Proceedings of the NATO Advanced Study Institute*, Melo, L. F., T. R. Bott and C. A. Bernardo, Eds., 1987. Kluwer Academic Publishers, Hingham, pp. 15–30,

Epstein, N., 1988b. "Particulate Fouling of Heat Transfer Surfaces: Mechanisms and Models," in *Fouling Science and Technology, Proceedings of the NATO Advanced Study Institute*, Melo, L. F., T. R. Bott and C. A. Bernardo, Eds., 1987. Kluwer Academic Publishers, Hingham, pp. 148–164.

Freeman, W. B., Middis, J., and Müller-Steinhagen, H., 1990. "Influence of Augmented Surfaces and of Surface Finish on Particulate Fouling in Double Pipe Heat Exchangers," *Chemical Engineering*, Elsevier, Vol. 27, pp. 1–11.

Garrett-Price, B. A., Smith, S. A., Watts, R. L., Knudsen, J. G., Marner, W. J., and Suitor, J. W., 1985. *Fouling of Heat Exchangers*, Noyes Publications, Park Ridge, NJ.

Katz, D. L., Knudsen, J. G., Balekjian, G., and Grover, S. S., 1954. "Fouling of Heat Exchangers," *Petroleum Refiner*, Vol. 33, No. 4, pp. 123–125.

Kern, D. Q., and Seaton, R. E., 1959, "A Theoretical Analysis of Thermal Surface Fouling." *British Chemical Engineering*, Vol. 14, No. 3, pp. 258–262.

Keysselitz, J., 1984. "Can Waterside Condenser Fouling be Controlled Operationally?," in *Fouling in Heat Exchange Equipment*, J. W. Suitor and A. M. Pritchard, Eds., ASME Symposium, Vol. HTD-Vol. 35, ASME, New York, pp. 105–111.

Kim, N-H., 1989. "A Theoretical and Experimental Study on the Particulate Fouling of Tubes Having Two-Dimensional Roughness," Ph.D. Thesis, Department of Mechanical Engineering, Penn State University.

Kim, N-H., and Webb, R. L., 1990. "Particulate Fouling Inside Tubes Having Arc-Shaped Two-Dimensional Roughness by a Flowing Suspension of Aluminum Oxide in Water," *Proceedings of the 9th International Heat Transfer Conference*, Jerusalem, Israel, pp. 139–146.

Kim, N-H., and Webb, R. L., 1991. "Particulate Fouling in Tubes Having a Two-Dimensional Roughness Geometry," *International Journal of Heat and Mass Transfer*, Vol. 34, pp. 2727–2738, 1991.

Kindlman, L., and Silverstrini, R., 1979. "Heat Exchanger Fouling and Corrosion Evaluation," AiResearch Mfg. Co. report 78–1516(2) on DOE contract DE-AC03–77ET11296, April 30.

Leitner, G. F., 1980. "Controlling Chiller Tube Fouling," *ASHRAE Journal*, Vol. 6, No. 2, pp. 40–43.

Lewis, M. J., 1975. "An Elementary Analysis for Predicting the Momentum and Heat Transfer Characteristics of Hydraulically Rough Surfaces," *International Journal of Heat Mass Transfer*, Vol. 97, pp. 249–254.

Marner, W. J., and Webb, R. L., 1983. "A Bibliography on Gas-Side Fouling," *Proceedings of the ASME-JSME Thermal Engineering Joint Conference*, Vol. 1, ASME, New York, pp. 559–570.

Melo, L. F., Bott, T. R., and Bernardo, C. A., 1987. *Fouling Science and Technology*, *Proceedings of the NATO Advanced Study Institute*, Kluwer Academic Publishers, Hingham, MA.

Moore, J., 1974. "Fintubes Foil Fouling for Scaling Services," *Chemical Processing*, August, pp. 8–10.

Müller-Steinhagen, H., Reif, F., Epstein, N., and Watkinson, A. P., 1988. "Influence of Operating Conditions on Particulate Fouling," *Canadian Journal of Chemical Engineering*, Vol. 66, pp. 42–50.

Owen, I., El-Cady, A. A., and Cleaver, J. W., 1987. "Fine Particle Fouling of Roughened Heat Transfer Surfaces," *1987 ASME-JSME Thermal Engineering Joint Conference*, P. D. Marto and I. Tanasawa, Eds., Vol. 3, ASME, New York, pp. 95–101.

Papavergos, P.G., and Hedley, A.B., 1984. "Particle deposition behavior from turbulent Flows." *Chemical Engineering Research and Development*, Vol. 62, pp. 275–295.

Rabas, T., Merring, R., Schaefer, R., Lopez-Gomez, R., and Thors, P., 1990. "Heat-Rate Improvements Obtained with the Use of Enhanced Tubes in Surface Condensers," presented at the EPRI Condenser Technology Conference, Boston, 1990.

Rabas, T. J., Panchal, C. B., Sasscer, D. S., and Schaefer, R., 1991. "Comparison of Power-Plant Condenser Cooling-Water Fouling Rates for Spirally-Indented and Plain Tubes," *Fouling and Enhancement Interactions*, T. J. Rabas and J. M. Chenoweth, Eds., ASME Symposium Vol. HTD-Vol. 164, ASME, New York, pp. 29–37.

Renfftlen, R. G., 1991. "On-line Sponge Ball Cleaning of Enhanced Heat Transfer Tubes," in *Fouling and Enhancement Interactions*, T. J. Rabas and J. M. Chenoweth, Eds., ASME Symposium Vol. HTD-Vol. 164, ASME, New York, pp. 55–60.

Roberts, P. B., and Kubasco, A. J., 1979. "Combined Cycle Steam Generator Gas-Side Fouling Evaluation: Phase 1 Final Report," Solar Turbines International report SR79-R-4557–20, July.

Sheikholeslami, P., and Watkinson, A.P., 1986. "Scaling of Plain and Externally Finned Heat Exchanger Tubes." *Journal of Heat Transfer*, Vol. 108, pp. 147–152.

Somerscales, E. F. C., and Knudsen, J. G., 1979. *Fouling of Heat Transfer Equipment*, Hemisphere Publishing Corp., Washington, D.C.

Somerscales, E. F. C., Ponteduro, A. F., and Bergles, A. E., 1991. "Particulate Fouling of Heat Transfer Tubes Enhanced on Their Inner Surfaces," in *Fouling and Enhancement Interactions*, T. J. Rabas and J. M. Chenoweth, Eds., ASME HTD-Vol. 164, ASME, New York, pp. 117–128.

Taborek, J., Aoki, T., Ritter, R. B., Palen, J. W., and Knudsen, J.W., 1972. "Fouling—The Major Unresolved Problem in Heat Transfer," *Chemical Engineering Progress,* Vol. 68, No. 2, pp. 59–67, and No. 7, pp. 69–78.

Watkinson, A. P., 1975. "Scaling of Spirally Indented Heat Exchanger Tubes," *Journal of Heat Transfer*, Vol. 108, pp. 147–152.

Watkinson, A.P., 1990. "Fouling of Augmented Heat Transfer Tubes," *Heat Transfer Engineering*, Vol. 11, No. 3, pp. 57–65.

Watkinson, A. P., 1991. "Interactions of Enhancement and Fouling," in *Fouling and Enhancement Interactions*, T. J. Rabas and J. M. Chenoweth, Eds., ASME Symposium Vol. HTD-Vol. 164, ASME, New York, pp. 1–7.

Watkinson, A. P., and Epstein, N., 1970. "Particulate Fouling of Sensible Heat Exchangers." *Proceedings of the 4th International Heat Transfer Conference*, Paper No. HE 1.6.

Watkinson, A. P., Louis, L., and Brent, R., 1974. "Scaling of Enhanced Heat Exchanger Tubes." *Canadian Journal of Chemical Engeering*, Vol 52, pp. 558–562.

Webb, R. L., and Chamra, L. M., 1991. "On-line Cleaning of Particulate Fouling in Enhanced Tubes," *Fouling and Enhancement Interactions*, T. J. Rabas and J. M. Chenoweth, Eds., ASME Symp. Vol. HTD-Vol. 164, ASME, New York, pp. 47–54.

Webb, R. L., and Kim, N-H., 1989. "Particulate Fouling in Enhanced Tubes," *Heat Transfer Equipment Fundamentals, Design, Applications, and Operating Problems*, R. K. Shah, Ed., ASME Symposium, Vol. HTD-Vol.108, ASME, New York, pp. 315–324.

Webb, R. L., Marchiori, D., Durbin, R. E., Wang, Y-J., and Kulkarni, A. K., 1983. "Heat Exchangers for Secondary Heat Recovery from Glass Plants," in *Fouling of Heat Exchange Surface*, R. W. Bryers, Ed., Engineering Foundation, New York, pp. 169–182.

Webber, W. O., 1960., "Does Fouling Rule Out Using Finned Tubes in Reboilers?," *Petroleum Refiner*, Vol. 39, No. 3, pp. 183–186.

Zhang, G., Bott, T. R., and Bemrose, C. R., 1990. "Finned Tube Heat Exchanger Fouling by Particles," *Proceedings of the 9th International Heat Transfer Conference*, Jerusalem, Vol. 6, pp. 115–120.

Zhang, G., Bott, T. R., and Bemrose, C. R., 1992. "Reducing Particle Deposition in Air-Cooled Heat Exchangers," *Heat Transfer Engineering*, Vol. 13, No. 2, pp. 81–87.

10.11 NOMENCLATURE

A	Heat transfer surface area ($= \pi d_i/L$), m or ft
B	Time constant defined by Equation 10.8 ($= \rho_f \tau_w k_f/\zeta$), s^{-1}
c_p	Specific heat, kJ/kg-K or Btu/hr-°F
C_b	Particulate concentration in bulk fluid, ppm
C_w	Particulate concentration at wall, ppm
d_e	Diameter at tip of fins for a finned tube; m or ft
d_i	Plain tube inside diameter, or diameter to the base of internal fins (or flutes) or roughness; m or ft
d_o	Tube outside diameter, fin root diameter for a finned tube, minor diameter of rectangular cross section tube; m or ft
d_p	Foulant particle diameter; m or ft
D	Diffusion coefficient ($= K_B T_{abs}/3\pi\mu d_p$), m^2/s or ft^2/s
e	Fin height or roughness height; m or ft
e^+	Roughness Reynolds number($= eu*/\nu$), dimensionless
f	Fanning friction factor, $\Delta p_f d/2LG^2$; dimensionless
G	Mass velocity based on the minimum flow area, kg/m^2-s or lbm/ft^2-s
h	Heat transfer coefficient, W/m^2-K or Btu/hr-ft^2-°F
k_w	Tube wall thermal conductivity, W/m-K or Btu/hr-°F
k_f	Foulant layer thermal conductivity, W/m-K or Btu/hr-ft-°F
K_B	Boltzmann constant ($= 1.38$ E $- 23$ J/kg)
K_m	Mass transfer coefficient, kg/m^2-s or lbm/ft^2-s
m_f	Foulant mass per unit area attached to wall, kg/m^2 or lbm/ft^2
$\dot{m}_d$	Foulant deposition rate, kg/m^2-s or lbm/ft^2-s
$\dot{m}_r$	Foulant removal rate, kg/m^2-s or lbm/ft^2-s
p	Axial pitch of surface or roughness elements, m or ft
Pr	Prandtl number ($= c_p\mu/k$), dimensionless
Re$_d$	Reynolds number based on the tube diameter ($= Gd/\mu$, $d = d_i$ for flow inside tube and $d = d_o$ for flow outside tube), dimensionless
R_f	Fouling resistance, m^2-K/W or ft^2-°F-hr/Btu
R_f^*	Asymptotic fouling resistance (Equation 10.10), m^2-K/W or m^2-°F-hr/Btu
dR_f/dt_o	Initial fouling rate, m^2-K/W-s or ft^2-°F/Btu
S	Sticking probability, dimensionless
Sc	Schmidt number ($= \nu/D$), dimensionless
t	Time, s or h
t_d	Delay time shown in Figure 10.1, s or h
t^+	Dimensionless particle relaxation time defined by Equation 10.11
T_{abs}	Absolute temperature, K or R
u	Flow velocity, m/s or ft/s
u*	Friction velocity, m/s or ft/s
U	Overall heat transfer coefficient, W/m^2-K or Btu/hr-ft^2-°F
U$_c$	Clean tube overall heat transfer coefficient, W/m^2-K or Btu/hr-ft^2-°F

U_f	Fouled tube overall heat transfer coefficient, W/m^2-K or Btu/hr-ft^2-°F
w	width of rib, m or ft
x_f	Foulant layer thickness at wall, m or ft

Greek Letters

α	Helix angle, degrees
μ	Dynamic viscosity, Pa-s or lbm/s-ft
ν	Kinematic viscosity, kg/m^2-s or lbm/ft^2-s
ξ	Deposit bond strength factor, dimensionless
ρ	Fluid density, kg/m^3 or lbm/ft^3
ρ_p	Density of a particle, kg/m^3 or lbm/ft^3
τ_a	Apparent wall shear stress based on pressure drop, N/m^2 or lbf/ft^2
τ_w	Wall shear stress, N/m^2 or lbf/ft^2

Subscripts

i	Designates inner surface of tube
o	Designates outer surface of tube
s	Smooth surface
p	Plain tube
w	Wall
τ^+	Nondimensional particle relaxation time

11

POOL BOILING

11.1 INTRODUCTION

Perhaps the most significant advances in enhanced heat transfer technology have been made in special surface geometries that promote high-performance nucleate boiling. These special surface geometries provide an increased heat transfer coefficient. Although the surface area may be increased, it is common to define the boiling coefficient in terms of the projected, or flat plate, surface area. The fact that "roughness" can improve nucleate boiling performance has been known for 60 years. For nearly 35 years this was regarded as an interesting but commercially unusable concept because the heat transfer improvement lasted for only a matter of hours before "aging" caused the performance to decay to the plain surface value. In the period 1955–1965, fundamental advances were made in understanding the character of nucleation sites and the shape necessary to form stable vapor traps. This understanding provided a basis for industrial research [1960–1975] that led to the development of commercially viable enhanced surface geometries. The first of these high-performance geometries was patented in 1968. In 1980 six nucleate boiling surface geometries were commercially available.

This chapter describes the fundamental and applied research which has brought about this major advance in the field of heat transfer. It also describes available correlations for prediction of the boiling coefficient and discusses current understanding of the boiling mechanism. The excellent book by Thome [1990] on enhanced boiling heat transfer is a recommended reference.

The reference list includes a number of patents. Industrial research, which favors publication via patents, has led the way in development of the enhanced boiling surfaces. Academic research has been important in explaining the physical mechanism of the enhanced surface.

11.2   EARLY WORK ON ENHANCEMENT (1931–1962)

In 1931, Jakob and Fritz investigated the effect of surface finish on nucleate boiling performance, as reported by Jakob [1949]. They boiled water from (1) a sandblasted surface and (2) a surface having a square grid of machined grooves (0.016-mm square with 0.48-mm spacing). The sandblasting provided no more than 15% improvement, which dissipated within a day. The grooved surface initially yielded boiling coefficients about three times higher than those of a smooth surface, but this performance also dissipated after several days.

The performance decay is described as an aging effect. The cavities formed by the roughening were not stable vapor traps and slowly degassed, leaving no vapor to sustain the nucleation process. Sauer's [1935] tests of a grooved tube (0.4-mm square grooves spaced at 1.6-mm) offered no additional promise. Interest in roughness was renewed about 1954, and a sustained effort has advanced the technology to the present level.

Between 1954 and 1962, three studies were performed using flat surfaces roughened with emery paper of different coarseness or lapping compound. Berenson [1962] observed nucleate boiling coefficient increases as large as 600% using lapped surfaces (Figure 11.1). The tests were not run long enough to establish the

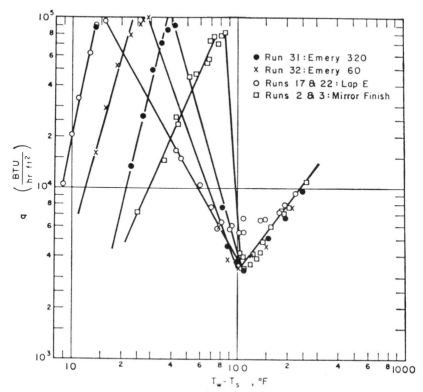

Figure 11.1   Effect of surface roughness for pentane boiling on copper. (From Berenson [1962].)

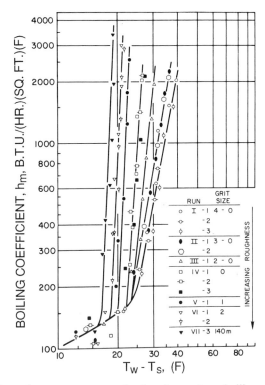

Figure 11.2 Effect of emery paper roughening for acetone boiling on copper. (From Kurihari and Myers [1960].)

long-term aging performance, although Corty and Foust's [1955] tests of grit-roughened surfaces did show some short-term aging. Kurihari and Myers [1960] boiled water and several organic fluids on copper surfaces roughened with different grades of emery paper. Figure 11.2 shows their acetone results for the different grit sizes. They clearly established that the increased boiling coefficient results from an increased area density of nucleation sites. Their results for five test fluids showed that $h \propto (n_s/A)^{0.43}$. These works demonstrated that boiling enhancement will occur if the site density of stable nucleation sites can be increased. The artificially formed sites must allow incipience to occur at a lower ΔT ($= T_w - T_s$) than for naturally occurring nucleation sites. These studies motivated later researchers to investigate other means of making artificial sites that showed incipience at low ΔT and functioned as stable vapor traps.

11.3 SUPPORTING FUNDAMENTAL STUDIES

Research on artificially formed nucleation sites was supported by parallel fundamental studies of the character of nucleation sites, the effect of cavity shape, and theoretical models of the nucleation process.

Particularly notable are the high-speed photographic studies of Clark et al. [1959], who identified naturally occurring pits and scratches, between 0.008- and 0.08-mm surface width, as active boiling sites for pentane. Griffith and Wallis [1960] showed that the cavity geometry is important in two ways. First, the mouth diameter determines the superheat needed to initiate boiling, and its shape determines its stability once boiling has begun. Figure 11.3a shows the reentrant cavity shape proposed by Griffith and Wallis. Figure 11.3b shows $1/r$ versus bubble volume, where $1/r$ is proportional to the liquid superheat required to maintain a vapor nucleus. When the vapor radius of curvature becomes negative (concave curvature of the liquid–vapor interface), the vapor nucleus may still exist in the presence of subcooled liquid. Therefore, a reentrant cavity should be a very stable vapor trap. Benjamin and Westwater [1961] were apparently the first to construct a reentrant cavity and demonstrate its superior performance as a vapor trap. This site remained active after surface subcooling, while the naturally occurring sites became flooded. In a later study, Yatabe and Westwater [1966] varied the interior dimensions and shape of reentrant cavities while maintaining the surface opening diameter at a constant value. These tests showed that the interior shape of a reentrant cavity is not important.

Hsu [1962] developed a model to define the size range of active cavities for incipient boiling on a heated surface. The model assumes that the cavity is of the proper shape to act as a stable vapor trap. Bankoff [1959] developed a dynamic

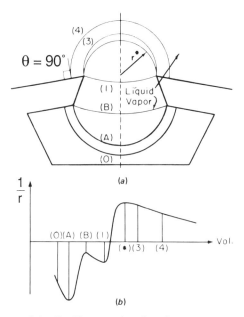

Figure 11.3 (a) States of the liquid–vapor interface in a reentrant cavity. (b) Reciprocal radius ($1/r$) versus vapor volume for liquid having a 90-degree contact angle. (From Griffith and Wallis [1960].)

model to predict the ability of a cavity to serve as a stable vapor trap (e.g., one that will not become liquid-filled). The analysis considered triangular and constant width circular cavities and grooves. He showed that grooves are poor vapor traps unless they are steep walled or poorly wetted. Bankoff's suggestion that grooves may function as vapor traps provided motivation for later research on cavities having a reentrant groove shape. Moore and Mesler [1971] showed that high vaporization rates occur across a "microlayer" of liquid at the base of the bubble.

These studies, which provided a strong basis for the later research on artificially formed nucleation sites, almost exclusively focused on nucleation in discrete cavities. However, the state-of-the-art enhanced surfaces generally consist of two dimensional reentrant grooves or interconnected cavities. Griffith and Wallis's suggestion of the reentrant cavity shape provided a major contribution.

11.4 TECHNIQUES EMPLOYED FOR ENHANCEMENT

A variety of surface treatments have been used to develop enhanced boiling surfaces. Survey articles on such treatments have been published by Webb [1981, 1983] and Dundin et al. [1990]. This provides an overview of the various techniques that have been investigated. Matijević et al. [1992] compare the enhancement for boiling water using a variety of the enhancement techniques described below. Their highest performance was obtained using a porous coating of sintered spherical particles. Note that they used a very large particle diameter, 3.0 mm being the smallest.

11.4.1 Abrasive Treatment

Because of the severe aging effects, there has been little sustained interest in this method. Chaudri and McDougall [1969] measured the long-term performance (up to 500 hr) of standard abrasive treated tubes. The abrasive treatment produced parallel scratches 0.025 mm or less in width. They boiled several organic fluids on copper and steel tubes and found only temporary benefits of the surface treatment. After several hundred hours the treated tubes showed essentially the same performance as the untreated tubes.

11.4.2 Open Grooves

Bonilla et al. [1965] formed parallel grooves by dragging a sharp-pointed scriber across a polished copper plate. The boiling coefficient of water was measured for scratch spacings of 1.6, 3.2, 6.4, 12.8, and 25.5 mm. Increased performance was observed for all groove spacings. Surprisingly, these authors found that the 6.4-mm spacing yielded the highest performance. They concluded that optimum performance occurs with a scratch spacing of 2 to 2.5 bubble diameters. The boiling coefficient was increased 100% at this optimum scratch spacing. Reentrant grooves have shown greater sustained-performance level than that of open grooves. These will be discussed later.

11.4.3 Three-Dimensional Cavities

Several tests of three-dimensional cavities have been performed. As previously discussed, Griffith and Wallis [1960] examined fundamental characteristics (incipience and stability) of a single 0.15-mm-diameter conical cavity. They showed that the addition of a nonwetting coating within the cavity allows incipience of water at lower superheat. Such nonwetting phenomena may be beneficial for high-surface-tension fluids, and will be discussed later.

Marto et al. [1968] measured the boiling performance of nitrogen on a 25.4-mm-diameter copper block. The nucleation sites were formed by pressing cylindrical or conical cavities in the surface. Figure 11.4 shows their results for 0.38-mm by 0.76-mm-deep cavities spaced at 3.7 mm. Significant enhancement was obtained. As previously noted, Griffith and Wallis [1960] have shown that a reentrant cavity shape (Figure 11.3) is preferred over a conical or cylindrical shape; lower superheat is required for incipience, and the reentrant shape provides a more stable vapor trap. Marto and Rohsenow [1966] boiled sodium (a nonwetting liquid) on a 75-mm-diameter block having 12 reentrant cavities. The cavity cross section is shown in Figure 11.5b. They also tested other surface treatments, which included 100-grit lapped finish (lap F) porous welds, a sintered coating, and reentrant cavities (12 cavities in 75-mm-diameter surface). Their results are summarized in Figure 11.5a, which shows the high-performance potential of reentrant cavities and porous coatings.

The formation of discrete cavities is difficult and of limited practicality. This has motivated innovators to develop commercially feasible concepts to produce a high area density of nucleation sites. Reentrant grooves and porous coatings have developed as favored methods.

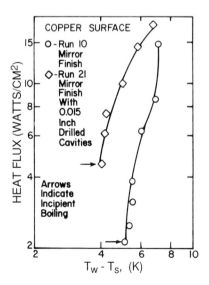

Figure 11.4 Nitrogen boiling at 101 kPa (1 atm) on a 25.4-mm-diameter copper surface. Run 21 surface has 13 circular cavities in the surface. (Adapted from Marto et al. [1968].)

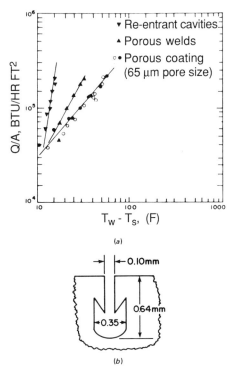

(a)

(b)

Figure 11.5   (a) Effect of several surface treatments on sodium boiling at 65 mm Hg. (b) Cross section of doubly reentrant cavities tested. (Adapted from Marto and Rohsenow [1966].)

11.4.4   Etched Surfaces

Mechanical roughness may be formed by chemical etching. Vachon et al. [1969] boiled water on a chemically etched stainless steel surface. This showed no better performance than a polished surface. Chu and Moran [1977] described the use of a laser beam to form cylindrical cavities having a reentrant shape (Figure 11.6).

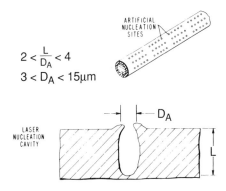

$$2 < \frac{L}{D_A} < 4$$
$$3 < D_A < 15\mu m$$

Figure 11.6   Reentrant cavities formed by laser beam.

11.4.5   Electroplating

Bliss et al. [1969] electroplated 0.13-mm layers of copper, chrome, nickel, cad-
mium, tin, and zinc on stainless steel tubes. Some platings increased the heat
transfer coefficient in the range of 200–300%. However, other coating materials
(tin, zinc, and nickel) produced a decreased heat transfer coefficient. They offer no
explanation for the observed results, but they do argue that the thermal properties of
the plating material do not, by themselves, account for the effect.

Albertson [1977] shows that electroplating at very high current densities causes
the formation of dendrites or nodules on the base surface. This copper porous
structure produced a large heat transfer increase for boiling R-12 on a 19-mm-
diameter tube. The performance is further enhanced by cold-rolling the tube to
partially compact the nodules.

11.4.6   Pierced Three-Dimensional Cover Sheets

Ragi [1972] describes a method of forming reentrant cavities using a three-
dimensional formed foil sheet, which is brazed to a flat base surface (Figure 11.7).
A 0.08-mm-thick foil sheet has a three-dimensional pattern of pyramids whose tops
are pierced, thereby forming reentrant cavities. Data are given for R-11 and water
with several geometric variations having cavity densities of 10–80/cm². Figure 11.7
shows the enhancement obtained with water at 101 kPa and 62 cavities/cm². At 60-
kW/m² heat flux, the boiling coefficient was increased a factor of 7.7; the corresponding enhancement level with R-11 was 2.5. Greater enhancement levels were
measured for increased cavity density (e.g. 180 cavities/cm²).

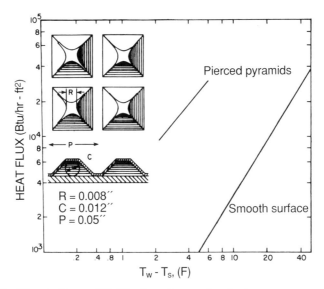

Figure 11.7   Water boiling at 101 kPa (1 atm) on surface having reentrant cavities formed
by brazing a pierced foil sheet to base surface.

Palm [1992] has done additional work using perforated metal foils. He has pierced 0.05-mm-thick flat sheets and investigated a range of hole densities and hole diameters. His R-22 data show earlier incipient boiling and show a higher heat flux than that of the commercial GEWA-TWX™ tube (Figure 11.12c) for $(T_w - T_s)$ < 3.5 K. Additional fabrication details are provided in a U.S. patent by Granryd and Palm [1988].

11.4.7 Attached Wire and Screen Promoters

Hasegawa et al. [1975] boiled water on a polished copper surface on which a screen was placed. The woven screens used in the experiments were 30, 50, 80, and 150 mesh, using either one or two layers of screening. The screens were held down by glass bars laid at the sides of the screens. They were not very effective as enhancement devices. The 150-mesh screen produced the greatest enhancement, about 50% increased boiling coefficient. The screens increased the critical heat flux about 35%. This improvement was generally greater for the larger mesh openings, and one screen layer was better than two layers. Cormon and McLaughlin [1976] performed similar experiments, using felt-metal nickel and copper wicking of various thicknesses (0.5–3 mm). At low heat flux (60 kW/m²), the boiling coefficient was improved about 70%, but the slope of the boiling curve was smaller than for the smooth surface. At increasing heat flux the wick surface becomes inferior; the crossover point occurs at 155 kW/m². In spite of the lower slope, the wick surface displayed a higher critical heat flux than the smooth surface. Improvements in the range of 50–70% were measured for the thinner wick structures.

Asakavicius et al. [1979] attached 8–12 layers of copper screening (0.07-mm open space) and boiled R-113, ethyl alcohol, and water. Enhancements in excess of 100% were observed at low heat fluxes, but little improvement was noted at high heat fluxes.

Figure 11.8a shows a loosely fitting metallic (or nonmetallic) wire-wrapped integral fin tube tested by Webb [1970]. A gap spacing of 0.013–0.05 mm provides substantial enhancement. Figure 11.8b, from Schmittle and Starner's [1978] patent, shows that ΔT is reduced approximately 60% by the wire wrap. This patent also discusses an interesting variation applicable to plain tubes (Figure 11.8b). This tube is wrapped with a 0.38-mm stranded nylon cord with 0.25-mm space between wraps. Figure 11.8c shows that the stranded nylon wrap substantially enhances the plain tube, but the wire-wrapped plain tube is ineffective.

11.4.8 Nonwetting Coatings

Griffith and Wallis [1960] were apparently the first to recognize that a thin coating of nonwetting material may improve the nucleating characteristics of water (a high-surface-tension fluid) on a boiling surface. They proposed that a nonwetting coating on the interior walls of a nucleating cavity will enhance its stability as a nucleating site. Thus, a lower temperature level will be needed to deactivate the cavity. The

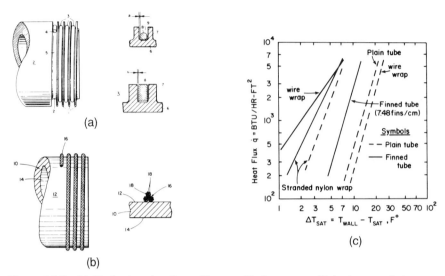

Figure 11.8   (a) Nucleation sites formed by metallic (or nonmetallic) wire-wrap in between the fins of an integral-fin tube. (b) Stranded nylon cord wrap on smooth tube. (c) R-11 boiling at 101 kPa (1 atm) on an integral fin tube (7.48 fins/cm) and enhancement provided by wire or stranded nylon wrap in space between fins.

nonwetting coating establishes a larger liquid–vapor interface radius within the cavity, which requires less superheat for its existence. However, they proposed that this coating would not affect the temperature at which a cavity will nucleate, because this is fixed by the cavity mouth radius. They performed experiments on single conical cavities 0.08 mm in diameter, formed by pressing a needle into the boiling surface. They found that it was easier to maintain boiling and get reproducible results from the paraffin cavity than from clean ones, and that unwetted cavities are more stable than wetted cavities.

Gaertner [1967] performed further work with artificial nucleation sites, covering the inside surface of the cavities with a nonwetting material (Figure 11.5b). He boiled water on a variety of surfaces having closely spaced nucleation sites formed by needle sharp punches and parallel scratches. The low-surface-energy material was coated over the surface containing the artificial sites, then removed from the flat face by abrasion, leaving a thin film of the material deposited in each cavity. The coated cavity surface boiled at a lower superheat and remained active for a much longer time. The heat transfer coefficient was considerably reduced if the coating was left on the entire surface. For this condition the bubbles spread over the surface and coalesced with bubbles formed at other sites, causing the entire surface to become vapor-blanketed.

Hummel [1965] describes a surface coated with small, nonwetted spots (Figure 11.9), which not only improves the boiling stability of water, but also provides a very significant heat transfer enhancement. Young and Hummel [1965] boiled water on a stainless steel surface that had been sprayed with Teflon, producing 30–60

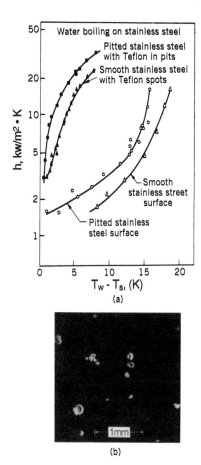

(a)

(b)

Figure 11.9 Enhancement for water boiling at 101 kPa (1 atm) on a stainless steel surface having minute nonwetted spots (30–60 spots/cm², 0.25 mm diameter or less). (b) Enlarged photo of Teflon-spotted smooth surface. (From Young and Hummel [1965].)

spots/cm² and 0.25-mm diameter or less. Figure 11.9a shows a dramatic enhancement, with significant nucleation occurring at $\Delta T < 0.5$ K. The theory of this augmentation concept is as follows: The Teflon spot surrounds a naturally occurring or a mechanically prepared nucleation site. When the bubble is in contact with the nonwetted surface, it has a much larger radius of curvature than for a wetted surface. Therefore, lower superheat is required for its existence and initial growth. Once the bubble grows in diameter past the area of the unwetted spot, it quickly reverts to the spherical shape and grows very rapidly because of evaporation of the thin liquid film at the base of the spherical bubble. It is undesirable to have the entire surface Teflon-coated, because the vapor would tend to blanket a large area of the surface, thereby reducing heat transfer. Figure 11.9a also shows data for a pitted surface with and without Teflon spots in the pits. The performance of the pitted surface is marginally better than the smooth surface when both have Teflon spots. Apparently, the mechanical pitting did not establish significantly more nucleation sites than those occurring naturally in the stainless steel strip.

The Teflon spotting method should be effective only for surface–liquid combinations that have large contact angles (e.g., water). Bergles et al. [1968] confirmed this in their tests with refrigerants, which have low surface tension and small contact angles. Their results showed that the Teflon spotting method did not favorably affect the boiling performance of the refrigerants.

Vachon et al. [1969] applied thin (0.008 mm) surface coatings of Teflon in an attempt to augment nucleate boiling of water on stainless steel. Gaertner [1967] and Hummel [1965] argue against the merits of a continuous surface coating because of the tendency of the surface to become vapor-blanketed. However, Vachon et al. did not observe vapor-blanketing. A considerable heat transfer enhancement was observed for the 0.008 mm thick coating, but a decrease of heat transfer occurred with a 0.04-mm coating because of the insulating effect of the thick coating.

11.4.9 Porous Surfaces

Milton's patents [1968, 1970, 1971] of a porous boiling surface provided a contribution of major practical significance. He describes a sintered porous metallic matrix bonded to the base surface. The coating is approximately 0.25 mm thick and has a void fraction of 50–65%. The optimum pore size depends on the fluid properties.

Figure 11.10a, which is based on data reported by Gottzmann et al. [1973], compares the performance between the High-Flux™ porous surface and a smooth surface. The ΔT required to produce a given heat flux was reduced by more than a factor of 10. The surface showed negligible decay of long-term performance because of aging. Gottzmann et al. [1971] also found that the critical heat flux for trichlorethylene is 80% higher than for a smooth surface.

Table 11.1, taken from Webb [1983], summarizes four different methods that have been used to make porous coatings. The Milton [1968] patent describes a five-step method for making a sintered particle coating. The particles are mixed into a liquid slurry, which contains a temporary binder. The surface is coated with this slurry, which is air-dried and then sintered in a baking oven. The sintering process takes place at a temperature slightly below the particle melting point. The sintered method may be disadvantageous if it anneals the base tube. The other methods listed in Table 11.1 are performed at lower temperatures, and the mechanical properties of the base tube are not adversely affected.

Table 11.1 shows that most of the published test data are for copper coatings using R-11 or R-113 as test fluids. The geometric variables of the porous coating are (d_p) the particle size (or size distribution), (2) the particle shape, (3) the coating thickness, and (4) the particle packing arrangement. The packing arrangement establishes the porosity of the matrix and depends on whether the particles are of uniform size and shape or whether a distribution of sizes and shapes is used. Figure 11.10b compares the R-11 boiling performance of the coatings listed in Table 11.1. Table 11.2 gives the geometric characteristics of each coating represented in Figure 11.10b, and are listed in order of increasing particle diameter. The particle diameters are in the range $0.044 < d_p < 1.0$ mm, and the coating thickness ranges from

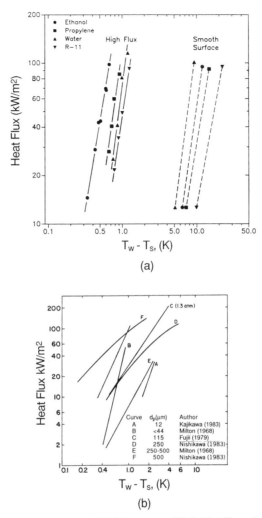

Figure 11.10 (a) Enhancement provided by porous High-Flux™ surface for three fluids boiling at 101 kPa (1 atm), as reported by Gottzmann et al. [1971, 1973]. From Thome [1990].) (b) Comparative performance of the Table 11.1 coatings for R-11 boiling at 1.0 atm, except as noted. (From Webb [1983].)

0.25 to 2.0 mm. Experimental error in measuring $T_w - T_s$ may be a major uncertainty, since its value is in the range of 1 K.

Nishikawa et al. [1983] and Nishikawa and Ito [1982] investigated the effect of particle size and coating thickness for R-11 and R-113 boiling on 18-mm-diameter, horizontal tubes. They used spherical particles of copper or bronze. The coating porosity ranged from 0.38 to 0.71. In Test 1 using 0.25-mm copper particle diameter (δ_p), they varied the coating thickness (δ) from 0.4 to 4 mm. This test showed that the maximum performance was obtained for $\delta/d_p = 4.0$. In Test 2, they used

TABLE 11.1 Porous Coatings Tested

Investigator	Particle Metal	Test Fluid	Coating Structure
O'Neill et al. [1972]	Cu, Ni, Al	R-11 and others	Sintered particles (nonuniforrn size and shape)
Nishikawa et al. [1983]	Cu, bronze	R-11, R-113, benzene	Sintered spherical particles
Fujii et al. [1979]	Cu	R-11, R-113	Sperical particles bonded by electroplating
Dahl and Erb [1976]	Al	R-113	Metal-sprayed powder
Janowski et al. [1978]	Cu	No published data	Electroplated polyurethane foam
Kajikawa et al. [1983]	Cu	R-11	Sintered steel fibers
Yilmaz and Westwater [1981]	Cu	Isopropanol, p-xylene	Metal spray with copper wire
Kartsounes [1975]	Cu	R-12	Metal-sprayed coating

Source: Webb [1983].

2.0-mm coating thickness (δ) and tested particle diameters of 0.10, 0.25, and 0.50 mm. The coating having $\delta/d_p = 4$ ($d_p = 0.25$ mm) again yields the highest boiling coefficient. Therefore, for spherical particles with $0.10 < d_p < 0.50$ mm, it appears that a coating having $\delta/d_p = 4$ is preferred for maximum performance. Tests 3 and 4 with bronze spherical particles show a much smaller effect of δ/d_p on the boiling coefficient; the $\delta/d_p = 4$ coating yields the highest boiling coefficient only for $q > 50$ kW/m². Below this heat flux, coatings having $2 < \delta/d_p < 6$ provided approximately equal performance. Kim and Bergles [1988] also report R-113 data on coatings made with $\delta/d_p = 4$ for particle diameters of 0.059, 0.096, and 0.247 mm.

Considerable data have been published on the sintered coating described by Milton [1968, 1970, 1971]—for example, see O'Neill et al. [1972] and Czikk and O'Neill [1979]. This is made from commercially produced powders, which are not spherical and have a range of particle sizes. About 45% of the particles are between 7 and 44 μm, with the rest smaller. The commercially produced tube is known as High-Flux™, and Figure 11.11a shows top and cross section views of the coating.

TABLE 11.2 Porous Copper Coatings Shown in Figure 11.10b

Investigator	d_p (μm)	δ (mm)	δ/d_p	ϵ
Kajikawa et al. [1983]	12	1.0	8	0.65
Milton [1968]	<44	0.25	6	
Fujii et al. [1979]	115	0.40	3–4	0.49
Nishikawa et al. [1983]	250	2.0	8	0.66
Milton [1968]	250–500	1.1	3	
Nishikawa et al. [1983]	500	2.0	4	0.42

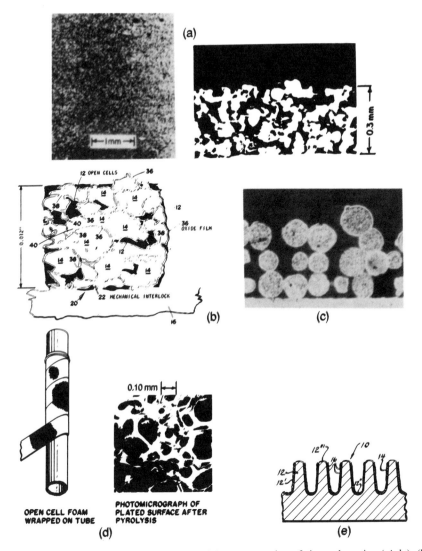

Figure 11.11 (a) Copper sintered surface (*left*); cross section of sintered coating (*right*). (b) Cross-section photo of aluminum flame-sprayed surface. (c) 0.115-mm-diameter spherical particles tested by Fujii et al [1979]. (d) Porous coating formed by plating copper on polyurethane foam wrapped tube. (e) Porous coating on fins formed by electroplating in a bath containing graphite particles. (From Webb [1983].)

Figure 11.11c shows a cross section of the spherical copper coating tested by Fujii et al. [1979], which has a nominal particle diameter of 0.115 mm and $\delta/d_p =$ 3. The coatings tested by Nishikawa et al. [1983] were also made of spherical, uniform-diameter particles. The pore dimensions and matrix porosity of the Figure 11.11a coating should be different from those for spherical particles of uniform diameter, since the smaller particles will tend to partially fill the voids between the larger particles. This raises a question concerning the important geometric parameters of the coating. Rather than the particle diameter, the pore size (the void space between particles) is probably a key characteristic dimension. Although the thermal conductivity of the particles is important, coatings made of aluminum and copper give comparable performance. Milton and Gottzman [1972] show that low-conductivity nickel particles produced considerable enhancement for boiling oxygen, although the boiling coefficient was only 30% as large as for a copper structure.

Because the independent parameters of the porous structure are the particle diameter (or particle size distribution), particle shape, and the coating thickness, one cannot calculate the pore diameter, unless the particle "stacking arrangement" is known. As seen by examining the porous structures of Figure 11.11, the "pore size" is not an explicitly measurable dimension. Czikk and O'Neill [1979] used a quantitative scanning microscope to examine cross-section slices of the coating in attempt to measure the pore size distribution. This resulted in a statistically defined pore size distribution, which was not wholly realistic. This is because the method could not differentiate between connected and closed pores. A more acceptable method involves use of a capillary rise test using a wetting fluid (near zero contact angle). By observing the height (H) that the wetting liquid rises on a porous coated plate, one may determine the "equivalent" pore size as $r = 2\sigma/\rho_l g H$. This method is used in the Milton patents [1968, 1970, 1971] to define the equivalent pore size.

The pores provide a high area density of relatively large cavities, which are believed to function as reentrant nucleation sites. Many of these pores are probably of a reentrant shape, which Griffith and Wallis [1960] have shown to be very stable vapor traps. The pores within the matrix are interconnected so vapor formed in one pore can activate adjacent pores. Gottzmann et al. [1971] theorize that essentially all vaporization occurs within the porous matrix, and the high performance is the result of two factors: (1) The porous structure entraps large radius vapor–liquid interfaces, compared to the very small nuclei in naturally occurring pits and scratches, which considerably reduces the theoretical superheat required for nucleation, and (2) the porous structure provides a much larger surface area for thin film or microlayer evaporation than exists with a flat surface. Thus, high evaporation rates can be accommodated with a very small film ΔT.

The sintering process described by Milton [1968] for copper powder on a copper base tube caused annealing and possible deformation of the base copper tube. Grant [1977] describes an improved metal powder, which allows a lower temperature sintering process (732–813°C). Use of a copper alloy base tube prevents full annealing.

Several other methods of forming a porous coating, which do not require high-temperature processing, have been proposed. These methods provide boiling performance comparable to that of the Milton patent [1968]. Fujii et al. [1979] bond several layers of fine copper particles to the base surface by electroplating with copper; this process is accomplished at room temperature. Shum [1980] forms a porous coating by electroplating in a bath containing graphite particles (see Figure 11.11e). The graphite particles are attracted to the surface and bonded by the plating operation. Dahl and Erb [1976] make an aluminum porous coating by flame-spraying the aluminum particles in an oxygen-rich flame atmosphere (see Figure 11.11b). The oxygen-rich flame forms an oxide film on the particle and prevents it from flattening upon impact with the surface. Flame-sprayed copper tends to flatten on impact, providing a less porous, lower performance surface. A U.S. patent by Modahl and Lukeroth [1982] describe aluminum flame-spray techniques that provide higher enhancement than obtained using that described by Dahl and Erb. [1976]. Sanborn et al. [1982] show that aluminum flame spray on the outer surface of the Figure 11.13f GEWA-T™ tube provides increased performance.

Liu et al. [1987] formed a porous coating by sintering multiple layers of 185 mesh copper screens to a copper surface. They tested R-113 and water for 1, 2, 3, 4, 5, 6, 7, 9, and 13 mesh layers. The 185-mesh wire screen was formed from 0.046-mm wires, provided 0.091-mm pore diameter, and had area and volumetric porosities of 44% and 56%, respectively. They found that the boiling coefficient increased with number of layers, up to five layers, and decreased for increasing number of layers for both R-113 and water. They provide an empirical, power law correlation of their data, whose maximum error was ±30%. The correlation assumes that the characteristic dimensions are pore diameter (not varied) and number of layers. They observed interesting hysteresis phenomena, which are discussed in Section 11.7.

Other innovative methods have been developed to make porous surfaces. Janowski et al. [1978] have developed a process for making a porous boiling surface (see Figure 11.11d). A tube is wrapped with an open-cell polyurethane foam, which is then copper-plated. The copper plating provides structural integrity and good thermal conduction. The polyurethane is removed by pyrolysis (300–520°C), which forms additional very small pores within the skeletal structure. The structure has 1.5-mm coating thickness, 97% void volume, 4 pores/mm, and 0.12-mm pore size, augmented by 0.02-mm pores within the skeletal structure. Sachar and Silvestri [1983] used vacuum deposition of aluminum in the presence of an inert gas to form a porous surface coating. Zohler [1990] flame-sprayed two dissimilar metals (e.g., zinc and copper) on a tube, and then etched the tube so that one of the metals was etched out. Voids are left where the metal has been etched away.

Porous coatings offer two potentially important advantages over enhancement formed by integral roughness. If tube-side enhancement is desired, it may be formed independent of the boiling-side enhancement. If corrosion requirements permit, the porous coating may be made using a less expensive material than required for base tube surface.

11.4.10   Structured Surfaces (Integral Roughness)

State-of-the-art concepts employ cold metal working to form a high area density of reentrant nucleation sites, which are interconnected below the surface. Typically, these are reentrant grooves or tunnels, rather than discrete cavities. Figure 11.12b shows five patented boiling surfaces of this type, which have significantly higher performance than the Figure 11.12a standard integral-fin tube. The surfaces were

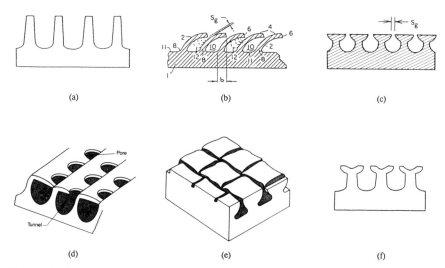

Figure 11.12 Illustration of six commercially available enhanced boiling surfaces. a) Integral-fin tube. b) Trane bent fin. c) Wieland GEWA-TW™. d) Hitachi Thermoexcel-E™, e) Wolverine Turbo-B™ or Furukawa ECR-40™, and f) Wieland GEWA-SE™, from Webb and Pais [1992].

developed for use in refrigeration evaporators, which use shell-side boiling on copper tubes. The first enhanced surface (Figure 11.12b) was patented in 1972. All of the enhanced surfaces shown in Figure 11.12 are commercially produced, and they are routinely used in refrigerant evaporators. Pais and Webb [1991] provide a survey of existing data for refrigerants on enhanced boiling surfaces. In single-tube pool boiling tests, the heat flux at fixed $T_w - T_s$ will be 2–4 times greater than that of the Figure 11.12a integral-fin tube.

The Figure 11.12b reentrant, single-grooved surface of Webb [1972] consists of integral-fin tubing (1300 fins/m, 0.8-mm fin height) in which the fins are bent to form a reentrant cavity. The groove opening at the surface is a critical dimension, and the recommended gap spacing is 0.0038–0.0089 mm for R-11. The performance is sharply reduced for gap widths outside this range. The preferred gap spacing is expected to be smaller for low-surface-tension fluids. All of the enhancements shown in Figure 11.12 are made by deforming the fins on integral-fin tubes. Saier et al. [1979] describes the Figure 11.12c tube, which is known as the Wieland GEWA-TW™ and has 750 fins/m. Figure 11.12d is the Wieland GEWA-SE™ tube having 1024 fins/m. The Figure 11.12e and 11.12f tubes are also made from integral-fin tubes, whose fins are further deformed to provide a three-dimensional structure. Figure 11.12e reported by Fujie et al. [1977], is formed from a low fin tube (1970 fins/m) which has small spaced cutouts at the fin tips. These saw-tooth fins are bent to a horizontal position to form tunnels having spaced pores at their top. This commercially available tube is called Thermoexcel-E™ and is further discussed by Arai et al. [1977] and Torii et al. [1978]. Figure 11.12f illustrates the Wolverine Turbo-B™. This tube was originally developed by Fujikake [1980]. Subsequent improvements have been made by Wolverine. The Wolverine version is described by Wolverine [1984] and has 1650 fins/m.

Figure 11.13 shows other structured surfaces reported in the literature. Kun and Czikk [1969] were awarded an early patent on a reentrant, cross-grooved surface (Figure 11.13a). A series of fine, closely spaced grooves (1100–9000 grooves/m) are cut in the surface, followed by a second set of cross-grooves, which are not as deep as the first set. The patent provides data on boiling nitrogen and water for a variety of groove densities. Kun and Czikk report that the opening width of the

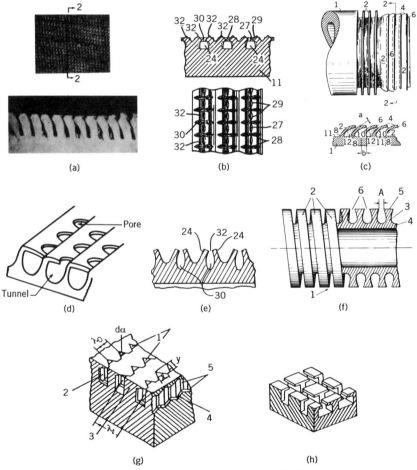

Figure 11.13 Illustrations of structured boiling surfaces. (a) Cross-grooved surface having reentrant cavities. (b) Deeply knurled Y-shaped fins. (c) Reentrant grooves formed by bending tips of integral-fin tube. (d) Reentrant grooves having minute spaced holes at top of tunnel. (e) Reentrant grooves formed by bending of fins of unequal height. (f) Reentrant grooves formed by flattening fin tips of integral-fin tube. (g) JK-2 tubes tested by Zhong et al. [1992]. (h) Geometry developed by Zhang and Dong [1992].

reentrant cavity is a critical dimension, as also reported by Webb [1972]. Maximum performance is obtained with a restricted opening dimension of 0.0013 mm for nitrogen and 0.006 mm for water.

11.4.11   Combination Structured and Porous Surfaces

Ma et al. [1986] have found quite high performance with water and methanol boiling on a copper plate having the following: (1) parallel rectangular grooves covered by fine mesh screens; (2) same as item 1, but with a 40-mesh screen covered by a thin brass, and plate having small pores (0.08 to 0.22 mm); (3) same as item 2, but varying the cross-sectional shape of the grooves (rectangular and vee). They varied the groove size and shape, the screen porosity, and the pore size in the item 2 variant. Their findings and preferred dimensions are as follows:

Variant 1: The best performance was obtained for water and methanol with 0.3-mm-wide, 0.4-mm-deep grooves spaced at 0.6 mm, which were covered with a 185-mesh copper screen (0.046-mm wire diameter). A 40-kW/m² heat flux was obtained at 1.0 K with methanol.

Variant 2: Grooved plates (0.60 mm groove pitch) covered by a 40 mesh screen, upon which a plate having 0.08-0.22 mm diameter pores at 277 per cm² was brazed. The best configuration provided 90 kW/m² heat flux at 1.0 K. This surface performance is competitive with the High-Flux™ surface for ethanol, as reported by Antonelli and O'Neill [1981].

Variant 3: Use of triangular grooves provided higher heat flux than rectangular grooves at heat fluxes above 80 kW/m².

The paper by Ma et al. [1986] is very interesting and is worthy of close reading. Although the use of screening above the grooves (or screening with a plate having pores) provides high performance, it may be relatively expensive. It appears that a standard porous surface (e.g., the High-Flux™) may provide competitive performance.

11.5   SINGLE-TUBE POOL BOILING TESTS OF ENHANCED SURFACES

Figure 11.14 shows the single-tube pool boiling performance of four commercially available enhanced surfaces measured by Yilmaz and Westwater [1981] using iso-propyl alcohol. These are the High-Flux™ (Figure 11.11a), the Thermoexcel-E™ (Figure 11.12d), the GEWA-TW™ (Figure 11.12c), and the ECR-40™ (Figure 11.12e), which is the same as an early version of the Wolverine Turbo-B. Figure 11.14 also shows the performance of a copper-metal sprayed surface (CSBS). The tubes were heated by steam condensing on the tube side, and data were taken for saturated isopropyl alcohol boiling at 103 kPa. The data span the nucleate boiling and transition regimes. The highest performance is given by the ECR-40 and High-Flux™ surfaces. However, at a typical design heat flux of 50 kW/m², the other

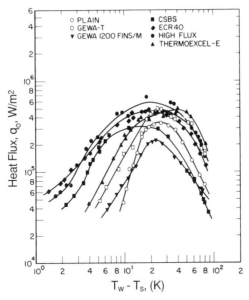

Figure 11. 14 Comparative single-tube pool boiling test results for isopropyl alcohol at 101 kPa (1 atm). (From Yilmaz and Westwater [1981].)

tubes provide considerable enhancement. The heat flux is defined in terms of the envelope area ($\pi d_o L$), except for the 1200-fin/m integral-fin tube (GEWA); the heat flux of this tube is defined in terms of the extended surface area. The enhanced surface tubes exhibit a critical heat flux (CHF) at least 40% greater than that of the plain tube. If the heat flux of the integral fin tube is defined in terms of the diameter over the fins, the heat flux will be 2.6 times greater. Then its CHF will be approximately equal to that of the High-Flux™ tube. More discussion on CHF is presented in Section 11.13. Li et al. [1992] tested a variant of the Thermoexcel-E™ tube shown in Figure 11.12d. The R-113 boiling coefficient of their best tube, having 0.14-mm pore diameter, was approximately 50% higher than that of the Thermoexcel-E™ tube.

Webb and Pais [1992] obtained saturated nucleate boiling data for five refrigerants (R-11, R-12, R-22, R-123, and R-134a) on a plain tube and geometries a, c, d, and f shown in Figure 11.12. The copper tubes had 17.5- to 19.1-mm diameter over the fins. Tests were performed at two saturation temperatures, 4.44°C and 26.7°C, using refrigerants R-11, R-12, R-22, R-123, and R-134a. The tubes were specially made to have a 9.53-mm smooth inside bore. This was done to facilitate wall thermocouple and electric cartridge heater installation. Figure 11.15 shows the R-22 boiling data at 4.44°C for the five tubes tested. Figure 11.16 shows the 4.44°C GEWA-SE™ data for five refrigerants. The complete data set are given by Webb and Pais [1992]. The data for the high-pressure refrigerants (R-12, R-22, and R-134a) are higher than that for the low-pressure refrigerants (R-11 and R-123). Note the difference in slope for the high- and low-pressure refrigerants, particularly for the GEWA-SE™. At 40-kW/m² heat flux, the boiling coefficient of the GEWA-SE™ relative to the 26 fin/in. tube is 1.58 for R-12 and 1.38 for R-134a, respectively.

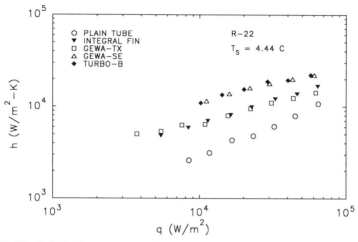

Figure 11.15   R-22 boiling data at 4.44°C for the five tubes tested by Webb and Pais [1992].

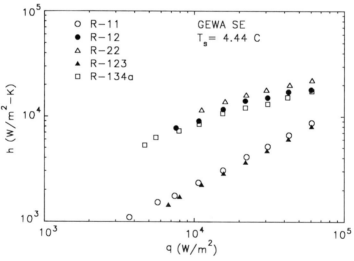

Figure 11.16   Boiling data for five refrigerants boiling at 4.44°C on the GEWA-SE™ tube. (Webb and Pais [1992].)

11.6   THEORETICAL FUNDAMENTALS

Small bubbles (e.g., d_b < 1.5 mm) are periodically grown and released from nucleation sites. After departure of a bubble, the nucleation sites must retain a small vapor embryo to facilitate the next bubble cycle. The cavities must be able to initially entrap vapor that will allow the nucleation process to take place. These phenomena must exist for both plain and enhanced surfaces. Carey [1992] provides an excellent discussion of these fundamental phenomena. We will summarize the key issues, and then discuss how the fundamental phenomena of boiling on enhanced surfaces differ from plain surfaces.

11.6.1 Liquid Superheat

The first item to understand is the thermodynamic requirement for a bubble to exist. Consider a static bubble of radius r, which exists in thermal equilibrium with a large volume of liquid. The pressure inside and outside the bubble are p_v and p_l, respectively. The net pressure force, $(p_v - p_l)\pi r^2$ is equal to the surface tension force, $2\pi r\sigma$. Equating the pressure and surface tension forces shows that

$$p_v - p_l = \frac{2\sigma}{r} \tag{11.1}$$

Thermodynamic tables (or an equation of state) give the saturation pressure (p_s) that corresponds to a given saturation temperature (T_s). The $p_s(T_s)$ is for a liquid–vapor interface having zero curvature. For the present case of a vapor bubble of radius r, Rohsenow [1985] shows that the pressure inside the bubble (p_v) is slightly lower than the pressure of a saturated liquid (p_s) at the same temperature. This difference is

$$p_v = p_s - \frac{\rho_v}{\rho_l - \rho_v}\frac{2\sigma}{r} \tag{11.2}$$

If $\rho_v \ll \rho_l$, then $p_v \simeq p_s$.

$$p_s - p_l \simeq \frac{2\sigma}{r} \tag{11.3}$$

It is frequently assumed that $p_v = p_s$ at the liquid temperature. Using this approximation, we may write Equation 11.1 as

$$T_l - T_s = \frac{2\sigma}{mr} \tag{11.4}$$

where the local slope of the saturation curve (p_s versus T_s) is $dp/dT \equiv m$. Frequently, writers use the Clasius–Clapeyron equation to approximate $dp/dT = \lambda/T_{abs}v_{fg}$. However, we will use the terminology of $m \equiv dp/dT$ for simplicity. Assuming $p_v \equiv p_{sat}$, Equation 11.4 shows that the liquid temperature must be superheated for the vapor bubble of radius r to exist.

Equation 11.4 was written for a vapor bubble (concave liquid shape). If a force balance is written on a liquid droplet (a convex liquid shape) of radius r, the resulting force balance will yield a negative sign on the right hand side of Equation

11.1. Then, the right-hand side of Equation 11.4 would be negative. In this case, the liquid must be subcooled an amount $(T_s - T_l)$ for a convex liquid surface of radius r to exist. Hence, we may write Equation 11.4 in a more general form:

$$T_l - T_s = \pm \frac{2\sigma}{mr} \qquad (11.5)$$

where the plus and minus signs are for concave and convex liquid curvatures, respectively.

11.6.2   Effect of Cavity Shape and Contact Angle on Superheat

Figure 11.17 shows vapor a liquid–vapor interface growing inside a conical cavity for three different values of liquid contact angle (θ). The interface curvature is larger for increasing contact angles. The liquid–vapor interface is spherical. Its radius (r) depends on the liquid contact angle (θ) and the angular opening (2ϕ) of the cavity. The interface radius changes with its location within the cavity. Figure 11.17a shows a highly wetting liquid ($\theta \ll \phi < 90$ degrees), and Figure 11.17c is for a nonwetting liquid ($\theta > \phi + 90$ degrees). Figure 11.17b is for nonwetting inside the cavity ($\theta > \phi + 90$ degrees), but wetting outside the cavity ($\theta < 90$ degrees). Figure 11.17d shows the variation of r_c/r within the conical cavity, for the three wetting characteristics. If $r_c/r < 1$, the required liquid superheat (see Equation 11.5) is less than that required at the cavity mouth. Figure 11.17d shows that a highly wetting surface requires higher superheat inside the cavity than at the mouth. Conversely, a poorly wetting surface requires smaller superheat inside the cavity than at the mouth.

The contact angle is not a thermodynamic property. It depends on the surface cleanliness and micro-roughness. Oxidized surfaces have smaller contact angles

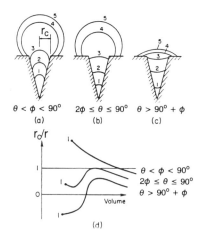

$\theta < \phi < 90°$ $2\phi \le \theta \le 90°$ $\theta > 90° + \phi$
   (a)   (b)   (c)

r_c/r

$\theta < \phi < 90°$
$2\phi \le \theta \le 90°$
$\theta > 90° + \phi$

Volume

(d)

Figure 11.17   Variation of bubble radius during growth for (a) $\phi < \theta < 90$ degrees, (b) $2\phi \le \theta \le 90$ degrees, (c) $\theta > \phi + 90$ degrees, and (d) Variation of r_c/r with cavity volume. (From Carey [1992].)

than do clean surfaces. For example, water deposited on a waxed surface will exist as beads, but it tends to spread on the surface after the wax has worn away. Chapter 3 of Carey [1992] provides a good discussion of contact angle.

The internal shape of the cavity has a significant influence on the liquid superheat required for a liquid–vapor interface to exist within a cavity. Webb [1983] gives equations for the interface radius for the five cavity geometries shown in Figure 11.18. These equations are developed using geometric relations for the cavity shape, as a function of the contact angle (θ) and the cavity angle (ϕ). Depending on θ and ϕ, the liquid either wets the cavity (concave liquid interface) or is nonwetting (convex liquid interface). The equations are as follows:

1. *Conical (Figure 11.18a)*

$$\frac{r}{x} = \frac{\sin \phi}{\sin (\theta - \phi \pm \pi/2)} \qquad (11.6)$$

The plus sign is used for a wetting fluid ($\theta < \pi/2 + \phi$) and the minus sign for a nonwetting fluid ($\theta > \pi/2 + \phi$).

2. *Cylindrical (Figure 11.18b)*

$$\frac{r}{r_c} = \pm \frac{1}{\cos \theta} \qquad (11.7)$$

The cylinder radius is r_c. The plus sign is used for a wetting fluid ($\theta < \pi/2$), and the minus sign is used for a nonwetting fluid.

3. *Reentrant Conical Cavity (Figure 11.18C and 11.18D)*

$$\frac{r}{x} = \frac{\sin \phi}{\sin [\pm(\theta + \phi - \pi/2)]} \qquad (11.8)$$

The minus sign is used for a wetting fluid ($\theta < \pi/2 - \phi$), and the plus sign is used for a nonwetting fluid.

4. *Doubly reentrant conical cavity (Figure 11.18E):* Same equation as for Figures 11.18C and 11.18D (Equation 11.8).

Equations 11.6 through 11.8 were applied to the cavity shapes in Figure 11.18 for $\theta = 15$ degrees and $r_c = 0.04$ mm to predict the liquid–vapor interface radius values. Then, from Equation 11.5, we may calculate the $T_l - T_s$ values required for a stable liquid–vapor interface. These results are shown in Figure 11.19. Figure 11.19 shows the value $1/r = m(T_w - T_s)/2\sigma$ versus y/r_c, where r_c is the cavity mouth radius. All of the cavity shapes require the same value of $T_l - T_s$ for a vapor bubble to exist at the cavity mouth. However, significantly different values of $T_w - T_s$ exist within the cavity. Figure 11.19 shows that cavity "A" requires the greatest superheat, and that the limiting superheat exists at the bottom of the cavity. The required superheat decreases with vapor volume in cavity "C," attaining a maximum

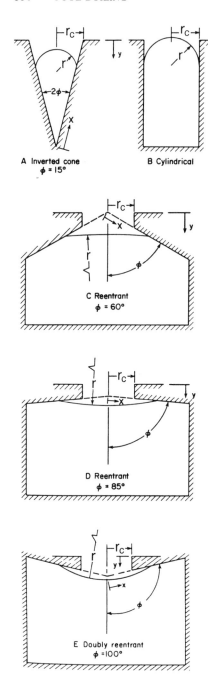

Figure 11.18 Illustration of vapor–liquid interface radius in cavities of different shapes for $r_c = 0.04$ mm and $\theta = 15$ degrees. (From Webb [1983].)

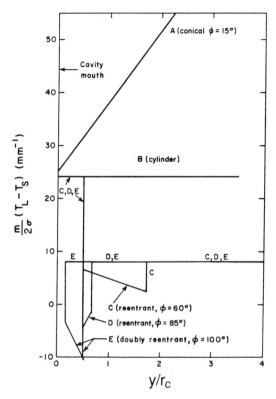

Figure 11.19 Required liquid superheat for stability of the Figure 11.18 cavity geometries having $R_c = 0.04$ mm and $\theta = 15$ degees. (From Webb [1983].)

at the cavity mouth, and decreases as the bubble grows outside the cavity. The required superheat within the cavity "B" is intermediate to that cavity "A" and cavity "C." The reentrant cavities (cavities C, D, and E) require the smallest $T_l - T_s$ within the cavity. Only the cavities in Figure 11.18D and 11.18E will be stable if a subcooled liquid is exposed to the cavity; this is because the liquid interface takes a convex shape along the conical side walls. The doubly reentrant cavity (Figure 11.18E) allows greater subcooling than does the cavity in Figure 11.18D.

Therefore, a reentrant cavity shape having a high contact angle will be very stable, and if $\theta > 90$ degrees $- \phi$, it will not flood in the presence of moderate subcooling of the liquid. For very small contact angles (e.g., $\theta = 4$ degrees) a reentrant cavity is susceptible to liquid flooding. However, the doubly reentrant cavity (Figure 11.18E) will establish a convex liquid shape just inside the lower cavity mouth for low-contact-angle fluids and provide a stable interface with a subcooled fluid. Low-surface-tension fluids (such as refrigerants and organics) have small contact angles (e.g., 5 degrees), and high-surface-tension fluids (such as water) have contact angles in the range of 70 degrees.

11.6.3   Entrapment of Vapor in Cavities

Section 11.6.2 addressed how the radius of the vapor interface changes with its movement within cavities of various shapes. Now, we must consider the means by which vapor (or air) is initially trapped within a cavity. Boiling systems are typically charged with subcooled liquid. After the system is charged, heat is applied and the subcooled liquid is heated to the boiling state. Some analysis has been done to explain how inert gas (air) can be trapped in cavities, when flooded by subcooled liquid. The fact that an inert gas is trapped in the cavity has implications on the boiling process, which are discussed in Section 11.6.4.

When the system is initially charged with liquid, a liquid front advances over the nucleation sites, as illustrated for a conical cavity in Figure 11.20a. The contact angle of the advancing liquid front is θ_a. If the liquid contacts the far wall of the cavity before it contacts the base, gas (air) will be trapped in the cavity. This will occur in Figure 11.20a if $\theta > 2\phi$. As discussed by Carey [1992], the advancing contact angle (θ_a) may be significantly greater than the static contact angle (θ). Hence, the possibility of trapping gas is influenced by the cavity geometry and the magnitude of θ_a.

Lorenz et al. [1974] have provided a simple method to calculate the size of the initially entrapped gas radius. If the cavity geometry and θ_a are known, one may calculate the volume of the trapped gas illustrated in Figure 11.20a. Assuming that this trapped gas volume reverts to the shape illustrated in Figure 11.20b, one may calculate the interface radius, shown in Figure 11.21. This may be an oversimplification, because Lorenz et al. assume that the static contact angle applies to both Figure 11.20a and Figure 11.20b. As suggested above, $\theta_a > \theta$. However, the Lorenz et al. concept is probably useful to explain how a finite liquid–vapor radius can exist in a cavity.

Some boiling systems (e.g., refrigerant systems) are evacuated before charging with liquid. Consider charging an evacuated system with R-22 at 300 K. At 300 K , R-22 will exist as a saturated liquid at 1097 kPa. Thus, R-22 enters the evacuated system as a vapor, and the system pressure increases during the filling process. This

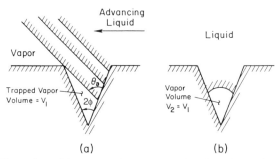

Figure 11.20   Illustration of how a cavity may trap air when the surface is flooded with liquid. (a) Advancing liquid front. (b) Liquid–vapor interface for the Figure 11.20a conditions. (From Carey [1992].)

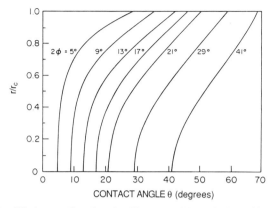

Figure 11.21 Equilibrium radius for the Figure 11.20 conical cavities with different ϕ and θ.

vapor will flood the cavities, and may be trapped by the mechanism described by Lorenz et al. [1974]. However, the vapor trapped in the cavity may not contain an inert gas. It is possible that the charged liquid may contain a dissolved gas, which comes out of solution during the process of raising the liquid to boiling temperature. This evolved gas may add to the trapped vapor volume.

11.6.4 Effect of Dissolved Gases

Equations 11.1 through 11.5 assume that the vapor does not contain inert gases. Should the system contain inert gases, the total pressure in the vapor bubble is $p_v + p_G$. Assuming $p_v \simeq p_s$, Equations 11.3 and 11.5 would then be written as

$$p_s - p_l \simeq \frac{2\sigma}{r} - p_G \qquad (11.9)$$

$$T_l - T_s = \pm \frac{1}{m}\left(\frac{2\sigma}{r} - p_G\right) \qquad (11.10)$$

Equation 11.10 shows that a concave vapor interface will require less liquid superheat when inert gas is present in the cavity. When boiling occurs from cavities containing inert gases, each departing bubble will remove a small amount of inert gas from the cavity. With increasing time, a higher $T_l - T_s$ will be required to sustain boiling, because the partial pressure of the inert gas will be reduced.

11.6.5 Nucleation at a Surface Cavity

Consider a bubble growing from a cavity, whose radius at the surface is r_c. Figure 11.22 shows that the cavity is contained in a surface, maintained at temperature T_w. How much liquid superheat must exist at the wall temperature for the bubble of

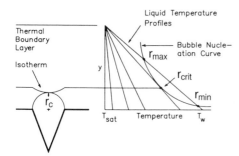

Figure 11.22 Necessary condition for bubble growing from cavity of radius r_c in a heated surface at T_w.

radius r_c to exist? This will define a necessary condition for nucleate boiling to occur at the surface. The liquid superheat at the top of the bubble, $T_l(r) - T_s$, must satisfy Equation 11.4 for $r = r_c$. Assuming steady-state conduction in the liquid, and a linear temperature profile in the liquid, the Fourier heat conduction equation gives

$$\frac{q}{A} = \frac{k_l[T_w - T_l(r_c)]}{r_c} \tag{11.11}$$

Writing $T_w - T_s = [T_w - T_l(r_c)] + [T_l(r_c) - T_s]$, the first and second terms on the right-hand side may be obtained from Equations 11.11 and 11.4, respectively. The resulting $T_w - T_s$ is

$$T_w - T_s = \frac{qr_c}{k_l} + \frac{2\sigma}{mr_c} \tag{11.12}$$

Hsu [1962] proposed a more refined analytical model to define the wall superheat necessary to support periodic bubble growth at a plain heated surface. The Hsu model assumes transient heat flux to the cold liquid layer that floods the surface after bubble departure.

If the cavity size is known, one may solve Equation 11.12 for the $T_w - T_s$ needed to support nucleation from the cavity at the specified heat flux. A problem in using the above model for plain surfaces is that the size of the natural nucleation sites is typically not known. For an enhanced surface, the bubbles grow from pores in the surface that are typically much larger than the microcavities associated with plain surfaces. Hence, substantially smaller superheat is required for the second term of Equation 11.11. More detail on nucleation on enhanced surfaces is given in Section 11.8.

11.6.6 Bubble Departure Diameter

Bubbles are periodically grown and released from the surface. The bubble departs when buoyancy force exceeds the surface tension force holding the bubble to the surface. Writing a force balance between buoyancy and surface tension forces on a departing bubble, with inertia forces neglected, we obtain

$$d_b = (6r_c \sin \theta)^{1/3} \left[\frac{2\sigma}{(\rho_l - \rho_v)g} \right]^{1/3} \tag{11.13}$$

where r_c is the surface cavity radius. This equation is not normally used for plain surfaces, since the cavity radius is typically not known. However, the surface pore radius is a known dimension for certain types of enhanced surfaces, which are discussed later. For plain surfaces, a different form of Equation 11.13 is typically used, and is based on the work of Fritz [1935]. The Fritz equation includes the empirical constant c_b ($= 0.0148\theta$ degrees) and does not use the cavity radius. This equation is

$$d_b = c_b \left[\frac{2\sigma}{(\rho_v - \rho_v)g} \right]^{1/2} \tag{11.14}$$

Other equations have been proposed to calculate the bubble departure diameter. Some of the equations also account for the inertia term in the momentum equation for the growing bubble. These equations are surveyed in Table 6.1 of Carey [1992].

11.6.7 Bubble Frequency

Several studies have shown that the nucleate boiling heat flux is proportional to the nucleation site density (n_s) on the heated surface. For example, the data of Kurihari and Myers [1960] for plain and emery-roughened surfaces show that $q \propto n_s^{0.33}$. Researchers who have modeled boiling on plain surfaces typically regard the bubble density as an empirical parameter in their models. This is the case for the Mikic and Rohsenow [1969] model, which is based on transient heat conduction to liquid during the bubble waiting period. The model assumes that the transient conduction occurs over a surface area twice the bubble departure diameter.

11.7 BOILING HYSTERESIS EFFECTS

The existence of a starting hysteresis effect has been observed by a number of researchers. This phenomenon occurs when heating is initially applied and the data are taken in order of increasing heat flux. As the heat flux is increased from zero, boiling will not commence until a certain surface superheat is attained. Then the surface suddenly "explodes" into nucleate boiling, at which point the surface temperature decreases. It is possible then to traverse up and down the boiling curve, q versus $(T_w - T_s)$, without significant hysteresis. Apparently, this phenomenon occurs when the nucleation sites contain few, or very small vapor nuclei, and it is necessary to evaporate the trapped liquid in order to activate the cavities. Bergles and Chyu [1982] have measured hysteresis effects of the High-Flux™ surface by boiling R-113 on a single electrically heated tube. Figure 11.23 shows test results from this study. The data were taken for different amounts of initial subcooling. This figure shows that the nucleation sites are suddenly activated when $2.5 < \Delta T < 8$K. In some cases, small boiling patches were observed and full surface activation did not occur until a higher heat flux was applied. They were able to shut off the test section heat input and restart the test from a low heat flux without the occurrence of the initial hysteresis effect. However, if the R-113 was sufficiently subcooled in the shutdown phase, the hysteresis would again occur. The possibility of cavity flooding is more likely for low-surface-tension, wetting fluids (e.g., R-113) than for a less-wetting fluid, such as water. The cavity geometry and the existence of dissolved gas should have a significant effect on hysteresis effects.

Kim and Bergles [1988] extended their evaluation of hysteresis effects on porous boiling surfaces. They boiled R-113 at 1.0 atm on horizontal, 4.7-mm square plates having porous coatings of spherical copper particles. They tested three different coatings having a thickness of four particle diameters (e.g., as in Figure 11.11c). The particle diameters were 0.059, 0.096, and 0.247 mm. A fourth coating had layers of different particle diameters. Incipient boiling occurred at 30 K $< \Delta T < 40$ K, which is higher than for the High-Flux™ tube tests of Bergles and Chyu [1982]. The superheat for incipient boiling was independent of the particle size.

Low-surface-tension fluids, such as R-113, are more susceptible to hysteresis effects than are high-surface-tension fluids, such as water. This is because the low-surface-tension fluids typically have smaller contact angles and are more likely to flood the cavities. Ko et al. [1992] measured hysteresis effects for water and R-113 on four different porous coatings having different particle sizes. Their data, on a given surface, shows that R-113 shows significantly higher incipient superheat than does water. They attempted to explain the incipient liquid superheat required for bubble existence by the equation $(T_w - T_s) = 2\sigma(\cos\theta)/r$. They argued that porous structures having small pores will have the largest incipient superheat. Unfortunately, they did not actually define the average pore size of the surfaces they tested. Hence, their conclusions are not well-justified. The tests of multiple layers

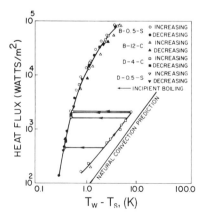

Figure 11.23 Starting hysteresis effects with R-113 boiling at 101 kPa (1 atm) on a horizontal High-Flux tube (Figure 11.10a) measured by Bergles and Chyu [1982]. (From Bergles and Chyu [1982].)

of sintered 185-mesh screen by Liu et al. [1987] showed decreasing hysteresis effects as the number of screen layers increased from one to five. Both R-113 and water showed similar hysteresis effects, although the hysteresis for water was a little less than for R-113.

Ma et al. [1986] note that combination structured/porous surfaces showed substantially less hysteresis with methanol than observed by Bergles and Chyu [1982] for R-113. Bar-Cohen [1992] provides a survey of hysteresis effects and Malyshenko and Styrikovich [1992] observed a peculiar hysteresis phenomenon for boiling on porous coatings made of low-thermal-conductivity particles. They found a significant difference between increasing and decreasing power boiling curves and attempted to explain the phenomenon.

The startup hysteresis effects suggest potential startup problems for equipment containing enhanced surfaces. This possibility will be discussed in Section 11.10. It appears that the only possible options for forming a vapor (or gas) nucleus in potentially vapor trapping cavities are:

1. Gas initially trapped in cavity.
2. Dissolution of dissolved gas from the liquid, because of liquid heating.
3. Blowing vapor or inert gas into the cavity, from an external source.
4. Heterogeneous nucleation.

The startup hysteresis effects suggest potential startup problems for equipment containing enhanced surfaces. This possibility will be discussed in Section 11.10.

11.8 BOILING MECHANISM ON ENHANCED SURFACES

11.8.1 Basic Principles Employed

The published literature shows that a complete understanding of the mechanism of boiling on enhanced surfaces does not exist. However, some very good recent progress has been made in theoretical modeling, which is discussed in Section 11.9. As previously discussed, there are two basic types of enhancement geometries, namely, porous and structured surfaces. The structured surfaces are more amenable to analysis and description of the boiling mechanism than are the porous surfaces. This is because the structured surfaces, as illustrated in Figure 11.12, contain well-defined dimensions. However, the porous surfaces of Figure 11.11 are made up of an irregular geometry of the porous structure.

In the author's opinion, four key concepts are employed to attain high performance:

1. The use of reentrant (or doubly reentrant) tunnel-type cavities is necessary.
2. The subsurface pores are substantially larger than those occurring on natural surfaces. This reduces the superheat requirement, as defined by Equation 11.4.

3. Liquid is supplied via the surface pores to subsurface capillary passages. The liquid is heated as it proceeds through the capillary passages; thus, in the vicinity of an active pore, $T_l > T_s$. Thin film evaporation occurs from liquid–vapor interfaces in the subsurface capillaries. When a bubble departs from the surface, the liquid–vapor interfaces in the subsurface capillaries are not subjected to the cold liquid temperature existing above the surface.

4. The subsurface capillaries are interconnected. Vapor produced within the subsurface structure at one site tends to activate other sites, because the subsurface sites are interconnected.

The six structured surfaces illustrated in Figure 11.12 are all made from an integral-fin tube. The Figure 11.12a fins are deformed to form reentrant tunnels in the Figure 11.12b and 11.12c geometries. A narrow gap exists at the top of the tunnel. The width of this gap has a strong effect on the boiling performance. A secondary operation is applied to the Figure 11.12d and 11.12e surfaces. The initially formed fins have a "sawtooth" configuration, as shown in Figure 12.6. In Figure 11.12d, the fins are bent so that there is no gap between adjacent fins. However, the sawtooth fin shape provides "pores" which communicate between the tunnel interior and the fluid above the surface.

Because both the Figures 11.12b, 11.12c and the Figures 11.12d, 11.12e geometries provide high performance, it is concluded that the pore structure of the Figure 11.12d geometry is not a necessary geometrical feature. However, if the pore structure is not provided, there is an "optimum" gap between adjacent fins. Ayub and Bergles [1988a, 1988b] boiled R-113 and water on the 748-fin/m (19-fin/in.) Figure 11.12c enhanced surface. They varied the gap spacing s_g at the fin tip. They found that the optimum gap dimension is 0.25 mm. However, Webb [1972] found the "optimum" fin gap is between 0.04 and 0.09 mm for R-11 boiling on the Figure 11.12b geometry.

Nakayama et al. [1980b] investigated the optimum pore diameter for the Figure 11.12d surface. The optimum pore diameter is 0.15 mm for water and 0.07 mm for nitrogen. The optimum pore diameter varies with saturation pressure.

11.8.2   Visualization of Boiling in Subsurface Tunnels

Nakayama et al. [1980a] were the first to perform a visualization study of boiling in a rectangular tunnel, with pores in a very thin copper sheet covering the top surface, as illustrated in Figure 11.24a. Liquid R-11 at 1.0 atm was contained above the cover plate. In one series of experiments, heat was provided from below by a cartridge heater in the base block. In other tests, the lid was electrically heated by dc current and was used as a resistance thermometer. The tunnel heights varied between 0.5 and 1.0 mm, and the width was held fixed at 1.0 mm. The diameters of the pores were uniform, from 0.05 to 0.5 mm in different lids tested. Boiling activity was recorded by a high-speed cine camera or a still camera. Their visual observations are summarized below.

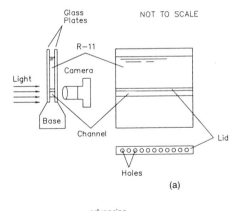

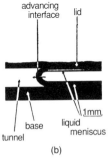

Figure 11.24 (a) Schematic of apparatus used by Nakayama et al. [1980a] to view boiling in tunnels. (b) Liquid–vapor interface observed in tunnel by Nakayama et al. [1980a]. (From Nakayama et al. [1980a].)

1. In the unheated state, a vapor plug existed in part of the tunnel. The volume of this plug reduced as the ambient temperature was decreased.

2. Upon addition of heat, the vapor region expanded along the tunnel, and the advancing liquid–vapor interface shown in Figure 11.24b was observed. A thin liquid film existed on the walls of the vapor region, with liquid holdup in the corners.

3. At wall superheats greater than 0.6 K, bubbles began to emerge from several pores at a frequency of about one every 8 seconds.

4. Incrementally increasing the wall superheat increased the bubble departure frequency and also the number of active pores.

These characteristics were observed for all test geometries except for the lid with the largest pores (0.5 mm). In this case, expulsion of the liquid from the tunnel was never complete, and liquid could be seen sloshing around in the tunnel while occupying roughly 10–50% of the tunnel volume.

Arshad and Thome [1983] performed a visualization study with water using an experimental apparatus similar to that of Nakayama et al. [1980a]. They soldered 0.005-mm-thick copper cover plates having grooves machined in the top surface. The cover plates contained spaced holes, whose diameter is between 0.15 and 0.25 mm. The ends of the tunnels were covered with thin glass plates. They tested triangular, rectangular, and circular cross-sectional groove shapes. Their apparatus was designed to allow observation along the axis of the tunnel, as opposed to the side viewing method of Nakayama et al. [1980a]. Figure 11.25 shows sketches of the activity observed for a triangular groove about 1 mm deep with 0.15-mm pore diameter. Nucleation occurred at the top right corner in this sequence, but in other tests it also occurred at the bottom or the top left corner. Once the vapor reached a pore in the cover plate, a bubble immediately grew in the liquid pool at this pore, and a thin liquid film could be observed covering the tunnel walls, taking their shape. At large heat fluxes, the liquid film progressively dried out.

Nucleation occurred at a bottom right corner for a rectangular tunnel. The vapor front reached the opposite wall before reaching the pore in the cover plate. Activation of circular tunnels was more difficult compared to the other two geometries. The following conclusions were drawn from the visualization study for the nucleate boiling mechanism on structured surfaces:

1. Thin film evaporation is the principal heat transport mechanism inside the tunnel.
2. The geometrical shape of the subsurface groove affects the shape and formation of the thin evaporating liquid film.
3. Dryout of the liquid film causes the surface to revert toward the performance of a plain surface.

Because Arshad and Thome [1983] viewed the end of the tunnels, their observations were unable to detect a meniscus moving down the tunnel, as was observed by Nakayama et al. [1980a]. However, both studies show the existence of alternate liquid vapor–liquid zones inside the tunnel, at least during the initial phase of boiling.

11.8.3 Boiling Mechanism in Subsurface Tunnels

Nakayama et al. [1982] described three possible evaporation modes, which may occur in a subsurface tunnel. Conceptual sketches of the three possible evaporation modes in the tunnel are shown in Figure 11.26. Nakayama et al. [1982] state that the

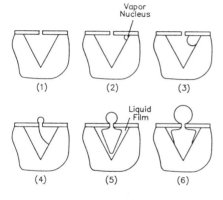

Figure 11.25 Sketches of boiling activity in a triangular cross-section tunnel. (From Arshad and Thome [1983].)

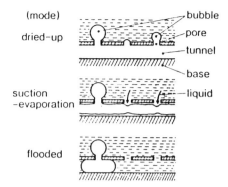

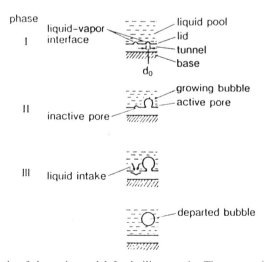

Figure 11.26 Heat transfer modes in tunnels of the Thermoexcel-E™ surface proposed by Nakayama et al. [1982]. (From Nakayama et al. [1982].)

"flooded mode" occurs at low heat flux, where most of the tunnel space is occupied by liquid, and an active pore acts as an isolated nucleation site. At higher heat flux, the "suction–evaporation mode" exists, where liquid is sucked in the tunnel space through inactive pores by pumping actions of bubbles growing at active pores. It then spreads along the tunnels and evaporates from menisci at the corners of the tunnel. With further heat flux increase, the "dried-up mode" exists, where the tunnel space is filled with vapor, and vaporization into bubbles takes place outside the tunnels.

Nakayama et al. [1980b] were the first to propose an analytical model to predict boiling performance of the Thermoexcel-E™ (Figure 11.12d) surface geometry. Their dynamic model described below, assumes that the bubble cycle occurs as illustrated in Figure 11.27:

Figure 11.27 Basis of dynamic model for boiling on the Thermoexcel-E™ surface proposed by Nakayama et al. [1980b]. (From Nakayama et al. [1980b].)

1. *Phase I (Pressure Build-up Phase):* Evaporation of liquid in the tunnel causes the internal vapor pressure to build up until the vapor nuclei located at the mouth of the pores attain a hemispherical shape, protruding outward into the external liquid pool.

2. *Phase II (Pressure-Reduction Phase):* At some pores, the vapor nuclei become active and grow into bubbles. Initially, the bubbles grow as a result of the pressure differential across the interface, which reduces the pressure in the tunnel as the vapor flows into the bubbles. They then continue to grow as a result of the inertial force imparted to the surrounding liquid.

3. *Phase III (Liquid-Intake Phase):* During the inertial stage of bubble growth the pressure inside the tunnel is reduced to less than that of the external liquid pool, and liquid is drawn into the tunnel at inactive pores. At the end of this phase the bubbles depart, all pores are closed by liquid menisci, and the liquid spreads along the angled corners of the tunnel by capillary forces. The cycle then repeats from this point.

11.9 PREDICTIVE METHODS FOR STRUCTURED SURFACES

11.9.1 Empirical Correlations

Very little has been done in the development of correlations for prediction of the boiling coefficient as a function of heat flux and fluid properties. This is largely because no workers have developed the required database for specific surface geometries. The database provided by Webb and Pais [1992] on five enhanced tube geometries is perhaps the largest existing database on any of the enhanced surfaces. Without a large database, a correlation has little chance of being very useful. Zhang and Dong [1992] developed a power-law correlation for the Figure 11.13h tube geometry, based on their data for R-113 and ethyl alcohol taken at pressures between 1.0 and 6.0 bar. The dimensionless parameters were selected, based on their speculation of the boiling mechanism. The data were correlated within ±20. Unfortunately, they do not give the dimensions of the enhancement geometry. Li et al. [1992] propose a power-law correlation for their 0.10-MPa R-11 and R-113 data on the Figure 11.13g tube. The database supporting the correlation is insufficient for generality.

11.9.2 Nakayama et al. [1980] Model

Nakayama et al. [1980b] developed an analytically based model to predict boiling performance of the Figure 11.28a surface, which was intended to simulate the Thermoexcel-E™ enhanced tube (Figure 11.12d). The Figure 11.28a test surface consisted of subsurface tunnels in a flat plate covered with pores, through which vapor bubbles escape into the bulk fluid. Their model is based on the "suction–evaporation" mode illustrated in Figure 11.26. Nakayama et al. assume that the total

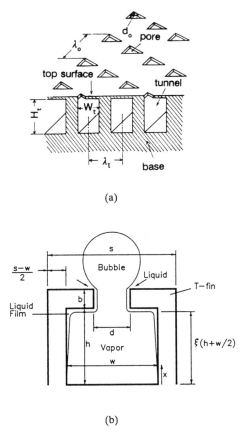

(a)

(b)

Figure 11.28 Schematic of the surface geometries modeled by (a) Nakayama et al. [1980b] (From Nakayama et al. [1980b]) and (b) Xin and Chao [1985] (From Xin and Chao [1985]).

heat flux is the sum of a latent and a single-phase (external convection) heat flux. This is expressed by

$$q = q_{tun} + q_{ex} \qquad (11.15)$$

The term q_{tun} results from evaporation in the tunnel, and q_{ex} is due to external convection to the bubble after it has emerged from the tunnel. The model for q_{tun} is based on Figure 11.27, and it assumes that wetting occurs only in the corners of the tunnel, the "suction–evaporation" mode of Figure 11.26. The model for q_{tun} required six empirical constants.

The external convection term is given by the following empirical expression:

$$q_{ex} = \left(\frac{T_w - T_s}{c_q} \right)^{5/3} n_s^{1/3} \qquad (11.16)$$

where c_q is an empirical constant that depends upon the fluid and the pressure. The c_q term was derived from experimental data for boiling data on a simulated Thermoexcel-E™ surface. Hence, one additional empirical constant is required to determine q_{ex}.

Nakayama et al. [1980a] measured q_{lat} heat flux fraction of the total heat flux for boiling on the Figure 11.28a surface and on a plain surface. This was determined using the measured bubble departure diameter and the nucleation site density. Figure 11.29 shows that q_{lat} contributes a much greater fraction of the total for the Figure 11.28a surface than for a plain surface. Hence, a significant fraction of the total evaporation occurs in the subsurface tunnels. A commercial R-11 refrigerant flooded evaporator operates at approximately 35-kW/m² heat flux.

The major deficiency of the Nakayama et al. [1980a] model is that it requires seven empirical constants. If contact angle is important, it is not directly accounted for in the analysis. Furthermore, the model is limited to the suction-evaporation mode of Figure 11.26.

11.9.3   Xin and Chao [1985] Model

Xin and Chao [1985] proposed a model for boiling on a planar (not tubular) GEWA-T™ (Figure 11.12c) surface. The model, illustrated in Figure 11.28b, assumes steady-state evaporation of a thin liquid film spreading over the inside surface of the tunnel. The vapor formed is ejected through the narrow opening. The T-shaped tunnels serve as stable reentrant cavities, which act as vapor traps. They considered this phenomenon as a counter-current two-phase flow; that is, vapor continuously flows upwards out of the tunnel through the center of the slit's opening, and liquid continuously flows downwards into the tunnel through the slit and along the fin wall. The dynamics of the thin evaporating film is controlled by the counter-current two-phase flow between the tunnel and the outside liquid pool. They assumed that the vapor and liquid flows are steady rather than cyclic. Therefore, the model does not take into account bubble growth and departure, dimension, or frequency in the model. Because the model assumes evaporation from the entire tunnel surface area, nucleation site density is not considered in the model. Heat was assumed to be

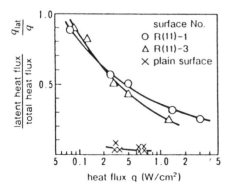

Figure 11.29 Contribution of latent heat transport to the total heat flux on the Figure 11.28a surface. (From Nakayama et al. [1980a].)

transferred as latent heat from inside the tunnel and by convection from the external surface to the liquid pool. Their data were predicted within ± 30% by the model, which requires eight empirical constants. This model appears to be based on a physically unrealistic assumption, since the bubble evolution process is periodic.

Wang et al. [1991] attempted to improve the Xin and Chao model. The Xin and Chao model was formulated for a plane T-finned surface but was assumed to be also applicable to tubular T-finned surface (e.g., Figure 11.12c). The basic assumptions of the Wang et al. [1991] model are the same as those of Xin and Chao [1985]. This includes Xin and Chao's physically unrealistic assumption of the countercurrent two-phase flow of vapor and liquid through the openings at the top of the tunnel. The additional assumptions of Wang et al. model regard the vapor flow path. Wang et al. considered that the heat conducted across the thin liquid film is carried away by two vapor streams: (1) the continuous vapor stream flowing inside the tunnel, and (2) the external vapor stream that has escaped from the tunnel. Both of them join at the top of the tube. They concluded with a multiple regression correlation having six empirical constants. Although their boiling data were predicted within ±20%, the presence of six empirical constants indicates the lack of physical realism in the model.

11.9.4 Ayub and Bergles [1987] Correlation

Ayub and Bergles [1987] obtained R-113 and water boiling data at 1.0 atm on the Figure 11.12c GEWA-TW™ tube. They used Equations 11.15 and 11.16 to predict their data. However, this required knowledge of the nucleation site density to calculate q_{ex}. They experimentally determined the nucleation site density on their surface and expressed it as a polynomial of $T_w - T_{sat}$. Following the suggestion of Nakayama et al. [1980b], they assumed that q_{tun} may be written as

$$q_{lat} = c_1 k(T_w - T_{sat}) \qquad (11.17)$$

They determined the constant c_1 for each fluid at one heat flux and assumed that it applied at all heat fluxes. Their q versus $(T_w - T_{sat})$ data correlated with a maximum deviation of 28% for water and 23% for R-113. Because they did not provide a method to predict n_s or the constant c_1 as a function of fluid properties, the correlation is of very limited value.

11.9.5 Evaluation of Models

Of the models discussed, only the Nakayama et al. [1980b] model appears to have a sound rational basis, which includes bubble dynamics. However, it is limited to the "suction–evaporation" mode illustrated in Figure 11.26, and it requires eight empirical constants.

Webb's [1972] observations of boiling on a flat plate having the Figure 11.12b surface geometry suggest that the tunnels contain alternating zones of liquid and vapor within the tunnel. A liquid meniscus exists at the intersection of each liquid and vapor zone. Visual observations show that the liquid meniscus moves back and forth along the length of the tunnel. Evaporation from this moving meniscus should be an important element of the boiling process. This corresponds to the "flooded mode" shown in Figure 11.26. It appears that a greater physical understanding is required to model the boiling process than is used by the models of Nakayama et al. [1980b] and Xin and Chao [1985].

Although a basic framework exists for a physically based model, an analytical basis is required to predict the effect of fluid properties on the several "constants" and "exponents." A key missing element in the prior work is a methodology to predict the nucleation site density as a function of heat flux. Assuming that the three evaporation modes illustrated in Figure 11.26 exist, separate analytical models would be required for each mode.

Webb and Haider [1992] have developed a more advanced model, which uses some of the features of the Nakayama et al. [1980a] model. This model requires two empirical constants and is described in the next section.

11.10   THE WEBB AND HAIDER MODEL

The work of Webb and Haider is applicable to the "flooded mode" illustrated in Figure 11.26. A dynamic model based on Equation 11.15 is used. Figure 11.30a shows a two-dimensional tunnel of circular cross section. A circular tunnel was chosen for development of the model, because of mathematical simplicity. The tunnels are two-dimensional, having a gap dimension s_g at their mouth. The basic model is also applicable to tunnels with rectangular cross section and with discrete pore openings, as shown in the Figure 11.12 or Figure 11.28a geometries.

Figure 11.30b illustrates the envisioned vaporization mechanism that occurs in the tunnels, which contain vapor filled regions between liquid slugs. A complete bubble formation and departure cycle is illustrated by Figure 11.31. Figures 11.30 and 1.31 show a thin liquid film on the wall between the menisci, which applies to highly wetting liquids that have very small contact angles (e.g., 2 degrees). For liquids such as water, whose contact angle is 25–40 degrees, the surface between the menisci would be dry. In this case, one must know the contact angle (θ) to define the shape of the meniscus. The Webb and Haider [1992] model assumes that the base surface is dry. Haider and Webb have developed a variant of the model that applies to highly wetting liquids, such as refrigerants.

Liquid evaporates from the thin film (if it exists) within the vapor plug, and from the meniscus at each end. The length of the vapor plug increases as evaporation occurs. The evaporation causes the pressure to increase in the tunnel, and causes translation of the meniscus within the tunnel. When the pressure in the tunnel reaches a critical value, the bubble is expelled, and the vapor plug shrinks to its minimum size, as shown by the left hand vapor plug in Figure 11.30b.

The evaporation rate in the tunnel (q_{tun}) and the heat flux to the expelled bubble (q_{ex}) are predicted by analytical models. It is necessary to predict the nucleation site density (n_s) and the bubble frequency (f) to obtain both q_{ex} and q_{tun}. The essence of

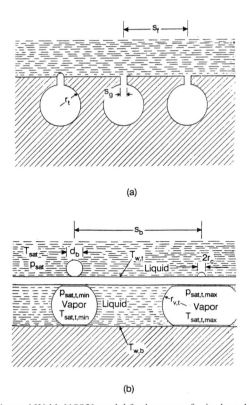

(a)

(b)

Figure 11.30 Haider and Webb [1993] model for heat transfer in the subsurface tunnels. (a) Illustration of subsurface tunnels of circular cross section. (b) Vaporization mechanism in subsurface tunnel.

the model is the methodology to predict the terms q_{tun}, q_{ex}, n_s, and f. Assuming a highly wetting liquid, the tunnel heat flux is written as $q_{tun} = (2Q_m + Q_f)/s_b s_f$, where Q_m is the heat transfer rate across the meniscus, Q_f is the heat transfer rate across the thin film, s_b is the axial pitch (in the tunnel) of the periodic vapor regions, and s_f is the transverse pitch of the tunnels, or "fins." The Q_m is found by calculating heat conduction rate across the meniscus over one bubble cycle. The instantaneous heat transfer rate is proportional to $T_w - T_{l,t}$, where $T_{l,t}$ is the local saturation temperature in the tunnel. The $T_{l,t}$ is a function of pressure in the tunnel. The Q_f is calculated as k_l/δ, where δ is the film thickness, which varies with position of the meniscus. The initial film thickness (δ_0) is calculated using lubrication theory, as described by Probstein [1989]. It is possible that the liquid film may become so thin that evaporation no longer occurs. Criteria for this condition are given by Mirzamoghadam and Catton [1988].

The bubble frequency ($f = 1/\tau$) is obtained by integrating the tunnel evaporation rate over one bubble cycle of period τ.

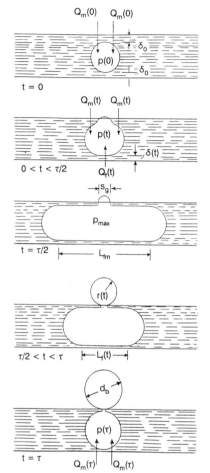

Figure 11.31 Bubble formation and departure cycle of the Haider and Webb model.

$$\int_0^{1/f} (2Q_m + Q_f)\, dt = \frac{\pi}{6}\, \rho_v \lambda d_b^3 \qquad (11.18)$$

A bubble is emitted when the buoyancy force exceeds the surface tension force holding the bubble to the fin gap spacing (s_g in Figure 11.30). The bubble departure diameter is calculated by Equation 11.14, which uses an empirical value for c_b.

The axial pitch of the tunnel nucleation sites (s_b) is obtained by solution of the momentum equation for the periodically oscillating liquids slug in the tunnel. The liquid oscillation is caused by the pressure difference between menisci at each end of the liquid slug. The pressures in alternate vapor plugs zones are assumed to be 180 degrees out of phase. The lowest pressure exists just before bubble departure, and the highest pressure occurs when a bubble of radius $s_g/2$ exists in the fin gap. The maximum pressure differential, which drives the periodically oscillating liquid slugs between the menisci, is given by

$$\Delta p_{max} = 2\sigma \left(\frac{2}{s_g} - \frac{2}{d_b} \right) \tag{11.19}$$

The pressure difference across a liquid slug is written as a periodic function, and the resulting momentum equation is

$$\frac{\partial u}{\partial t} = \frac{\Delta p_{max} \cos \omega t}{\rho x_l} + \frac{v}{r} \frac{\partial}{\partial r} \left(r \frac{\partial u}{\partial r} \right) \tag{11.20}$$

Solution of the momentum equation gives the length of the oscillating liquid slug, x_l. The axial spacing between nucleation sites, which is shown as s_b on Figure 11.30b, is given by

$$s_b = x_l + \frac{L_{f,m}}{2} \tag{11.21}$$

where $L_{f,m}$ is the maximum length of the vapor plug (Figure 11.31) and is given by $L_{f,m} = 2u_0/\pi f$, with u_0 being the initial velocity of the moving meniscus. With s_b and the fin pitch (s_f) known, the nucleation site density is given by

$$n_s = \frac{1}{s_f s_b} \tag{11.22}$$

Two methods are available to predict q_{ex}. The first is the empirical equation (Equation 11.16) developed by Nakayama et al. [1980a]. The second method is strictly analytical and is based on the model of Mikic and Rohsenow [1969] for boiling on plain surfaces. In the latter model, the heat flux is calculated assuming transient conduction to the cold liquid layer (at $T = T_s$) that replaces the superheated liquid layer upon bubble departure. Presently unpublished work by Haider and Webb modifies the Mikic and Rohsenow [1969] model to account for convection in the liquid. Using the known bubble frequency and nucleation site density, one may directly calculate q_{ex}.

The Haider and Webb model was used to predict the data of Nakayama et al. [1980b], who provide data for R-11 boiling at 26.7°C on the Figure 11.28a surface, which has 1818 fins/m (46 fins/in.). The predicted values included the tunnel heat flux (q_{tun}), the external heat flux (q_{ex}), the total heat flux (q), bubble frequency (f), and nucleation site density (n_s). The predictions show reasonable agreement with the Nakayama et al. data. The reader should consult recent publications of Haider and Webb for the more advanced model.

11.11 BOILING MECHANISM ON POROUS SURFACES

11.11.1 O'Neill et al. [1972] Thin Film Concept

O'Neill et al. [1972] theorized that vapor bubbles exist within the pores formed by
the void space between the stacked particles. Thin liquid films exist on the surface
of the particles. Heat is transferred by conduction through the particle matrix, and
then by conduction across the thin liquid film, where evaporation occurs. The pores
within the matrix are interconnected so that: (a) liquid can be supplied to the pores
and (b) vapor can pass through the matrix to the free liquid surface. As vapor is
generated within a pore, the pressure in the vapor bubble increases. When the
pressure is sufficiently high, it overcomes the surface tension retention force, and
the vapor is forced through the interconnected pores to the liquid surface. O'Neill et
al. assumed that each pore contains a vapor bubble and is therefore an active
nucleation site. They outlined an analytical model based on the thin film concept,
which is described in Section 11.12.1 In a later publication, Czikk and O'Neill
[1979] proposed that not all pores are active. They give the following pore classification:

1. *Active Pores.* These are reentrant pores that always contain vapor bubbles.
2. *Intermittent Pores.* These contain vapor as the result of vapor flow generated
   in interconnected active pores; these are envisioned as larger, non-reentrant
   pores.
3. *Liquid-Filled Pores.* These are small non-reentrant pores, whose opening
   radius is too small to permit bubbles at the prevailing $T_w - T_s$. They are
   permanently liquid-filled and supply superheated liquid to the active pores.
4. *Nonfunctional Pores.* These are "closed" pores, which do not contain the
   boiling fluid.

11.11.2 Kovalev et al. [1990] Concept

The static model of O'Neill et al. [1972] is not realistic, because it does not account
for the dynamics of fluid flow within the matrix and does not include bubble
dynamics (bubble departure, frequency and site density). Kovalev et al. [1990]
proposed a more rationally based model, which accounts for: (a) evaporation from
menisci within the matrix and (b) the dynamics of fluid flow within the porous
matrix. Figure 11.32 illustrates the concepts on which their model is based. Figure
11.32a shows that evaporation takes place from stationary menisci of radius r_m
within the matrix of thickness δ. The meniscus radius changes over the matrix
thickness. The pressure difference between the vapor and the liquid is $(p_v - p_l) =
2\sigma/r_m$ at a subsurface meniscus of radius r_m. The temperature of the porous matrix
is greatest at the base of the matrix ($x = 0$), and it decreases in the direction of the
surface. The local liquid temperature within the matrix (T_l) is equal to the local
matrix temperature. Thermodynamic equilibrium requires that T_l must be superheated an amount $(T_l - T_s) = 2\sigma/mr_m$ for a meniscus of radius r_m to exist. Hence, at

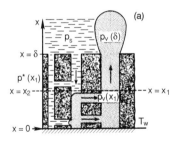

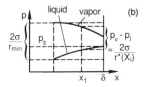

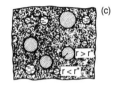

Figure 11.32   (a) Concept of Kovalev et al. [1990] model for porous boiling surface. (b) Distribution of pressure in vapor and liquid channels over matrix depth. (c) Top view of matrix showing vapor- and liquid-filled channels.

distance x from the base of the matrix, all pores having a radius $r_m \leq 2\sigma/m(T_l - T_s)$ are assumed to be liquid-filled, and pores having $r > r_m$ are assumed to be vapor-filled. Because the matrix temperature decreases with distance from its base (x), the value of r_m increases in the direction from the base. Hence, $r_m = r_m(x)$.

Evaporated vapor formed at the menisci within the matrix flows to the surface via the larger vapor-filled, interconnected pores. Bubbles are emitted at the surface from the vapor filled capillaries. Vapor flow within the matrix is caused by "capillary pressure." Because $(p_v - p_l) = 2\sigma/r_m$ and r_m increases in the direction of the surface, a pressure gradient exists within the matrix. This pressure gradient forces the vapor through the capillaries to the surface. Figure 11.32b shows a pressure gradient in the vapor, due to friction losses. The liquid exists at saturation pressure p_s at the matrix surface. At this point, the vapor pressure is greater than p_s by the amount $2\sigma/r_m(\delta)$. The liquid flow in the small liquid-filled capillaries causes a pressure gradient within the liquid. Some detail of the mathematical model is given in Section 11.12.2.

11.12   PREDICTIVE METHODS FOR POROUS SURFACES

Several models or correlations have been proposed for prediction of porous boiling surfaces. The first model was developed by O'Neill et al. [1972] and is described in detail by Webb [1983]. The O'Neill et al. model assumes thin film evaporation on the particles. A different model was proposed by Kovalev et al. [1990], who assume that evaporation occurs at menisci within the porous structure.

11.12.1   O'Neill et al. [1972] Model

O'Neill et al. [1972] developed an analytical model to predict the heat flux for this envisioned boiling mechanism. The model, described in detail by Webb [1983], assumes that the matrix consists of precisely stacked spherical particles and that each pore contains a vapor bubble. Although the O'Neill et al. [1972] model is a static model, it contains several concepts of fundamental importance and may serve as the starting point for the development of a more sophisticated dynamic model. The model, described in detail by Webb [1983], assumes that $T_w - T_s$ may be written as the sum of two terms:

1. The temperature drop across the thin liquid film between the surface of the particle and the liquid $(T_w - T_s)$ at the surface of the bubble.
2. The liquid superheat $(T_l - T_s)$ required for the existence of a vapor bubble within the pore of radius r_b.

$$T_w - T_s = (T_w - T_l) + (T_l - T_s) \qquad (11.23)$$

The second term of Equation 11.23 is given by Equation 11.4, and the first term is calculated using the Fourier equation for heat conduction across the liquid film of thickness δ. The heat flux q is based on the area of the base surface rather than the total liquid film surface of the matrix (S). This gives

$$T_w - T_l = \frac{qA\delta}{k_l S} \qquad (11.24)$$

The key assumptions of the model are as follows:

1. The uniform diameter spherical particles are interconnected with liquid or vapor to flow between them.
2. The pore diameter is defined as the diameter of the largest sphere that may be contained within the void space. The geometric packing arrangement of the spherical particles is known. Hence, the pore diameter is calculable, as illustrated in Figure 11.33 for two possible packing arrangements.
3. Each pore is active and functions as a stable vapor trap.
4. The matrix is a perfect heat conductor; thus $k_m = \infty$, and there is no temperature drop within the coating.

Based on the particle packing geometry and the coating thickness, one may show that

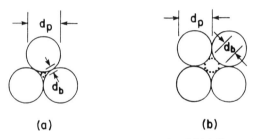

Figure 11.33   Definition of d_b and d_p for (a) staggered and (b) inline packings as used in the O'Neill et al. [1972] analytical model.

$$T_w - T_s = \frac{\beta q r_b^2}{k_l} + \frac{2\sigma}{mr_b} \qquad (11.25)$$

where β is a geometry factor given by

$$\beta = \left[1 - (1 - \epsilon) \left(1 + \frac{V_b}{V_T} \right) \right] \left(\frac{V_T}{Sr_b} \right)^2 \qquad (11.26)$$

The geometry factor is calculable for given values of the particle diameter d_p and the particle packing arrangement.

Comparisons of porosity measurements of the High-Flux™ coating show $0.50 < \epsilon < 0.65$, which approximately agrees with the calculated porosity for an inline packing arrangement ($\epsilon = 0.48$). Predictions by O'Neill et al. [1972] have used $\beta\delta = 1.04$, which is for the inline packing arrangement.

The matrix packings tested by Nishikawa and Ito [1982] and Nishikawa et al. [1983] had $0.38 < \epsilon < 0.71$, which differs substantially from the $\epsilon = 0.48$ associated with the inline packing. It is not possible to perform measurements on actual matrix packings to determine their geometric parameter, $\beta\delta$. Hence, the geometric parameters required for this model are indeterminate for actual packings.

11.12.2   Kovalov et al. [1990] Model

The evaporation concept proposed by Kovalev et al. [1990] was described in Section 11.11.2 and illustrated in Figure 11.32. The pores sizes are distributed between $r_{min} \leq r \leq r_{max}$. At distance x from the base surface, a meniscus of radius r_m will exist in a pore of radius r. At this x-location, pores of $r < r_m$ are liquid-filled, and pores of $r > r_m$ are vapor-filled. The pressure difference across the meniscus interface at location x is given by

$$p_v(x) - p_l(x) = \frac{2\sigma}{r_m(x)} \qquad (11.27)$$

The vapor evaporated at a meniscus passes through the interconnected vapor-filled pores and is emitted at the surface of the porous coating. Equation 11.27 is satisfied at each distance x from the base of the coating.

The momentum equation is written for parallel cylindrical capillary channels illustrated in Figure 11.32c. The small channels are liquid-filled, and the large channels are vapor-filled. Prior to describing the model, one must understand how the porosity and permeability for the liquid- and vapor-filled channels are defined. The term $f(r)$ is the pore size distribution function of the porous structure, as shown in Figure 11.34 from Kovalev and Lenkov [1981] for a particular porous coating. Figure 11.34 shows that the dominant pore size, maximum $f(r)$, is approximately 55 μm. The term $\phi(r)$ is defined as the integral pore distribution and is shown in Figure 11.34. It is defined by

$$\phi(r) = \int_{r_{min}}^{r} f(r) \, dr \tag{11.28}$$

The total porosity of the matrix (ϵ) is equal to $\phi(r_{max})$. If all pores between r_{min} and r_m are liquid-filled, the porosity of these liquid-filled pores is $\epsilon_l = \phi(r_m)$. Similarly, if all pores having $r > r_m$ are vapor-filled, their porosity is $\epsilon_v = \phi(r_{max}) - \phi(r_m)$.

The frictional pressure drop in the liquid-filled pores is calculated using the Darcy equation,

$$\frac{dp}{dx} = \frac{G v_l}{K_l} \tag{11.29}$$

where the permeability of the liquid-filled pores is K_l. The permeability of the porous matrix for single-phase flow is defined as $K = CF(r_{max})$, where $F(r_{max})$ is defined by

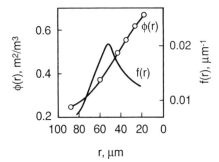

Figure 11.34   Pore size distribution, $f(R)$ and integral distribution function, $\phi(R)$ for a 1.1-mm-thick sintered coating made of 0.1-mm spherical particles.

$$F(r_{\max}) = \frac{\displaystyle\int_{r_{\min}}^{r_{\max}} r^2 f(r)\, dr}{\displaystyle\int_{r_{\min}}^{r_{\max}} f(r)\, dr} \tag{11.30}$$

and C is an empirical constant obtained by test data for single-phase flow in the matrix. It is necessary to subdivide the permeability (K) into the permeability for the liquid-filled channels (K_l) and the vapor-filled channels (K_v). The permeability of the matrix having liquid-filled pores of $r_{\min} < r < r_m$ is $K_l = CF(r_m)$, where $F(r_m)$ is given by Equation 11.30 with the upper limit of the integral in the numerator set as r_m, rather than $r_{\max}$. The permeability of the matrix for vapor flow is $K_v = K - K_l = K - CF(r_m)$.

Kovalev et al. [1987, 1990] analytically model the flow and heat transfer within the porous structure by writing momentum and energy equations for the liquid and vapor flow within the liquid- and vapor-filled capillaries. The two-phase momentum equation is given by

$$\frac{dr_m}{dx}\left[\frac{2\sigma}{r_m^2} - G^2 f(r_m)\left(\frac{1}{\rho_l \epsilon_l^3} + \frac{1}{\rho_v \epsilon_v^3}\right)\right] =$$
$$G^2 \frac{dG}{dx}\left(\frac{1}{\rho_v \epsilon_v^2} - \frac{1}{\rho_l \epsilon_l^2}\right) + \frac{v_l G}{K_l} + \frac{v_v G}{K_v} + \frac{2 f_v G^2}{\rho_v \epsilon_v^2} \tag{11.31}$$

The pore distribution functions $f(r_m)$, $\phi(r_m)$ and $F(r_m)$ have been previously defined. Solution of the equations requires (1) experimental data for the porous matrix that define the pore distribution functions, (2) the pressure drop characteristic versus mass velocity for simultaneous liquid and vapor flow in the porous structure containing liquid-filled small pores, and (3) the permeability of the matrix. Kovalev and Ovodkov [1992] describe an experimental technique to determine the item (2) information. Kovalev et al. [1990] show the measured pressure drop characteristics for a porous matrix.

The energy equation for the porous matrix is

$$k_l \frac{d^2 T_m}{dx^2} = h_v (T_m - T_s) \tag{11.32}$$

where subscript m refers to the matrix properties, and h_v is the volumetric heat transfer coefficient for evaporation from the stationery menisci. The analytical solution for h_v is given by Kovalev et al. [1990]. The momentum and energy equations must be numerically solved over the matrix thickness. The model is simplified if one assumes that the matrix temperature is constant over its thickness. Thus, all menisci are of a constant radius.

Kovalev et al. [1990] shows that the model reasonably predicts the boiling coefficient for water at 1.0 atm boiling on 1.0-mm-thick matrix made of 0.3- to 0.4-mm-diameter particles. It is important to note that their model also predicts the critical heat flux. They provide predictions which describe the performance of porous structures having different matrix thermal conductivity, permeability, and thickness.

The model appears to be an important contribution to understanding of the boiling mechanism in porous structures. The Kovalev et al. [1990] paper does not provide sufficient detail and definitions necessary to easily understand their analysis method. Polezhaev and Kovalev [1990] provide some additional description of the model.

11.12.3 Nishikawa et al. [1983] Correlation

Nishikawa et al. [1983] developed an empirical correlation of their data on their matrix geometries, which include copper and bronze particles having $0.10 < d_p < 1.0$ mm and $0.4 < \delta < 4.0$ mm. Boiling data were taken at 1 atm for R-11, R-113, and benzene. A linear, multiple regression technique was employed, and Figure 11.35 shows the correlated data. Most of the data are correlated within $+30\%$.

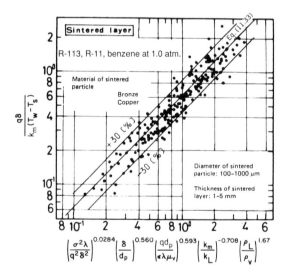

R-113 and R-11 Tests Conducted Using Spherical Particles

Test	Mat'l.	d_p (μm)	δ (μm) (mm)	(mm)	δ/d_p
1	Cu	250		0.4,1,2,4	1.6,4,8,16
2	Cu	100, 250[a]			4[a],8[a],20
		500		2	
3	Bronze	500		1,2,3,4	2,4,6,8
4	Bronze	250[a],500[a]			2,2.7,4,8[a]
		750,1000		2	

[a] Also with R-11.

Figure 11.35 Nishikawa et al. [1983] correlation of their R-11 and R-113 data for coatings made of uniform spherical particles. (From Nishikawa et al. [1983].)

The algebraic equation of the correlation is

$$\frac{q\delta}{k_m \Delta T} = 0.001 \left(\frac{\sigma^2 \lambda}{q^2 \delta^2} \right)^{0.0284} \left(\frac{\delta}{d_p} \right)^{0.56} \left(\frac{q d_p}{\epsilon \lambda \mu_v} \right)^{0.593} \left(\frac{k_m}{k_l} \right)^{-0.708} \left(\frac{\rho_l}{\rho_v} \right)^{1.67}$$

(11.33)

The left-hand side of Equation 11.33 is a Nusselt number based on the coating thickness, and the third term on the right is a Reynolds number based on the total vapor flow in the matrix, using the particle diameter as the characteristic dimension.

Webb [1983] used the O'Neill et al. [1972] and Nishikawa et al. [1983] correlations to predict the boiling coefficient for several porous structures.

11.12.4   Zhang and Zhang [1992] Correlation

This power-law correlation is based on dimensionless parameters assumed to control the boiling process. Zhang and Zhang [1992] tested 12 bronze-sintered coatings on a flat surface, which span $0.11 \le d_p \le 0.53$ mm and $0.94 \le \delta \le 4.6$ mm using water, ethyl alcohol, and R-113 at approximately 100 kPa. The data were correlated within $\pm 25\%$.

11.13   CRITICAL HEAT FLUX

As noted in Section 11.5, enhanced surfaces may have a higher critical heat flux than plain surfaces. The present understanding of critical heat flux in nucleate boiling on a plain, horizontal surface is based on the theory of hydrodynamic instability of the closely spaced vapor jets, as described by Zuber [1958]. These vapor jets carry the latent heat away from the surface. When they break down, the surface can no longer support nucleate boiling. The CHF is given by

$$q_{CHF} = \rho_v \lambda u_v \left(\frac{A_j}{A} \right)$$

(11.34)

where u_v is the velocity of the vapor jet, and A_j is the jet cross-sectional flow area. The jet velocity is given by

$$u_v = \left(\frac{2\pi\sigma}{\rho_v \lambda_H} \right)$$

(11.35)

where λ_H is the pitch of one-dimensional Taylor instability waves. λ_H is given by $A_j/A = \pi/16$. The model shows that the CHF is determined by λ_H, which is determined by the balance of surface tension against inertia and buoyancy forces on a plain surface. Lienhard [1987] shows that the CHF on other smooth surface shapes (e.g., a horizontal tube) is influenced by the shape of the surface.

An explanation of higher CHF on an enhanced surface of the same basic shape (e.g., a horizontal plate) would require that the spacing of the vapor jets is different from that on a smooth surface. Polezhaev and Kovalev [1990] have proposed such an explanation for a porous boiling surface. They propose that the vapor jet velocity will still be controlled by Equation 11.35. However, the spacing of the bubble columns is fixed by the spacing of the large pores, from which vapor is emitted. Therefore, the Taylor wavelength used in Equation 11.35 does not control the spacing of these bubble columns. They observed that the spacing of the vapor columns on a porous boiling surface was one-tenth that occurring on a plain surface:

$$q_{\text{CHF}} = 0.52\epsilon^{2.28}\lambda \left(\frac{\sigma\rho_v}{r_{\text{br}}} \right)^{1/2} \tag{11.36}$$

where r_{br} is defined as the "breakdown" pressure differential and is approximately 30% larger than the smallest vapor-filled pores. Their correlation shows that q_{CHF} is governed by the porosity of the surface.

Corresponding work has not been done for the structured boiling surfaces, as illustrated in Figure 11.13. However, it appears that the spacing of the vapor jets would be controlled by the subsurface structure, rather than by the Taylor wavelength.

11.14 CONCLUSIONS

This survey has described the evolution of special nucleate boiling surfaces employed on the outer surface of a tube. Two basic types of surface structures exist, namely, porous coatings and reentrant grooves. Rapid advances have been made since 1968, and several high-performance tubes are commercially employed. These have found extensive use in the refrigeration industry and limited use in process applications.

The key to the high performance of the porous and reentrant grooved structures is attributed to three factors: (1) a pore or reentrant cavity within a critical size range, (2) interconnected cavities, and (3) nucleation sites of a reentrant shape. When the cavities are interconnected, one active cavity can activate adjacent cavities. It appears that the dominant fraction of the vaporization occurs at very thin liquid films within the subsurface structure. The reentrant cavity shape provides a stable vapor trap, which will remain active at very low liquid superheat values.

The porous coated surfaces offer the opportunity for a duplex tube material construction. Thus, the boiling surface may be made of less expensive material than required to meet the corrosion characteristics of the tube-side fluid.

Significant progress has been made in analytical models to explain the mechanism of boiling in enhanced surfaces. However, more work is required.

Enhanced tubes may require a substantial liquid superheat for cavity activation. Start-up problems are not observed in refrigeration systems, since the compressor startup provides the necessary liquid superheat.

Future research and development activity is expected to focus on the following:

1. The development of analytical models and design correlations applicable to a wide range of fluid properties.
2. Application studies to define where such enhanced boiling surfaces may be effectively used.
3. Improved methods of manufacture which yield lower product cost. The most cost-effective surface geometry is yet to be established. Most of the surfaces produced to date are of copper. Other materials are needed if wide-scale application is to be achieved.
4. Finally, doubly enhanced tubes (inside and outside enhancement) require extensive development effort. The preferred boiling-side enhancement may be dependent on the method of manufacture.

11.15 REFERENCES

Albertson, C. E., 1977. "Boiling Heat Transfer Surface and Method," U.S. Patent 4,018,264.

Antonelli, R., and O'Neill, P. S., 1981. "Design and Application Considerations Heat Exchangers with Enhanced Boiling Surfaces," in *Heat Exchanger Sourcebook*, J. Palen, Ed., Hemisphere Publishing Corp., Washington, D.C., pp. 645–661.

Arai, N., Fukushima, T., Arai, A., Nakajima, T., Fujie, K. and Nakayama, Y., 1977. "Heat Transfer Tubes Enhancing Boiling and Condensation in Heat Exchangers of a Refrigerating Machine," *ASHRAE Transactions*, Vol. 83, No. 2, pp. 58–70.

Arshad, J., and Thome, J. R., 1983. "Enhanced Boiling Surfaces: Heat Transfer Mechanism Mixture Boiling," *Proceedings of the ASME-JSME Thermal Engineering Joint Conference*, Vol. 1, pp. 191–197.

Asakavicius, J. P., Zukauskav, A. A., Gaigalis, V. A., and Eva, V. K., 1979. "Heat Transfer from Freon-113, Ethyl Alcohol and Water with Screen Wicks," *Heat Transfer—Soviet Research*, Vol. 11, No. 1, pp. 92–100.

Ayub, Z. H., and Bergles, A. E., 1987. "Pool Boiling from GEWA Surfaces in Water and R-113," *Wärme-und Stoffübertragung*, Vol. 21, pp. 209–219.

Ayub, A. H. and Bergles, A. E., 1988a. "Pool Boiling Enhancement of a Modified GEWA-T Surface in Water," *Transactions of the ASME, Journal of Heat Transfer*, Vol. 10, pp. 266–267.

Ayub, Z. H., and Bergles, A. E., 1988b. "Nucleate Pool Boiling Curve Hysteresis for GEWA-T Surfaces in Saturated R-113," *ASME Proceedings of the National Heat Transfer Conference*, Vol. 2, ASME Symposium, Vol. HTD-Vol. 96, ASME, New York, pp. 515–521.

Bankoff, S. G., 1959. "Entrapment of Gas in the Spreading of a Liquid Over a Rough Surface," *AIChE Journal*, Vol. 4, No. 1, pp. 24–26.

Bar-Cohen, A., 1992. "Hysteresis Phenomena at the Onset of Nucleate Boiling," in *Pool and External Flow Boiling*, V. J. Dhir and A. E. Bergles, Eds., ASME, New York, pp. 1–14.

Benjamin, J. E., and Westwater, J. W., 1961. "Bubble Growth in Nucleate Boiling of a Binary Mixture," *International Developments in Heat Transfer*, ASME, New York, pp. 212–218.

Berenson, P. J., 1962. "Experiments on Pool Boiling Heat Transfer," *International Journal of Heat and Mass Transfer*, Vol. 5, pp. 985–999.

Bergles, A. E., and Chyu, M. C., 1982. "Characteristics of Nucleate Pool Boiling from Porous Metallic Coatings," *Journal of Heat Transfer*, Vol. 104, pp. 279–285.

Bergles, A. E., Bakhru, N., and Shires, J. W., Jr., 1968. "Cooling of High-Power Density Computer Components," Massachusetts Institute of Technology, EPL Report 70712–60.

Bliss, F. E., Hsu, S. T., and Crawford, M., 1969. "An Investigation into the Effects of Various Platings on the Film Coefficient During Nucleate Boiling from Horizontal Tubes," *International Journal of Heat Mass Transfer*, Vol. 1, pp. 1061–1072.

Bonilla, C. F., Grady, J. J., and Avery, G. A., 1965. "Pool Boiling Heat Transfer from Scored Surfaces," *Chemical Engineering Progress Symposium Series*, Vol. 61, No. 57, pp. 281–288.

Brothers, W. S., and Kallfelz, A. J., 1979. "Heat Transfer Surface and Method of Manufacture," U.S. Patent 4,159,739.

Carey, V. P., 1992. *Liquid–Vapor Phase-Change Phenomena*, Hemisphere Publishing Corp., Washington, D.C.

Chaudri, I. H., and McDougall, I. R, 1969. "Aging Studies in Nucleate Pool Boiling of lsopropyl Acetate and Perchlorethylene," *International Journal of Heat and Mass Transfer*, Vol. 12, pp. 681–688.

Chu, R. C., and Moran, K. P., 1977. "Method for Customizing Nucleate Boiling Heat Transfer from Electronic Units Immersed in Dielectric Coolant," U.S. Patent 4,050,507.

Clark, H. B., Strenge, P. H., and Westwater, J. W., 1959. "Active Sites for Nucleate Boiling," *Chemical Engineering Progress Symposium Series*, Vol. 55, No. 29, pp. 103–110.

Corman, J. C., and McLaughlin, M. H., 1976. "Boiling Augmentation with Structured Surfaces," *ASHRAE Transactions*, Vol. 82, No. 1, pp. 906–918.

Corty, C., and Foust, A. S., 1955. "Surface Variables in Nucleate Boiling," *Chemical Engineering Progress Symposium Series*, Vol. L20 51, No. 17, pp. 1–12.

Czikk, A. M., and O'Neill, P. S., 1979. "Correlation of Nucleate Boiling from Porous Metal Films," in *Advances in Enhanced Heat Transfer*, J. M. Chenoweth, J. Kaellis, J. W. Michel, and S. Shenkman, Eds., ASME, New York, pp. 103–113.

Dahl, M. M., and Erb, L. D., 1976. "Liquid Heat Exchanger Interface and Method," U.S. Patent 3,990,862.

Dundin, V. A., Danilova, G. N., and Tikhonov, A. V., 1990. "Enhanced Heat Transfer Surfaces for Shell-and-Tube Evaporators of Refrigerating Machines," *Refrigerating Machines*, Series XM-7, pp. 1–46 (in Russian).

Fritz, W., 1935. "Berechnung des Maximalvolume von Dampfblasen," *Physikalische Zeitschrift*, Vol. 36, pp. 379–388.

Fujie, K., Nakayama, W., Kuwahara, H. and Kakizakci, K., 1977. "Heat Transfer Wall for Boiling Liquids," U.S. Patent 4,060,125.

Fujii, M., Nishiyama, E., and Yamanaka, G., 1979. "Nucleate Pool Boiling Heat Transfer from Microporous Heating Surface," in *Advances in Enhanced Heat Transfer*, J. M. Chenoweth, J. Kaellis, J. W. Michel, and S. Shenkman, Eds., ASME, New York, pp. 45–51.

Fujikake, J., 1980. "Heat Transfer Tube for Use in Boiling Type Heat Exchangers and Method of Producing the Same," U.S. Patent 4,216,826.

Gaertner, R. F., 1967. "Methods and Means for Increasing the Heat Transfer Coefficient between a Wall and Boiling Liquid," U.S. Patent 3,301,314.

Gottzmann, C. F., Wulf, J. B., and O'Neill, P. S., 1971. "Theory and Application of High Performance Boiling Surfaces to Components of Absorption Cycle Air Conditioners," *Proceedings of the Conference on Natural Gas Resource Technology*, Session V, Paper 3, Chicago, IL.

Gottzmann, C. F., O'Neill, P. S., and Minton, P. E., 1973. "High Efficiency Heat Exchangers," *Chemical Engineering Progress*, Vol. 69, No. 7, pp. 69–75.

Granryd, E., and Palm, B., 1988. "Heat Transfer Element," U.S. Patent 4,787,441.

Grant, A. C., 1977. "Porous Metallic Layer and Formation," U.S. Patent 4,064,914.

Griffith, P. and Wallis, J. D., 1960. "The Role of Surface Conditions in Nucleate Boiling," *Chemical Engineering Progress Symposium Series*, Vol. 56, No. 49, pp. 49–63.

Han, C. Y., and Griffith, P., 1965. "The Mechanism of Heat Transfer in Nucleate Pool Boiling—Parts I and II," *International Journal of Heat and Mass Transfer*, Vol. 8, pp. 887–917.

Hasegawa, S., Echigo, R., and Irie, S., 1975. "Boiling Characteristics and Burnout Phenomena on a Heating Surface Covered with Woven Screens," *Journal of Nuclear Science Technology*, Vol. 12, No. II, pp. 722–724.

Hsu, Y. Y., 1962. "On the Size Range of Active Nucleation Cavities on a Heating Surface," *Journal of Heat Transfer*, Vol. 84, pp. 207–216.

Hummel, R. L., 1965. "Means for Increasing the Heat Transfer Coefficient Between a Wall and Boiling Liquid," U.S. Patent 3,207,209.

Jakob, M., 1949. *Heat Transfer*, John Wiley & Sons, New York, pp. 636–638.

Janowski, K. R., Shum, M. S., and Bradley, S. A., 1978. "Heat Transfer Surface," U.S. Patent 4,129,181.

Kajikawa, T., Takazawa, H., and Mizuki, M., 1983. "Heat Transfer Performance of the Metal Fiber Sintered Surfaces," *Heat Transfer Engineering*, Vol. 4, No. 1, pp. 57–66.

Kartsounes, G. T., 1975. "A Study of Surface Treatment on Pool Boiling Heat Transfer in Refrigerant 12," *ASHRAE Transactions*, Vol. 81, pp. 320–326.

Kim, C. J., and Bergles, A. E., 1988. "Incipient Boiling Behavior of Porous Boiling Surfaces Used for Cooling Microelectronic Chips," in *Particulate Phenomena and Multiphase Transport*, Vol. 2, T. N. Veziroğlu, Ed., Hemisphere Publishing Corp., New York, pp. 3–18.

Ko, S-Y. Liu, L., and Yao, Y-Q., 1992. "Boiling Hysteresis on Porous Metallic Coatings," in *Multiphase Flow and Heat Transfer, Second International Symposium*, Vol. 1, X-J. Chen, T. N. Veziroğlu, and C. L. Tien, Eds., Hemisphere Publishing Corp., New York, pp. 259–268.

Kovalev, S. A., and Lenkov, V. A., 1981. "Mechanism of burnout with Boiling on a Porous Surface," *Thermal Engineering*, Vol. 28, No. 4, pp. 201–203.

Kovalev, S. A., and Ovodkov, O. A., 1992. "A Study of Gas–Liquid Counterflow in Porous Media," *Experimental Thermal and Fluid Science*, Vol. 5, pp. 457–464.

Kovalev, S. A., Solov'yev, S. L., and Ovodkov, O. A., 1987. "Liquid Boiling on Porous Surfaces," *Heat Transfer—Soviet Research*, Vol. 19, No. 3, pp. 109–120.

Kovalev, S. A., Solov'yev, S. L., and Ovodkov, O. A., 1990. "Theory of Boiling Heat Transfer on a Capillary Porous Surface," *Proceedings of the 9th International Heat Transfer Conference*, Vol. 2, pp. 105–110.

Kun, L. C., and Czikk, A. M., 1969. "Surface for Boiling Liquids," U.S. Patent 3,454,081. (Reissued August 21, 1979 Re. 30,077.)

Kurihari, H. M. and Myers, J. E., 1960. "Effects of Superheat and Roughness on the Boiling Coefficients," *AIChE Journal*, Vol. 6, No. 1, pp. 83–91.

Li, Z., Tan, Y., and Wang. S., 1992. Investigation of the Heat Transfer Performance of Mechanically Made Porous Surface Tubes with Ribbed Tunnels, *Multiphase Flow and Heat Transfer, Second International Symposium*, Vol. 1, X-J. Chen, T. N. Veziroğlu, and C. L. Tien, Eds., Hemisphere Publishing Corp., New York, pp. 700–707.

Lienhard, J. H., 1987. *A Heat Transfer Textbook*, 2nd edition., Prentice–Hall, Englewood Cliffs, NJ.

Liu, X., Ma, T., and Wu, J., 1987. "Effects of Porous Layer Thickness of Sintered Screen Surfaces on Pool Nucleate Boiling Heat Transfer and Hysteresis Phenomena," *Heat Transfer Science and Technology*, B-X. Wang, Ed., Hemisphere Publishing Corp., New York, pp. 577–583.

Lorenz, J. J., Mikic, B. B., and Rohsenow, W. M., 1974. "The Effect of Surface Conditions on Nucleate Boiling Characteristics," *Proceedings of the 5th International Heat Transfer Conference*, Vol. 4, 35–49.

Ma, T., Liu, X., Wu, J., and Li, H., 1986. "Effects of Geometrical Shapes and Parameters of Reentrant Grooves on Nucleate Pool Boiling Heat Transfer from Porous Surfaces," *Heat Transfer 1986, Proceedings of the 8th International Heat Transfer Conference*, Vol. 4, pp. 2013–2018.

Malyshenko, S. P., and M. A. Styrikovich, 1992. Heat Transfer at Pool Boiling on Surfaces with Porous Coating, *Multiphase Flow and Heat Transfer, Second International Symposium*, Vol. 1, X-J. Chen, T. N. Veziroğlu, and C. L. Teen, Eds., Hemisphere Publishing Corp., New York, pp. 269–284.

Marto, P. J., and Rohsenow, W. M., 1966. "Effects of Surface Conditions on Nucleate Pool Boiling of Sodium," *Journal of Heat Transfer*, Vol. 88, pp. 196–204.

Marto, P. J., Moulson, J. A., and Maynard, M. D., 1968. "Nucleate Pool Boiling of Nitrogen with Different Surface Conditions," *Journal of Heat Transfer*, Vol. 90, pp. 437–444.

Matijević, M., Djurić, Zavargo, Z., and Novaković, 1992. "Improving Heat Transfer with Pool Boiling by Covering of Heating Surface with Metallic Spheres," *Heat Transfer Engineering*, Vol. 13, No. 3, pp 49–57.

Mikic, B. B., and Rohsenow, W. M., 1969. "A New Correlation of Pool-Boiling Data Including the Effect of Heating Surface Characteristics," *Journal of Heat Transfer*, Vol. 91, pp. 245–250.

Milton, R. M., 1968. "Heat Exchange System," U.S. Patent 3,384,154.

Milton, R. M., 1970. "Heat Exchange System," U.S. Patent 3,523,577.

Milton, R. M., 1971. "Heat Exchange System with Porous Boiling Layer," U.S. Patent 3,587,730.

Milton, R. M. and Gottzmann, C. F., 1972. "High Efficiency Reboilers and Condensers," *Chemical Engineering Progress.*, Vol. 68, No. 9, pp. 56–61.

Mirzamoghadam, A., and Catton, I., 1988. "A Physical Model of the Evaporating Meniscus," *Journal of Heat Transfer*, Vol. 110, pp. 201–207.

Modahl, R. J., and Lukeroth, V. C., 1982. " Heat Transfer Surface for Efficient Boiling of Liquid R-11 and Its Equivalents," U.S. Patent 4,354,550.

Moore, F. D., and Mesler, R. B., 1971. "The Measurement of Rapid Surface Temperature Fluctuations during Nucleate Boiling of Water," *AIChE Journal*, Vol. 7, pp. 620–624.

Nakayama, W., Daikoku, T., Kuwahara, H., and Nakajima, T., 1980a. "Dynamic Model of Enhanced Boiling Heat Transfer on Porous Surfaces Part I: Experimental Investigation," *Journal of Heat Transfer*, Vol. 102, pp. 445–450.

Nakayama, W., Daikoku, T., Kuwahara, H., and Nakajima, T., 1980b. "Dynamic Model of Enhanced Boiling Heat Transfer on Porous Surfaces Part II: Analytical Modeling," *Journal of Heat Transfer*, Vol. 102, pp. 451–456.

Nakayama, W., Daikoku T., and Nakajima, T., 1982. "Effects of Pore Diameters and System Pressure on Saturated Pool Nucleate Boiling Heat Transfer From Porous Surfaces," *Journal of Heat Transfer*, Vol. 104, pp. 286–291.

Nishikawa, K., and Ito, T., 1982. "Augmentation of Nucleate Boiling Heat Transfer by Prepared Surfaces," in *Heat Transfer in Energy Problems*, T. Mizushina and W. J. Yang, Eds., Hemisphere Publishing Corp., New York, pp. 111–118.

Nishikawa, K., Ito, T., and Tanaka, K., 1983. "Augmented Heat Transfer by Nucleate Boiling at Prepared Surfaces," *Proceedings of the 1983 ASME-JSME Thermal Engineering Conference*, Vol. 1, pp. 387–393.

O'Neill, P. S., Gottzman, C. F., and Terbot, J. W., 1972. "Novel Heat Exchanger Increases Cascade Cycle Efficiency for Natural Gas Liquefaction," in *Advances in Cryogenic Engineering*, K. D. Timmerhaus, Ed., Plenum, New York, pp. 420–437.

Pais, C., and Webb, R. L., 1991. "Literature Survey of Pool Boiling on Enhanced Surfaces," to be published in *ASHRAE Transactions*, Vol. 97, Pt. 1, pp. 79–89.

Palm, B., 1992. "Heat Transfer Enhancement in Boiling by Aid of Perforated Metal Foils," in *Recent Advances in Heat Transfer*, B. Sunden and A. Zukauskas, Eds., Elsevier Science Publishers, New York.

Polezhaev, Y. U., and Kovalev, S. A., 1990. "Modelling Heat Transfer with Boiling on Porous Structures," *Teploenergetika*, Vol. 37, No. 12, pp. 5–9 (in Russian). Also in *Thermal Engineering*, Vol. 37, No. 12, pp. 617–620.

Probstein, R. F., 1989. *Physicochemical Hydrodyynamics—An Introduction*, Butterworths, Boston, pp. 280–289.

Ragi, E. G., 1972. "Composite Structure for Boiling Liquids and Its Formation," U.S. Patent 3,684,007.

Rohsenow, W. M., 1985. "Boiling," Chapter 12 in *Handbook of Heat Transfer Fundamentals*, McGraw–Hill, New York, pp. 12–15.

Sachar, S. S., and Silvestri, V. J., 1983. "Porous Film Heat Transfer," U.S. Patent 4,381,818.

Saier, M., Kastner, H. W., and Klockler, R., 1979. "Y- and T-Finned Tubes and Methods and Apparatus for Their Making," U.S. Patent 4,179,911.

Sanborn, D. F., Holman, J. L. M., and Ware, C. D., 1982. "Heat Exchange Surface with Porous Coating and Subsurface Cavities," U.S. Patent 4,359,086.

Sauer, E. T., 1935. M.S. Thesis, Department of Mechanical Engineering, Massachusetts Institute of Technology, Cambridge, MA.

Schmittle, K. V., and Starner, T. E., 1978. "Heat Transfer in Pool Boiling," U.S. Patent 4,074,753.

Shum, M. S., 1980. "Finned Heat Transfer Tube with Porous Boiling Surface and Method for Producing Same," U.S. Patent 4,182,412.

Szumigala, E. T., 1971. "Manufacturing Method for Boiling Surfaces," U.S. Patent 3,566,514.

Thome, J. R., 1990. *Enhanced Boiling Heat Transfer*, Hemisphere Publishing Corp., Washington, D.C.

Torii, T., Hirasawa, S., Kuwahara, H., Yanagida, T., Fujii, M., and Ito, T., 1978. "The Use of Heat Exchangers with Thermoexcel's Tubing in Ocean Thermal Energy Power Plants," ASME paper 78-WA-HT-65.

Vachon, R. I., Nix, G. H., and Tanger, G. E., 1968. "Evaluation of Constants for the Rohsenow Pool-Boiling Correlation," *Journal of Heat Transfer*, Vol. 90, pp. 239–247.

Vachon, R. I., Nix, G. H., Tanger, G. E., and Cobb, R. E., 1969. "Pool Boiling Heat Transfer from Teflon-Coated Stainless Steel," *Journal of Heat Transfer*, Vol. 91, pp. 364–370.

Wang, D. Y., Cheng, J. G., and Zhang, H. J., 1991. "Pool Boiling Heat Transfer from T-Finned Tubes at Atmospheric and Super-atmospheric Pressures," in *Phase Change Heat Transfer*, E. Hensel, V. K. Dhir, R. Greif, and J. Fillo, Eds., ASME Symposium Vol., HTD-Vol. 159, ASME, New York, pp. 143–147.

Webb, R. L., 1970. "Heat Transfer Surface which Promotes Nucleate Boiling," U.S. Patent 3,521,708.

Webb, R. L., 1972. "Heat Transfer Surface Having a High Boiling Heat Transfer Coefficient," U.S. Patent 3,696,861.

Webb, R. L., 1981. "The Evolution of Enhanced Surface Geometries for Nucleate Boiling," *Heat Transfer Engineering*, Vol. 2, No. 3–4, pp. 46–69.

Webb, R. L., 1983. "Nucleate Boiling on Porous Coated Surfaces," *Heat Transfer Engineering*, Vol. 4, No. 3–4, pp. 71–82.

Webb, R. L., and Haider, S. I., 1992. "An Analytical Model for Nucleate Boiling on Enhanced Surfaces," in *Pool and External Flow Boiling*, V. J. Dhir and A. E. Bergles, Eds., ASME, New York, pp. 345–360.

Webb, R. L., and Pais, C., 1992. "Nucleate Boiling Data for Five Refrigerants on Plain, Integral-Fin and Enhanced Tube Geometries," *International Journal of Heat and Mass Transfer*, Vol. 35, No. 8, pp. 1893–1904.

Wolverine, 1984. *Engineering Data Book II*, K. J. Bell and A. C. Mueller, Eds., Wolverine Tube Corp., Decatur, AL.

Xin, M. D., and Chao, Y. D., 1985. "Analysis and Experiment of Boiling Heat Transfer on T- Shaped Finned Surfaces," AIChE paper, 23rd National Heat Transfer Conference, Denver, Colorado.

Yatabe, J. M., and Westwater, J. W., 1966. "Bubble Growth Rates for Ethanol-Water and Ethanol-Isopropanol Mixtures," *Chemical Engineering Progress Symposium Series*, Vol, 62, No. 64, pp. 17–23.

Yilmaz, S., and Westwater, J. W., 1981. "Effect of Commercial Enhanced Surfaces on the Boiling Heat Transfer Curve," in *Advances in Enhanced Heat Transfer 1981*, R. L. Webb, T. C. Carnavos, E. F. Park, Jr., and K. M. Hostetler, Eds., ASME Symposium Vol., HTD-Vol. 18, ASME, New York, pp. 73–92.

Young, R. X., and Hummel, R. L., 1965. "Improved Nucleate Boiling Heat Transfer," *Chemical Engineering Progress Symposium Series*, Vol. 61, No. 59, pp. 264–470.

Zhang, H., and Dong, L., 1992. Analysis and Experiment of Pool Boiling Heat Transfer from CIT-Shaped Finned Tube Above Atmospheric Pressure, *Multiphase Flow and Heat Transfer, Second International Symposium*, Vol. 1, X-J. Chen, T. N. Veziroğlu, and C. L. Tien, Eds., Hemisphere Publishing Corp., New York, pp. 384–392.

Zhang, Y., and Zhang, H., 1992. "Boiling Heat Transfer from a Thin Powder Porous Layer at Low and Moderate Heat Flux," in *Multiphase Flow and Heat Transfer, Second International Symposium*, Vol. 1, X-J. Chen, T. N. Veziroğlu, and C. L. Teen, Eds., Hemisphere Publishing Corp., New York, pp. 358–366.

Zhong, L., Tan, Y., and Wang, S., 1992. "Investigation of the Heat Transfer Performance of Mechanically Made Porous Surface Tubes with Ribbed Channels," *Multiphase Flow and Heat Transfer, Second International Symposium*, Vol. 1, X-J. Chen, T. N. Veziro—lu, and C. L. Teen, Eds., Hemisphere Publishing Corp., New York, pp. 700–707.

Zohler, S. R., 1990. "Porous Coating for Enhanced Tubes," U.S. Patent 4,890,669.

Zuber, N., 1958. "On Stability of Boiling Heat Transfer," *Transactions of the ASME*, Vol. 80, pp. 711–720.

11.16 NOMENCLATURE

A	Base area, m^2 or ft^2
A_j	Cross-sectional flow area of vapor jet, m^2 or ft^2
c_b	Empirical constant in Equation 11.14
c_p	Specific heat of the liquid, J/kg-K or Btu/lbm-°F
d_b	Bubble departure diameter, m or ft
d_t	Diameter of subsurface tunnel, m or ft
f	Bubble formation and departure cycle frequency, 1/s
G	Mass velocity, kg/s-m^2 or lbm/s-ft^2
g	Acceleration due to gravity, m/s^2 or ft/s^2
h	Depth of the fin gap spacing, m or ft
h_{nb}	Nucleate boiling heat transfer coefficient, W/m^2-K or Btu/hr-ft^2-°F
k	Thermal conductivity, W/m-K or Btu/hr-°F
L_f	Length of thin film region shown on Figure 11.31, m or ft
m	Average slope of the P–T curve for the working regime, Pa/K or lbf/ft^2-°F
n	Exponent on q versus $(T_w - T_{sat})$ equation, dimensionless
n_s	Nucleation site density, 1/m^2 or 1/ft^2
p	Pressure, Pa or lbf/ft^2
p_{cr}	Critical pressure, Pa or lbf/ft^2
p_G	Partial pressure of inert gas, Pa or lbf/ft^2
Δp	Pressure differential across a liquid slug, Pa or lbf/ft^2
Δp_{max}	Maximum pressure differential across a liquid slug, Pa or lbf/ft^2
Q	Heat transfer rate, W or Btu/hr
Q_{ex}	$q_{ex}A$, W or Btu/hr
Q_f	Heat transfer rate to thin film region of Figure 11.31, W or Btu/hr
Q_m	Heat transfer rate to one meniscus of Figure 11.31, W or Btu/hr
q	Total heat flux, W/m^2 or Btu/hr-ft^2
q_{CHF}	Critical heat flux, W/m^2 or Btu/hr-ft^2
q_{ex}	Heat flux from the external surface, W/m^2 or Btu/hr-ft^2
q_{tun}	Heat flux from the tunnel surface, W/m^2 or Btu/hr-ft^2
r	Liquid-vapor interface radius, m

r_b	Pore radius used in Equation 11.25, m or ft
r_c	Cavity radius at surface, m or ft
r_m	Meniscus radius shown on Figure 11.32, m or ft
r_t	Radius of subsurface circular tunnel, m or ft
R	Gas constant for saturated water vapors, J/kg-K or lbf-ft/lbm-R
s_b	Axial bubble pitch along the tunnel, m or ft
s_f	Transverse tunnel pitch, m or ft
s_g	Fin gap spacing, m or ft
S	Surface area of liquid film in porous structure, m² or ft²
T_l	Temperature at liquid–vapor meniscus, K or °F
T_s	Temperature of saturated pool liquid, K or °F
T_w	Wall temperature, K or °F
ΔT	$T_w - T_s$, K or °F
u	Velocity, u_o (initial velocity of translating meniscus), m/s or ft/s
v	Specific volume, $v_{lv} = v_v - v_l$, m³/kg or ft³/lbm
V	Volume, V_T (total volume of porous surface), V_b (total bubble volume), m³ or ft³
x	Length coordinate, x_l (length of the liquid slug, Figure 11.30), m or ft

Greek Letters

β	Geometry factor defined by Equation 11.26, dimensionless
δ	Liquid film thickness, δ_0 (initial film thickness) m or ft
ϵ	Porosity, dimensionless
θ	Liquid contact angle, θ_a (advancing contact angle), θ_r (receding contact angle), degrees
λ	Latent heat, J/kg or Btu/lbm
λ_H	Pitch of one-dimensional Taylor instability waves, m or ft
ν	Kinematic viscosity of the liquid, m²/s or ft²/s
ρ	Density, kg/m³ or lbm/ft³
σ	Surface tension of the liquid, N/m or lbf/ft
τ	Time period of the bubble departure and formation cycle, s
τ_w	Wall shear stress, N/m² or lbf/ft²
ϕ	Angle between cavity centerline and wall, degrees
ω	Angular frequency of oscillation of the meniscus, rad/s

Subscripts

CHF	Critical heat flux
l	Liquid phase
m	Porous matrix
s	At saturated condition
t	In subsurface tunnel
v	Vapor phase
w	At tube wall

12

VAPOR SPACE CONDENSATION

12.1 INTRODUCTION

This chapter is concerned with enhancement of vapor space condensation. Geometries include plates and tubes (horizontal and vertical). If the vapor flows in the direction of the draining condensate film, the interfacial shear stress will enhance condensation. Condensation with significant vapor velocity is called "convective condensation," and is discussed in Chapter 14. Because the surface orientation will affect condensate drainage characteristics, one must distinguish between horizontal and vertical tube orientations.

The majority of enhancement techniques of practical interest are limited to the passive types. These include special surface geometries for enhancement of film condensation and nonwetting coatings or additives for promotion of dropwise condensation. Electric field enhancement of film condensation appears to be very promising, and is discussed in Chapter 15.

Condensation will occur on a surface whose temperature is below the vapor saturation temperature. The condensed liquid formed on the surface will exist either as a wetted film or in droplets. Droplets are formed, if the condensate does not wet the surface. Although dropwise condensation yields a very high heat transfer coefficient, it cannot be permanently sustained. Dropwise condensation may be promoted by liquid additives or surface coatings that inhibit surface wetting. As the surface slowly oxidizes, the surface will eventually become wetted, and the process will revert to filmwise condensation. Hence, filmwise condensation is currently the more important process.

Study of the literature shows that surface tension effects are an important phenomenon in enhancement of film condensation. The technique classification in

Table 1.1 lists "rough" and "extended" surfaces separate from "surface tension devices." However, surface tension forces are probably the dominant mechanism for enhancement on rough and extended surfaces, in the absence of vapor velocity. The importance of surface tension forces in enhancement of film condensation was first described by Gregorig [1954]. However, its importance in affecting condensation on extended surfaces (e.g., integral-fin tubes) was not recognized until the late 1970s (e.g., see Karkhu and Borovkov [1971]). Consequently, the separate classifications of "extended surfaces" and "surface tension devices" applied to film condensation is somewhat ambiguous. We will more precisely define "surface tension devices" as concepts that do not increase the surface area of the base surface. However, they may include loosely attached (poor thermal contact) devices, such as wires.

Since 1981, significant advances have been made in understanding the importance of surface tension in enhancement of film condensation on finned surfaces—for plates and horizontal tubes. The key advances involve understanding the role that surface tension force plays in draining the condensate from the fins, and in retaining condensate within the interfin region of finned tubes. This understanding has culminated in the development of analytical models for predicting the condensation rate on horizontal integral-fin tubes and on banks of integral-fin tubes. The models have been validated for both low- and high-surface-tension fluids. In addition, data have been obtained to establish the "row effect" on several high-performance integral-fin tube geometries. High-performance fin geometries have been identified, as well as techniques to model currently used fin geometries. Models have been developed to predict condensate retention and to account for the effect of drainage strips.

12.1.1 Condensation Fundamentals

The Nusselt [1916] analysis provided the foundation for our present understanding of laminar film condensation. By neglecting the convection terms in the energy equation, the thermal resistance across the condensate film, of thickness δ, is given by the Fourier heat conduction equation. Defining the condensation coefficient in terms of $T_{sat} - T_w$, we obtain

$$q = h(T_{sat} - T_w) = \frac{k_l(T_{sat} - T_w)}{\delta} \qquad (12.1)$$

Hence, $h = k_l/\delta$. This equation is very simple. The difficult part is in determining the film thickness. The Nusselt analysis developed the equation for δ on a vertical, gravity-drained plate, as illustrated in Figure 12.1. This analysis, for zero interfacial shear, is given in heat transfer textbooks. Neglecting the inertia terms and assuming $u = u(y)$, the momentum equation is

$$(\rho_l - \rho_v)g + \mu_l \frac{d^2u}{dy^2} = 0 \qquad (12.2)$$

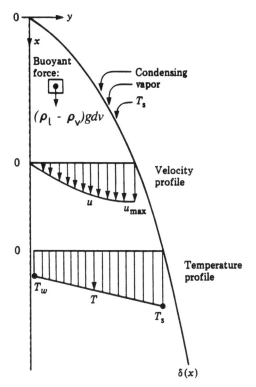

Figure 12.1 Illustration of gravity-drained condensation on a vertical plate.

Combining Equation 12.2 with an energy balance on the condensate film, the Nusselt result for $1/\delta$ may be written as

$$\frac{1}{\delta} = \left(\frac{\lambda F_g}{4\nu k_l x \Delta T_{vs}}\right)^{1/4} \tag{12.3}$$

where $F_g = (\rho_1 - \rho_v)g$, is the gravity force per unit volume. Substitution of Equation 12.3 in Equation 12.1 gives the local condensation coefficient at location x from the top of the plate:

$$h = 0.707 \left[\frac{k_l^3 g(\rho_l - \rho_v)\lambda}{\nu_l \Delta T_{vs} x}\right]^{1/4} \tag{12.4}$$

The average condensation coefficient $(\bar{h})$ on the plate of length L is obtained by integrating Equation 12.4 over $0 \leq x \leq L$, to obtain

$$\bar{h} = 0.943 \left[\frac{k_l^3 g(\rho_l - \rho_v)\lambda}{\nu_l \Delta T_{vs} L}\right]^{1/4} \tag{12.5}$$

Equation 12.5 may be written in terms of the condensate Reynolds number (Re_l = $4\Gamma/\mu_L$), as shown in heat transfer textbooks (e.g., Incropera and DeWitt [1990]). Γ is the condensate mass flow rate per unit wetted perimeter (P_w). The Reynolds number form of Equation 12.5 is

$$\bar{h} = 1.47 \left[\frac{k_l^3 \rho_l (\rho_l - \rho_v)g}{\mu_l^2} \right]^{1/3} Re_L^{-1/3} \qquad (12.6)$$

Equation 12.6 is more convenient for a sizing problem, where the required tube length (L) is unknown.

Interfacial shear stress (τ_i) will cause ripples in the condensate film, thereby yielding a higher condensation coefficient. For $\tau_i = 0$, the condensate film will transition to a turbulent film. Incropera and DeWitt [1990] give equations for rippled and turbulent films. For $\tau_i \gg 0$, transition to a turbulent film may occur at Re_l as small as 300. Collier [1981] provides equations for these more complex situations.

Nusselt also derived equations for laminar film condensation on inclined plates and horizontal tubes. For an inclined plate, the gravity force component, in the direction of the plate, is $g \sin \theta$, where θ is the plate angle from the vertical. Thus, the average condensation coefficient on an inclined plate is given by Equation 12.5 with g replaced by $g \sin \theta$. To obtain the equation for horizontal tubes, Nusselt integrated the inclined plate equation over $0 \le \theta \le \pi$ to obtain

$$\bar{h} = 0.728 \left[\frac{k^3 g (\rho_l - \rho_v) \lambda}{v_l \Delta T_{vs} d} \right]^{1/4} \qquad (12.7)$$

Nusselt extended his analysis of horizontal tubes to predict the "row effect." For a vertical rank of horizontal tubes, each successive tube row will receive condensate generated on the upper tube rows. Nusselt showed that the average condensing coefficient on N-tube rows ($\bar{h}_N$) related to the condensing coefficient on a single horizontal tube by the relation

$$\frac{\bar{h}_N}{\bar{h}_1} = N^{-1/4} \qquad (12.8)$$

Equation 12.8 shows that the condensation coefficient on a gravity-drained bank of horizontal tubes should decrease as $N^{-1/4}$. Equation 12.8 assumes no interfacial shear and no mixing of the condensate, which is unrealistic for actual tube bundles. The "row effect" in actual tube banks may be substantially less than that predicted by the Nusselt model. This is because of splashing and mixing in the liquid film. More discussion of the row effect is given in Section 12.4.8.

12.1.2   Basic Approaches to Enhanced Film Condensation

As shown by Equation 12.1, the thermal resistance in film condensation is that of conduction across the condensate film. The local film thickness is determined by the force that drains the condensate. Equations 12.3 through 12.8 assume that gravity force drains the condensate film. Other possible drainage forces are surface tension, suction, and centrifugal force. *Any technique that yields a reduced film thickness will enhance the film condensation coefficient.*

Therefore, a surface geometry which promotes reduced film thickness will provide enhancement. Short, vertical fins on horizontal tubes will have a smaller film thickness than on the base tube, thus providing enhancement.

An alternate to gravity-drained films is the use of surface tension forces for condensate removal. An example is the vertical fluted tube proposed by Gregorig [1954]. This tube has axial fins of a special shape, and will be discussed later. In a gravity-drained condenser, the lower portions of the bundle suffer from condensate inundation or flooding. A number of mechanical means may be envisioned to remove the accumulated condensate, which would allow reduced condensate film thickness in the lower portion of the condenser. Similarly, when condensation occurs inside a serpentine tube circuit, one may envision possibilities for condensate removal.

In vapor-shear-controlled condensation, high vapor velocity will provide positive effects due to interfacial shear or condensate entrainment. This contributes an exponent in Equation 12.8 less than the 1/4 value predicted by the Nusselt theory, which assumes zero interfacial shear. In shear-controlled flow, enhancement may be provided by reducing the cross sectional vapor flow area as condensation proceeds to lower vapor qualities. This is possible for tube-side condensation in a multipass design by reducing the number of tubes in parallel in each succeeding pass. Similar techniques are possible for shell-side condensation in large tube bundles. These concepts are discussed in detail in Chapter 14.

When noncondensibles are present, an additional thermal resistance is introduced in the vapor at the vapor–liquid interface. Mixing of the gas film will substantially reduce this thermal resistance. The maintenance of high vapor velocities, or special surface geometries, which promote a higher heat transfer coefficient in this gas film will substantially alleviate the performance deterioration due to noncondensibles.

Surface roughness may also provide mixing within the condensate film. However, this will not be effective if the film is laminar.

12.2 DROPWISE CONDENSATION

If surface wetting can be prevented, high-performance dropwise condensation will occur. Griffith [1985] reviews recent advances and presents an excellent discussion of the expected performance and practical aspects of applying dropwise condensation to steam condensers. Carey [1992] provides an analytical treatment of the proposed mechanisms of dropwise condensation. Two basic techniques for promoting dropwise condensation exist, namely, nonwetting surface coatings and chemical additives. Because low-surface-tension fluids more easily wet a surface than do high-surface-tension fluids, steam is a much more viable candidate for promotion of dropwise condensation than are the refrigerants or many organics. This is unfortunate, since the condensation coefficients for the second group of fluids are substantially less than that for steam.

Iltscheff [1971] condensed R-22 on several coated horizontal tubes (intended to promote dropwise condensation) and found that the condensation coefficients were below or approximately equal to those predicted by the Nusselt model for film condensation. Iltscheff implies that he observed condensation in the dropwise mode. This raises a question that has plagued many experimental studies of dropwise condensation: Were noncondensible gases present? The presence of noncondensible gases may substantially offset the potential enhancement provided by dropwise condensation.

The majority of the experimentation has been performed with steam. Successful surface coatings include noble metals and plastic (e.g., PTFE) and chemical additives (e.g., oleic acid). Chemical additives may be effective for up to 1000 hours, and then require surface cleaning and re-promotion for further effectiveness. When plastic coatings are used, one must account for the additional thermal resistance of the coating. This may be 60% as large as the resistance associated with film condensation. For a typical steam condenser design, Hanneman [1977] shows that dropwise condensation may provide an order of magnitude smaller steam-side resistance, resulting in a 40% surface area reduction for titanium tubes. If the thermal resistance of a 0.0015-mm-thick PTFE coating is included, the surface area saving is reduced to only 10%.

Most dropwise condensation studies have been performed on vertical plates or single horizontal tubes. If applied to a tube bundle, the effects may be substantially reduced due to inundation and vapor shear effects which cause reversion to film condensation.

12.3 SURVEY OF ENHANCEMENT METHODS

This section discusses film condensation on vertical plates and tubes and on horizontal tubes. Each enhancement technique is separately discussed.

12.3.1 Coated Surfaces

Figure 12.2 shows the cross section of a vertical condensing surface on which nonwetting strips are attached. Theoretical predictions for this geometry have been performed by Cary and Mikic [1973] and Brown and Matin [1971]. Because the condensate film is thinned near the nonwetting strips, high heat transfer coefficients should prevail. Brown and Matin show that the enhancement is dependent on the liquid contact angle and the thermal conductivity of the base surface. Cary and Mikic reason that an additional enhancement mechanism will be present due to a surface-tension-induced secondary flow. They show that the liquid surface temperature will be reduced near the nonwetting strip, resulting in a surface tension gradient causing the secondary flow (e.g., the "Marangoni effect"). The predicted results of Cary and Mikic, and of Brown and Matin do not appear to be consistent. Accounting

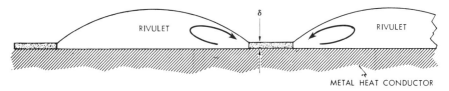

Figure 12.2 Cross section of a vertical condensing surface on which nonwetting strips are attached. (From Brown and Matin [1971].)

for the secondary flows, Cary and Mikic predict much lower enhancement levels than those of Brown and Matin. Cary and Mikic conclude that at best, "modest" enhancement levels can be expected (e.g., 30% or less). Glicksman et al. [1973] measured the effect of nonwetting Teflon strips for steam condensing on a horizontal tube. They helically wrapped 3.2-mm-wide, 0.16-mm-thick tape on a 12.7-mm-diameter copper tube and tested two strip spacings, namely, $p/d_o = 3$ and 6. The $p/d_o = 3$ wrap gave a 35% increased condensation coefficient. However, greater enhancement was obtained with a simple axial tape strip along the bottom of the tube. This yielded a 50% enhancement level. Addition of the helical wrap with the bottom axial tape strip did not provide further enhancement.

A U.S. Patent by Notaro [1979] describes a coated surface geometry for enhanced film condensation. It consists of an array of small-diameter metal particles bonded to the tube surface. The particles are 0.25 to 1.0 mm high, covering 20–60% of the tube surface. Condensation occurs on the particle array and drains along the smooth base surface. High condensation rates occur on the convex surfaces of the particles, due to surface tension forces which maintain very thin condensate films on the particles. Figure 12.3 shows a photograph of the surface and illustrates the thinned condensate films on the particles. The patent provides performance data for several fluids condensing on vertical tubes. For a given particle height, there is an optimum particle spacing. A 6.0-m-long vertical tube having a 50% area density of 0.5-mm-diameter particles yielded a steam condensation coefficient 17 times that predicted by the Nusselt equation for an equal-length smooth tube.

12.3.2 Roughness

Medwell and Nicol [1965] and Nicol and Medwell [1966] investigated enhancement due to a closely knurled roughness for a condensate film flowing down a vertical surface. The knurled roughness provides mixing in the condensate film, and hence increases the condensing coefficient. Nicol and Medwell [1966] present a theoretical treatment of the problem. They apply the heat–momentum analogy for roughness developed by Dipprey and Sabersky [1963], which was discussed in Chapter 9. Their theory shows that the benefits of roughness are characterized by the "roughness Reynolds number," $e^+ = eu^*/\nu_l$. For constant film thickness (δ), increasing roughness height increases e^+, which reduces the thermal resistance of the viscous influenced region. This effect continues for larger roughness sizes up to $e^+ = 55$, at which viscosity no longer influences the thermal resistance. As the condensate film

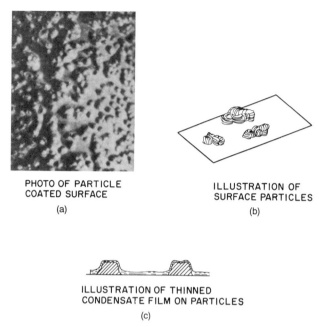

PHOTO OF PARTICLE
COATED SURFACE
(a)

ILLUSTRATION OF
SURFACE PARTICLES
(b)

ILLUSTRATION OF THINNED
CONDENSATE FILM ON PARTICLES
(c)

Figure 12.3   Enhanced condensation surface formed by small-diameter metal particles bonded to the base surface. (a) Photograph of actual surface. (b) Illustration of particles bonded to surface. (c) Illustration of thin condensate film on particles and thick condensate film on base surface. (From Notaro [1979])

thickness increases, for constant roughness height, the e^+ will decrease, causing viscous effects to influence the thermal resistance. The film may become so thick that the roughness characteristic approaches the hydraulically smooth condition, (e^+ = 5) where no effective mixing occurs in the roughness zone. Thus, the enhancement produced by a given roughness size is directly related to the condensate film thickness. Their theory shows that the maximum possible enhancement is approximately 100%. The steam condensation tests of Nicol and Medwell [1966] offer good support for the theory. The largest roughness (e = 0.5 mm) yielded an enhancement of approximately 90%.

12.3.3   Horizontal Integral-Fin Tubes

Horizontal integral-fin tubing illustrated in Figure 12.4 has found wide commercial acceptance for condensation on horizontal tubes. These tubes are commercially available with 433–1575 fins/m (11–40 fins/in.). Integral-fin tubes provide substantial performance improvement over plain tubes, particularly for low-surface-tension fluids which use 748–1575 fins/m. This occurs because of (a) the area increase provided by the fins and (b) the thin condensate films formed on the short fins. The thin condensate films are primarily due to surface tension effects, which

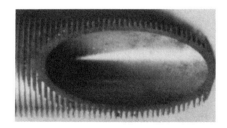

26 FPI   **Figure 12.4**   Horizontal integral-fin tube.

are discussed in detail in Section 12.4. Table 12.1 shows the enhancement ratio determined by Webb et al. [1985] for R-11 condensing on 19-mm-diameter (d_e) commercial integral-fin tubes. The surface area of the integral-fin tubes is based on the envelope diameter over the fins, $A/L = \pi d_e$. The table shows that the enhancement ratio increases with increasing fins/m. The 1378-fin/m tube provides an enhancement ratio of 5.28 compared to a 19-mm-diameter plain tube.

Substantial data have been reported for condensation on integral-fin tubes. Much of the existing data are chronologically listed in Table 12.2. Marto [1988] presents an excellent survey of much of these data.

12.3.3.1 Condensate Retention   The integral-fin tubes listed in Table 12.1 are routinely used for condensation of low-surface-tension fluids. However, they will not be effective for high-surface-tension fluids, such as water (steam). This is because capillary (surface tension) force retains condensate between the fins on the lower side of the tube, where the condensate thickness is $\delta = e_o$ (fin height). Because $h = k_l/\delta$, the condensation coefficient in this condensate flooded zone is very small. Figure 12.5 shows that the fins are condensate flooded over an angle β. Rudy and Webb [1983, 1985] and Honda et al. [1983] show that the condensate retention angle for fins of rectangular cross section is given by

$$\beta = \pi c_b = \cos^{-1}\left(1 - \frac{4\sigma}{d_o \rho_l g s}\right) \qquad (12.9)$$

Table 12.1   Enhancement Ratio for R-11 Condensing on Integral-Fin Tubes (d_e = 19 mm, T_{sat} = 35 °C, ΔT_{vs} = 9.5 K)

Fins/m (m^{-1})	e_o (mm)	h (mm)	h/h_p (W/m² K)
748	1.5	8,070	2.64
1024	1.5	11,970	3.91
1378	0.9	16,140	5.28

Table 12.2   Published Data for Condensation on Integral-Fin Tubes

Fins/m	Fluids	Reference
608	Methyl chloride, sulfur chloride, R-22, n-pentane, propane	Beatty and Katz [1948]
630	R-22, n-butane, acetone, water	Katz and Geist [1948]
396–770	R-12	Henrici [1961]
748–1024	R-22	Pearson and Withers [1969]
748–1024	R-22	Takahashi et al. [1979]
1060–1610	R-11	Carnavos [1980]
1020–2000	R-113, Methanol	Honda et al. [1983]
95–1000	Water	Yau et al. [1985]
100–667	Water	Wanniarachchi et al. [1986]
400–1000	R-113	Masuda and Rose [1987]
200–1333	Water	Marto et al. [1988]
1417	R-11	Sukhatme et al. [1990]
400–667	R-113	Briggs et al. [1992]
1000	R-113	Wang et al. [1990]

Equation 12.9 is based on a force balance between gravity and surface tension forces on the condensate. Equation 12.9 shows that the condensate retention angle (β) increases with increasing surface tension (σ) and with decreasing fin spacing (s). Condensation of steam on a 19-mm-diameter tube having 0.25-mm fin thickness would result in total flooding if the fin density were greater than 1000 fins/m. Jaber and Webb [1993] show that integral-fin tubes for steam condensation should use no more than 630 fins/m (16 fins/in.).

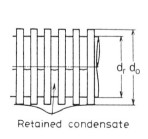

Retained condensate

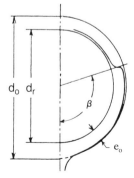

Figure 12.5   Illustration of condensate flooding angle (β) on horizontal integral-fin tubes.

The amount of condensate retained between the fins may be reduced by attaching "drainage strips" to the bottom of the tube. A drainage strip is simply a thin plate of height H_d that hangs from the bottom of the tube. This alters the balance between surface tension and gravity, which establishes the condensate retention angle. Addition of the drainage plate adds a surface tension force in the downward direction. Because of the added downward surface tension force, less condensate is retained on the tube. Honda et al. [1983] present theoretical relations to predict the condensate retention angle for finned tubes with drainage strips. Their theory shows that if the drainage strip is made of a porous material, the capillary condensate removal force will be increased. Table 12.3 shows the effect of drainage strips for R-113 condensing on an 18.9-mm-diameter tube (diameter over fins) having 2000 fins/m and 1.13-mm fin height, as measured by Honda and Nozu [1987b]. Two drainage strip materials were used. The first was polyvinyl chloride (PVC), and the second was porous nickel having a permeability of $1.7E-9$ m². For a strip height (H_d) of 12.6 mm, the PVC strip reduced c_b (fraction circumference flooded) 9%, and the porous strip reduced c_b by 51.6%. The shorter 4.0-mm porous strip reduced c_b by 33%. As shown in Table 12.3, the 14.6-mm porous strip provided a 65% increase in the condensation coefficient, as opposed to only 11% for the PVC strip of the same height. The improvements are solely due to reducing the condensate retention angle.

Although significant enhancement is provided by porous strips, it is difficult to envision how such drainage strips may be employed in an actual condenser tube bundle. They would be added after the tube is inserted in the bundle, which is cumbersome. Sufficient vertical clearance between tubes must be provided to accommodate the strip height. These strips would act as barriers to horizontal vapor flow. Hence, the vapor must flow downward within the bundle. Vapor shear would occur on the drainage strips, which may increase the vapor pressure drop. However, the strip may act as a "splitter plate," which acts to decrease the vapor pressure drop.

Yau et al. (1986) have measured the effect of drainage strips on steam condensation. Using an 8.0-mm-high copper drainage strip on a 15.9-mm-outer-diameter tube having 667 fins/m, they found a 30% condensation enhancement. They also showed that the same drainage strip provided little benefit for a plain tube.

Table 12.3 Effect of Drainage Strips for Methanol Condensing on a 2000-fin/m Tube at $(T_s - T_w) = 5$ K

Strip	H_d (mm)	c_b	h (W/m²-K)
None	—	0.62	6200
PVC	12.6	0.57	6900
Porous	12.6	0.32	10200
Porous	4.0	0.42	7500

12.3.3.2 Advanced Surface Geometries   Integral-fin tubes having a saw-toothed fin shape have been developed, which have higher condensing coefficients than the standard integral-fin tubes. Figures 12.6a, 12.6c, and 12.6d show three such saw-tooth fin geometries, which are commercially used in refrigerant condensers. Figure 12.4 is the standard integral-fin tube (1024 fins/m, 26 fins/in.), and Figure 12.6b has a Y-shaped fin tip. The geometries of these tubes are given in Table 12.4.

The Figure 12.6d Tred-26D™ fin has the same height and fin pitch as the Figure 12.4 standard 1024-fin/m (26-fin/in.) integral-fin tube. The notch depth in the Tred-26D™ fin tip is approximately 40% of the fin height. Test data have been reported for refrigerants condensing on copper tubes for the Figure 12.6 and Figure 12.7 geometries. Figure 12.7a shows the R-11 condensation coefficient data of Webb et al. [1985] for three commercially used integral-fin tubes, which have 748, 1024, and 1378 fins/m. Figure 12.7b shows the R-11 data of Sukhatme et al. [1990] for four 1417-fins/m tubes having fin heights between 0.46 and 1.22 mm. Note that the condensation coefficient is based on the envelope area over the fins, $\pi d_e L$. Figure 12.7a shows that the condensation coefficient increases with decreasing fin pitch. For the same fin pitch, Figure 12.7b shows that the condensation coefficient increases with increasing fin height.

Webb and Murawski [1990] measured the R-11 condensation coefficient for the Figure 12.6b–d and Figure 12.4 tube geometries, which are shown in Figure 12.8. Figure 12.8 shows that the highest single tube performance is provided by the Turbo-C™ tube, with the lowest given by the standard 1024-fin/m tube. The condensation coefficient of the standard 1024-fin/m tube is 60% that of the Turbo-C™. The second and third best performance is given by the GEWA-SC™ (80% of Turbo-C™) and Tred-26D™ (72% of Turbo-C™) tubes, respectively.

Wang et al. [1990] report the R-113 performance of a unique finned tube having lateral ripples in the fins. The tube had 1000 fins/m with 1.47-mm-high fins. Their data show that the condensation coefficient is 30–40% higher than that of the

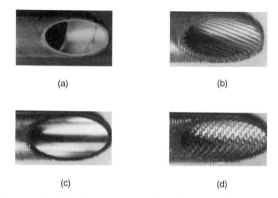

(a)   (b)

(c)   (d)

Figure 12.6   Photographs of enhanced condensing tubes. (a) Hitachi Thermoexcel-C™. (b) Wieland GEWA-SC™. (c) Wolverine Turbo-C™. (d) Sumitomo Tred-26D™.

Table 12.4 Dimensions of the Figure 12.4 and 12.6 Tubes

Tube Type	Figure	d_o (mm)	e_o (mm)	Fins/m
Integral-fin	12.4	18.9	1.3	1024
Thermoexcel-C™	12.6a	18.9	1.2	1378
GEWA-SC™	12.6b	18.9	1.3	1024
Turbo-C™	12.6c	18.9	1.1	1575
Tred-26D™	12.6d	18.9	1.3	1024

Thermoexcel-C™ tube. They propose that surface tension force pulls the condensate into the valleys of the ripples.

No analytical models have been specifically developed to predict the effect of the notched fin tip geometry. However, surface tension drainage from the sawtooth fins is expected to account for the higher performance of the fin. A small tip radius establishes a surface tension gradient to affect condensate drainage.

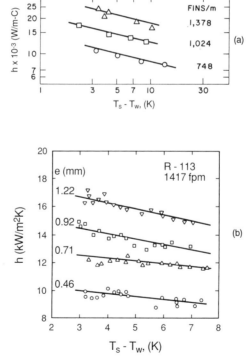

Figure 12.7 R-11 condensation coefficient for standard copper integral-fin tubes (748, 1024, and 1378 fins/m) as reported by Webb et al. [1985]. (From Webb et al. [1985].) (b) R-11 condensation coefficient on 1417-fin/m copper integral-fin tubes with fin heights between 0.46 and 1.22 mm reported by Sukhatme et al. [1990].

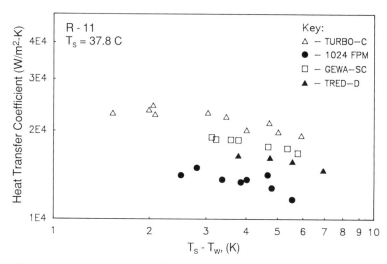

Figure 12.8   R-11 condensation coefficient for copper enhanced horizontal tubes. (From Webb and Murawski [1990].)

12.3.3.3   Steam Condensation   Figure 12.9 shows six tube geometries consciously developed for steam condensation. The first four are integral-fin tubes, and the fifth is the attached particle tube shown in Figure 12.3. The sixth tube (Figure 12.9f) is discussed by Wildsmith [1980]; it has a corrugated internal surface ($p = 6.4$ mm, $e_i = 0.7$ mm) and has fine triangular threads ($e_o = 0.3$ mm, $p_f = 1.0$ mm) on the outer surface. Wildsmith reports that the Figure 12.9f tube provides $UA/U_p A_p = 1.47$ for the same water-side pressure drop. The Figure 12.9 tube dimensions are listed in Table 12.5. Note that they typically have a larger fin pitch than the tubes listed in Table 12.4. Only one of the Table 12.5 tubes are made of copper. Electric utility steam condensers do not use copper tubes. Rather, they use 90/10 Cu/Ni, stainless steel, or titanium. These materials have much lower thermal conductivity than copper, so fin efficiency becomes a limiting factor, especially for steam condensation. As shown in Table 12.5, quite small fin heights are required. The Table 12.5 tubes were designed to have a thicker fin thickness at the base (t_b) than at the tip (t_t) to increase the fin efficiency. The thermal conductivities of the tube materials are: Cu/Ni ($k_w = 45$ W/m-K), stainless steel ($k_w = 14$ W/m-K), and titanium ($k_w = 22$ W/m-K).

The last column of Table 12.5 shows the outside enhancement ratio for steam condensation at $(T_s - T_w) = 2.3$ K and $T_s = 54°C$. The E_o of the copper NW-11C is much higher than that of the NW-11C/N tube, which has the same fin geometry. The difference is a result of the fin efficiency. The 11C/N, 16SS, and A/P-50 tubes all provide condensation enhancement in the range of 1.64–1.75.

All of the tubes in Tables 12.4 and 12.5, except the standard integral-fin, have tube-side enhancement. The tube-side enhancement is for the tube-side water flow. Adding the shell-side enhancement results in a reduction of the inside diameter. For

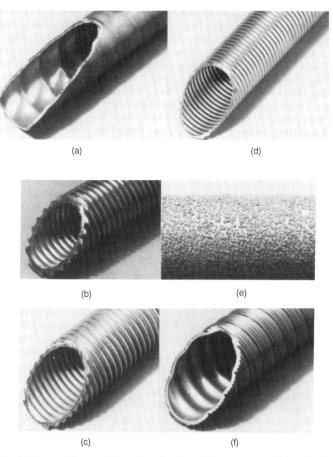

(a)   (d)

(b)   (e)

(c)   (f)

Figure 12.9 Horizontal integral-fin tubes developed for steam condensation by Jaber and Webb [1993]. (a) Wolverine Korodense™. (b) Copper Wieland 11-NW™. (c) Copper–nickel Wieland 11-NW™. (d) Stainless steel Wieland NW-16. (e) UOP attached particle tube. (f) Yorkshire MERT (Multiply Enhanced Roped Tube).

tube-side flow at fixed velocity, $\Delta p/L \propto (d_i)^5$. Hence, any reduction of the tube inner diameter will significantly increase the tube-side pressure drop. Efforts to reduce the external fin height will benefit the tube-side pressure drop. Note that the outside fin heights are typically lower in Table 12.5 than in Table 12.4.

12.3.4 Corrugated Tubes

Figure 12.10 shows three basic types of commercially available "corrugated" tubes. These tubes also provide tube-side enhancement. Figure 12.10a has axially spaced helical grooves pressed in the outer surface, which form helical ridges in the inner surface. Surface tension forces are responsible for the condensation enhancement.

Table 12.5   Dimensions of the Figure 12.9 Tubes

Tube Type	Figure	Fins/m	Material	d_o (mm)	e_o (mm)	t_b (mm)	t_t (mm)	E_o
Wolverine Korodense™	12.9a	None	Cu/Ni	22.2	NA[a]	NA	NA	NA
Wieland NW-11C	12.9b	433	Cu	19.0	1.1	0.9	0.3	2.80
Wieland NW-11C/N	12.9c	433	Cu/Ni	22.2	1.1	0.9	0.3	1.75
Wieland NW-16SS	12.9d	630	SS	18.9	0.3	1.2	0.8	1.70
UOP A/P-50	12.9e	—	Cu/Ni	22.2	0.5	—	—	1.70
Yorkshire MERT	12.9f	1000	Cu/Ni	25.4	0.3	0.3	0.05	2.00 (est)

[a]NA, not applicable.
[b], stainless steel.

Surface tension force pulls the condensate into the outer helical grooves, which act as condensate drainage channels. Thin condensate films exist on the convex profile between the drainage grooves.

Mehta and Rao [1979] condensed steam on the Figure 12.10a tubes for systematically varied groove pitch (p) and the groove depth (e_i) with a constant diameter aluminum tube (19.5-mm outside diameter and 1.8-mm wall thickness). Figure 12.11 shows their results for $p = 6.35$ mm with e_i increasing from 0.13 mm to 1.0 mm. The enhancement (h/h_p) initially increases with increasing groove depth and attains a maximum of 1.38 at $e_i = 0.35$ mm, after which h/h_s decreases with increasing groove depth. The h/h_p decreases for increasing groove pitch (e_i = constant) as one would expect. Mehta and Rao also measured the water-side heat transfer and friction characteristics for their tubes. These data are reported by Rao [1988].

Withers and Young [1971] show that the Figure 12.10a corrugated tube will provide a 30–50% tubing material reduction, compared to a smooth-tube steam condenser designed for equal water-side pressure drop. Their data show that the steam-side enhancement is 35–50%, relative to a smooth tube.

Marto et al. [1979] performed comparative tests of all of the Figure 12.10

(a)   (b)   (c)

Figure 12.10   Doubly enhanced tube geometries for condensation on horizontal tubes tested by Marto et al. [1979]. (a) Helically corrugated tube. (b) Turbotec™ tube. (c) Corrugated tube formed by rolling a corrugated sheet and seam welding. (From Marto et al. [1979].)

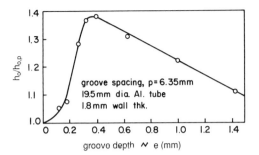

Figure 12.11 Steam condensation data on corrugated tubes reported by Mehta and Rao [1979]. (From Webb [1981].)

geometries with steam condensation (100°C) which included variation of the geometric parameters for each tube type. Their results for $p = 9.58$ mm, $d = 0.4$–0.6 with a 15.9-mm-diameter tube appear to be in conflict with those of Mehta and Rao. Marto et al. found that the condensing coefficient is approximately 10% below the smooth tube value. The low steam-side enhancement may be due to an error in their use of the Wilson plot method to separate the water- and steam-side resistances. Only the Figure 12.10a tube provided significant steam-side enhancement, approximately 35%. The dominant enhancement occurred on the water side, where they obtained h/h_p values in the range of 2–4 for most tubes. The water-side enhancement was accompanied by substantial increased pressure drop. These results imply that such tubing would be of value only if the water-side offers the controlling thermal resistance. Furthermore, the steam-side enhancement is marginal, compared to that possible with other surface geometries.

Yorkshire [1982] describes corrugated tubes having different corrugation pitches and depths. Their brochure presents data and a design correlation for steam-side condensing performance as a function of surface geometry. Their data are generally consistent with the Figure 12.11 data of Mehta and Rao [1979].

12.3.5 Surface Tension Drainage

The use of surface tension forces to affect condensate drainage is a very important and effective enhancement technique. We have previously noted enhancement by surface tension forces. In this section, our discussion will briefly describe data for various geometries. Section 12.4 provides detailed discussion of the mechanism and theory of surface tension enhancement. Equations are provided in Section 12.4 to select preferred surface shapes. The geometries described in this section are limited to those which do not provide significant increased surface area. Such surface tension drainage may be invoked by loosely attached wires, which have essentially no thermal contact with the base heat transfer surface. In some cases, a surface area increase may be involved, which technically should be discussed under "extended surfaces." However, it is discussed in this section because the enhancement mechanism is caused by surface tension force.

12.3.5.1   Vertical Fluted Tubes   Gregorig [1954] was the first to propose use of surface tension forces to enhance laminar film condensation on a vertical surface. Figure 12.12a illustrates a horizontal cross section through the wall of a vertical fluted tube, and Figure 12.12b is a photograph of a doubly fluted tube. Because of the surface curvature, the liquid pressure in the convex film is greater than that of the vapor. The combination of convex and concave surfaces establishes a surface tension force which draws the condensate from the convex surface into the concave region. A high condensation rate occurs on the convex portions of the fluted surface due to nearly horizontal drainage by the surface tension force. The concave portions serve as vertical, gravity drainage channels. The resulting heat transfer coefficients averaged over the total surface area are substantially higher than for a uniform film thickness on a smooth tube. The size of the flutes should be selected such that the drainage channel will be filled to capacity at the bottom of the vertical surface. Therefore, longer tubes would require larger drainage channels. This is discussed by Webb [1979] and by Adamek and Webb [1990a].

Mori et al. [1979] also investigated the effect of placing circular disks (Figure 12.13a) at spaced intervals to remove the condensate flowing down the tube, thereby exposing new condensing surfaces below the disk. Figure 12.13b shows their experimental and predicted results for R-113 condensing on a vertical surface 50-mm height. This figure shows that the 25-mm disk spacing improves the performance of the 50-mm-high surface. Their accompanying theory allows prediction of an optimum disk spacing. Combs and Murphy [1978] also tested the effect of spaced runoff

(a)

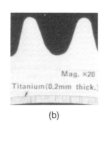

(b)

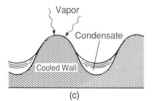

(c)

Figure 12.12   Vertical fluted tubes. (a) Cross section of fluted tube. (b) Photograph of doubly fluted tube. (Courtesy of Sumitomo Light Metal, Ltd.) (c) Detail of fin cross section.

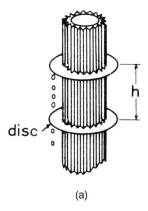

(a)

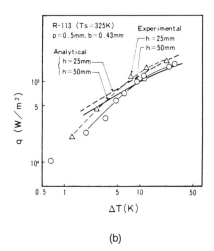

(b)

Figure 12.13 (a) Vertical fluted tube fitted with drainage skirt. (b) Predicted and experimental results on tube having drainage skirt. (From Mori et al [1979].)

disks for ammonia condensation on 1.2-m-long fluted tubes. They show that the disks should be effective, provided that the tube is sufficiently loaded with condensate.

Fluted tubes have received considerable attention for vertical tube condensers used in desalination and are commercially available. Thomas [1968] and Carnavos [1974] give performance data for single and doubly fluted tubes (see Figure 12.12). They show enhancement ratios for h/h_p (total area basis) in the range of 4–8 for tube lengths between 0.50 and 0.60 m. Combs and Murphy [1978] report similar enhancement ratios for condensing ammonia on 1.2-m-high tubes. The flute geometry of their tubes was not selected based on theoretical relations. In fact, some of the tubes they tested are actually "curtain rods" they purchased from a department store!

Newson and Hodgson [1973] tested 32 tubes of the types illustrated in Figure 12.14a. They condensed atmospheric steam on a 1.13-m-long vertical tube against

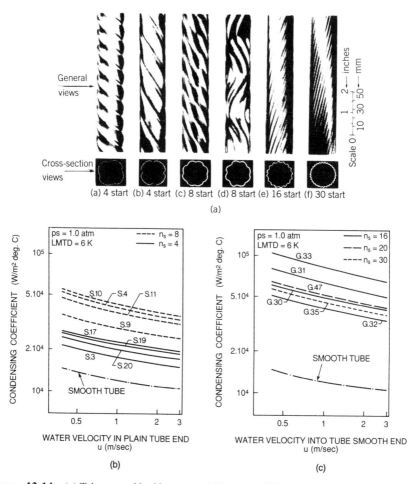

Figure 12.14 (a) Tubes tested by Newson and Hodgson [1973] and reported by Butterworth and Hewitt [1978]. (b) condensation coefficient for 4 and 8-start tubes, (c) condensation coefficient for 16, 20 and 30-start tubes. (From Newson and Hodgson [1973].)

cooling water inside the tubes. Each tube provides inside and outside enhancement, causing a swirl flow to the tube-side coolant. Their tubes have 4, 8, 16, 20, or 33 flutes (n_f). All tubes were made from a 31.8-mm-outside-diameter, 0.89-mm-wall plain tube. The geometric enhancement parameters are the number of flutes, the groove depth, and the helix angle. They propose that the tubes provide enhancement by draining off the condensate after each gravity-drained, vertical length between the grooves, thus acting as a short vertical tube. We propose that surface tension drainage is the actual condensation enhancement mechanism. Figure 12.14b shows the condensation coefficient for a selected group of the 4- and 8-flute tubes. Figure 12.14c shows the condensation coefficient for a selected group of the 16-, 20-, and 30-flute tubes. The tubes having 16, 20, or 30 flutes have a higher condensation coefficient than do the 4- or 8-flute tubes. The condensation and water-side heat transfer coefficients were separated using the Wilson plot method. Table 12.6 lists the geometry details, E_o ($= h_o/h_{op}$), E_i ($= h_i/h_{ip}$), U/U_p, and the water-side pressure drop ratio ($\Delta p/\Delta p_p$) at 1.52-m/s water velocity. In contrast to the condensation-side enhancement, the lower number of starts and higher helix angles (α) provide the higher E_i values. The relatively large water-side $\Delta p/\Delta p_p$ are partially caused by the reduced cross-sectional flow area of the fluted tubes. The U/U_p values are dependent on the split of thermal resistances between the outside and inside of the tube. At the reference test condition (93°C water temperature, 60°C log-mean temperature difference, and 1.52-m/s water velocity), 46% of the total thermal resistance of the smooth tube was on the water-side.

12.3.5.2 Axial Wires on Vertical Smooth Tubes   Loosely attached, spaced vertical wires on a vertical surface (Figure 12.15) can also provide surface tension condensation enhancement. If the wire diameter (e) is appreciably larger than the condensate film thickness (δ) and the wires are wetted by the condensate, surface tension force draws the condensate into a rivulet at the wire (region A). This produces film thinning in the space between the wires (region B). The enhancement occurs due to the thinned film in region B, and the condensate drains down along the wires. Thomas [1967] provides experimental data for this enhancement technique. In a second publication, Thomas [1968] shows that square wires are more effective than circular wires of the same height. This is because square wires have a greater condensate carrying capacity. Butizov et al. [1975] and Rifert and Leont'yev [1976] have also worked with loosely attached vertical wires. Their work with steam condensation supports conclusions advanced by Thomas. Using 1.0-mm-diameter circular wires spaced at nine wire diameters, they measured enhancement levels in the range of 3–6 on a 1.3-m-long test section. The enhancement decreases with increasing heat flux because the wires become more quickly loaded with

Table 12.6 Test Results of Newson and Hodgson [1973] for Stream Condensation on Vertical Sprially Fluted Tubes

Tube	n_f	e/D_h	α (degrees)	E_o	E_i	U/U_p	$\Delta p/\Delta p_p$
S.3	4	0.078	56	1.45	1.79	1.69	4.54
S.17	4	0.068	65	1.90	2.21	1.99	6.97
S.11	8	0.055	20	3.00	1.19	1.61	3.0
S.10	8	0.056	27	3.50	1.26	1.70	3.5
S.4	8	0.050	45	3.20	1.47	1.82	3.2
S.9	8	0.033	56	2.35	1.91	2.02	3.56
G.33	16	0.051	14	6.75	1.23	1.90	1.74
G.31	16	0.070	21	5.00	1.34	1.87	1.91
G.30	16	0.063	27	4.00	1.47	1.93	2.15
G.32	16	0.053	36	6.20	1.66	2.04	2.52
G.35	20	0.048	21	3.60	1.37	1.85	2.03
G.47	30	0.047	14	4.10	1.19	1.61	1.47

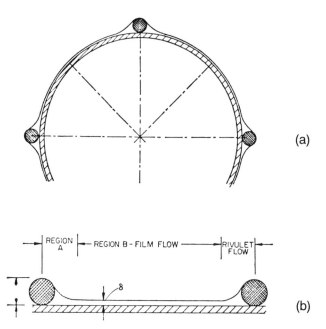

Figure 12.15 (a) Vertical tube fitted with loosely attached wires for enhancement of condensation as reported by Thomas [1968]. (b) Detail showing condensate drainage. (From Thomas [1968[.)

condensate. The Kun and Ragi [1981] patent describes 1.4-mm wires spaced at 2.0 mm on a vertical tube. The vertical wires were held in place by a second set of wires helically wrapped at 28-mm axial pitch. For the same LMTD between the condensing steam and the water-side coolant, the attached wire design increased the heat flux at the bottom of the tube by a factor of 23.

Although the attached wire concept is effective, the fluted tube concept of Figure 12.14 provides enhancement on both sides, is probably cheaper, and is more practical for tube insertion in a tube sheet. The fluted tubes are made with plain tube ends.

12.3.5.3 Wire Warp on Horizontal Tubes
Thomas et al. [1979] tested ammonia condensation on a smooth, horizontal tube wrapped with a spaced wire in a helical manner. A 38-mm-diameter aluminum tube was wrapped with 2.4-mm-diameter wire spaced at 9.5 mm. The condenser contained 147 tubes. The measured condensing coefficient was approximately three times that predicted by the Nusselt equation for a smooth tube. Surface tension forces draw the condensate to the base of the wires, which act as condensate runoff channels. On the basis of this test, it appears that external ridges (wire) are more effective for condensation enhancement than are circumferential grooves. Other studies of wire-wrapped horizontal tubes are reported by Marto and Wanniarachchi [1984] and Fujii et al. [1987]. Fujii et al. condensed R-11 and ethanol on an 18-mm-diameter tube and varied the wire pitch (p_f) for wire diameters (e) of 0.1, 0.2, and 0.3 mm. They found maximum enhancement for

closely spaced wires, $p_f/e \simeq 2.0$, which provided $h/h_p \simeq 3.5$. For fixed wire pitch, the larger-diameter wires produced higher enhancement.

Practical problems are associated with this enhancement method for shell-and-tube heat exchangers. It would be necessary to expand the tube end, in order to slip the wire-wrapped tube into the tube sheets.

12.3.5.4 Surface Tension Effects in Zero Gravity In a U.S. patent, Staub [1966] describes surface geometries intended to operate in a zero gravity environment, where gravity forces do not exist to drain the condensate. His geometry consists of an array of circular spine-fins having rounded fin tips and a capillary wicking material on the base surface. Surface tension force drains condensate from the rounded spine-fin tip and the condensate is removed from the base surface by the capillary wicking. Such a surface may effectively operate in any orientation, regardless of whether a gravity field is present. No data are presented in the patent.

12.3.6 Electric Fields

Recently, considerable work has been done using electric fields to enhance film condensation. Application of an electric field to a dielectric fluid is termed an "electrohydrodynamic" (EHD) effect. Yabe [1991] enhanced R-113 condensation on a vertical plain tube by a factor of 4.5 using an electric field strength of 4 MV/m. However, the enhancement did not exceed that provided by a horizontal Thermoexcel-C™ tube (Table 12.4). Section 15.4 discusses recent research.

12.4 SURFACE-TENSION-DRAINED CONDENSATION

12.4.1 Fundamentals

The phenomenon takes advantage of the fact that the pressure difference across a liquid–vapor interface is influenced by the local radius of the interface. By forming an interfacial shape that has a changing radius, one can establish a pressure gradient in the liquid film, which drains condensate from the surface.

To understand the basic principles of the pressure difference across an interface, consider a liquid droplet of radius r. The liquid droplet exists in a saturated vapor environment, whose pressure is p_v. The liquid pressure in the droplet is p_l. Surface tension force acts on the liquid–vapor interface. If the droplet is cut in half, and a force balance is written on the pressure and surface tension forces, we obtain

$$(p_l - p_v)\pi r^2 = 2\pi r\sigma \qquad (12.10)$$

Solving for $p_l - p_v$, we obtain

$$p_l - p_v = \frac{2\sigma}{r} \qquad (12.11)$$

which shows that the pressure in the liquid is greater than that in the vapor. The value of $p_l - p_v$ increases as the droplet radius, r, is decreased. Next consider a two-dimensional fin having semicircular-shaped fin tip, of constant radius r, illustrated in Figure 12.16. Because of surface tension force, the pressure in the liquid is higher than that of the vapor ($p_v = p_{sat}$). A force balance between pressure and surface tension forces shows that

$$p_l = p_{sat} + \frac{\sigma}{r} \qquad (12.12)$$

$p_l - p_{sat}$ is equal to σ/r, rather than the $2\sigma/r$ of Equation 12.10, because the droplet considered by Equation 12.10 is a surface of revolution, whereas Figure 12.16 involves a two-dimensional shape. Now, consider the fin profile shown in Figure 12.17. This profile has a small radius at the tip, and the local radius increases with increasing distance from the tip. At the fin base ($\theta_m = \pi/2$), $r = \infty$. As shown by Equation 12.12, the liquid pressure will decrease with increasing distance from the fin tip, because the local radius of the condensate increases. This means that condensate formed on the fin would be drained from the fin tip toward the fin base. The pressure gradient in the liquid film is obtained by differentiating Equation 12.12 and is given by

$$\frac{dp}{ds} = \sigma \frac{d(1/r)}{ds} \qquad (12.13)$$

where $dp/ds < 0$. The largest pressure gradient exists at the fin tip (where r is the smallest) and decreases along the fin length. The fin tip radius on Figure 12.16 is constant, and suddenly changes to $r = \infty$ at the tangent point. This radius discontinuity will result in an intense local surface tension gradient at the tangent point. However, gravity force will control after this point.

Figure 12.16 considered a convex liquid profile. Next, we wish to consider the concave liquid profile that exists at the base of the fin shown in Figure 12.17. To understand the effects of the concave interface shape, we first consider the forces

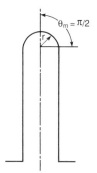

Figure 12.16   Fin having constant radius fin tip.

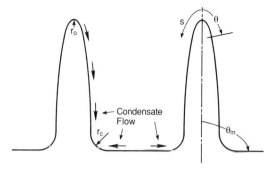

Figure 12.17 Fin having small tip radius with increasing radius from fin tip.

acting on a vapor bubble, which involves a concave liquid shape. The pressure in the bubble is p_v. If the bubble is cut in half, and a force balance is written on the pressure and surface tension forces, we obtain

$$(p_v - p)\pi r^2 = 2\pi r\sigma \tag{12.14}$$

Solving for $p_l - p_v$, we obtain

$$p_l - p_v = -\frac{2\sigma}{r} \tag{12.15}$$

Thus, for a vapor bubble, or a concave interface, the liquid pressure is less than the vapor pressure. Similar to the development for Equation 12.14, the pressure gradient for a two-dimensional concave liquid film is given by

$$\frac{dp}{ds} = -\sigma\frac{d(1/r)}{ds} \tag{12.16}$$

Hence, for a concave interface shape, the liquid pressure decreases in the direction of *decreasing* radius. Consider condensation on the base surface between the fins shown in Figure 12.17. The condensate interface radius at the centerline between the fins is very large as compared to the corner radius r_c at the root of the fin ($r \cong \infty$). Hence, condensate in the root region would be pulled into the corner. Similarly, condensate is pulled from the region near the fin base into the corner.

The larger the pressure gradient, the smaller the condensate film thickness (δ). Assuming that the film is laminar, the local condensation coefficient is given by $h = k_f/\delta$. Thus a strong surface-tension-induced pressure gradient will provide a high condensation coefficient. A unique profile shape is not required. Virtually any profile shape that has small tip radius, with increasing local radius along the surface (the s direction), will establish a surface tension drainage pressure gradient.

As previously noted, Gregorig [1954] was the first to propose that surface tension can enhance film condensation. His work addressed condensation on a convex surface profile. Later work by Adamek [1981] and Kedzierski and Webb [1990] developed advanced profile shapes. Webb et al. [1982] performed measurements on finned plates having rectangular fin shape and showed that special fin profile shapes are not required to obtain surface tension drainage. Hence, the prior work on profile shapes can be applied to both vertical and horizontal finned surfaces. Figure 12.18 shows a vertical finned plate consisting of convex surface-tension-drained profiles and concave condensate drainage channels. If the plate is vertical, surface tension force pulls the condensate (generated on the convex profile) horizontally to the drainage channel, where it is drained downward by gravity. Adamek and Webb [1990a] developed equations to predict the condensation coefficient in the flat-sided drainage channels. Both surface tension and gravity forces may be important in the drainage channel. Adamek and Webb [1990b] extended their work to predict condensation on horizontal integral-fin tubes. Honda and Nozu [1987a] have also developed prediction theory for horizontal integral-fin tubes. We will develop the theoretical equations for surface-tension-drained condensation on convex and flat-sided fins and in the drainage channel.

A number of quantitative, convex profile shapes have been described. Gregorig [1954] described a profile shape that gives constant film thickness. Smaller film

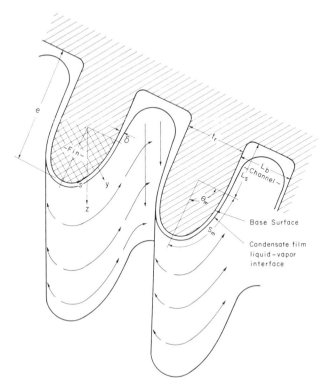

Figure 12.18   Illustration of finned plate with convex fin profiles designed for surface-tension-drained condensation, and rectangular condensate drainage channels.

thickness is given by smaller tip radius (r_o). Zener and Lavi [1974] defined a profile shape that gives constant pressure gradient. Consideration of the profile shape described by Zener and Lavi provides important fundamental understanding of the theory of surface-tension-drained condensation. Figure 12.19 shows a convex base profile, and a thin condensate film on the surface. The coordinate directions along the curved surface and normal to the surface are s and η, respectively. The condensate film (thickness δ) is surface-tension-drained at velocity $u(\eta)$ in the s direction. The Zener and Lavi profile has radius r_o at $s = 0$, and $r = \infty$ at $s = S_m$, where $\theta = \pi/2$. If profile shape is chosen, such that dp/ds = constant, Equation 12.12 shows that

$$\frac{dp}{ds} = \sigma \left[\frac{1/r(S_m) - 1/r(0)}{S_m - 0} \right]$$

(12.17)

$$= -\frac{\sigma}{r_o S_m}$$

Writing the momentum equation in the s direction of Figure 12.19, and assuming $dp/ds \gg (\rho_l - \rho_v)g$, and using the boundary layer approximations, the momentum equation in the s direction is

$$-\frac{dp}{ds} + \mu_l \frac{\partial^2 u}{\partial \eta^2} = 0$$

(12.18)

Comparison of Equation 12.18 with the momentum equation for gravity-drained condensation (Equation 12.2) shows that $-dp/ds$ in Equation 12.18 corresponds with the term $F_g = (\rho_l - \rho_v)g$ in Equation 12.2. The solution of Equation 12.2 for gravity-drained condensation is given by Equations 12.3, 12.4, and 12.5. The solution of

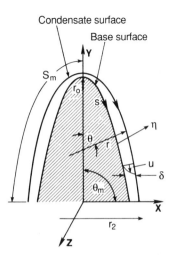

Figure 12.19 (a) Condensation on a profile having small tip radius, with increasing radius along the arc length.

the momentum equation for dp/ds = constant may be obtained from Equations 12.3, 12.4, and 12.5 simply by replacing $(\rho_l - \rho_v)g$ by $\sigma/r_o S_m$, and L by S_m. Thus, the average condensation coefficient over the arc length S_m for the Zener and Lavi profile is obtained from Equation 12.5 as

$$\bar{h} = 0.943 \left(\frac{k_l^3 \lambda \sigma}{v_l \Delta T_{vs} r_o S_m^2} \right)^{1/4} \tag{12.19}$$

The convex profile shape having dp/ds = constant leads to a simple solution for surface-tension-drained condensation. Examination of Equation 12.19 shows that a higher condensation coefficient may be obtained simply by reducing the tip radius, r_o.

Other profile shapes have been defined, which do not have dp/ds = constant. If $dp/ds \neq$ constant, the local condensate film thickness varies over the profile length. For this case, Adamek [1981] shows that the local film thickness is given by

$$\delta^4(s) = \frac{F_p}{\sigma} \left[\frac{d(1/r)}{ds} \right]^{-4/3} \int \left[\frac{d(1/r)}{ds} \right]^{1/3} ds \tag{12.20}$$

12.4.2 Adamek's Generalized Analysis

Adamek [1981] defined a family of convex interface profiles that support surface tension drainage. The family includes both the Gregorig [1954] and the Zener and Lavi [1974] profile shapes. The geometrical shape of this family is given by the equation

$$\frac{1}{r} = \left(\frac{\theta_m}{S_m} \right) \left(\frac{\zeta + 1)}{\zeta} \right) \left[1 - \left(\frac{s}{S_m} \right)^\zeta \right] \tag{12.21}$$

which is valid for $-1 \le \zeta \le \infty$. The relation between the tip radius (r_o), the total profile length (S_m), and the angle θ_m is given by

$$\frac{S_m}{r_o \theta_m} = \frac{\zeta + 1}{\zeta} \tag{12.22}$$

The local pressure gradient in the film is given by applying Equation 12.16 to Equation 12.21. Adamek used Equation 12.21 in Equation 12.20 to obtain the local film thickness, $\delta(s)$. The result is

$$\frac{1}{\delta} = 0.538 \left[\left(\frac{\sigma \theta_m}{F_p \Delta T_{vs}} \right) \left(\frac{(\zeta + 1)(\zeta + 2)}{S_m^{\zeta + 1} s^{2 - \zeta}} \right) \right]^{1/4} \tag{12.23}$$

By integrating Equation 12.23 over $0 < s < S_m$, Adamek obtained the average condensate film thickness. Using $\bar{h} = k/\delta_{av}$ he obtained

$$\bar{h} = 2.149 \frac{k_l}{S_m} \left[\left(\frac{F_p \theta_m S_m}{\Delta T_{vs}} \right) \left(\frac{\zeta + 1}{(\zeta + 2)^3} \right) \right]^{1/4} \tag{12.24}$$

If $\zeta = 1$ and $\theta_m = \pi/2$, one obtains the result for $dp/ds = $ constant, Equation 12.19. If $\zeta = 2$ and $\theta_m = \pi/2$, the result is for the $\delta = $ constant case described by Gregorig [1954].

Figure 12.20a shows the average condensation coefficient as a function of arc length (S_m), for $\theta = \pi/2$ and $-0.5 < \zeta < 4$. Figure 12.20b shows the shape of the surface for different values of ζ. The ζ parameter characterizes the aspect ratio of the fin cross section (height/thickness at S_m). The aspect ratio increases as ζ decreases.

In a mathematical sense, the term r in Equations 12.20 or 12.21 is the *radius of curvature*. The term *curvature* (κ) is the rate of change in direction of the local tangent to the curve per unit arc length. The curvature is the inverse of the radius of curvature, $\kappa = 1/r$. Thus, a curve shape having small local r will have large curvature. All of the profiles start with high curvature at the fin tip, which requires a small tip radius. The curvature decreases along the length of the profile, and has zero curvature (infinite radius) at θ_m. For a constant θ_m and S_m, Adamek shows that the maximum h will occur when $\zeta = -0.5$. Inspection of Equation 12.21 shows that when $\zeta \leq 0$, the initial curvature is as follows: $\kappa_o = 1/r_o = \infty$. The physical significance of $\kappa_o = \infty$ is that $dp/ds = \infty$ at the fin tip. The practical realization of $r_o = 0$ for $\zeta < 0$ is a physical impossibility—although it is mathematically possible.

Adamek's family of profiles has no special physical significance, other than the fact that the curves promote surface tension drainage from the tip to the fin base. It

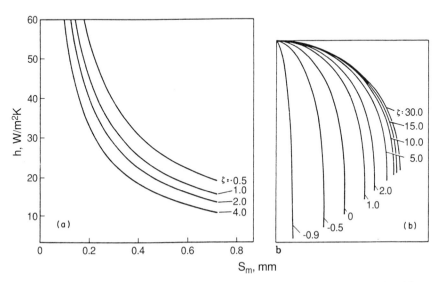

Figure 12.20 (a) Average condensation coefficient versus S_m. (b) Profile shape for different ξ values in Equation 12.19. (From Adamek [1981].)

is possible to identify other curve shapes. More discussion of alternate fin profiles will be presented in a later section.

Adamek's analysis assumes that surface tension dominates over gravity force over the entire profile length (e.g., Bo ≪ 1). This condition may not be satisfied over the total arc length, S_m. The Bond number of the Adamek profiles is given by

$$\text{Bo} = \frac{\rho_l g S_m^{\zeta+1}}{\sigma \theta_m (\zeta + 1) s^{\zeta-1}} \tag{12.25}$$

For dp/ds = constant, the Bond number is given by

$$\text{Bo} = (\rho_l - \rho_v) g \frac{r_o S_m}{\sigma} \tag{12.26}$$

The Bond number will decrease with increasing direction from the fin tip. At the end of the profile, Bo approaches infinity. This means that the condensate streamline will have a component in the vertical direction.

Kedzierski and Webb [1987] performed experiments to validate the theoretical predictions for the Equation 12.24 profile shapes. They used the electrostatic discharge machining (EDM) method with numerical control of the machining head, to make finned plates with profiles defined to high accuracy. They made finned plates similar to that illustrated in Figure 12.18 having the Figure 12.20b Gregorig and Adamek profiles (ζ = 2 and −0.05, respectively) machined into the face of a 102-mm-long, copper test section and tested with R-11 at 1.3 kPa. The theoretical models for the Gregorig and Adamek profiles are based on the curvature of the liquid–vapor interface. Hence, the shape of the base surface is slightly different from the interface shape. The Gregorig and Adamek theories were used to subtract the condensate film thickness from the interface surface to obtain the shape of the base surface. This profile shape was produced by the EDM method. The profiles consist of the convex arc (S_m) and a flat condensate drainage channel of depth L_s and width L_b. Because the Gregorig or Adamek theories are applicable only to the convex arc length (S_m), it was necessary to subtract the condensation rate in the drainage channel from the total. The model developed by Adamek and Webb [1990a] allows prediction of the condensation coefficient for both the convex profile and the drainage channels. Adamek and Webb showed very good agreement between the theoretically predicted and the experimental condensation rates on the finned blocks.

12.4.3 "Practical" Fin Profiles

There are two primary applications for surface-tension-drained condensation. The first is for a vertical tube (Figure 12.12) or for horizontal, integral-fin tubes (Figure 12.4).

Prior to the Webb et al. [1982] tests on plates having rectangular fins, designers generally did not recognize that surface tension governed the condensate drainage from the fin surfaces on horizontal integral-fin tubes. Figure 12.21 shows the cross section of the fins used on commercial integral-fin tubes. This figure shows that the fins are of a trapezoidal shape, and that the sides are flat. Table 12.7 shows the dimensions of the fins, as reported by Webb et al. [1985].

The basic geometry of the fins is defined by specifying the fin pitch (p_f), the fin height (e), and the fin shape. For the fins in Table 12.7, the shape is defined by the fin thickness at the base (t_b) and the tip (t_t). Generally, one will specify the minimum values of t_b and t_t that are manufacturable. Assume that we wish to make an integral-fin tube having 748 fins/m with 1.53-mm-high fins. If the Gregorig profile is used, its aspect ratio (fin height/fin base) is 0.75. Hence, the t_b would be 2.04 mm. This thickness is substantially greater than the 0.42-mm value of Table 12.7. Use of greater thickness values will waste material and cause greater condensate retention. For $e = 1.53$, the Gregorig fins are so thick that it would not be possible to have 748 fins/m. An Adamek $\zeta = -0.86$ profile would provide $e = 1.53$ and t_b

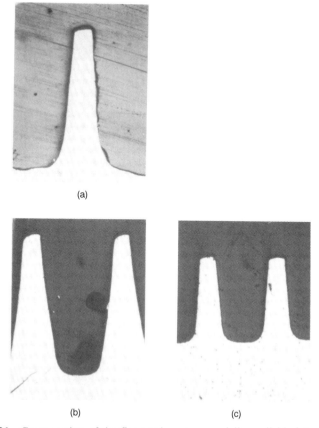

(a)

(b)   (c)

Figure 12.21   Cross section of the fins used on commercially available integral-fin tubes; (a) 748 fins/m, (b) 1024 fins/m, (c) 1378 fins/m.

Table 12.7 Dimensions of Commercially Available Integral-Fin Tubes

Fins/meter (fins/m):	748	1024	1378
Outside diameter, d_o (mm):	19.0	19.0	19.0
Area ratio, $A_o/(\pi d_o L)$:	2.91	3.60	3.18
Fin height, e (mm):	1.53	1.53	0.89
Fin thickness at tip, t_t (mm):	0.20	0.20	0.20
Fin thickness at base, t_b (mm):	0.42	0.52	0.29
Aspect ratio, e/t_b:	3.6	2.9	3.1

$= 0$. In the previous discussion, we have concluded that any Adamek profile having $\zeta \leq 0$ is not a practical choice, since it requires a zero-radius liquid–vapor interface at the fin tip. Practical minimum tip radii are probably limited to 0.05 mm. Hence, we quickly conclude that none of the existing theoretical profiles are practical for our design conditions!

Webb and Kedzierski [1990] have defined an alternate set of condensate surface profiles that allow one to independently specify the fin tip radius (r_o), the fin base thickness (t_b), the fin height (e), and the angle (θ_m). For the previously specified parameters, the fin profile shape is a function of the parameter Z, which describes the "fatness" of the profile. The equation describing the fin profiles is presented and discussed in Appendix A. If the fin profile is a continuous function, one may develop an analytical solution for the condensation coefficient, following the same procedure used by Adamek [1981] for his ζ profiles. The local condensate thickness is given by integrating Equation A.1 in Appendix A with respect to the angle θ for the profile of interest. The integration is

$$\delta^4(s) = \frac{F_p}{\sigma} \left[\frac{d(1/r)}{d\theta} \right]^{-4/3} \int_0^{\theta_m} \left[\frac{d(1/r)}{d\theta} \right]^{1/3} ds \qquad (12.27)$$

Equation 12.27 may be analytically integrated. Jaber and Webb [1993] give the result of the analytical integration of Equation 12.27. Then, the average condensation coefficient over the profile length, S_m, is given by $\bar{h} = k/\delta_{av}$, where

$$\delta_{av} = \frac{1}{S_m} \int \delta(s) \, ds \qquad (12.28)$$

Kedzierski and Webb [1990] have compared the conductance ($\bar{h}S_m$ for $\theta_m = \pi/2$ for several profiles. The results are shown in Table 12.8. For a fin base thickness of $t_b = 0.356$ mm, the value of $\bar{h}S_m$ of the Webb and Kedzierski profile is within 10% of the Adamek profile, which has zero tip radius. The $t_b = 0.356$-mm Gregorig profile has a small $\bar{h}S_m$ because of its small fin height. If the height of the Gregorig fin is set at 1.45 mm, $\bar{h}S_m$ will increase, but the base thickness will increase to 1.88 mm, which wastes a considerable amount of fin material and limits the fin pitch to considerably larger values than for the $t_b = 0.356$-mm fins.

Table 12.8 Comparison of $\bar{h}S_m$ on Three Profile Types

Profile	Parameter	e (mm)	t_b (mm)	$\bar{h}S_m$ (W/m-s)
Gregorig	$\xi = 2$	1.45	1.88	8.04
Gregorig	$\xi = 2$	0.28	0.356	5.31
Adamek	$\xi = -0.78$	1.45	0.356	9.45
Webb and Kedzierski ($r_o = 0.025$ mm)	$Z = 10$	1.45	0.356	8.45

An additional question of interest is: How precisely must the convex profiles be machined? Or, what variation in condensation coefficient will occur (for fixed e and t_b) as r_o or the shape of the profile (Z, θ_m) is varied? These questions were evaluated using the Kedzierski and Webb profiles. Table 12.9 shows the calculated results for a 1.07-mm fin height (e) and 0.356-mm base thickness (t_b). Lines 1–3 show that changing Z from 100 to 10 increases h by 10%. Increasing the tip radius from 0.025 to 0.051 mm decreases h by 3.0% (lines 2 and 4), and increasing the tip radius from 0.051 to 0.102 mm decreases h by 7.6% (lines 2 and 5). Decrease of θ_m from 90 to 85 degrees causes a 21% decrease (lines 1 and 6). Small r_o and Z provide higher performance. Manufacture of 0.051-mm fin tip radius is a practical possibility. Moderate variation of the tip radius and profile shape (Z) will have little effect on h. However, it is important to maintain θ_m close to 90 degrees.

12.4.4 Prediction for Trapezoidal Fin Shapes

It is important to be able to predict condensation on trapezoidal (or rectangular) fin shapes. These fins shapes cannot be written in terms of continuous functions as used by Adamek [1981] or by Kedzierski and Webb [1990]. Such fin shapes may have both surface-tension- and gravity-drained regions. We will describe both approximate and quite precise, but also complex, models.

12.4.4.1 Adamek and Webb Model   Figure 12.22a illustrates the Adamek and Webb [1990a] model, which is simple in concept. The model segments the fin into

Table 12.9 Effect of Parameters Z, Γ, and θ_m on $\bar{h}$ ($e = 1.07$ mm, $t_b = 0.356$ mm)

Line	Z	r_o (mm)	θ_m (degrees)	S_m (mm)	$\bar{h}$ (W/m²–K)
1	10	0.025	90	1.12	7520
2	50	0.025	90	1.14	7080
3	100	0.025	90	1.14	6790
4	50	0.051	90	1.15	6880
5	50	0.102	90	1.12	6360
6	50	0.051	85	1.11	5940

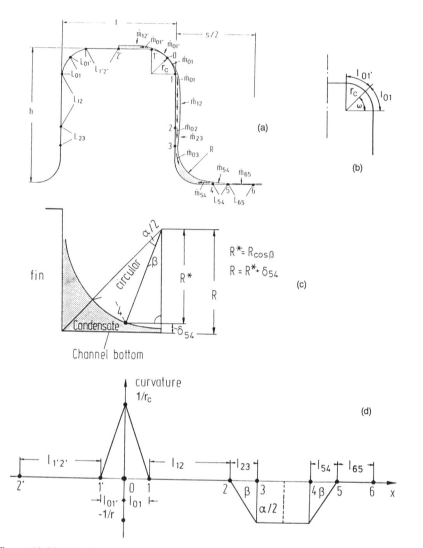

Figure 12.22 (a) Cross section of fin and root illustrating the defined condensate drainage regions (not to scale). (b) Detail of a fin tip corner for rectangular fin. (c) Detail of drainage in the corner of a drainage channel. (d) Curvature of the condensate film in the drainage regions defined in Fig. 12.21a. (From Adamek and Webb [1990b].)

several surface-tension- or gravity-drained zones. We will illustrate the basic concept of the model for condensation on a vertical plate having rectangular fins of pitch p_f, thickness t_f, and height e. The various drainage regions are noted on Figure 12.22a. Linear surface tension gradients are assumed for the surface-tension-drained zones. The fin tip has a corner radius of r_c. Equation 12.3, which gives the local film thickness for gravity-drained condensation at location x, may be written in the general form

$$\delta(x) = \left(\frac{F_p x}{F_d}\right)^{1/4} \tag{12.29}$$

where x is the coordinate in the drainage length, and F_d is the drainage force. If the region is gravity-drained, $F_d = g(\rho_l - \rho_v)$, and if it is surface-tension-drained, $F_d = \sigma(1/r_1 - 1/r_2)/\Delta x$. The r_1 and r_2 refer to the local film radius at the beginning and end of the region, respectively. Surface tension pulls the condensate from the tip radius r_o and into the drainage corner, containing a concave film of radius R. Regions that have no change of film radius are gravity-drained. Gravity drains regions $L_{1'2'}$, L_{12}, L_{34}, and L_{65} of Figure 12.22a. Table 12.10 defines the drainage length and the drainage force (F_d) for the several drainage regions illustrated in Figure 12.22a. Figure 12.22b shows detail of a fin tip corner for a rectangular fin, Figure 12.22c shows detail of drainage in the corner of a drainage channel, and Figure 12.22d shows the curvature $(1/r)$ of the condensate film for each of the regions denoted on Figure 12.22a.

Knowledge of the local film thickness permits calculation of the local condensation coefficient, since $h(x) = k_l/\delta(x)$. Regions $L_{1'2'}$, L_{12}, L_{23}, and L_{54} do not start from zero film thickness, so Equation 12.29 must be modified to account for the film thickness at the beginning of the zone. Consider a region L_{jk} that bounds upstream region L_{ij}, on which condensate flow $\dot{m}_{ij}$ was generated. Adamek and Webb [1990a] show that

$$\delta_k(x) = \left(\frac{3v_l\dot{m}_{ij}}{f_{ij}} + \frac{F_p x}{f_{jk}}\right)^{1/4} \tag{12.30}$$

Equation 12.30 may be appropriately used to calculate the film thickness of regions $L_{1'2'}$, L_{12}, L_{23} and L_{54}. Readers are referred to Adamek and Webb [1990a,

Table 12.10   Definition of Drainage Regions for Figure 12.22

Region	x	F_d
1'2'	$t - r_o$	ρg
01'	$\pi/4r_o$	$\sigma/r_o l_{01'}$
01	$\pi/4r_o$	$\sigma r_o l_{01}$
12	$e - r_o - l_{23}$	ρg
23	$4[F_p \Delta T R_c/\sigma]^{1/6}$	$\sigma/R l_{23}$
34	$\pi/4R$	None
54	l_{23}	$\sigma/R l_{54}$
65	$p_f - t_f - R - \delta_{54}$	ρg

1990b] for the calculation details. The average condensation coefficient for the surface is given by

$$\bar{h} = \frac{\Sigma h_{ij} l_{ij} \Delta T_{ij}}{l_{ij} \Delta T_{ij}} \tag{12.31}$$

where the summation refers to all of the L_{ik} drainage zones shown in Figure 12.22a. Adamek and Webb [1990a, 1990b] have validated the theory with good success for vertical finned plates and horizontal finned tubes.

12.4.4.2 Approximate Model   It is desirable to have a simple model to predict the condensation coefficient on such trapezoidal fins such as those of Figure 12.21. A good approximation is illustrated in Figure 12.23a. A K–W profile of tip radius r_c is fitted to the fin sides. The flat top on the fin is of width $(t_t - 2r_c)$. The base thickness of the K–W profile is $[t_b - (t_t - 2r_c)]$. The condensation coefficient on the fin side is predicted for this K–W profile. The condensation coefficient on the fin tip is predicted using the same procedure as Adamek, and is described in Section 12.2.4.1. Figure 12.23b shows the nature of the approximation at the fin corner. The dashed line part of the K–W profile is shown only for illustration purposes and is not part of the model. Figure 12.23c shows the change of curvature over the fin,

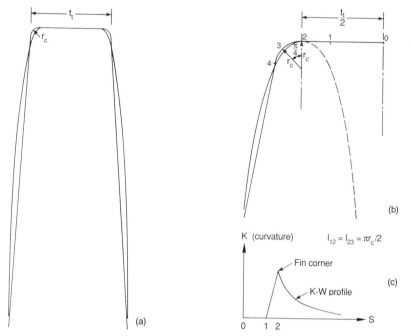

Figure 12.23   (a) Kedzierski–Webb profile fitted to the fin side. (b) Approximation at fin tip. (c) Change of curvature over fin surface.

where point 0 is the center of the fin tip. Region L_{01} is gravity-drained, and regions L_{12} and L_{2s} (fin side) are surface-tension-drained.

An alternate approach is to fit the Adamek [1981] ζ profile having the aspect ratio, $e/(t_b - t_t + 2r_c)$, to the fin side. This is probably not as good an approach as use of the K–W profile, because the Adamek profile results in a zero radius at the fin tip, rather than the actual corner radius, r_c.

More precise, but more complex models have been developed for prediction of the condensation coefficient on fins of trapezoidal and rectangular cross section. These models are due to Honda and Nozu [1987a] and Adamek and Webb [1990a, 1990b].

12.5   HORIZONTAL INTEGRAL-FIN TUBE

The first analytical model for horizontal integral-fin tubes was developed by Beatty and Katz [1948]. This model assumes that gravity force drains the condensate from the fins. Later work by Webb et al. [1982] showed that surface tension force, rather than gravity force, drains the condensate from the fins. This observation led the way for application of surface-tension-drained models to integral-fin tubes. The first simple surface tension drainage model was proposed by Karku and Borovkov [1971]. Rudy and Webb [1983, 1985] showed that the lower part of the fin circumference is flooded by condensate, and that the model should account for this condensate flooding. The first model to incorporate surface tension drainage and condensate retention was developed by Webb et al. [1985]. Later work by Honda and Nozu [1987a] and Adamek and Webb [1990b] bring us to the most advanced models for integral-fin tubes. This section will survey these models and advances.

12.5.1   The Beatty and Katz Model

Beatty and Katz [1948] proposed that the heat transfer may be modeled by use of the Nusselt [1916] equations for condensation on horizontal tubes and vertical plates. The model assumes that gravity force drains the condensate from the fins, and that no condensate retention occurs on the lower side of the tube. The area-weighted condensation coefficient (h) is calculated as the average on the finned surface (h_f), and on the tube surface between the fins (h_h) as given by

$$h\eta = h_h \frac{A_r}{A} + \eta_f h_f \frac{A_f}{A} \tag{12.32}$$

where $A = A_f + A_r$. They used Equations 12.5 and 12.7 to calculate h_f and h_h, respectively. Beatty and Katz assumed that the characteristic length (L) needed in Equation 12.5 is given by

$$L = \frac{\pi(d_e^2 - d_o^2)}{4d_e} \tag{12.33}$$

Their test data for six low-surface-tension fluids condensing on several finned tube geometries (433 to 633 fins/m) were predicted within ±10% using Equation 12.32.

The Beatty and Katz [1948] model is fundamentally incorrect, for two reasons: First, it does not account for the condensate flooded zone illustrated by Figure 12.5. Second, surface tension, rather than gravity, drains the condensate from the fins. Webb et al. [1982] condensed R-12 on the face of a 51-mm-diameter cylinder, in which fins of rectangular cross section were machined (Figure 12.24a). The fins were 1.0 mm high and 0.30 mm thick at 0.84-mm pitch. The finned surface was tested in the three geometric orientations listed in Table 12.11. Figure 12.24b shows the test results, and the predicted condensation coefficient using the Nusselt equa-

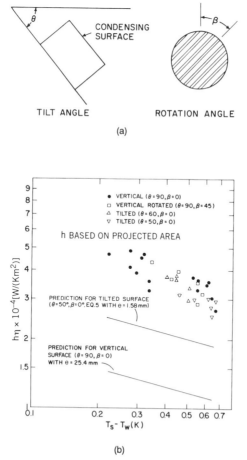

Figure 12.24 Finned plate tests conducted by Webb et al. [1982] which prove that surface tension forces drain the condensate from integral-fins. (a) Illustration of test surface. (b) Test data. (From Webb et al. [1982].)

tion for gravity drained condensation (Equation 12.5). The different tilt and rotation angles change the characteristic length (L). The characteristic length associated with the several plate orientations is listed in Table 12.11. Figure 12.24b shows the following:

1. The measured condensation coefficient is essentially independent of geometric orientation.
2. The data are underpredicted 70% by Equation 12.5, as applied to a finned plate.

These results show that the condensate drainage from the fin surface is not controlled by gravity. Rather, it is controlled by surface tension force. Thus, the assumption of gravity drainage in the Beatty and Katz [1948] model is invalid. The Figure 12.24 data argue that the theoretical relations for condensation on surface-tension-drained fin profiles may be applied to the fins on integral-fin tubes.

12.5.2   Precise Surface-Tension-Drained Models

Honda and Nozu [1987a] describe a more complex model, which is applicable to the trapezoidal cross-section fins illustrated in Figure 12.21. Figure 12.25 defines the geometry, on which their model is based. The condensation coefficient is predicted for two regions: the "u region" (unflooded circumferential fraction) and the "f region" (flooded fraction) shown in Figure 12.25a. These regions are illustrated by Figures 12.25b and 12.25c, respectively. The overall structure of their model is

$$h = \frac{(1 - c_b)h_u\eta_u\Delta T_u + c_b h_f\eta_f\Delta T_f}{(1 - c_b)\Delta T_u + c_b\Delta T_f} \tag{12.34}$$

where the first term in the numerator is for the u region, and the second term is for the f region. Equation 12.34 recognizes that the fin efficiency and the $(T_s - T_w) \equiv \Delta T_{vs}$ in the u and f regions are different (e.g., $\Delta T_{vs,u}$ and $\Delta T_{vs,f}$). Equations for calculation of the fin efficiency in the two regions are given by the authors. The heat transfer coefficient is based on the nominal area, $\pi d_e L$. The model predicts the condensation coefficient, based on surface tension drained theory, for the

Table 12.11   Summary of Tests Conducted by Webb et al. [1982] on the Figure 12.24 Finned Plate

θ (degrees)	β (degrees)	L (m)
90	0	25.4
90	45	25.4
50	0	1.58
60	0	1.60

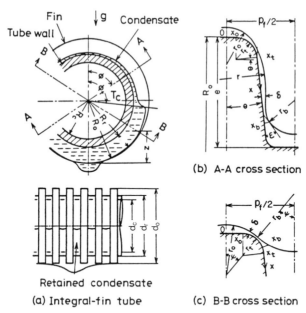

(b) A-A cross section

(a) Integral-fin tube

Retained condensate

(c) B-B cross section

Figure 12.25 Physical model of Honda, Nozu and Takeda [1987a]. (From Honda et al. [1987a].)

condensate–vapor interface surfaces illustrated in Figures 12.25b and 12.25c. Note that this includes the interface in the root area between the fins.

The prediction of h in the u region breaks the interface into four length increments: the fin tip, the corner of the fin tip, and the upper and lower regions of the fin side. A two-region model is used for the f region. Condensation is neglected on the condensate surface in the flooded region, but it is evaluated at the fin tips. The reader should consult Honda and Nozu's [1987b] paper for the details of the predictions. The model recognizes that the tube surface temperature is different in the u and f regions, due to circumferential heat conduction in the tube wall thickness.

The values of the wall temperatures $T_{w,u}$ and $T_{w,f}$ were determined by solving the circumferential heat conduction equation in the tube wall, assuming constant heat transfer coefficients in the u and f regions, and neglecting radial conduction. Analytical solutions were obtained, and are given by Honda and Nozu [1987a]. Because the solution depends on the fin efficiency and the heat transfer coefficient, an iterative solution is necessary. Once converged values of $T_{w,u}$ and $T_{w,f}$ have been obtained, the mean tube wall temperature, $T_{w,m} = (T_{w,u} + T_{wf})/2$ is calculated. The local heat flux is calculated using Equation 12.34 with temperature difference $T_v - T_{w,m}$. Honda et al. [1987b] have improved the model to account for (a) surface tension drainage in the root region (important for wide fin spacing) and (b) the wall temperature variation between the fin root and the tube wall between the fins. The improved model slightly reduces the error band, relative to the Honda and Nozu [1987a] model.

12.5.3 Approximate Surface-Tension-Drained Models

Because of the complexity of the precise models described in Section 12.5.2, it is desirable to have a simpler model for use, where the complex computer models are not available. This section describes a recommended approximate model. A valid model must account for the effect of condensate retention. It should also account for surface tension drainage from the fins. In currently unpublished work, Jaber and Webb have developed an improvement on the model of Webb et al. [1985]. The basic structure of the present model is given by

$$\eta hA = (1 - c_b)[h_h A_r + \eta_{fu}h_{fu} A_f] + c_b\eta_{ff} h_{ff} A_{ft} \qquad (12.35)$$

where $A = A_{fs} + A_{ft} + A_r$ (fin sides, fin tip, and root region), and the finned area is $A_f = A_{fs} + A_{ft}$.

The improvements relative to the Webb et al. [1985] model relate to calculation of the terms h_{fu} and h_{ff}. The first term in square brackets on the right hand side is for the unflooded fraction of the tube circumference, and the second term is for the condensate flooded region. The condensate bridged surface fraction is $c_b = \alpha/\pi$, and the unflooded fraction is $1 - c_b$. The c_b is calculated using Equation 12.9. The condensation coefficients in the unflooded region are h_r (the root tube) and h_{fu} (the fin). The h_{fu} term is composed of condensation contributions on the fin side (h_{fus}) and the fin tip (h_t). The composite h_u is given by

$$h_{fu}l_u = h_{fus}e + h_t t_t \qquad (12.36)$$

h_{fus} is predicted for a Kedzierski–Webb profile of height e and base thickness $t = t_b - t_t$. This methodology was discussed in Section 12.4.4.2. The condensation coefficient on the fin tip (h_t) is calculated per Adamek and Webb [1990b] and is described in Section 12.4.4.1.

If the fin pitch is relatively small (e.g., 1.3 mm), condensation in the root region will be gravity-dominated. Then, the condensation coefficient on the base tube (h_h) is calculated from the Nusselt equation for horizontal tubes (Equation 12.7), taking account of the additional condensate drained from the fin profile to the base tube. Equation 12.7 may be written in terms of the condensate Reynolds number as

$$h_h = 1.51 \left(\frac{\mu_l^2 \mathrm{Re}_l}{k_l^3 \rho_l^2 g} \right)^{-1/3} \qquad (12.37)$$

and the condensate Reynolds number is

$$\mathrm{Re}_l = \frac{4\dot{m}_r}{\mu_l(p_f - t_b)} \qquad (12.38)$$

The term $\dot{m}_r$ in Equation 12.38 is the sum of the condensate formed on the fin ($\dot{m}_f$) and that generated on the base tube between the fins ($\dot{m}_h$). Typically, $\dot{m}_f \gg \dot{m}_h$ so that small error would occur by neglecting $\dot{m}_h$. However, a rigorous solution is possible, and is described by Webb et al. [1985]. For wider fin pitches (e.g., greater than 1.3 mm) one may account for surface tension drainage in the fin root region.

Condensation occurs in the flooded region on the fin tip and on the fin corners. The h_{ff} in Equation 12.34 is given by

$$h_{ff}\, l_{ff} = h_{ffs}(\pi r_c/2) + h_t(t_t - \pi r_c/2) \qquad (12.39)$$

where h_{ffs} and h_t are given by k/δ. For h_{ffs}, δ is calculated using Equation 12.29 with $x = \pi/(4r_o)$ and $F_d = \sigma(1/r_o + 2/s)$. For h_t, δ calculated with Equation 12.29 and Table 12.10 using $l_{1,2}$, and l_{01}.

The fin efficiency for the unflooded region (h_{fu}) is calculated using a standard fin efficiency equation. The fin efficiency in the flooded region (h_{ff}) is calculated as described in Appendix B.

Webb et al. [1985] predicted h_f using an Adamek ζ profile, which fit the aspect ratio of the actual fin cross section (approximately, $\zeta = 0.8$). They neglected condensation on the fin tip, in the flooded region ($h_{ff} = 0$). Their R-11 data for condensation on 748, 1024, and 1378-fin/m tubes were predicted within $\pm 20\%$. Using the more refined model, Jaber and Webb [1993] predicted the same data within $\pm 13\%$. The model predicted the condensing coefficient of steam, refrigerant R-11, and refrigerant R-113, a variety of tubes, within $\pm 15\%$. The tube materials included copper and the Figure 12.9a–e Cu–Ni, stainless steel finned tubes.

12.5.4   Comparison of Theory and Experiment

Honda and Nozu [1987a] predicted the condensation coefficients for 22 different integral-fin tubes and 12 different fluids using their model and the Webb et al. [1985] model. The test fluids and tube geometries include many of those listed in Table 12.2. The Honda and Nozu model predicts most of the data (including steam) within $\pm 20\%$. The Rudy and Webb [1985] model also predicted most of the low-surface-tension fluid data within $\pm 20\%$. The steam data were over predicted 30–40%. The Adamek and Webb (1990b) model was used to predict data for seven different fluids (both low surface tension and steam) on 80 different finned tube geometries. The data were predicted within $\pm 15\%$. Hence, the Adamek and Webb [1990b] and the Honda and Nozu models have comparable predictive ability. Sukhatme et al. [1990] tested the ability of the above models to predict his R-11 data for four 1417-fin/m tubes having fin heights between 0.46 and 1.22 mm. They found that the simple Webb et al. [1985] model was as good as the Honda and Nozu [1987a] model for their particular test conditions.

12.6   HORIZONTAL TUBE BANKS

Tube banks are subject to effects of condensate inundation and vapor velocity. Condensate inundation data account for the drainage of condensate from the upper tube rows, without vapor velocity effects. Condensate inundation reduces the condensation coefficient. Vapor velocity generally exerts a shear stress on the liquid film, which acts to thin the film. This will enhance the condensation coefficient, unless the vapor shear opposes gravity. Vapor shear data may be provided for a single horizontal tube, without inundation. An actual condenser tube bundle experiences both inundation and vapor shear effects. Problems involving *both* condensation and vapor shear are more properly classified under "convective condensation," which is discussed in Chapter 14.

Webb [1984] provides a survey of inundation and vapor velocity effects for refrigerant condensers. A recent paper by Michael et al. [1989] provides an update of recent work on enhanced tubes.

12.6.1 Condensate Inundation without Vapor Shear

Much work has been done to study condensation on single, horizontal enhanced tubes. However, very few studies have been undertaken to determine the "row effect" exponent for enhanced tubes. The row effect is defined as

$$\frac{\overline{h_N}}{h_1} = N^{-m} \tag{12.40}$$

where $\overline{h}$ is the average condensation coefficient for N-tube rows, and h_1 is for a single tube. The exponent m determines the row effect. Kern [1958] has recommended $m = 1/6$ as an approximate value for plain tubes. Katz and Geist [1948] measured the row effect for six rows of finned tubes having 590 fins/m and 1.6-mm fin height using R-12, n-butane, acetone, and water. They found $m = 0.04$. Figure 12.26 shows row effect data for steam condensation on plain and enhanced tubes. The plain tube data of Brower [1985] reasonably agree with the Kern recommendation of $m = 1/6$. Marto and Wanniarachchi [1984] measured the row effect for a wire-wrapped plain 16-mm-diameter tube (1.6-mm wire diameter, 8.0-mm wire pitch) and found $n = 0.025$. Marto [1986] also found $m = 0.04$ for an integral-fin tube (400 fins/m, with 1.0-mm fin spacing and fin height).

Webb and Murawski (1990) have measured the row effect for R-11 condensing on four enhanced tube geometries (1024 standard integral fin, Tred-26D™, Turbo-B™, and GEWA-SC™). The R-11 was condensed on a vertical rank of five horizontal tubes. The tube geometries are shown in Figure 12.6 and described in Table 12.4. The single tube performance of these tubes was shown on Figure 12.8. Figure 12.27 shows the row effect plotted in the form of the individual tube condensation coefficient versus condensate Reynolds number leaving the tube. The average condensation coefficient on N-tube rows is given by

$$\overline{h_N} = a\,\mathrm{Re}_l^{-n} \tag{12.41}$$

where a/Re_l^n is the equation of the lines in Figure 12.27. Figure 12.27 shows that the data for all tube rows of a given tube geometry fall on a single curve, when plotted in the h_N versus Re_l format. Because the h_N depends only on Re_l, and not on

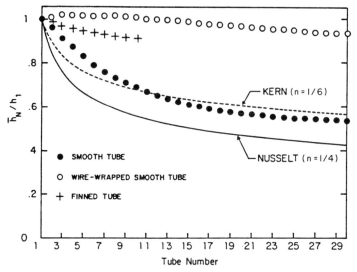

Figure 12.26 Condensation row effect exponent for steam condensation at 1 bar. (From Marto and Wanniarachchi [1984].)

tube rows, Figure 12.27 may be interpreted for any number of tube rows within the Re_l range of the data. Table 12.12 gives curve fits of straight lines drawn through the h_N versus Re_l curves of Figure 12.27.

The standard 1024-fin/m tube shows no row effect within the Re_l range tested. Honda et al. [1987a] condensed R-113 on a rank of standard 1024-fin/m tubes, and

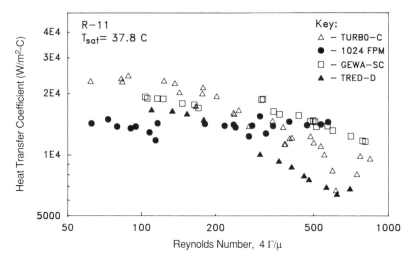

Figure 12.27 Row effect data for the Figure 12.8 enhanced tube geometries. (From Webb and Murawski [1990].)

Table 12.12   Exponent _n_ and Constant _a_ in eq. 12.41

Geometry	a × 10⁻³	n
1024 fins/m	12.90	0.000
GEWA-SC™	54.14	0.220
Turbo-C™	257.80	0.507
Tred-D™	269.90	0.576

also found zero row effect. The GEWA-SC™ tube had the second best single-tube performance and the second best row effect ($n = 0.22$). The greatest row effect was displayed by the Tred-D tube, having a Reynolds number exponent of $n = 0.58$. Although the Turbo-C™ tube showed the highest single-tube performance, its row effect ($n = 0.51$) is only a little better than that of the Tred-D. These row effect exponents may be compared to that for a plain tube, which has $n = 1/3$, according to the Nusselt theory and given by Holman [1986].

It appears that the superior row effect of the integral-fin tube occurs because the high, solid fins prevent axial spreading of the condensate. Hence, the condensate is channeled over the tube, leaving a region free of condensate between the condensate drainage columns. The poorest row effect was yielded by the geometries having a sawtooth fin shape. It appears that the sawtooth cuts in the fins allow axial spreading of the condensate, which drips from the tube row above.

If the condensate Reynolds number leaving the last row of a bank of tubes is known, one may find the average condensation coefficient on the bank of N-tube rows by integrating Equation 12.41. The result is given by

$$\overline{h_N} = \frac{a(\mathrm{Re}_{l,N}^{-n} - \mathrm{Re}_{l,1}^{-n})}{(1 - n)(\mathrm{Re}_{l,N} - \mathrm{Re}_{l,1})} \tag{12.42}$$

where $\mathrm{Re}_{l,N}$ and $\mathrm{Re}_{l,1}$ are the condensate Reynolds numbers leaving the Nth and the first tube rows, respectively. An example is given later, which shows how the bundle average condensation coefficient may be calculated using Equation 12.42.

It is interesting to compare the condensate flow rates for steam and R-11 condensing at the same heat flux. For condensation at 38°C, the R-11 condensate Reynolds number will be 25 times greater, and the condensate volume flow rate will be 9.5 times greater. Hence, the condensate flow patterns may be considerably different in R-11 and steam condensers for the same heat flux.

12.6.2   Condensate Drainage Pattern

Honda et al. [1987a] have studied the condensate flow pattern for a bank of integral-fin tubes ($d_o = 15.8$ mm, $e = 1.4$ mm, 1063 fins/in). Condensate was pumped over a rank of three tubes, with vertical tube pitches of 22 or 44 mm using R–113, methanol and n-propanol. They defined four condensate flow patterns: droplet mode, column mode, column and sheet mode, and sheet mode. Figure 12.28

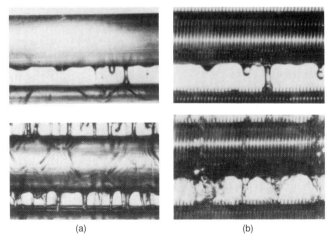

(a)   (b)

Figure 12.28 Flow patterns observed by Honda et al. [1987a] for R-113 condensating at 0.12 MPa on 15.9 diamter plain and 1060 fins/m integral-fin tubes. The top and bottom figures are for the first and 13th rows, respectively. (a) Plain tubes: First row ($Re_1 = 19$), 13th row ($Re_1 = 150$), (b) 1024 fins/m integral-fin tubes: First row ($Re_1 = 50$), 13th row ($Re_1 = 550$). (Photographs courtesy of Professor H. Honda).

illustrates the droplet and columnn modes on plain and 1060 integral-fin tubes. They analyzed their results to define the flow pattern transitions and the characteristics of the condensate flow. Figure 12.29 shows their flow pattern map for R–113 at 22-mm tube pitch. Assuming that the parameters Γ, σ, ρ, and g are important, dimensional analysis shows that the following dimensionless group is significant:

$$K = \left(\frac{\lambda}{\sigma^{3/4}}\right)\left(\frac{g}{\rho_l}\right)^{1/4} \qquad (12.43)$$

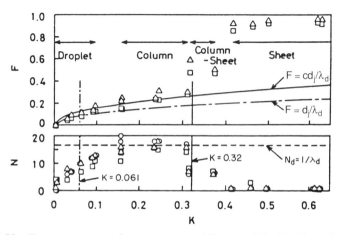

Figure 12.29 Flow pattern map of parameters F and N versus K for R-113 condensing on a 1060-fin/m tube. (From Honda et al. [1987b].)

The parameter N in Figure 12.29 is the number of axial droplet release locations along the tube length (160 mm). The parameter F is defined as d_j/λ_d, as illustrated in Figure 12.29. The λ_d parameter is the one-dimensional Taylor instability wavelength, which is defined in the Nomenclature. As shown by Figure 12.30, the F parameter is the fraction of the tube length that is totally condensate-flooded. This fraction of the tube length will have a very small condensation coefficient. They show that $F = cd_j/\lambda_d$, where d_j is the column diameter and $c = 1.5$ for R-12. $F = cd_j/\lambda_d$ is plotted as the upper solid line on the F versus K curve of Figure 12.29. The authors also present an analytical equation to predict the column diameter, d_j.

Considerable work has been done to determine the effect of vapor velocity for condensation on plain tubes. Much of this work is summarized by Fujii [1991]. Several studies have been done for integral-fin tubes. Recent work includes Lee and Rose [1984], and Michael et al. [1989] used R-113 vapor flow normal to a single horizontal tube. Honda et al. [1991, 1992] studied R-113 condensing in downward flow over inline and staggered finned tube bundles, respectively. Michael et al. tested downward vapor Reynolds number ($Re_v = d_o u_\infty/\mu_l$) for R-113 and steam. The plain and finned tubes had a 19-mm outer diameter. The finned tubes had 1.0-mm fin height and fin thickness, and three finned tubes were used (200, 400, and 800 fins/m). They found that vapor velocity benefits both finned and plain tubes. Figure 12.31a shows how the U value is influenced by vapor velocity (u_∞) for 3-m/s water-side velocity. Figure 12.31b shows how the enhancement ratio (E_o) is affected by Re_v for R-12. Figure 12.31b shows that the enhancement ratio tends to decrease with increasing vapor velocity. The reason for the decrease of E_o is that the smooth tube benefits more from vapor velocity than does the finned tube.

12.6.3 Prediction of the Condensation Coefficient

Having developed their analytical model to predict the condensation coefficient on a single-finned tube, and established the condensate flow pattern for a bank of tubes,

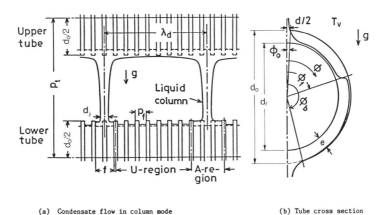

(a) Condensate flow in column mode   (b) Tube cross section

Figure 12.30 Illustration of liquid flow in column region and definition of flow pattern parameters shown in Figure 12.29. (From Honda et al. [1987b].)

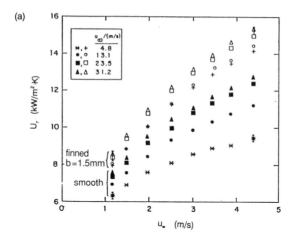

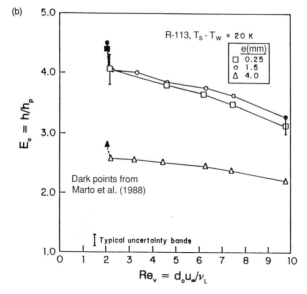

Figure 12.31 (a) Effect of vapor velocity (u_∞) on U_r for 3-m/s water-side velocity (b) Variation of the enhancement ratio with the vapor Reynolds number for R-113 condensing on finned tubes having 1.0-mm-thick square fins with fin spacings of 0.25, 1.5, and 4 mm. (From Michael et al. [1989].)

Honda et al. [1987b] proceeded to predict the condensation coefficient for each tube row. Their predictive model is presently limited to the integral-fin tubes. The basic structure of their model is as follows. The tube length between the draining condensate columns/droplets (λ_d) is divided into the U and A regions illustrated on Figure 12.30. The A region is that receiving condensate from the draining columns. The U region does not receive condensate from the tube above, and essentially acts as it were in the top tube row. The condensate behavior on the tube circumference is influenced by the condensate retention angle. Hence, the U region is divided into region U_u (unflooded) and the U_f (flooded region), where flooding is defined by the condensate retention angle. The A region similarly contains regions A_u (unflooded) and A_f (flooded). Their model assumes that the condensate retention angle is the same in the U and A regions. Each of these four regions is modeled as one of the four cases illustrated in Figure 12.32. In the U_f and A_f regions, only case D of Figure 12.32 is possible. Case A, B, or C would typically apply to the U_u and A_u regions, depending on the fin height and condensation rate.

Having defined the four tube regions, they applied the model of Honda et al. [1987c] to the U_u and U_f regions. The same model was applied to the A_u and A_f regions, with correction for condensate inundation. They used the model to predict the condensation coefficient for a multi-row case (6–9 rows) of R-12 and 12 rows of steam condensation. The R-12 data were predicted within 12%, and the steam data were 5–20% lower than the experimental value. The Adamek and Webb [1990b] model may be equally applied to predicting the condensation coefficient on a bank of integral-fin tubes.

The Honda et al. [1987c] model for a bank of tubes is presently limited to

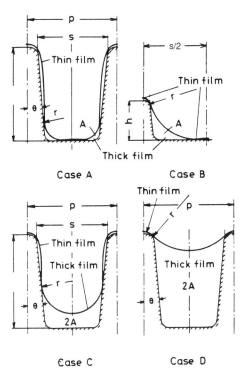

Figure 12.32 Four possible cases of condensate profile on a bank of integral-fin tubes. (From Honda et al. [1987b].)

integral-fin tubes, which neatly channel the condensate and which have two-dimensional condensate profiles within the fin region. No model exists for tubes such as the Turbo-C™ tube (cf. Figure 12.6) which have sawtooth fin shapes. Although the GEWA-SC™ tube of Figure 12.6 will have a two-dimensional condensate profile, it may not channel the condensate in the same manner as occurs on the integral-fin tube. Hence, flow pattern studies would be required to characterize this tube. It should be possible to develop a single-tube model, but such work has not yet been done.

12.7 CONCLUSIONS

Surface tension forces are dominant in draining the condensate from the fins on plates and tubes having typical commercial fin geometries. With the geometry of the condensate–vapor interface defined, models exist to predict the condensation coefficient.

Finned horizontal tubes experience condensate retention, which fills the interfin region around a fraction of the tube circumference. Analytical models have been developed to predict the condensate retention angle for typically used integral-fin tubes.

Experiments have been performed to measure the "row effect" on enhanced horizontal tube geometries. For integral-fin tubes, these experiments show that the row effect loss is smaller than that of a plain tube. The enhanced tube geometries having a sawtooth fin shape show a greater row effect loss than do integral-fin tubes.

Work has been done to define the flow patterns that exist on a bank of integral-fin tubes. Honda et al. [1989] have shown excellent success in predicting the condensation coefficient on a bank of integral-fin tubes.

Future research is needed to characterize and predict the condensation coefficient for the more complex surface geometries (e.g., those having a sawtooth fin shape).

12.8 REFERENCES

Adamek, T. A., 1981. "Bestimmung der Kondensationgrossen auf feingewellten Oberflachen zur Ausle-gun aptimaler Wandprofile," *Wärme-und-Stoffübertragung*, Vol. 15, pp. 255–270.

Adamek, T. A., and Webb, R. L., 1990a. "Prediction of Film Condensation on Vertical Finned Plates and Tubes—A Model for the Drainage Channel," *International Journal of Heat and Mass Transfer*, Vol. 33, No. 8, pp. 1737–1749.

Adamek, T. A., and Webb, R. L., 1990b. "Prediction of Film Condensation on Horizontal Integral-fin Tubes," *International Journal of Heat and Mass Transfer*, Vol. 33, No. 8, pp. 1721–1735.

Beatty, K. O., Jr., and Katz, D. L., 1948. "Condensation of Vapors on Outside of Finned Tubes," *Chemical Engineering Progress*, Vol. 44, No. 1, pp. 908–914.

Briggs, A., Wen, X. L., and Rose, J. W., 1992. "Accurate Measurements of Heat-Transfer

Coefficients for Condensation on Horizontal Integral-Fin Tubes," *Journal of Heat Transfer*, Vol. 114, pp. 719–726.

Brower, S. K., 1985. "The Effects of Condensate Inundation on Steam Condensate Heat Transfer in a Tube Bundle," M.S. Thesis, Naval Postgraduate School, Monterey, CA.

Brown, C. D., and Matin, S. A., 1971. "The Effect of Finite Metal Conductivity on the Condensation Heat Transfer to Falling Water Rivulets on Vertical Heat Transfer Surfaces," *Journal of Heat Transfer*, Vol. 93, pp. 69–76.

Butizov, A. I., Rifert, V. G., and Leont'yev, G. G., 1975. "Heat Transfer in Steam Condensation on Wire-Finned Vertical Surfaces," *Heat Transfer—Soviet Research*, Vol. 7, No. 5, pp. 116–120.

Carey, V. P., 1992. "Liquid–Vapor Phase-Change Phenomena," Hemisphere Publishing Corp., Washington, D.C.

Cary, J. D., and Mikic, B. B., 1973. "The Influence of Thermocapillary Flow on Heat Transfer in Film Condensation," *Journal of Heat Transfer*, Vol. 95, pp. 20–24.

Carnavos, T. C., 1974. Chapter 17 in *Heat Exchangers: Design and Theory Sourcebook*, Ed., N. Afgan and E. U. Schlünder, McGraw-Hill, New York.

Carnavos, T. C., 1980. "An Experimental Study: Condensing R-11 on Augmented Tubes," ASME paper 80-HT-54.

Collier, J. G., 1981. *Convective Boiling and Condensation*, McGraw–Hill, New York, pp. 332–340.

Combs, S. H., and Murphy, R. W., 1978. "Experimental Studies of OTEC Heat Transfer Condensation of Ammonia on Vertical Fluted Tubes," *Proceedings of the 5th OTEC Conference.*, Feb. 20–22, 1978, Miami Beach, FL, Vol. l, Sect. 6, pp. 111–122.

Dipprey, D. F., and Sabersky, R. H., 1963. "Heat and Momentum Transfer in Smooth and Rough Tubes at Various Prandtl Numbers," *International Journal of Heat and Mass Transfer*, Vol. 6, pp. 329–353.

Fujii, T., 1991, *Theory of Laminar Film Condensation*, Springer-Verlag, New York, p. 71.

Fujii, T., Wang, W. C., Koyama, S., and Shimizu, Y., 1987. "Heat Transfer Enhancement for Gravity Controlled Condensation on a Horizontal Tube by Coiling a Wire," in *Heat Transfer Science and Technology*, B-X. Wang, Ed., Hemisphere Publishing Corp., Washington, D.C., pp. 773–780.

Glicksman, L. A., Mikic, B. B., and Snow, D. F., 1973. "Enhancement of Film Condensation on the Outside of Horizontal Tubes," *AIChE Journal*, Vol. 19, pp. 636–637.

Gregorig, R., 1954. "Film Condensation on Finely Rippled Surfaces with Consideration of Surface Tension," *Zeitschrift für Angewandte Mathematik und Physik*, Vol. V, pp. 36–49.

Griffith, P., 1985. "Condensation, Part 2 Dropwise Condensation," Chapter 11 in *Handbook of Heat Transfer Fundamentals*, W. M. Rohsenow, H. P. Hartnett, and E. N. Ganic, Eds., McGraw-Hill, New York, pp. 11.37–11.50.

Hanneman, R. J., 1977. "Recent Advances in Dropwise Condensation Theory," ASME paper 77-WA/HT-21.

Henrici, K., 1961. "Kodensation von Frigen 12 und Frigen 22 an glatten und berippten Rohren," Dissertation, TU Karlsruhe.

Holman, J.P., 1986. *Heat Transfer*, 6th edition, McGraw–Hill, New York.

Honda, H., and Nozu, S., 1987a. "A Prediction Method for Heat Transfer During Film Condensation on Horizontal Low Integral-Fin Tubes," *Journal of Heat Transfer*, Vol. 109, pp. 218–225.

Honda, H., and Nozu, S., 1987b. "Effect of Drainage Strips on the Condensation Heat Transfer Performance of Horizontal Finned Tubes," in *Heat Transfer Science and Technology*, Wang, B-X., Ed., Hemisphere Publishing Corp., Washington, D.C., pp. 455–462.

Honda, H., and Nozu, S., 1987c. "A Theoretical Model of Film Condensation in a Bundle of Horizontal Low Finned Tubes, *Journal of Heat Transfer*, Vol. 111, pp. 525–532.

Honda, H., Nozu, S. and Mitsumori, K., 1983. "Augmentation of Condensation on Horizontal Finned Tubes by Attaching a Porous Drainage Plate, *Proceedings of the ASME-JSME Thermal Engineering Conference*, Honolulu, Vol. 3, pp. 289–295.

Honda, H., Nozu, S., and Takeda, Y., 1987a. "Flow Characteristics of Condensate on a Vertical Column of Horizontal Low Finned Tubes," *Proceedings of the 2nd ASME-JSME Thermal Engineering Joint Conference*, Vol. 1, pp. 517–524.

Honda, H., Nozu, S., and Uchima, B., 1987b. "A Generalized Prediction Method for Heat Transfer During Film Condensation on a Horizontal Low Finned Tube," *Proceedings of the ASME-JSME Thermal Engineering Joint Conference*, Vol. 4, pp. 385–392.

Honda, H., Uchima, B., Nozu, S., Nakata, H., and Torigoe, E., 1991. "Film Condensation of Downward Flowing R-113 Vapor on In-line Bundles of Horizontal Finned Tubes," *Journal of Heat Transfer*, Vol. 113, pp. 479–486.

Honda, H., Uchima, B., Nozu, S., Torigoe, E., and Imai, S. 1992. "Film Condensation R-113 on Staggered Bundles of Horizontal Finned Tubes," *Journal of Heat Transfer*, Vol. 114, pp. 442–449.

Iltscheff, S., 1971. "Some Experiments Concerning the Attainment of Drop Condensation with Flourinated Refrigerants," *Kaltetechnik Klimatisierung*, Vol. 23, pp. 237–241.

Incropera, F. P., and DeWitt, D. P., 1990. *Fundamentals of Heat and Mass Transfer*, John Wiley & Sons, New York, pp. 615–619.

Karkhu, V. A., and Borovkov, V. P., 1971. "Film Condensation of Vapor at Finely-Finned Horizontal Tubes," *Heat Transfer—Soviet Research*, Vol. 3, No. 2, pp. 183–191.

Katz, D. L., and Geist, J. M., 1948. "Condensation of Six Finned Tubes in a Vertical Row," *Transactions of the ASME*, Vol. 70, pp. 907–914.

Kedzierski, M. A., and Webb, R. L., 1987. "Experimental Measurements of Condensation on Vertical Plates with Enhanced Fins," in *Boiling and Condensation in Heat Transfer Equipment,* ASME Symposium Vol., HTD-Vol. 85, E. G. Ragi, T. M. Rudy, T. J. Rabas, and J. M. Robertson, Eds., ASME, New York, pp. 87–95.

Kedzierski, M. A., and Webb, R. L., 1990. "Practical Fin Shapes for Surface Tension Drained Condensation," *Journal of Heat Transfer*, Vol. 112, pp. 479–485.

Kern, K. Q., 1958. "Mathematical Development of Loading in Horizontal Condensers," *AIChE Journal*, Vol. 4, pp. 157–160.

Kun, L. C., and Ragi, E. G., 1981. "Enhancement for Film Condensation Apparatus," U.S. Patent 4,253,519.

Lee, W. C., and Rose, J. W., 1984. "Forced Convection Film Condensation on a Horizontal Tube with and without Non-condensing Gases," *International Journal of Heat and Mass Transfer*, Vol. 27, pp. 519- 528.

Marto, P. J., 1986. "Recent Progress in Enhancing Film Condensation Heat Transfer on Horizontal Tubes," *Heat Transfer 1986—Proceedings of the Eighth International Heat Transfer Conference*, Vol. 1, pp. 161–170.

Marto, P. J., 1988. "An Evaluation of Film Condensation on Horizontal Integral-Fin Tubes," *Journal of Heat Transfer*, Vol. 110, pp. 1287–1305.

Marto, P. J., and Wanniarachchi, A. S., 1984. "The Use of Wire-Wrapped Tubing to Enhance Steam Condensation in Tube Bundles," in *Heat Transfer in Heat Rejection Systems*, ASME Symposium Vol. HTD-Vol. 37, S. Sengupta and Y. G. Mussalli, Eds., ASME, New York, pp. 9–16.

Marto, P. J., Reilly, D. J., and Fenner, J. H., 1979. "An Experimental Comparison of Enhanced Heat Transfer Condenser Tubing," in *Advances in Enhanced Heat Transfer*, J. M. Chenoweth et al., Eds., ASME, New York, pp. 1–10.

Marto, P. J., Zebrowski, D., Wanniarachchi, A. S., and Rose, J. W., 1988. "Film Condensation of R-113 on Horizontal Finned Tubes," in *Proceedings of the 1988 National Heat Transfer Conference*, Vol. 2, H. R. Jacobs, Ed., ASME Symposium, HTD-96, ASME, New York, pp. 583–592.

Masuda, H., and Rose, J. W., 1987. "An Experimental Study of Condensation of Refrigerant R113 on Low Integral-Fin Tubes," in *Heat Transfer Science and Technology*, Hemisphere Publishing Corp., Washington, D.C., pp. 480–487.

Medwell, J. O., and Nicol, A. A., 1965. "Surface Roughness Effects on Condensing Films," ASME paper 65-HT-43.

Mehta, H. H., and Rao, M. R., 1979. "Heat Transfer and Frictional Characteristics of Spirally Enhanced Tubes for Horizontal Condensers," in *Advances in Enhanced Heat Transfer*, J. M. Chenoweth et al., Eds., ASME, New York, pp. 11–22.

Michael, A. G., Marto, P. J., Wanniarachchi, A. S., and Rose, J. W., 1989. "Effect of Vapor Velocity During Condensation on Horizontal Smooth and Finned Tubes," in *Heat Transfer with Change of Phase*, ASME Symposium, Vol. HTD-Vol. 114, I. S. Habib and R. J. Dallman, Eds., ASME, New York, pp. 1–10.

Mori, Y., Hijikata, H., Hirasawa, S. and Nakayama, W., 1979. "Optimized Performance of Condensers with Outside Condensing Surface," in *Condensation Heat Transfer*, J. M. Chenoweth et al., Eds., ASME Symposium Vol., ASME, New York, pp. 55–62.

Newson, I. H., and Hodgson, T. D., 1973. "The Development of Enhanced Heat Transfer Condenser Tubing," *Desalination*, Vol. 14, pp. 291–323.

Nicol, A. A., and Medwell, J. O., 1965.

Nicol, A. A., and Medwell, J. O., 1966. "The Effect of Surface Roughness on Condensing Steam," *Canadian Journal of Chemical Engineering*, Vol. 66, No. 3, pp. 170–173.

Notaro, P., 1979. "Enhanced Condensation Heat Transfer Device and Method," U.S. Patent 4,154,294.

Nusselt, W., 1916. "Die Oberflachenkondensation des Wasserdampfes," *Zeitschrift des Vereines der Deutschen Ingenieur*, Vol. 60, pp. 541–569.

Pearson, J. F., and Withers, J. G., 1969. "New Finned Tube Configuration Improves Refrigerant Condensing," *ASHRAE Journal*, pp. 77–82.

Rao, M. R., 1988. "Heat Transfer and Friction Correlations for Turbulent Flow of Water and Viscous Non-Newtonian Fluids in Single-Start Spirally Corrugated Tubes," *Proceedings of the 1988 National Heat Transfer Conference*, Vol. 1, ASME Symposium Vol. HTD-96, New York, pp. 677–683.

Rifert, V. G., and Leont'yev, G. G., 1976. "An Analysis of Heat Transfer with Steam Condensing on a Vertical Surface with Wires to Promote Heat Transfer," *Teploenergetika*, Vol. 23, No. 4, pp. 74–80.

Rudy, T. M., and Webb, R. L., 1983. "An Analytical Model to Predict Condensate Retention on Horizontal Integral-Fin Tubes,"*ASME-JSME Thermal Engineering Joint Conference*, Vol. 1, ASME, New York. pp. 373–378,

Rudy, T. M., and Webb, R .L., 1985. "An Analytical Model to Predict Condensate Retention on Horizontal Integral-Fin Tubes," *Journal of Heat Transfer*, Vol. 107, pp. 361–368.

Staub, P. W., 1966. "Condensing Heat Transfer Surface Device," U.S. Patent 3,289,752.

Sukhatme, S. P., Jagadish, B. S., and Prabhakaran, P., 1990. "Film Condensation of R-11 Vapor on Single Horizontal Enhanced Condenser Tubes," *Journal of Heat Transfer*, Vol. 112, pp. 229–234.

Takahashi, A., Nosetani, T., and Miyata, K., 1979. "Heat Transfer Performance of Enhanced Low Finned Tubes with Spirally Integral Inside Fins," Sumitomo Light Metal Technical Report, Vol. 20, pp. 59–65.

Thomas, D. G., 1967. "Enhancement of Film Condensation Rates on Vertical Tubes by Vertical Wires," *Industrial and Engineering Chemistry Fundamentals*, Vol. 6, No. 1, pp. 97–102.

Thomas, D. G., 1968. "Enhancement of Film Condensation Rate on Vertical Tubes by Longitudinal Fins," *AIChE Journal*, Vol. 6, No. 1, pp. 644–649.

Thomas, A., Lorenz, J. J., Hillis, D. A., Young, D. T., and Sather, N. P., 1979. "Performance Tests of the 1 MWt Shell and Tube Exchangers for OTEC," *Proceedings of the 6th OTEC Conference*, Paper lc.

Wang, S. P., Hijikata, K., and Deng, S. J., 1990. "Experimental Study on Condensation Heat Transfer by Various Kinds of Integral-finned Tubes," *Proceedings of the 2nd International Symposium on Condensation and Condensers*, University of Bath, Bath, England, pp. 397–406.

Wanniarachchi, A. S., Marto, P., and Rose, J., 1986. "Filmwise Condensation of Steam on Horizontal Tubes with Rectangular-Shaped Fins," in *Multiphase Flow and Heat Transfer*, V. K. Dhir, J. C. Chen, and O. C. Jones, Eds., ASME Symposium Vol. HTD. Vol. 47, ASME, New York, pp. 93–99.

Webb, R. L., 1979. "A Generalized Procedure for the Design and Optimization of Fluted Gregorig Condensing Surfaces," *Journal of Heat Transfer*, Vol. 101, pp. 335–339.

Webb, R. L., , 1981. "The Use of Enhanced Heat Transfer Surface Geometries in Condensers," *Power Condenser Heat Transfer Technology: Computer Modeling, Design, Fouling*, P. J. Marto, and R. H. Nunn, Eds. Hemisphere Pub. Corp., Washington, DC, 287–324.

Webb, R. L., 1984. "The Effects of Vapor Velocity and Tube Bundle Geometry on Condensation in Shell-Side Refrigeration Condensers," *ASHRAE Transactions*, Vol. 90, Part 1B, pp. 39–59.

Webb, R. L., and Kedzierski, M. A., 1990. "Practical Fin Shapes for Surface Tension Drained Condensation," *Journal of Heat Transfer*, Vol. 112, pp. 479–485.

Webb, R. L., and Murawski, C. G., 1990. "Row Effect for R-11 Condensation on Enhanced Tubes," *Journal of Heat Transfer*, Vol. 112, pp. 768–776.

Webb, R. L., Keswani, S. T., and Rudy, T. M., 1982. "Investigation of Surface Tension and Gravity Effects in Film Condensation," *Heat Transfer 1982* (*Proceedings of the Seventh International Heat Transfer Conference*), Vol. 5, Hemisphere Publishing Corp., New York, pp. 175–181.

Webb, R. L., Rudy, T. M., and Kedzierski, M. A., 1985. "Prediction of the Condensation Coefficient on Horizontal Integral-Fin Tubes," *Journal of Heat Transfer*, Vol. 107, pp. 369–376.

Wildsmith, G., 1980. "Open Discussion" section in *Power Condenser Heat Transfer Technology*, P. J. Marto and R. H. Nunn, Eds., Hemisphere Publishing Corp., Washington, D.C., pp. 463–468.

Withers, J. G., and Young, E. H., 1971. "Steam Condensing on Vertical Rows of Horizontal Corrugated and Plain Tubes," *Industrial and Engineering Chemistry—Process Design and Development*, Vol. 10, No. 1, pp. 19–30.

Yabe, A., 1991. "Active Heat Transfer Enhancement by Applying Electric Fields," *ASME/ JSME Thermal Engineering Proceedings*, Vol. 3, pp. xv–xxiii.

Yau, K. K., Cooper, J. R., and Rose, J. W., 1985. "Effect of Fin Spacing on the Performance of Horizontal Integral-Fin Condenser Tubes," *Journal of Heat Transfer*, Vol. 107, pp. 377–383.

Yau, K. K., Cooper, J. R., and Rose, J. W., 1986. "Horizontal Plain and Low-Finned Condenser Tubes—Effect of Fin Spacing and Drainage Strips on Heat Transfer and Condensate Retention," *Journal of Heat Transfer*, Vol. 108, pp. 946–950.

Yorkshire, 1982. *YIM Heat Exchanger Tubes: Design Data for Horizontal Rope Tubes in Steam Condensers*, Technical Memorandum 3, Yorkshire Imperial Metals, Ltd., Leeds, England.

Zener, C., and Lavi, A., 1974. Drainage Systems for Condensation," *Journal of Heat Transfer*, Vol. 96, pp. 209–205.

12.9 NOMENCLATURE

A	Heat transfer surface, A_f (fin area), A_r (root region), A_{ft} (fin tip), A_{fs} (fin side), m² or ft²
Bo	Bond number, ratio of gravity to surface tension force, dimensionless
c_b	Fraction of tube circumference condensate bridged ($= \beta/\pi$), dimensionless
d_j	Column jet diameter (Figure 12.29), m or ft
d_e	Diameter over fins of finned tube, m or ft
d_o	Tube outside diameter, m or ft
D_h	Hydraulic diameter, m or ft
D_c	Diameter of helix, m or ft
D_m	Mean diameter of fluted tube, m or ft
e	Fin height, roughness height, or corrugation depth; m or ft
e^+	Roughness Reynolds number (eu^*/v_l), dimensionless
E_i	Tube-side enhancement ratio, compared to plain tube (h_i/h_{ip}); dimensionless
E_o	Outside enhancement ratio, compared to plain tube h_o/h_{op}, dimensionless
f_{ik}	Force per unit volume within region L_{ik}, N/m³ or lbf/ft³
F	Parameter in Figure 12.29 ($= d_j/\lambda d$), dimensionless
F_d	Drainage force $F_d = \sigma(1/r_1 - 1/r_2)/\Delta x$ (surface tension drainage), N/m³ or lbf/ft³
F_g	Gravity drainage force $(\rho_l - \rho_v)g$, N/m³ or lbf/ft³
F_p	Property group $4k_l v_l \Delta T_{vs}/\lambda$, m⁻⁶ or ft⁻⁶
g	Gravitational constant, m/s² or ft/s²
G	Mass velocity in tube, $G_v = xG$, $G_l = (1-x)G$, kg/m²–s or lbm/ft²-s
$\bar{h}$	Average condensation coefficient, W/m²–K

h	Heat transfer coefficient, h_1 (on top row), h_t (on fin tip), h_h (horizontal plain tube), h_f (on fins), W/m²-K or Btu/hr-ft²-F
h_N	(on Nth tube row), $\bar{h}_N$ (average over N tube rows), W/m²-K or Btu/hr-ft²-F
h_{Nu}	Condensation coefficient calculated from Nusselt's laminar film model, W/m²-K or Btu/hr-ft²-F
h_s	Heat transfer coefficient on smooth surface, W/m²-K or Btu/hr-ft²-F
h_w	Water-side heat transfer coefficient, W/m²-K or Btu/hr-ft²-F
H_d	Height of drainage strip, m or m
H_L	Helix length per 360-degree revolution in the Figure 12.14 tubes, m or ft
k_l	thermal conductivity of condensate, W/m-K or Btu/hr-ft-F
k_w	Tube thermal conductivity, W/m-K or Btu/hr-ft-F
K	Parameter in Equation 12.43, dimensionless
l_{ik}	Length between points $x = i$ and $x = k$ (Figure 12.22), m or ft
L_{ik}	Region between points $x = i$ and $x = k$ (Figure 12.22)
L	Tube or surface length, m or ft
L_b	Width of drainage channel (Figure 12.18), m or ft
L_s	Depth of drainage channel (Figure 12.18), m or ft
LMTD	Logarithmic mean temperature difference, K or F
m	Row effect exponent (dimensionless)
$\dot{m}_{ik}$	Condensate flow per unit width rate on region L_{ik}, in Figure 12.22, kg/s-m or lbm/s-ft
$\dot{m}$	Condensate flow rate: $\dot{m}_l$ (from tube), $\dot{m}_h$ (on base tube), $\dot{m}_r$ (sum of $\dot{m}_f$ and $\dot{m}_h$), kg/s or lbm/s
N	Number of flutes on tube
N	Number of horizontal tube rows in depth
p	Pressure, p_{sat} (saturation pressure), kPa or lbf/ft²
p	Axial pitch of surface elements, m or ft
p_f	Fin pitch, m or ft
P	Pumping power, W or HP
P_w	Wetted perimeter of inner tube surface, m or ft
Nu	Nusselt number, dimensionless
Pr	Prandtl number, dimensionless
q	Heat flux, W/m² or Btu/hr-ft²
Q	Heat transfer rate, W or Btu/hr
r	Local radius of liquid–vapor interface, m or ft
r_c	Radius at corner of fin tip on trapezoidal fin, m or ft
r_o	Tip radius of convex profile, m or ft
R	Radius of drainage interface at fin root, m or ft
$R*$	$R - \delta_{54}$, projection of R on side wall from point 4 of Figure 12.22, m or ft
Re_l	Condensate Reynolds number ($4\Gamma/\mu_l$), dimensionless
Re_v	Vapor Reynolds number ($d_o u_\infty/\mu_l$), dimensionless
s	Spacing between fins ($= p_f - t$), m or ft

s	Coordinate distance along curved condensing profile, m or ft
S_m	Length of convex profile, m or ft
t	Fin thickness: t_b (at base), t_t (at tip); m or ft
T_s	Saturation temperature, also T_{sat}, K or °F
T_w	Wall temperature, K or °F
ΔT	ΔT_u (unflooded region), ΔT_f (flooded region), K or °F
ΔT_{vs}	$T_s - T_w$, K or °F
u_∞	Vapor velocity, m/s or ft/s
u^*	Frictional shear velocity $(\tau/\rho)^{1/2}$, m/s or ft/s
U	Overall heat transfer coefficient, U_r (based on root diameter), U_p (plain tube) W/m²-K or Btu/hr-ft²-F
V	Volume of heat exchanger tubing material, m³ or ft³

Greek Symbols

α	Angle in Figure 12.22, radians
β	Condensate retention angle (Figure 12.5), radians
Γ	Condensate mass velocity on horizontal tube, $\dot{m}_l/L$; kg/s-m or lbm/s-ft
Γ	Condensate mass velocity leaving vertical tube, $\dot{m}_l/\pi d_o$; kg/s-m or lbm/s-ft
δ	Condensate film thickness, m or ft
ζ	Parameter in Adamek profile shape equation, dimensionless
η_f	Fin efficiency; η_{fu} (unflooded region), η_{ff} (flooded region), dimensionless
θ_m	Rotation angle from fin tip to fin base, radians
λ	Latent heat, J/kg or Btu/lb
λ_d	One-dimensional Taylor instability wavelength, m or ft
μ	Dynamic viscosity: μ_l (of liquid), μ_v (of vapor), kg/s-m² or lbm/s-ft²
π	Constant (3.14159)
ρ	Density: ρ_l (of liquid), ρ_v (of vapor), kg/m³ or lbm/ft³
σ	Surface tension, N/m or lbf/ft
τ_i	Interfacial shear stress, N/m² or lbf/ft²
τ_w	Wall shear stress, N/m² or lbf/ft²

Subscripts

av	Average value
f	Flooded region
i	Inner surface of tube
m	Mean value
o	Outer surface of tube
p	Plain tube
s	Smooth tube
u	Unflooded region

Unsubscripted variables refer to enhanced tube.

APPENDIX A: THE KEDZIERSKI AND WEBB [1990] FIN PROFILE SHAPES

The key geometric parameters that define the basic shape of the profile are:

1. The fin height (e) and base thickness (t_b)
2. The fin tip radius (r_o) and the "rotation angle," θ_m

Equation A.1 defines the profile shape as a function of e, t_b, r_o, and Z, for $\theta_m = \pi/2$. The term Z is a "shape factor" parameter:

$$r = c_1 + c_2 e^{Z\theta} + c_3 \theta \tag{A.1}$$

The constants c_1, c_2, and c_3 are given by Equations A.2, A.3, and A.4:

$$c_1 = r_o - c_2 \tag{A.2}$$

$$c_2 = \frac{0.5 t_b(\sin \theta_m - \theta_m \cos \theta_m - e(\cos \theta_m + \theta_m \sin \theta_m - 1)}{c_4 - [2(1 - \cos \theta_m) - \theta_m \sin \theta_m]}$$

$$\tag{A.3}$$

$$+ \frac{r_o[\theta_m \sin \theta_m - 2(1 - \cos \theta_m)]}{c_4 - [2(1 - \cos \theta_m) - \theta_m \sin \theta_m]}$$

$$c_3 = \frac{0.5 t_b - c_1 \sin \theta_m - c_2 Z \exp(Z\theta_m)\cos \theta_m/(Z^2 + 1)}{\cos \theta_m + \theta_m \sin \theta_m - 1}$$

$$\tag{A.4}$$

$$- \frac{c_2[\exp(Z\theta_m)\sin\theta_m - Z]/(Z^2 + 1)}{\cos \theta_m + \theta_m \sin \theta_m - 1}$$

$$c_4 = [\exp(Z\theta_m)[1 - \cos\theta_m + Z(\sin\theta_m - \theta_m)]$$

$$\tag{A.5}$$

$$+ (Z\theta_m - 1)\cos\theta_m - (Z + \theta_m)\sin\theta_m +)/(Z^2 + 1)$$

The constants defined by Equations A.2 to A.5 are obtained by applying the following boundary conditions to Equation A.1:

$$r = r_o \qquad \text{at } \theta = 0 \tag{A.6}$$

$$t_b = 2 \int_0^{\pi/2} r \cos \theta \, d\theta \tag{A.7}$$

$$e = \int_0^{\pi/2} r \sin \theta \, d\theta \tag{A.8}$$

APPENDIX B: FIN EFFICIENCY IN THE FLOODED REGION

In the flooded fraction of the tube, condensation occurs on the fin tip. Because the fins are quite short, a one dimensional conduction model is acceptable. Assuming no heat transfer from the fin sides, heat conduction in a fin of constant thickness, t, is given by

$$Q = k_w A_b (T_{w,t} - T_{w,r})/e \tag{B.1}$$

where $T_{w,r}$ is the fin base temperature, $T_{w,t}$ is the fin tip temperature, and h is the fin height. The heat transferred by convection from the fin tip is

$$Q = h_t A_t (T_{sat} - T_{w,t}) \tag{B.2}$$

Combining Equations B.1 and B.2 and solving for $T_{w,t}$ gives

$$T_{w,t} = (T_{w,r} + Nu_t T_{sat})/(Nu_t + 1) \tag{B.3}$$

where $Nu_t = h_t e/k_w$. Substitution of Equation B.3 in Equation B.2 gives

$$Q = h_t A_t (T_{sat} - T_{w,r})/(Nu_t + 1) \tag{B.4}$$

Let Q_{max} be the heat transfer rate for a fin, having $T_{w,t} = T_{w,r}$, which is

$$Q_{max} = h_{t,\infty} A_t (T_{sat} - T_{w,r}) \tag{B.5}$$

where $h_{t,\infty}$ is the condensation coefficient if $T_{w,t} = T_{w,r}$. The fin efficiency is defined as q/q_{max}, and is the ratio of Equations B.4 and B.5, giving

$$\eta_f = h_t/[h_{t,\infty}(Nu_t + 1)] \tag{B.6}$$

The fin tip efficiency may be iteratively determined by assuming a fin tip temperature and using Equations B.1 and B.2 to check for convergence of the fin tip temperature.

13

CONVECTIVE VAPORIZATION

13.1 INTRODUCTION

Chapter 11 addressed pool boiling on plates and circular tubes. In pool boiling, the generated vapor is removed from the surface by the departing bubbles. However, this is not the case for convective vaporization. For example, for vaporization within a tube the generated vapor is contained within the tube, and it affects the local heat transfer coefficient, via the flow pattern. The combination of total flow rate and local vapor velocity establishes a flow pattern. The flow pattern changes with vapor quality. Horizontal tubes may behave differently from vertical tubes, because gravity force acts to stratify the liquid. Hence, enhancement requirements for horizontal tubes may be different from those of vertical tubes. Circular tubes are not the only "channel flow" geometry of interest. Brazed aluminum heat exchangers may use flat, extruded aluminum tubes having rectangular flow passages, or they may be of the plate-and-fin construction. Another "channel flow" geometry of interest is the annulus. Boiling on the outside of tubes in a bundle also involves convective effects. These geometries are also addressed.

Prior to addressing the subject of enhancement, we will provide information relating to fundamental understanding of two-phase flow and convective heat transfer.

13.2 FUNDAMENTALS

This section is concerned with fundamentals of vaporization in channels (e.g., in horizontal or vertical plain tubes). Understanding these processes in plain tubes provides a foundation for understanding the performance of enhanced surfaces. We will discuss only the aspects of two- phase flow and heat transfer that are particularly relevant to enhanced heat transfer. For a more detailed discussion of the general topic, the reader should consult a text on the subject area, such as Carey [1992].

When vaporization or condensation occurs in a tube, convective effects occur, which do not exist in pool boiling or vapor condensation, without vapor velocity. The convective effects are associated with the influence of shear stress on the liquid–vapor interface. Furthermore, if gravity force exceeds vapor shear forces, the liquid phase will tend to stratify in horizontal tubes.

As the vapor quality increases in vaporization, the vapor velocity increases causing increased vapor shear. In condensation the converse exists. A two-phase flow experiences different "flow patterns" as vapor quality changes along the tube length. The heat transfer coefficient is significantly affected by the flow pattern. Hence, it is important to understand how flow pattern affects the heat transfer coefficient in two-phase heat transfer.

13.2.1 Flow Patterns

Consider vaporization in a tube having subcooled liquid entering and 100% leaving vapor quality. Figures 13.1 and 13.2 illustrate the flow patterns that will exist in vertical and horizontal evaporator tubes, respectively. Gravity leads to stratification

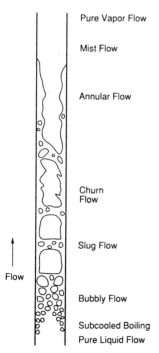

Pure Vapor Flow

Mist Flow

Annular Flow

Churn
Flow

Slug Flow

Flow

Bubbly Flow

Subcooled Boiling
Pure Liquid Flow

Figure 13.1 Flow patterns for complete vaporization for upward flow in a vertical tube. (From Carey [1992].)

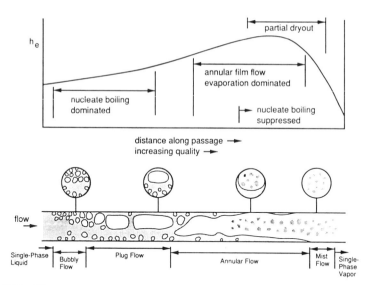

Figure 13.2   Flow patterns for complete vaporization in a horizontal tube. (From Carey [1992].)

in the Figure 13.2 horizontal tube. Stratification effects are more severe at lower vapor qualities, because vapor shear forces are smaller at low vapor quality.

Consider the situation for complete condensation in tubes. Figure 13.3 illustrates the flow patterns for condensation in a horizontal tube. These flow patterns are similar to the reverse of Figure 13.2, with one exception. Vapor may be generated by nucleate boiling at the tube wall, as illustrated in Figure 13.2. Condensation may occur in co-current downward flow. The flow patterns would appear similar to those in Figure 13.1, except vapor bubbles (caused by nucleate boiling) would not exist at the tube wall.

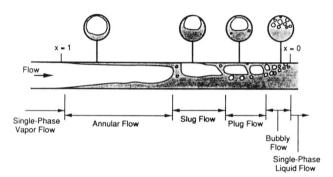

Figure 13.3   Flow patterns for complete condensation in a horizontal tube. (From Carey [1992].)

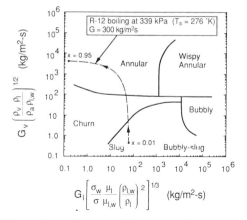
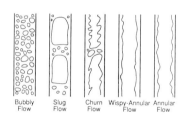

Figure 13.4 Hewitt and Roberts [1969] flow pattern map for vertical co-current upward flow. The figure is annotated to show the flow patterns encountered evaporating by R-12. (From Carey [1992].)

Work has been done to characterize the flow pattern as a function of mass velocity and vapor quality. Figure 13.4 shows the map defined by Hewitt and Roberts [1969] for vertical upward flow. Figure 13.5 show the map defined by Baker [1954] for flow in a horizontal tube. The Taitel and Dukler [1976] map is more accurate, but requires too much space to explain here. See Carey [1992] for details.

The dashed line in Figure 13.4 shows the flow patterns experienced by R-12 evaporating from $0.01 \leq x \leq 0.95$ in a vertical tube. This figure shows that the flow patterns are very sensitive to the mass velocity of the liquid and the vapor. The

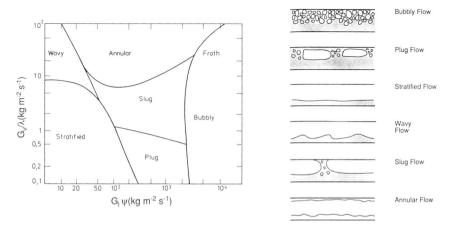

Figure 13.5 (a) Flow pattern map of Baker [1954] for a horizontal tube. (b) Defined flow patterns. (From Carey [1992].)

phase mass velocities are defined as $G_l = G(1 - x)$ and $G_v = G_x$, where G is the total flow rate divided by the pipe cross-sectional area. The reverse set of flow patterns would be experienced by a condensing flow, which traverses the same vapor qualities.

13.2.2   Convective Vaporization in Tubes

Figure 13.2 shows that three different heat transfer mechanisms can exist, depending on the flow pattern. These are nucleate (subcooled, bubbly, and plug flow), thin film evaporation (annular flow), and single-phase heat transfer to gas in the drywall region, downstream from the annular flow region. The third drywall region may or may not exist, depending on whether the "critical heat flux" has been exceeded in the annular flow region. Chen [1966] proposed that liquid velocity tends to suppress nucleate boiling. He proposed that the nucleate boiling contribution is $h_{nb} = Sh_{nbp}$, where h_{nbp} is for nucleate pool boiling and S is the suppression factor. If nucleate boiling is totally suppressed in the annular film region, heat transfer occurs by thin film evaporation (convective evaporation). If the film flow is laminar, the heat transfer coefficient is given by $h = k_l/\delta$, where δ is the film thickness. In the drywall region, the heat transfer coefficient is given by the appropriate equation for heat transfer to a gas.

It is possible that both nucleate boiling and convective evaporation heat transfer will exist at the same time. If so, one way to account for both components is via the superposition model, which was proposed by Chen [1966]. This model assumes that the total heat flux (q) is the sum of the "nucleate boiling" component and the "convective evaporation" component. This is written as

$$q = q_{nb} + q_{cv} \tag{13.1}$$

The terms q_{nb} and q_{cv} are calculated at the value of wall superheat [$\Delta T_{ws} \equiv (T_w - T_s)$] that exists in the system. One may write the superposition model in terms of the component heat transfer coefficients by dividing Equation 13.1 by the wall superheat to obtain

$$h = h_{nb} + h_{cv} \tag{13.2}$$

where h_{nb} is the nucleate boiling contribution and h_{cv} is the convective contribution. One calculates $h_{cv} = Fh_l$, where h_l is calculated from an appropriate equation for single-phase flow of the liquid phase alone. For vaporization inside a tube, one may calculate h_l using the Dittus–Boelter equation as given by

$$h_l = \frac{0.023k_l}{d_i} \left[\frac{d_i G(1 - x)}{\mu_l} \right]^{0.8} \mathrm{Pr}^{1/3} \tag{13.3}$$

Figure 13.6 illustrates the convective vaporization curve, (frequently called the "flow boiling" curve) as a log–log plot of heat flux versus wall superheat. The

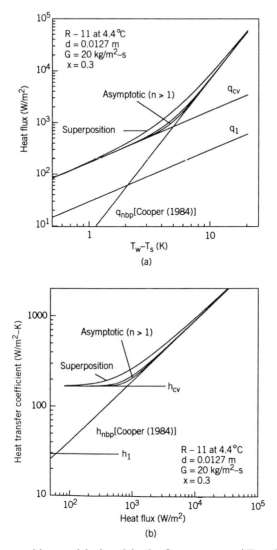

Figure 13.6 Superposition model plotted in the form q versus ΔT_{ws}. (From Webb and Gupte [1992].)

constant parameters are mass velocity (G) and vapor quality (x). The convective vaporization curve is asymptotic to q_{cv} at low ΔT_{ws}, and asymptotic to q_{nb} at high ΔT_{ws}. The suppression factor, S, is assumed to be unity in Figure 13.6 for the purpose of illustration. The q_{cv} asymptote has unity slope, because h_{cv} is independent of ΔT_{ws}. The objective of a convective vaporization correlation is to predict the heat flux in the curved region between the asymptotes. Steiner and Taborek [1992] have proposed that an asymptotic model gives a better fit of the data in the region between q_{nb} and q_{cv} than does the superposition model (Equation 13.1). This model is written as

$$q^n = q_{nb}^n + q_{cv}^n \qquad (13.4)$$

At low heat flux, the vaporization curve is asymptotic to h_{cv} (two-phase convection). At high heat flux, the vaporization curve is asymptotic to h_{nb} (nucleate boiling). The line below and parallel to h_{cv} in Figure 13.6b is that for the liquid phase flowing alone (h_l). The corresponding heat flux and heat transfer coefficient is q_l (liquid phase convection) and h_l, respectively. The ratio, F, is defined as the "two-phase convection multiplier" and is equal to q_{cv}/q_l or h_{cv}/h_l. The curve labeled "superposition" ($n = 1$) illustrates the superposition model. Using the superposition model, one would derive F from experimental data by the equation

$$F = \frac{q - q_{nb}}{q_l} = \frac{h - Sh_{nbp}}{h_l} \qquad (13.5)$$

When $n > 1$, the curved portion of the convective vaporization curve is asymptotic with the limiting straight lines q_{cv} and q_{nb} (Figures 13.6a) and h_{cv} and h_{nb} (Figure 13.6b). One asymptote of the prediction is the line of fully convective vaporization coefficient, where the nucleate boiling contribution is insignificant. The other asymptote is the nucleate boiling coefficient, where the contribution due to convective heat transfer is insignificant. The curved region between the two asymptotes contracts as the value of n increases. As $n \to \infty$, the curved region shrinks to zero length. Thus, the contributions of nucleate boiling and convective evaporation are added, to some extent, in the curved region. If $n = 1$, the total heat flux is the sum of the two components, which is the superposition model.

Enhancement of convective vaporization may be approached by increasing either h_{nbp} or h_{cv} (cf. Equation 13.2). Using a porous boiling surface on the inner surface of a plain tube will cause $h_{nbp} \gg h_{cv}$, and the tube performance will be "nucleate boiling dominated" up to quite high vapor qualities. In this case, the total heat transfer coefficient will be quite sensitive to heat flux, but relatively insensitive to flow rate. However, if the enhancement consists of internal fins, which should not significantly enhance the nucleate boiling component, the tube performance will be "convection dominated." Here, the total heat transfer coefficient will be sensitive to flow rate, but relatively insensitive to heat flux.

Most tube-side enhancements work to enhance the convective term. Examples are corrugated roughness, internal fins, the "star insert," and twisted tapes. The use of fine grooves at a helix angle on the inner surface may use capillary forces to transport liquid to the upper tube wall. Any approach that thins the film without causing dry spots will provide enhancement of the h_{cv} term in Equation 13.2. Roughness provides enhancement by turbulent mixing of the liquid film. Such a roughness may promote droplet entrainment at moderate vapor qualities, which causes a thinner film on the tube wall.

It is important to recognize that a low heat transfer coefficient will exist if the wall becomes dry. As shown in Figures 13.2 and 13.5, horizontal gravity forces may act to stratify the flow, causing the upper part of the tube to be dry or intermittently dry.

13.2.3 Two-Phase Pressure Drop

Refrigerant pressure drop affects the local saturation temperature in the bundle. In addition, the shell-side heat transfer coefficient correlation is also directly related to the frictional pressure gradient. The refrigerant pressure drop is given by

$$\Delta p = \Delta p_F + \Delta p_g + \Delta p_a \tag{13.6}$$

The three terms on the right-hand side of Equation 13.6 account for the friction, gravity, and acceleration contributions. For in-tube flow, the friction contribution, Δp_F, is calculated by

$$\Delta p_F = \frac{4 f_f L}{d_i} \frac{[G(1-x)]^2}{2\rho_l} \phi_l^2 \tag{13.7}$$

where f_f is the friction factor for liquid flow at mass velocity $G(1-x)$ and N is the number of tube rows crossed. The term ϕ_l^2 is a function of $X_{tt} \equiv (\Delta p_{F,l}/\Delta p_{F,v})^{0.5}$ is called the *Martinelli parameter*. The Martinelli parameter for turbulent flow of the gas and liquid phases in a circular tube is

$$X_{tt} = \left(\frac{1-x}{x}\right)^{0.9} \left(\frac{\rho_v}{\rho_l}\right)^{0.5} \left(\frac{\mu_l}{\mu_v}\right)^{0.1} \tag{13.8}$$

For flow in tubes, Chisholm [1967] has shown that ϕ_l^2 may be expressed in terms of X_{tt} by

$$\phi_l^2 = 1 + \frac{C}{X_{tt}} + \frac{1}{X_{tt}^2} \tag{13.9}$$

where C is an empirical constant that depends on the flow geometry. Chisholm [1967] shows that use of $C = 20$ in Equation 13.9 is applicable for flow in smooth tubes. Ishihara et al. (1980) proposes $C = 8$ for flow normal to plain tube bundles. Webb et al. [1990] propose that $C = 8$ is also applicable to enhanced tubes, which have a relatively smooth outside diameter. Because no correlations have been developed for the integral fin geometry, a possible choice is to assume that the Ishihara correlation is also applicable.

Prediction of Δp_g and Δp_a require knowledge of the void fraction, α. See Carey [1992] or other books on two-phase flow for void fraction correlations.

Tube-side enhancement techniques for certain enhanced geometries may increase the friction component (Δp_F). However, they should not affect the gravity or acceleration components of the pressure drop. Hence, a significant increase of Δp_F may not have a significant effect on the total pressure drop. This of course depends on what fraction of the total pressure drop is due to Δp_g and Δp_a.

13.2.4   Effect of Flow Orientation on Flow Pattern

For complete vaporization the flow pattern may be significantly different for the following situations:

1. *Vertical Tubes with Vapor and Liquid Flowing Down:* For co-current down flow in a vertical tube, the liquid is confined to the tube wall, except for possibly entrained droplets. In this case, there is essentially no change of flow pattern, except for the amount of entrained liquid. Typically, such evaporators are supplied with a greater liquid rate than is evaporated. The ratio of the inlet-to-evaporated liquid is called the *recirculation ratio*. Typically, a recirculation ratio of 2–4 is used to prevent dryout of the liquid film. This case is called "falling film," or "thin film" evaporation. For plain surfaces, nucleate boiling probably would not occur. However, nucleate boiling may provide a significant contribution, if an enhanced nucleate boiling surface is used. Except for enhanced nucleate boiling tubes, convective effects are important, and vapor shear forces may influence the convective contribution.

2. *Vertical Tubes with Vapor and Liquid Flowing Up:* This situation experiences the flow patterns shown in Figure 13.1, and convective effects are important. Nucleate boiling may be important at low-to-moderate vapor qualities. The heating surface may become dry at high vapor qualities.

3. *Horizontal Tubes with Co-current Vapor–Liquid Flow:* This situation is quite similar to item 2 above. However, gravity-influenced stratification of the liquid, combined with heat flux, may act to dry the upper tube surface. Hence, an internal enhancement geometry that acts to provide surface wetting may be more important than for vertical tubes.

Refrigerant pressure drop is an important consideration in evaluating tube-side enhancements for vaporization, particularly for fluids having high dT/dp. This concern is addressed in Chapter 4.

13.2.5   Convective Vaporization in Tube Bundles

Webb and Gupte [1992] survey models and correlations for convective vaporization in tubes and tube banks. They give recommended equations for calculation of the suppression factor, S, and the two-phase multiplier, F. The superposition model (Equation 12.1) or the asymptotic model (Equation 12.4) may be used for tube banks. One would use (1) an appropriate equation for pool boiling on horizontal tubes (h_{nbp}), (2) the applicable equation for single-phase heat transfer in tube banks (h_l), and (3) correlations for F and S that are applicable to tube banks. Several F-factor correlations have been proposed for tube banks. These are discussed by Webb and Gupte [1992].

The flow pattern that exists in a tube bundle depends on the bundle orientation, and how the bundle is circuited on the tube side. For vertical up-flow in a vertical thermosyphon reboiler, one would expect a flow pattern grossly similar to that in a vertical tube. For a horizontal tube bundle, the flow pattern would depend on whether it were a natural circulation reboiler or a flooded refrigerant evaporator. Relatively low vapor qualities (e.g., 10%) exist at the top tube row in a reboiler). However, complete evaporation occurs in a refrigerant evaporator. Polley et al. [1980] provide some description of flow patterns for shell-side vaporization.

13.2.6   Critical Heat Flux

The term *critical heat flux* (CHF) refers to critical conditions that may occur in convective vaporization, at which the heat transfer coefficient precipitously drops to a much lower level. The result of this phenomenon depends on the thermal boundary condition. If a heat flux boundary condition exists, the tube wall temperature will suddenly increase to compensate for the lower heat transfer coefficient. If a temperature boundary condition exists for a two-fluid heat exchanger, the local heat flux will decrease in response to the reduced overall heat transfer coefficient. A destructive condition may exist for a heat flux boundary condition, since the tube wall temperature may increase to the tube melting temperature. Such a condition may exist in a nuclear reactor or a fossil fuel boiler, whose walls are heated by radiant flames. However, this would not be the case in a two-fluid heat exchanger, where the second fluid fixes the tube wall temperature.

A vaporizing flow is subject to CHF at two different flow conditions depending on the heat flux and enthalpy of the flowing fluid. These are explained below:

1. At subcooled or low-vapor-quality conditions, nucleate boiling is the dominant heat transfer mechanism. If the heat flux is high enough at the particular local flow enthalpy, the boiling may suddenly revert to film boiling, for which the heat transfer coefficient drops to a much smaller value. This condition is called *departure from nucleate boiling* (DNB).
2. At moderate or high vapor qualities, the flow pattern is annular with entrained droplets. If the liquid film on the tube wall dries out, the heat transfer coefficient will precipitously drop. This condition is described as *dryout*. The dryout condition occurs at lower vapor qualities as the mass velocity and local $T_w - T_s$ increase.

Both types of CHF conditions may exist in shell-side vaporization as well as tube-side vaporization. Certain enhancement techniques are effective in alleviating both CHF conditions. They are discussed in Section 13.4.

13.3   ENHANCEMENT TECHNIQUES IN TUBES

Enhancement geometries used for pool boiling on the outside of circular tubes are generally not applicable to vaporization inside tubes. This is because the flow pattern and the convective effects significantly alter the vaporization process. A variety of enhancement techniques have been investigated. These include twisted-tape inserts, internal fins, and integral roughness. This chapter will address enhancements applicable to co-current flow in horizontal and vertical tubes.

13.3.1 Internal Fins

Work on internally finned tubes was slow to develop, because practical manufacturing techniques to form internal fins did not exist until the late 1960s. Figure 13.7 shows four basic types of internally finned tubes that have been investigated. Figure 13.7a shows axial and helical internal fins made in copper tubes by a swaging process. A 19-mm-diameter tube may have 12–30 fins, typically 1.5–3 mm in height, and are of the same generic type tested by Carnavos [1980], discussed in Section 8.3.2. Such copper finned tubes are made by a swaging process, which is slow and relatively expensive. Axial internal fins may be made at a higher speed using a hot extrusion process. One may add a helix angle by twisting the extruded fins. Figure 13.7b shows internally finned (or grooved) tubes made of steel, which are used in steam power boilers. The fin height is approximately 1.1 mm and is used to increase the departure from nucleate boiling. Figure 13.7c shows an aluminum extended surface area tube made by first extruding an aluminum plate with the corrugated pattern, then rolling the strip in a helix, followed by seam welding. This tube has also been made in thin wall titanium. Here, the titanium sheet is corru-

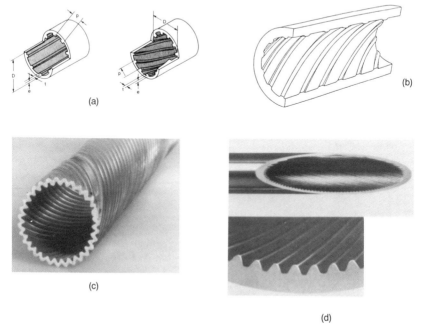

Figure 13.7 (a) Axial and helical internal fins. (b) Grooved, axial steel tube for power boilers. (c) General Atomics spirally fluted tube. (d) Wieland micro-fin tube.

gated, then rolled in the helical form and welded. Figure 13.7d shows a modern *micro-fin* tube which has a small-scale fin structure, and is used for refrigerants. A 9.0-mm tube may have 60 fins, 0.2–0.25 mm high at a small helix angle (e.g., 15 degrees). These micro-fins may be formed in a copper tube at high speed by drawing a the tube over a grooved slug.

The earliest work on internal fins was reported by Boling et al. [1953], who worked with refrigerant evaporation. Lavin and Young [1965] were the first to test an integral internal fin. They tested the four tube geometries shown in Figure 13.8 using R-12 and R-22 in horizontal and vertical flow. This very interesting paper provides data for specific flow regimes (subcooled liquid, subcooled nucleate boiling, annular flow, and mist flow). Kubanek and Miletti [1979] tested R-22 in copper internal fin and aluminum insert tubes (see Figure 8.1c). They measured the average heat transfer coefficient in a 0.8-m- and 2.0-m-long test sections with $\Delta x = 0.7$. Figure 13.9 shows their data, for which the heat transfer coefficient (h) is based on the nominal plain tube area ($A/L = \pi d_i$). Tubes 22, 25, and 30 have either 30 or 32 fins, 0.51–0.64 mm high. Tube 30 has axial fins, and the fins in tubes 22 and 25 are spiraled. Tube 25, having the tightest fin spiral, provides the highest enhancement— approximately 200% at $G = 100$ kg/m²-s. The aluminum insert device (tube 24C) gave the highest enhancement at low mass velocity, but is below that of the internal fin tubes at high velocity.

Schlünder and Chawla [1969] tested R-11 in 14-mm-diameter tubes having aluminum inserts (of the generic type illustrated by Figure 8.1c) inserted in the tube. The copper tube was "shrink fitted" to the insert to minimize thermal contact

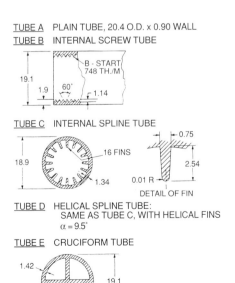

TUBE A   PLAIN TUBE, 20.4 O.D. x 0.90 WALL
TUBE B   INTERNAL SCREW TUBE

TUBE C   INTERNAL SPLINE TUBE

TUBE D   HELICAL SPLINE TUBE:
         SAME AS TUBE C, WITH HELICAL FINS
         $\alpha = 9.5°$

TUBE E   CRUCIFORM TUBE

Figure 13.8 Tube geometries tested by Lavin and Young [1965] with R-12 and R-22 in vertical and horizontal flow. (From Lavin and Young [1965].)

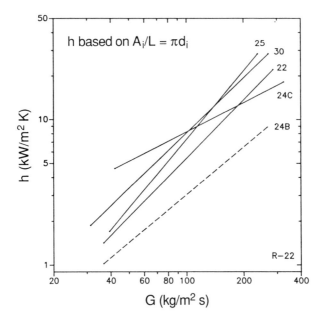

Tube type and number	Number of fins	Tube internal diameter, mm	Tube hydraulic diameter, mm	Fin height, mm	Fin pitch, mm	Wetted area per unit length, mm²/m×10⁻³	Nominal area per unit length, mm²/m×10⁻³	Area ratio	Heated length, m
Plain, 24B		14.4	14.4			45.3	45.3	1.00	0.80
Insert, 24C	5	14.4	4.09		610	90.8	45.2	2.00	0.80
Finned, 22	32	14.7	7.57	0.635	305	87.0	46.2	1.88	0.80
Finned, 25	32	14.7	7.57	0.635	152	87.2	46.2	1.89	0.80
Finned, 30	30	11.9	6.30	0.508	102	68.0	37.4	1.82	0.80

Figure 13.9 R-22 heat transfer vaporization coefficient (h based on $A_i/L = \pi d_i$) data of Kubanek and Miletti [1979] at $T_s = 4.4°C$ in 0.8-m-long tube with $\Delta x = 0.7$. The table defines the tube geometries.

resistance. They tested inserts having two, four, or eight legs. These inserts act as full-height fins, and also reduce the hydraulic diameter of the flow passage. They provide empirical correlations for the heat transfer and pressure drop data. Pearson and Young [1970] provide R-22 data on the Figure 13.10 aluminum insert having five legs. Their data are taken for complete evaporation in a four-pass evaporator, with heat transfer coefficients measured for each pass.

Panchal et al. [1992] have tested the Figure 13.7c aluminum tube with upflow of R-11. Tests were conducted for subcooled and saturated vaporization. Figure 13.11 shows the saturated vaporization results for exit vapor qualities between 0.4 and 0.9. The heat transfer coefficient is based on the total internal area (1.64 times that

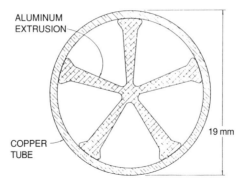

Figure 13.10 Tube having five-leg aluminum insert with shrink fit.

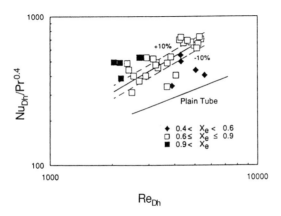

Tube Specifications

Parameter	Value
Tube material	Aluminum, Al 6063
Wall thickness	1.65 mm
Tube flow area	563.2 mm^2
Mean inside diameter	26.8 mm
Inside perimeter	137.54 mm
Outside perimeter	136.42 mm
Equivalent diameter	16.38 mm
Flute angle to tube axis	30°
Flute spacing	2.63 mm
Flute depth	1.54 mm
Effective tube length	4.45 m

Figure 13.11 Convective vaporization coefficient for R-11 (296–1010 kPa) in vertical spirally fluted tube. (From Panchal et al. [1992].)

of a plain tube of the same inside diameter). The Nusselt and Reynolds number are defined as hD_h/k_l and $\mathrm{Re_{Dh}} = D_hG(1 - x)/\mu_l$, respectively. Figure 13.11 shows the heat transfer coefficient averaged over $0 < x < x_e$, where x_e is the exit vapor quality. The internal enhancement ($E_{hi} = h/h_p$) is between 1.25 and 1.6. If the enhancement were based on the nominal plain tube area, $2.05 \leq E_{hi} \leq 2.6$. The solid line through the data set is the prediction based on the Chen [1966] convective vaporization model.

Ito and Kimura [1979] and Shinohara and Tobe [1985] describe a tube having very small triangular-shaped fins, which has become known as the "micro-fin" tube. This tube is used in virtually all presently manufactured refrigerant condensers, either air- or water-cooled. Because of the importance of this tube, it is separately described in Section 13.3.2.

13.3.2   The Micro-Fin Tube

The tube was first developed by Fujie et al. [1977] of Hitachi Cable, Ltd. and is described by Tatsumi (1979). An improved Hitachi design is described by Shinohara and Tobe [1985] and by Shinohara et al. [1987]. The version described by Shinohara and Tobe (1985) is close to that now made by tube manufacturers in Japan, Europe, and the United States. It is also used for condensation of refrigerants, which is discussed in Chapter 14. Figure 13.7d shows Wieland's [1991] version of the micro-fin tube.

The tube is presently made in diameters between 4 and 16 mm. Depending on the manufacturer, the micro-fin tube provides an R-22 evaporation coefficient 100–300% higher than that of a plain tube. Table 13.1, from Yasuda et al. [1990], shows (a) the geometry changes that have been made by Hitachi Cable in their 9.52-mm-outside-diameter Thermofin™ tube and (b) the resulting performance improvements. The 9.52-mm-outside diameter Thermofin-HEX™ tube has 60 fins with 0.20-mm height (e) and 40-degree fin apex angle (β), spaced at 2.32 times the fin height (p/e) and at 18-degree helix angle. The ratio values are stated relative to a plain tube (subscript p). $A_i/A_{i,p}$ is the total internal surface area ratio, and $\mathrm{Wt/Wt}_p$ is the tube weight ratio, per unit tube length. h/h_p is the evaporation coefficient (based on $A_i/L = \pi d_i$ relative to a plain tube operated at the same mass velocity).

Figure 13.12 shows the cross-section fin geometries associated with Table 13.1. Figure 13.13 shows the R-22 evaporation coefficient of the Table 13.1 tubes as a

Table 13.1   Chronological Improvements in the Hitachi Thermofin™ Tube
(d_o = 9.52 mm)

Year	Geometry	e	p/e	α	β	n	$A_i/A_{i,p}$	$\mathrm{Wt/Wt}_p$	h/h_p
1977	Original	0.15	2.14	25	90	65	1.28	1.22	2.0
1985	EX	0.20	2.32	18	53	60	1.51	1.19	2.6
1988	HEX	0.20	2.32	18	40	60	1.60	1.19	3.2
1989	HEX-C	0.25	2.32	30	40	60	1.73	1.28	3.2

Source: Yasuda et al. [1990].

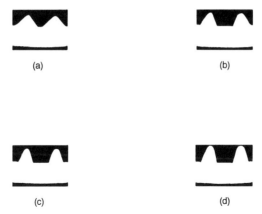

Figure 13.12   Cross sections of Hitachi Thermofin™ tubes. (a) Thermofin™, (b) Thermofin EX™, (c) Thermofin HEX™, and (d) Thermofin HEX-C™. (From Yasuda et al. [1990].)

function of mass velocity. Table 13.1 shows that the evaporating coefficient of the Thermofin HEX-C™ tube is 3.1 times that of a plain tube at the same mass velocity. The improved versions have higher fins (e) and sharper apex angles (β). Figure 13.14, from Ito and Kimura [1979], shows how the vaporization coefficient of the original Thermofin tube (Figure 13.12a) is influenced by helix angle. This figure

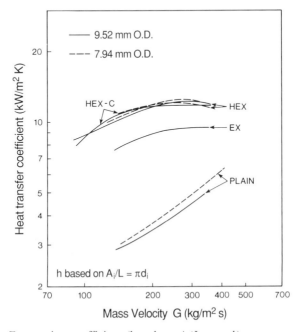

Figure 13.13   Evaporation coefficient (based on $A_i/L = \pi d_i$) versus mass flux for the Hitachi Thermofin tubes.

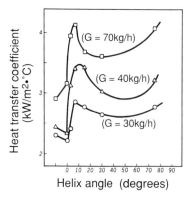

Figure 13.14 Effect of groove angle on evaporation coefficient for the original Hitachi Thermofin™ tube. (From Ito and Kimura [1979].)

shows that the optimum performance occurs at a helix angle of approximately 10 degrees. The vaporization coefficient significantly decreases for helix angles greater than 20 degrees.

A very interesting feature of the tube performance is its low refrigerant pressure drop. Figure 13.15, from Ito Kimura [1979], shows that the R-22 pressure drop of the Figure 13.12a Thermofin tube is only 10% higher than that of a plain tube. Yasuda et al. [1990] report that the R-22 pressure drop of the 9.52-mm Thermofin HEX-C™ tube (Figure 13.12d) is approximately 35% greater than that of a plain tube at 200 kg/s-m².

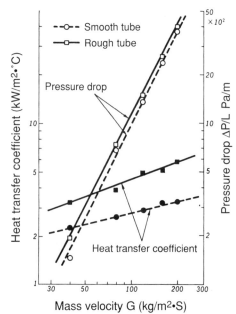

Figure 13.15 Comparison of evaporation coefficient and pressure drip versus mass velocity for original Hitachi Thermofin™ tube. (From Ito and Kimura [1979].)

Yoshida et al. [1987] performed a detailed experimental study attempting to explain the enhancement mechanism of the micro-fin tube. They measured local R-22 heat transfer coefficients on the top, side, and bottom of the tube for a range of vapor qualities and mass velocities. Their measurements were done for the three micro-fin geometries listed in Table 13.2. Tube C has primary grooves 0.24 mm deep at +30 degrees, and secondary grooves 0.15 mm deep at −15 degrees. Figure 13.16 shows the enhancement ratio, $E_h = h/h_p$, versus vapor quality for low and high mass velocities. At G = 300 kg/m²-s, all three tubes give approximately E_i = 1.5 for $0.2 \leq x \leq 0.9$. However, at G = 100 kg/m²-s, the tubes give much higher enhancement ratios. Tube B with 30 degree helix angle is better than tube A with 10 degree helix angle. Figure 13.17 explains why greater enhancement occurs at the lower G than at the higher value. Figure 13.17 shows the local heat transfer coefficients on the top, side, and bottom of tube B and the plain tube. At G = 300 kg/m²-s, there is little difference between the h-values at the three circumferential locations. However, at G = 100 kg/m²-s, the figure shows that much higher h-values are obtained on the top and sides of the tube than exist for the plain tube. Yoshida et al. conclude that the narrow grooves carry liquid to the sides and top of the tube by capillary wetting. Thus, thin films are provided around the entire tube circumference. Low heat transfer coefficients exist on the top and sides of the plain tube at low G, because the tube surface is dry.

Although Figure 13.14 shows an optimum helix angle of approximately 8 degrees, there is no uniform agreement on this optimum. The Shinohara et al. [1987] patent shows a relatively flat optimum of 13 degrees for R-22 evaporation in a 9.5-mm-outside-diameter tube.

Eckels et al. [1992] provide R-22 evaporation coefficients and pressure drop at 2.0°C for five currently used micro-fin tube geometries. The reported evaporation coefficient is the average value for 0.10 entering and 0.85 leaving vapor quality. Schlager et al. [1990] tested three 12.7-mm-outside-diameter micro-fin tubes having different helix angles (15, 18, and 25 degrees) with R-22. However, their tubes also had different fin heights ($0.15 \leq e_i \leq 0.3$ mm) and pitch, so they could not define the effect of specific geometry factors on the performance differences. Their test results for all three geometries agree very closely.

Schlager et al. [1988a, 1988b] and Eckels and Pate [1991] report the effect of oil on the performance of R-22, R-134, and R-12, respectively. Oil slightly reduces the evaporation coefficient for typically used oil concentrations.

Table 13.2 Micro-fin Tubes (15.8-mm Outer Diameter) Tested by Yoshida et al. [1987]

Tube	e	α	N	A/A_p
A	0.24	10	60	1.28
B	0.24	30	60	1.35
C	0.24/0.15	30/15	60	1.35

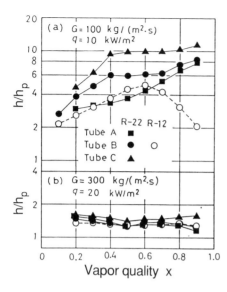

Figure 13.16 Circumferentially averaged R-22 evaporation coefficients in 12-mm-inside-diameter micro-fin tubes described in Table 13.2. (From Yoshida et al. [1987].)

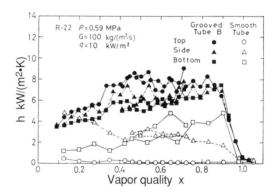

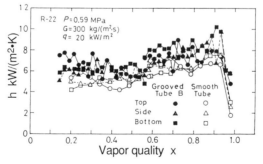

Figure 13.17 Local R-22 evaporation coefficients on top, side and bottom of 12 mm inside diameter micro-fin tubes described in Table 13.2. (a) G = 100 kg/m²-s, b) G = 300 kg/m²-s. (From Yoshida et al. [1987].)

13.3.3   Swirl Flow Devices

Twisted-tape inserts were investigated quite early in the search for enhancement techniques. This is because technology to internally fin or roughen a tube was not yet developed, and because twisted tapes were easy to make. However, their use is now limited, except for special situations discussed below. The reported test data include water (subcooled and saturated boiling), refrigerants, and liquid metals. The twisted-tape data are typically taken using electrical heating in the tube wall with a loosely fitting tape. The tape geometry is usually described as $y = H/d_i$, where H is the axial length for one-half tape revolution. The twist geometry may also be described in terms of the helix angle of the tape, which is $\tan \alpha = \pi d_i/2H = \pi/2y$.

Because the flow pattern changes along the tube length, it is probable that a given enhancement technique may work better for one flow pattern than for others. This is true for the twisted tape. It is particularly effective in the drywall region, where liquid flows as entrained droplets. The tape throws the liquid back on the tube wall. The vapor quality at which the critical heat flux occurs decreases for increasing flow rate and heat flux. Hence, improvements found at high heat fluxes and flow rates would not be expected at low heat fluxes and flow rates. Bergles et al. [1971] provide data on twisted tapes ($\alpha = 10$ and 22 degrees) to enhance heat transfer in the drywall region for boiling nitrogen at 140–170 kPa. Their tests with an electrically heated tube wall showed that the tapes approximately doubled the heat transfer coefficient in the post-critical vapor quality region, relative to a plain tube. However, the tapes did not move the critical vapor quality to a significantly lower value.

Jensen and Bensler [1986] tested R-113 in a vertical upflow using an electrically heated 8.1-mm-inside-diameter tube. Figure 13.18a shows the effect of vapor quality for $q = 46$ kW/m². The figure shows that h increases with x, and that h increases with increasing tape helix angle (α). Note that the $\alpha = 6.4$ degree tape provided equal or smaller heat transfer coefficients than did the plain tube. The maximum measured enhancement for the best tape ($\alpha = 38.6$ degrees) is 50%. Figure 13.18b shows the effect of heat flux for $x = 0.46$. Increasing heat flux benefits both the plain and enhanced tubes. This is apparently because of the nucleate boiling component. The highest-performing twisted tape ($\alpha = 38.6$ degrees) benefits least, because of its larger convective contribution. Jensen [1985] reported pressure drop data for the same tapes tested by Jensen and Bensler [1986].

Agrawal et al. [1982, 1986] tested R-12 in a 10-mm-inside-diameter horizontal tube with electric heating. Agrawal et al. [1986] tested helix angles between 8.8 and 22.7 degrees at 138 kPa with heat fluxes between 8 and 13.6 kW/m². Figure 13.19 shows the enhancement ratio, $E_h = h/h_p$, for the low and high heat fluxes. At the lowest heat flux (Figure 13.19a), negligible enhancement was provided for $x < 0.6$. Figure 13.19b shows that increasing the heat flux to 13.6 kW/m² provided greater enhancement for the tapes having $\alpha \geq 15.7$ degrees. Surprisingly, the $\alpha \geq 15.7$-degree tapes show poor enhancement for $x > 0.65$. This behavior is not expected, and confirmation is desirable. Note that the highest heat flux in Figure 13.19 is less than the smallest heat flux in Figure 13.18. Hence, convection should play a larger role in Figure 13.19 than in Figure 13.18. Blatt and Alt's [1963] tests for R-11 in a

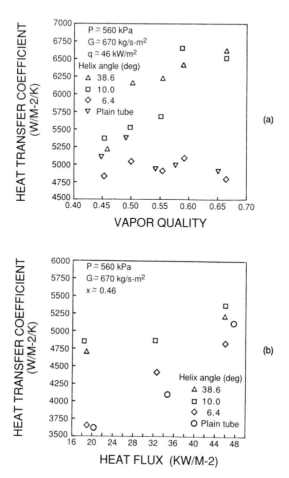

Figure 13.18   R-113 evaporation coefficient in 8.1-mm-inside-diameter vertical tubes. (a) Effect of vapor quality for $q = 46$ kW/m². (b) Effect of heat flux for $x = 0.46$.

6.4-mm-inside-diameter tube with condensing steam on the outer surface showed much less sensitivity to heat flux than observed by Agrawal et al. Agrawal et al. [1982] measured the adiabatic, friction pressure drop characteristics of the twisted-tape tubes. The friction pressure drop of the $y = 3.76$ tube was 1.7–3.7 times that of a plain tube. However, they do not define how the pressure drop ratio is influenced by vapor quality. One concludes that the frictional pressure drop ratio substantially exceeds the heat transfer enhancement ratio.

It is concluded that the twisted tape is a fairly low-performance enhancement device for refrigerants, as compared to the micro-fin tube. Furthermore, the pressure drop increases are significant, compared to the heat transfer increases. Its main application appears to be in situations where CHF may exist. Then, the swirl effect may delay the inception of CHF to a higher vapor quality. Jensen [1985] presents a

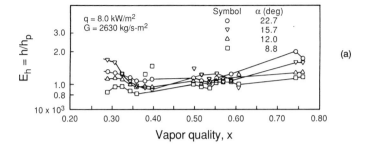

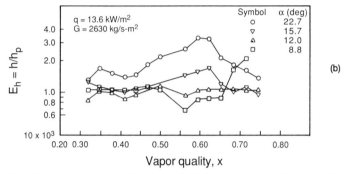

Figure 13.19 Enhancement ratio (h/h_p) for R-12 evaporating in 10-mm-inside-diameter horizontal tubes with twisted tapes. (a) Effect of vapor quality for $q = 8.0 \text{ kW/m}^2$. (b) Effect of vapor quality for $q = 13.6 \text{ kW/m}^2$.

parametric theoretical analysis of R-12 in a tube having a twisted tape for $3 \leq y \leq 21$, with different flow rates and external heat transfer coefficients. For conditions below the CHF, he shows that the R-12 side enhancement is between 20% and 40%. A twist ratio of three provides very little benefit over $y = 12$. The Jensen analysis includes calculation of pressure drop and flow power. He concludes that suppression of the CHF is the main attractive feature of a twisted-tape insert. More information on the CHF is given in Section 13.4.

13.3.4 Roughness

Among the earliest enhancement devices studied is the wire coil insert shown in Figure 1.2c. These devices typically provide a given enhancement at much higher pressure drop than is obtained by the integral roughness described below. A recent experimental program using wire coil inserts (1.24-mm wire diameter) for R-22 vaporization is described by Varma et al. [1991].

Withers and Habdas [1974] investigated R-12 vaporization at 0°C in 19.1-mm-outside-diameter (16.3-mm-inside-diameter) single-helix corrugated tubes of the type illustrated in Figure 9.3c. Their data were taken for entering saturated liquid and 100% exit vapor quality, with heat input from high-velocity water in an annulus. A Wilson plot method was used to derive the average vaporization coefficient in the 3.96-m-long test section. The geometry parameters of the corrugated tube are the dimensionless roughness height (e/d_i), and roughness spacing (p/e). Data were obtained for (a) six roughness geometries having $0.4 \leq e \leq 1.6$ mm and $3.2 \leq p \leq 15.9$ mm and (b) a plain tube. The authors proposed that the combined effect of e/d_i and p/e may be described by the *severity factor*, which they define as $\phi \equiv (e/d_i)/(p/e)$. Figure 13.20 shows the tube performance as a function of the severity factor. Note that the refrigerant flow rate is not constant on the figure. The R-12 flow rate is determined by the entering water temperature (15.6°C) and the vaporization coefficient. The geometry with the highest vaporization coefficient will have the highest flow rate. The highest performance is provided by $\phi = 0.004$, for which the vaporization coefficient is 2.9 times that of the plain tube ($\phi = 0$) with a 1.8 times higher flow rate. The pressure drop (friction plus momentum) ratio $(\Delta p/\Delta p_p)$ and mass flow rate ratio (W/W_p) are also shown in Figure 13.20. At $\phi = 0.004$, $W/W_p = 2.8$, which means that the evaporation rate in the enhanced tube is 2.8 times that in the plain tube. The large $\Delta p/\Delta p_p$ is partially a result of the substantially higher flow rate in the enhanced tube. Withers and Habdas recommend use of $\phi = 0.00234$ ($e = 0.8$ mm and $p = 15.9$ mm) because the R-12 pressure drop is lower. This tube is offered by Wolverine as a commercial product, known as Koro-Chil™.

Shinohara and Tobe [1985] describe an improvement on the corrugated tube tested by Withers and Habdas. This tube is a corrugated micro-fin tube called TFIN-CR™ and is shown in Figure 13.21d. Table 13.3 compares the dimensions and performance of a 16-mm-outside-diameter TFIN-CR™ tube with the standard corrugated tube (CORG™), a micro-fin tube (TFIN™) and five-leg star fin insert. The heat transfer coefficients are based on the outer diameter (16 mm). The micro-

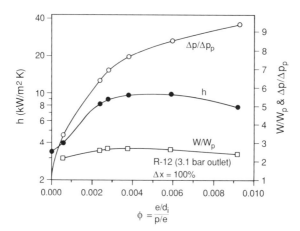

Figure 13.20 R-22 vaporization performance of 19.1-mm-diameter corrugated tubes of Withers and Habdas [1974] plotted versus severity factor (ϕ).

Figure 13.21 Photographs of evaporator tubes tested by Shinohara and Tobe [1985]. (a) Star-fin insert, (b) corrugated tube (CORG™), (c) micro-fin (TFIN™), and (d) corrugated micro-fin (TFIN-CR™). (From Shinohara and Tobe [1985].)

grooves in the TFIN™ tubes are 0.2 mm, and the helix angle is 8 degrees. The 5.7-m-long test section was operated at 2°C. Note that the enhancement level of the TFIN tube is not as high as listed in Table 13.1. Furthermore, the enhancement ratio of the CORG™ tube is not as high as measured by Withers and Habdas [1974], although its value of ϕ is higher ($\phi = 0.0042$).

Akhanda and James [1988] provide data for convective vaporization of sub-cooled water ($p = 11,000$ kPa) in a rectangular cross-section channel containing 0.5-mm-high ribs with axial pitches between 0.5 and 3.50 mm.

13.3.5 Coated Surfaces

The High-Flux™ porous coating may be applied to the inner tube surface. Whereas internal fins, corrugated tubes, and twisted tapes provide enhancement by affecting the convective term of Equation 13.2, the porous internal coating primarily affects

Table 13.3 Tube Comparison for R-22 Evaporation at $G = 100$ kg/h

Geometry	e_i (mm)	p (mm)	h (kW/m²)	Δp (KPa)	h/h_p	$\Delta p/\Delta p_p$
Plain	0	NA[a]	1.63	20.6	1.00	1.00
Star insert	0	NA	2.06	35.8	1.26	1.74
CORG™	0.7	8.0	2.15	32.4	1.32	1.57
TFIN™	0	NA	2.19	26.0	1.34	1.26
TFIN-CR™	0.7	8.0	2.99	32.4	1.83	1.57

[a]NA, not applicable

the nucleate boiling term. This provides very high performance for convective vaporization. The performance of the porous coated tube is quite sensitive to heat flux, because of the large nucleate boiling contribution. Figure 13.22 shows the data of Czikk et al. [1981] for the High-Flux™ tube with vertical up- flow of oxygen (101 kPa) in an 18.7-mm-diameter tube and is taken from Thome [1990]. The performance is compared with that of a plain tube, whose performance was pre- dicted by Thome [1990] using the Chen [1966] superposition model (Equation 13.2). The exit vapor quality is shown by the numbers in parentheses. At the same heat flux, the $T_w - T_s$ of the porous tube is approximately one-tenth that of the plain tube. Also shown in Figure 13.22 is the High-Flux™ pool boiling data of Antonelli and O'Neill [1981] at 101 kPa. The convective boiling points are in reasonable agreement with the pool boiling data, which show that the performance of the tube is dominated by the pool boiling term of Equation 13.2. Note that vapor quality and mass velocity have little effect on the High-Flux™ tube performance, which is another indication that nucleate boiling dominates the tube performance. The very low values of ΔT_{ws} are probably responsible for the data scatter.

The High-Flux™ tube shows an effect of mass velocity for horizontal flow. The ammonia data of Czikk et al. [1981] shows increasing performance with mass velocity for the range tested ($30 \leq G \leq 450$ kg/m2-s). This is apparently because the mass velocity is not high enough to produce annular flow. Hence, the upper

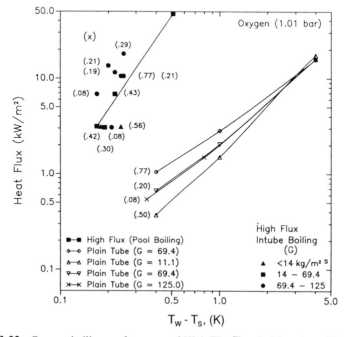

Figure 13.22 Oxygen boiling performance of High-Flux™ and plain tubes. (From Thome [1990].)

surface of the tube is partially dry. Wetting improves with mass velocity, which explains the mass velocity effect. Although some dryout may exist, the High-Flux™ performance was still 10 times better than that of a plain tube.

Ikeuchi et al. [1984] provide R-22 vaporization data in a horizontal porous coated tube. Their tests were performed with 15% inlet vapor quality and 70–95% exit vapor quality or with 5 K superheat. Figure 13.23 shows their data, as reported by Thome [1990]. Note that the data for 5 K superheat are lower than those for incomplete evaporation for both the plain and the High-Flux™ tube. The reduced performance at 5 K superheat is a result of CHF dryout at high vapor quality. These data clearly show the superiority of the High-Flux™ tube at all vapor qualities.

13.3.6 Perforated Foil Inserts

Palm [1990] has tested the perforated foil inserts discussed in Section 11.4.6 for convective vaporization of R-22 in a 15-mm-inside-diameter tube. Eight different foil inserts were tested; their hole diameters were between 0.10 and 0.22 mm, and their hole densities were between 0.5 and 1.5 holes/mm². At 50% vapor quality and heat fluxes below 10 kW/m², the best foil increases the local heat transfer coefficient by 50%. The enhancement is low, relative to the porous coated tube (Section 13.3.5) and the micro-fin enhancement. Poor performance was obtained at high vapor quality, presumably because the space between the tube wall and the foil was dried out. Conklin and Vineyard [1992] found little performance benefit of a foil insert having 0.5 holes/mm² for R-22.

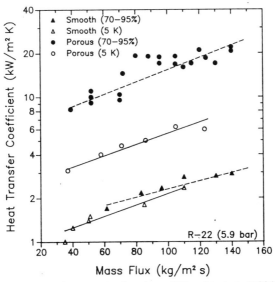

Figure 13.23 R-22 convective vaporization data of Ikeuchi et al. [1984] for porous and plain tubes, as reported by Thome [1990]. (From Thome [1990].)

13.3.7   Coiled Tubes and Return Bends

Coiled tubes are used in a variety of situations, including chemical reactors, steam generators, agitated vessels, and storage tanks. The centrifugal force imposed on the two-phase flow imposes a secondary flow imposed on the two-phase flow regimes. Liquid droplets are thrown to the outer wall, and the liquid film spirals along the tube wall to the inner surface of the coil. This effect causes an increase of the dryout heat flux. A number of publications exist on vaporization in coiled tubes. Jensen and Bergles [1981] provide a recent literature update.

Crain and Bell [1973] tested two helical coils with steam and an electric heated tube wall at approximately 170 kPa with $0.45 \leq x \leq 1.0$ and developed a correlation with an average error of 44%. The correlation, which does not account for the effect of pressure, is

$$\frac{hd_i}{k_l} = 0.0587 \left[\frac{d_i G (1 - x)}{\mu_l} \right]^{0.85} Pr_l^{0.4} \left(\frac{d_i}{D_c} \right)^{0.1} x^{-7.6} \qquad (13.10)$$

Equation 13.10 shows that little enhancement is provided by a coil, since the $(d_i/D_c)^{0.1}$ contribution is small. The -7.6 exponent on the vapor quality makes the correlation appear questionable.

Campolunghi et al. [1976] tested a large (1 MW) coil having $d_i = 15.5$ mm and $D_c = 836$ mm with steam over $80 \leq p \leq 170$ bar. Their correlation shows that $h \propto \exp(0.0132p)$. Hughes and Olson [1975] measured the DNB and the CHF for R-113 in a planar coil. They found that the CHF is higher on the concave surface than on the convex surface. Jensen and Bergles [1981] measured the DNB and dryout CHF conditions in several electrically heated coils using R-113. They provide correlations for both CHF conditions.

Gu et al. [1989] tested a subcooled fluorocarbon fluid (FC-72) in a single, glass rectangular cross section channel, curved over 180 degrees. The channel radius was 50.8 mm, and the channel cross section was 5.6 mm × 27.0 mm. The channel was locally heated on the outer radius by electrical elements that were (a) 9.5 mm long upstream of the bend and (b) 135 degrees around the bend. Local heat transfer data were obtained at the location of the two heater elements. They observed higher heat transfer coefficients and higher CHF in the curved channel. They provided a correlation to account for the effect of centrifugal acceleration. Note that the data and correlation do not account for effects that occur on the inner wall of the channel, at which the limiting conditions exist.

13.4   CRITICAL HEAT FLUX (CHF)

The literature review in Chapter 11 showed that structured and porous enhanced surfaces increase the CHF for nucleate pool boiling. Section 11.13 describes a mechanism by which the CHF is increased for a porous boiling surface. Equation 13.1 applies to convective vaporization and states that the total heat flux is the sum of the nucleate boiling and convective contributions. Hence, increasing the convective component can also lead to an increase of the CHF. As noted in Section 13.2.6, there are two CHF regions which occur at two different flow conditions: (1) DNB, which occurs at subcooled or low vapor quality and high heat flux, and (2) dryout, which occurs at higher vapor quality.

13.4.1 CHF in Tubes

13.4.1.1 Twisted Tape The twisted tape can increase the DNB and the CHF for subcooled boiling. Gambill et al. [1961] performed extensive tests with subcooled boiling of water. At high velocities, the CHF was increased as much as 200%. At lower velocities, the increase was only 20%. The velocity effect is explained by Equation 13.2. At increased velocity, convective contribution supports a higher heat flux at a given $T_w - T_s$. Gambill [1963, 1965] also speculates on reasons for the increased CHF. Bergles et al. [1971] investigated dryout CHF for nitrogen (140–170 kPa) in a 10-mm-diameter vertical tube having twisted tapes ($y = 4.1$ and 8.5). They show that the tape delays the vapor quality at which dryout occurs. They develop a superposition-based correlation to predict the heat transfer coefficient in the evaporation and the drywall regions. Cumo et al. [1974] also provide data for R-12 in vertical tubes ($y = 4.4$).

Jensen [1984] developed a correlation to predict the CHF for twisted tapes, relative to the CHF in a plain tube. The correlation is

$$\frac{q_{cr}}{q_{cr,p}} = (4.597 + 0.0925y + 0.004154y^2)\left(\frac{\rho_l}{\rho_v}\right) + 0.09012 \ln\left(\frac{a}{g}\right) \qquad (13.11)$$

where $q_{cr,p}$ is the critical heat flux in a plain tube operated at the same inlet flow conditions, and a is the radial acceleration given by

$$a = \left(\frac{2}{d_i}\right)\left(\frac{u_a \pi}{2y}\right)^2 \qquad (13.12)$$

13.4.1.2 Grooved Tubes Steel grooved tubes of the type shown in Figure 13.7b are used in vertical "wet wall" power boilers. These boilers operate with subcooled or low vapor quality water. Swensen et al. [1962] show that the grooved tubes permit operation at higher heat flux and lower mass flow rate than is possible with plain tubes. Watson et al. [1974] also investigated the CHF with water and defined the minimum mass velocity needed to produce swirl flow. In the "wet wall" boiler, a radiant heat flux is applied to one side of the tube. Kitto and Wiener [1982] investigated the effect of nonuniform circumferential heat flux and tube inclination angle on the CHF. For a peak-to-average heat flux ratio of 1.9, they observed that the CHF was higher than that for a uniform circumferential heat flux. The CHF was 30% smaller for 30-degree inclination angle (from the vertical), relative to the vertical orientation. Their tests showed that the CHF for the ribbed tube was as much as three-times that of a plain tube for the same flow conditions and orientation. Additional data are provided by Chen et al. [1992].

13.4.1.3 Corrugated Tubes   Withers and Habdas [1974] evaporated saturated R-12 in 19-mm-diameter, horizontal corrugated tubes (Figure 9.3b), as described in Section 13.3.4. The tests investigated the dryout CHF. They found that the dryout condition occurred at much higher vapor quality in the corrugated tubes, relative to a plain tube at the same operating conditions. They were able to obtain 100% evaporation in the corrugated tube at flow rates up to three times that attainable in the plain tube. The supportable heat flux increased in direct proportion to the flow rate increase.

13.4.1.4 Mesh Inserts   Mergerlin et al. [1974] showed that steel mesh and brush-type inserts increased the CHF by several hundred percent. However, the pressure drop increase was horrendous!

13.5 PREDICTIVE METHODS FOR IN-TUBE FLOW

Predictive equations have been developed for several of the enhancement techniques. These equations are typically based on modifications to vaporization in plain tubes. Schlager et al. [1990] provide a literature survey of correlations for vaporization inside tubes. Although these correlations may contain fluid property groups, their development may have been based on data for only one fluid and at one pressure. Hence, they cannot be reliably used for other fluids, or even other pressures with the same fluid. Frequently, the modification to account for the enhancement geometry is derived from terms previously developed to account for single-phase flow in that particular geometry. Only heat transfer correlations are presented here. Refer to the references for the accompanying pressure drop correlations.

There are a variety of correlations for vaporization in plain tubes. Thus, the geometry modifier is written in dimensionless terms and is derived by linear multiple regression methods. Such methods have no rational basis and use assumed dimensionless parameters and statistical correlating methods. Furthermore, there is no guarantee that the dimensionless values used are general. In fact, some of the plain tube correlations are based on similar statistical methods. What makes the correlations for the enhanced surfaces more risky is that the enhanced tube correlation contains more geometric variables than does the plain tube correlation.

There are a variety of correlations for vaporization in plain tubes. Thus, the modified correlation developed for the enhanced tube is as strong (or weak) as the base correlation. One of the most popular correlations is the Chen [1966] model defined by Equation 13.2. This model adds a convective term and a nucleate boiling term, which depends on heat flux. An early in-tube vaporization correlation developed by Pierre [1964] for refrigerants is

$$\frac{h d_i}{k_l} = B \left[\left(\frac{\Delta x \lambda}{L} \right) \left(\frac{d_i G}{\mu_l} \right)^2 \right]^n \tag{13.13}$$

when $x \leq 0.9$ ($B = 0.0009$, $n = 0.5$) and for $x > 0.9$ ($B = 0.0082$, $n = 0.4$). In contrast to the Chen model, the Pierre correlation does not contain heat flux. If the enhancement is expected to have a significant nucleate boiling contribution, one would not expect that modification of the Pierre correlation would work very well. Conversely, if there is no nucleate boiling contribution (no heat flux dependence), a correlation based on the Pierre correlation would be acceptable. With this introduction, we will briefly discuss the correlations that have been developed for the various tube-side enhancements.

13.5.1 High Internal Fins

The Azer and Sivakumar [1984] correlation uses the correlating parameters of the Pierre [1964] equation, multiplied by geometry factors to account for the internal fin geometry. This correlation is

$$\frac{hd_i}{k_l} = B\left[\left(\frac{J\Delta x\lambda}{L}\right)\left(\frac{d_iG}{\mu_l}\right)^2\right]^n(1 + 0.0024F_1^{3.72} F_2^{-8.88}) \qquad (13.14)$$

where $B = 12.24$, $n = 0.146$, and F_1 and F_2 account for the fin geometry. The parameters F_1 and F_2 were used by Carnavos [1980] to correlate single-phase flow data in internally finned tubes and are discussed in Chapter 8. These parameters are defined as follows: $F_1 = A_{fa}/A_{fc}$ (actual flow area/core flow area between fin tips), and $F_2 = A_n/A_a$ (plain tube surface area/total surface area). Because the correlation was based only on R-113 data, one should use the correlation with caution!

Schlünder and Chawla [1967] provide an empirical correlation for aluminum inserts (Figure 13.10) based on their R-11 test data. The correlation is not expected to apply to different fluids.

13.5.2 Micro-fins

Very little work has been done to develop predictive methods for the micro-fin tube (Figure 13.7d). Cui et al. [1992] developed an empirical correlation of their R-502 data on nine tube geometries. The correlation uses the parameters of the Pierre [1964] plain tube correlation, plus a parameter to account for the micro-fin geometry.

13.5.3 Twisted-Tape Inserts

The Agrawal et al. [1986] correlation is also based on modifying the Pierre [1964] correlation. It is

$$\frac{h}{h_{Pie}} = 0.00188Re_s^{2.23}Bo^{1.62}y^{-0.357} \qquad (13.15)$$

where h_{Pie} is the value calculated by the Pierre equation (Equation 13.13) for a plain tube. The swirl Reynolds number (Re_s) is defined in Chapter 7, and the term Bo is the Boiling number ($q/G\lambda$) which includes a heat flux dependency. The correlation is based on R-12 data for three twisted tapes. As noted in Section 13.3.3, one would not expect the twisted tape to show a heat flux dependency unless it were operated at quite high heat flux.

13.5.4 Corrugated Tubes

No correlations are reported in the literature. However, Withers and Habdas [1974] provide a curve-fit of their R-12 data on two corrugation geometries using the Pierre [1964] correlation, Equation 13.13.

13.5.5 Porous Coatings

No correlations have been developed and tested. However, Ikeuchi et al. [1984] curve-fit their R-22 data in the form of the Pierre [1964] correlation, Equation 13.13. It would be reasonable to predict the porous tube heat transfer coefficient using the Chen [1966] correlation or a later variant. This approach would add the pool boiling and convective components.

13.6 TUBE BUNDLES

Chapter 11 provided information for pool boiling on enhanced tubes. Because the enhanced tubes are intended for use in tube bundles, it is important to know whether tube bundle data will differ from single-tube pool boiling data. Two effects can be active in a tube bundle, which may alter the performance from single-tube results. These are: (1) static head effects and (2) convective effects.

For fluids having a large change of saturation temperature with change of pressure (dT/dp), static liquid head will cause increased saturation temperature in the lower part of a tube bundle, and thus reduce the local driving temperature difference. This was discussed in Chapter 4, and Figure 4.2 shows dT/dp for several fluids. Because of the two-phase mixture in the tube bundle, however, the static head effect will be less than indicated in the nonboiling condition for the same liquid level.

Boiling in tube bundles may span a wide range of flow conditions. In kettle reboilers, this typically involves an entering subcooled liquid and a vapor quality change of 15% or less over the bundle depth. However, a flooded refrigerant evaporator is essentially a once-through device, with 15% entering vapor quality and 100% leaving vapor quality. The mass velocity in a kettle reboiler is higher than that in a flooded refrigerant evaporator.

Early work to simulate boiling in tube bundles was unable to simulate this wide range of flow conditions. Hence, one should be careful to generalize about such simulations. Examples of early laboratory simulations of boiling in tube bundles are Fujita et al. (1986) and Muller (1986). Muller used saturated R-11 at 101 kPa in a bundle having 18 finned tubes in six rows with three tubes in each row. Fujita et al.

(1986) obtained data for R-113 boiling on an 11-tube bundle (7 rows) of 25.4-mm-diameter plain tubes. The effects of tube position in the bundle, distribution of heat flux, and system pressure were studied. They found that the boiling coefficient on the bottom tube row was approximately equal to that for a single tube. However, substantial enhancement was observed for the upper tube rows at moderate heat flux. In the experiments of Muller (1986) and Fujita et al. (1986), the refrigerant entered the boiling vessel as saturated liquid with zero vapor quality. Although these papers give an insight regarding the effect of tube position and intertube spacing on the boiling heat transfer, they involve quite low mass velocity and leaving vapor quality.

As shown by Equation 13.2, the composite heat transfer coefficient will be significantly affected by the use of an enhanced boiling surface. Later studies will be discussed that allow a more perceptive understanding of the effect of enhanced tubes in tube bundles.

13.6.1 Convective Effects in Tube Bundles

As noted in Section 13.2.5, the performance of a tube, when used in a tube bundle, will depend on the enhancement level for pool boiling. Depending on the pool boiling performance level of the tube, it will show one of the following two characteristics in convective vaporization:

1. *High Pool Boiling Coefficient:* The tube will not be very sensitive to convective effects. However, it will be quite sensitive to heat flux, which indicates that the tube performance is dominated by nucleate boiling.
2. *Low Pool Boiling Coefficient:* This tube will benefit substantially from convective effects. However, its performance in convective vaporization will not be very sensitive to heat flux. This will occur if the performance is convection-dominated.

Nakijima and Shiozawa [1975] obtained test data for an R-11 evaporator having 185 19-mm-diameter tubes (748 fins/m, 1.5-mm fin height) with four water passes on the tube side. Water passes 1–2 were in the lower part of the tube bundle, and passes 3–4 were in the upper part of the bundle. By measuring the water temperature leaving pass 2, they were able to determine the evaporation coefficient for the bottom and top halves of the tube bundle. For 15% inlet vapor quality, the evaporation coefficient in the lower half of the bundle was 4–7 times higher than that of a single tube in pool boiling. The evaporation coefficient in the upper half of the bundle was 1.5 times higher than in the lower half of the tube bundle. Convective effects cause the higher evaporation coefficient in the upper half of the bundle.

Webb and Gupte [1992] provide description and discussion of correlations for convective vaporization in tubes and in tube bundles. Webb and Gupte also provide convective vaporization data for the Figure 11.12a, e, and f tube geometries using R-11, and R-123 at operating conditions typical of refrigerant evaporators (5°C). The data were taken on 19.0-mm-diameter tubes on 23.8-mm triangular pitch in a tube bundle simulator, which allowed independent control of heat flux, mass velocity, and vapor quality. Pool boiling data of these geometries for R-22 are shown in Figure 11.15. Figure 11.15 shows that the GEWA-SE™ tube provides a substantially higher boiling coefficient than the integral-fin tube. The importance of convective affects is demonstrated by comparing the GEWA-SE™ tube with the lower-performance integral-fin tube.

Figure 13.24a shows the 1024-fin/m integral-fin tube data of Gupte [1992] at $G \simeq 13$ kg/m²-K. Figure 13.24a shows that the heat transfer coefficient is a strong function of vapor quality, whereas heat flux has a small effect. This means that the performance is dominated by convection, rather than by nucleate boiling. As the vapor velocity increases, the convective contribution increases because of the increased vapor velocity. Data for different mass velocities (not shown here) show that the integral-fin tube performance is significantly affected by mass velocity.

Figure 13.24b, from Gupte [1992], shows the convective boiling coefficient of the GEWA-SE™ tube with R-134a at 26.7°C (48 kPa) and mass velocity $G \simeq 13$ kg/m²-s. In contrast to Figure 13.24a, the figure shows that the GEWA-SE™ performance is highly sensitive to heat flux, which indicates that the performance is relatively insensitive to convective effects. Examination of the data at different mass velocities (not shown here) shows negligible effect of mass velocity. Note that the GEWA-SE™ tube maintains its high nucleate-boiling-dominated performance up to the highest vapor qualities tested (0.95).

Figure 13.25 shows the ratio of the pool boiling heat transfer coefficient to the convective vaporization coefficient for the two tubes. Figure 13.25a shows that $h_{nbp}/h < 1$ for the integral-fin tube. The h_{nbp}/h ratio decreases with increasing vapor quality, which indicates the importance of convection. The h_{nbp}/h ratio is smaller at low heat flux, where the nucleate boiling contribution is smallest. Figure 13.25b shows that the h_{nbp}/h ratio is approximately 1.0 for the GEWA-SE™ tube, and that the ratio is insensitive to vapor quality. Furthermore, the ratio is relatively insensitive to mass velocity, which indicates that the tube performance is nucleate-boiling-dominated.

This example shows that a tube having high pool boiling performance may also be dominated by nucleate boiling in tube bundle applications. However, the performance improvement (relative to integral-fin tubes) in an evaporator tube bundle is not as high as is measured in single tube pool boiling tests. This is because the forced convection enhancement that occurs in the tube bundle is more beneficial to standard integral-fin tubes than it is to the enhanced nucleate boiling tubes. Webb et al. [1990] compare the performance of large flooded refrigerant evaporators having integral-fin and enhanced tubes.

Jensen and Hsu [1987] reported R-113 data for boiling in plain tube bundles. The data were taken on 8-mm-diameter tubes at much higher temperatures (71–110°C) and mass velocities (100–675 kg/m²-s) than studied by Gupte and Webb [1992]. Because of the high mass flow velocities tested by Jensen and Hsu [1987], the exit vapor qualities were quite low,—that is, less than 0.35. These data showed little effect of vapor quality, possibly because of the low vapor quality range tested.

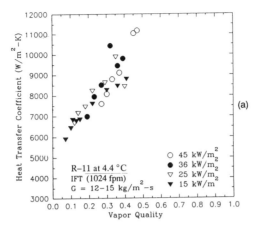

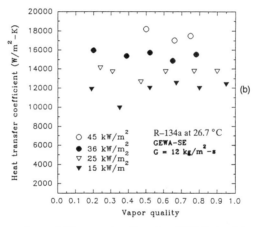

Figure 13.24 Comparison of 19-mm-outer-diameter. 1024-fin/m integral-fin and GEWA-SE™ tubes in a simulated tube bundle with 23.8-mm equilateral pitch. (a) 1024-fin/m integral-fin tube with R-11 at 4.4°C. (b) GEWA-SE™ tube with R-134a at 26.7°C.

Jensen et al. [1992] also provide tube bundle data on plain, Turbo-B™ and High-Flux™ tubes. For p = 206 kPa, 80 kW/m², and G = 217 kg/m²-s, the negligible effect of vapor quality was observed for the High-Flux™ and Turbo-B™ tubes, as expected. The plain tube at q = 30 kW/m² inexplicably showed no effect of vapor quality for exit vapor qualities as high as 0.74.

Yilmaz et al. [1981] and Arai et al. [1977] also show that plain or integral-fin tubes yield a substantially higher boiling coefficient in the tube bundle geometry because of the forced convection contribution. However, the boiling coefficient of enhanced tubes is so high that the forced convection effects are not expected to materially enhance their performance in a tube bundle configuration. Because of the

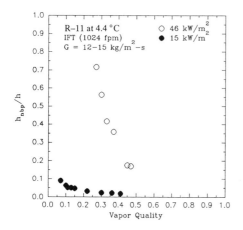

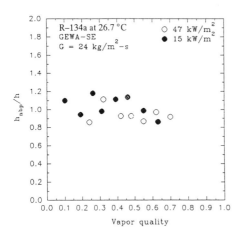

Figure 13.25   Ratio of pool boiling-to-convective vaporization coefficients for the Figure 13.24 simulated tube bundle data. (a) 1024-fin/m integral-fin tube with R-11 at 4.4°C. (b) GEWA-SE™ tube with R-134a at 26.7°C.

higher performance of the plain or integral-fin tubes in the bundle configuration, the designer is cautioned not to expect the same boiling coefficient improvement, as measured in pool boiling.

Figure 13.26 shows the R-11 performance of the High-Flux™ surface measured in pool boiling and tube bundle tests reported by Czikk et al. [1981]. All surface coatings were made from the same production run. This figure shows that the tube bundle performance is as high as or higher than that measured in pool boiling tests. Arai et al. [1977] report similar results for R-12 boiling on the Thermoexcel-E™ surface.

Bukin et al. [1982] measured convective vaporization in small tube bundles for 13 different types of porous coatings. Their porous surfaces were made by flame spraying, by metallic deposition, by sintering, and by wrapping the tube with layers

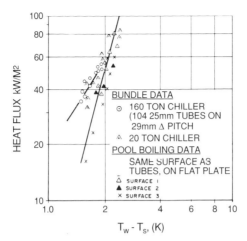

Figure 13.26   R-11 boiling at 39 kPa on the High-Flux™ surface for pool boiling and a 106 tube bundle.

of glass or stainless steel screening. The sintered porous coatings provided the best performance, and much better than that of plain or integral-fin tubes. Single tube data were not provided.

13.6.2   Starting Hysteresis in Tube Bundles

Lewis and Sather [1978] boiled a subcooled ammonia feed from a High-Flux™ tube bundle having 279 tubes. Heat was supplied by warm water on the tube side. The authors observed that a shutdown caused apparent deactivation of the nucleation sites resulting from flooding. When the boiler was restarted after a shutdown period of several hours and operated at the same heat flux (17.4 kW/m²), a smaller overall heat transfer coefficient was observed. About 100 hr of operation was required for the U value to climb from the low initial value (U_o = 600 W/m²-K) to the stable value of U_o = 785 W/m²-K. It appears that this phenomenon results from the starting hysteresis discussed by Bergles and Chyu [1982]. Apparently, the temperature boundary condition did not provide sufficient liquid superheating to immediately activate all of the nucleation sites. Thus, their activation was a gradual process. Lewis and Sather describe a special start-up process that allows vaporization of the liquid in the flooded pores and hence avoidance of the 100-hr period for site activation. Bergles and Chyu's analysis of the Lewis and Sather data suggests that the boiling coefficient corresponding to the stable U_o = 785 W/m²-K is in close agreement to that measured in a single-tube boiling test.

Boiling tests in R-11 and R-12 evaporators by Arai et al. [1977] have not shown the initially low performance on start-up as observed by Lewis and Sather. This results from a different start-up procedure of the refrigeration cycle. In the shutdown condition, the refrigerant liquid is saturated or very close to being saturated. When the compressor is started, the evaporator pressure is lowered, which acts to evaporate the liquid. Because the refrigerant is throttled to the evaporator from the higher condenser pressure, a two-phase mixture enters at the bottom of the tube bundle. This vapor may tend to displace liquid from deactivated cavities. This suggests that there may be some advantage in site activation by bringing a two-phase mixture in at the bottom of the tube bundle.

The author's research has shown that impingement of a two-phase jet on a superheated surface will activate nucleation sites. Czikk et al. [1981] and O'Neill [1981] state that the low initial performance measured by Lewis and Sather has not been noted in process applications of the High-Flux™ boiling surface. O'Neill has observed no advantage in introducing a two-phase feed at the bottom of the tube bundle. The possibility of lower initial performance on start-up may be of little significance in process applications. This equipment is usually brought on line and maintained in continuous operation for very long time periods (e.g., months or years). The existence of lower initial performance for the first hundred hours of operation would be negligible significance. Bergles and Chyu [1982] provide interesting speculations on tube bundle versus single-tube performance and possible start-up problems due to hysteresis.

13.7 PLATE-FIN HEAT EXCHANGERS

Brazed aluminum, plate fin heat exchangers are used in cryogenic gas-processing operations (air separation, and natural gas liquefaction) and for the separation of light hydrocarbons in ethylene plants. Used in an air separation plant, they will condense nitrogen against evaporating liquid oxygen. Figure 13.27 shows a brazed aluminum plate-fin heat exchanger for cryogenic service. Typical fin geometries used in this heat exchanger are shown in Figure 5.2. The same fin geometries are used for both single- and two-phase heat transfer.

The same convective vaporization phenomena exist in plate-fin heat exchangers as exist in tube bundles. Equation 13.2 still applies. Typical brazed plate-fin exchangers generally do not use enhanced boiling surfaces. Hence, convective effects will tend to be quite important. References treating convective vaporization in plate-fin heat exchangers include Robertson [1980], Carey and Shah [1988], Chapter 13 of Thome [1990], and Chapter 13 of Carey [1992]. Thome [1990] summarizes the available data on plate-fin geometries, and he also summarizes the empirical correlations that have been developed.

Mandrusik and Carey [1989] describe a method to predict the boiling coefficient in a plate-fin exchanger containing the offset strip fin (Figure 5.2d). The prediction is based on use of Equation 13.2. The authors use an accepted plain surface correlation to calculate h_{nb}, and they account for the fin efficiency. The h_{cv} term in Equation 13.2 is predicted using $h_{cv} = Fh_l$ with the Chen model to predict the F factor. The single-phase component (h_l) may be predicted using the correlations for the offset strip fin given in Chapter 6. A correlation for the F factor was developed based on their convective boiling data in a laboratory model of the offset strip fin array using R-113 as the working fluid. The authors obtained the F factor from their

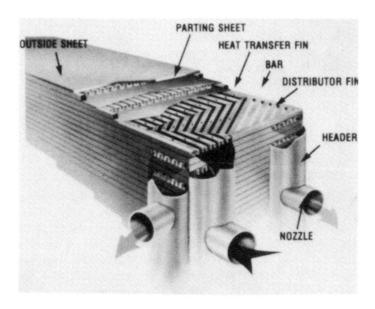

- ASME design pressure—Vacuum to 1,200 psig
- ASME design temperatures— –450 to +150 F
- Cross section—Up to 45" x 72"
- Length—Up to 240"
- Materials—Aluminum alloys 3003, 3004, 5083, 6061

Figure 13.27 Brazed aluminum plate-and-fin heat exchanger for cryogenic service. (Courtesy of Altec, LaCrosse, WI.)

data using Equation 13.5, and they correlated it as a function of X_{tt} as is accepted practice. They obtained

$$F = \left(1 + \frac{28}{X_{tt}^2} \right)^{0.372} \tag{13.16}$$

They used the suppression factor of Bennett and Chen [1980]. Their method follows accepted procedures for convective vaporization, as described, for example, by Carey [1992] or by Webb and Gupte [1992]. The reader should be careful about applying the F factor correlation developed for the offset strip fin to other plate-fin surface geometries. The F factor may be geometry-sensitive. Work has not been done to develop F factor correlations for other enhanced plate-fin geometries.

The surface geometry used in plate-fin heat exchangers is typically aluminum sheet metal, which should exhibit the nucleate boiling characteristics of a plain surface. A surface geometry that gives high single-phase convective heat transfer will give a high convective evaporation coefficient (cf. Equation 13.2). The offset strip fin is popular for convective boiling, because of its high single-phase heat transfer performance. It is conceivable that an enhanced boiling surface can be used in plate-fin geometries. In fact, this was the intent of Kun and Czikk [1969], who developed the Figure 11.13a enhanced surface for proprietary use in plate-fin exchangers. It is also feasible to use a porous coating, such as the Figure 11.11a sintered surface. This should have a significant effect on the nucleate boiling term of Equation 13.2.

13.8   THIN FILM EVAPORATION

Thin film evaporation is frequently used in the refrigeration, process, desalination, and cryogenic industries. It may be used inside vertical tubes, or on the outside of a bundle of horizontal tubes. For a horizontal tube bundle, one uses spray nozzles to distribute the film over the bundle frontal area. Thin film evaporation is particularly valuable for pressure-sensitive fluids. If pool boiling were used for such fluids, the elevation of saturation pressure over the bundle depth decreases the local $T_w - T_s$.

Typically, the liquid feed rate is several times the amount evaporated. This is done to prevent de-wetting of the tube surface. The recirculation ratio (R_R) is defined as the total flow rate divided by the rate evaporated. If any fraction of the tube is dry, the tube performance of the dry region is essentially "lost." Heated films are quite susceptible to rupture. When falling films are used in a bundle of tubes, entrainment effects will also act to diminish the liquid film thickness. If $R_R > 1$, a pump must be used to return the unevaporated liquid to the top of the evaporator.

13.8.1   Horizontal Tubes

Chyu et al. [1982] measured the evaporation coefficient on single, horizontal enhanced boiling tubes operated in the thin film evaporation mode. A liquid distributor tube located above the horizontal test tube dripped saturated liquid on the tube. Data for distilled water at 101 kPa were obtained on a plain tube and three enhanced tubes. The enhanced tubes are the Figure 11.11a High-Flux™ porous coated tube, the Figure 11.12a 1024 fin/m integral-fin tube, and the Figure 11.12c Wieland GEWA-T™ tube. Figure 13.28 summarizes their results. The operating conditions are defined in Table 13.4. Figure 13.28 shows that plain tubes provide higher performance in thin film evaporation than in pool boiling for $\Delta T_{ws} > 6$ K. Nucleate boiling occurred for $q > 20$ kW/m², which accounts for the upward slope of the curve at high heat flux. For a laminar film ($d_o G_l / \mu_l < 1600$), the evaporation coefficient is approximately given by $h = k_l / \delta$, where δ is the film thickness. Hence, the plain tube curve for $q < 20$ kW/m² should have unity slope, and the q versus ΔT_{ws} curve should be higher at lower values of G_l. If the G_l is reduced below a critical value, surface de-wetting will occur, and the evaporation coefficient will decrease. Hence, it is important that the surface remain wetted.

Figure 13.28 shows that the slope of the q versus ΔT_{ws} curve for the High-Flux™ surface is the same for thin film evaporation (curve 3) as for pool boiling (curve 4). Nucleate boiling was observed to occur in the film for $q > 8$ kW/m². This explains why the thin film and pool boiling curves have the same slope. Curve 3 is taken with

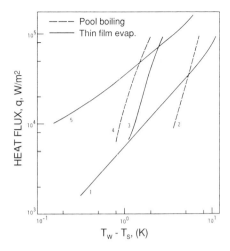

Figure 13.28 Data for thin film evaporation (*solid lines*) and pool boiling (*dashed lines*). Film evaporation: curve 1 (plain tube), curve 3 (High-Flux™), and curve 5 (GEWA-T™ and 1024-fin/m integral-fin tube). Pool boiling: curve 2 (plain tube) and curve 4 (High-Flux™).

increasing power steps. The data for decreasing power steps is approximately the same as curve 4 for pool boiling.

The GEWA-T™ surface (curve 5) did not show any nucleation in the film for the tested levels of q. Note that the GEWA-T™ surface gives the highest performance for $q < 40$ kW/m². The performance of the 1024-fin/m integral-fin surface was nearly identical to that of the GEWA-T™ surface. As noted in the discussion of condensation row effect in Chapter 12, the integral-fin tube "channels" the flow, which prevents axial spreading of the film on the surface. Hence, it is unlikely that the 1024-fin/m tube would provide high performance in a tube bundle, especially at lower values of G_l.

As noted above, nucleation (boiling) may or may not occur in the film. Nucleate boiling would not be expected at normal operating conditions for low-performance surfaces, such as integral-fin tubes. We use the terminology of "evaporation" to define cases where boiling does not occur in the film. Owens [1978] and Conti [1978] measured thin film evaporation and thin film boiling of ammonia in the laminar and turbulent flow regimes on horizontal tubes. Their tests were done with heat transfer from a two-tube array located below the distributor tube. This allowed

Table 13.4 Operating Conditions for Figure 13.28

Curve	Geometry	G_l (kg/s-m)	Heat Transfer Mode
1	Plain	78	Thin film
2	Plain	0	Pool boiling
3	High-Flux™	48	Thin film
4	High-Flux™	0	Pool boiling
5	GEWA-T™	38	Thin film

determination of the effect of the vertical spacing (H) on the performance. Owens' data are for a 50.8-mm-diameter (d_o) plain tube, which span $120 \leq \text{Re}_l \leq 10,000$, $5 \leq q \leq 55$ kW/m², and $0.25 \leq H/d_o \leq 2.1$. He provides useful design correlations for the complete range of conditions studied. Conti tested 25.4-mm-diameter integral-fin tubes spaced at $H/d_o = 1.25$. The first tube had 1339 fins/m and 1.3-mm fin height. The second tube had 787 fins/m, and 1.5-mm fin height. The fins on the 787-fin/m tube were machined to also provide data for 0.5- and 0.94-mm fin height. The highest performance was provided by the 787-fin/m tube with 1.5-mm fin height. It provided an enhancement (based on $A/L = \pi d_e$) 3.2 times that of a plain tube. The lower fin heights provided smaller enhancement.

Lorenz and Yung [1979] provide a semiempirical model for thin film boiling on plain horizontal tubes. Parken et al. [1990] provide additional data for water on 25.4- and 50.8-mm-diameter plain tubes, and they present correlations for the evaporation and boiling regions.

13.8.2   Vertical Tubes

The correlating parameters for falling film flow (nonboiling) are generally based on the semiempirical correlation developed by Chun and Seban [1971], who studied laminar and turbulent film flow on a vertical plate. They give correlations for the laminar and turbulent regions and include the effect of Prandtl number. This work was done without vapor shear. Later work by Kosky [1971] defined the effect of co-current vapor shear on the film thickness and on the transition Reynolds number.

Three studies report tests using the enhanced boiling discussed in Chapter 11 for application to refrigeration evaporators. These are Fagerholm et al. [1985], Nakayama et al. [1982], and Takahashi et al. [1990]. All studies attempted to use R_R close to 1.0. Fagerholm et al. tested a plain tube and three enhanced tubes— Thermoexcel-E™ (Figure 11.12d), GEWA-T™ (Figure 11.12c) and High-Flux™ (Figure 11.11a) geometries with R-114 on a single 2.0-m-long vertical tube as the inner tube of an annulus. The enhanced tubes experienced boiling in the film at all heat fluxes tested, $4 \leq q \leq 18$ kW/m². Entrainment caused partial dewetting when $R_R < 3$. The entrainment-induced dewetting increases with increasing heat flux. The Thermoexcel-E™ and GEWA-T™ tubes were more prone to dewetting than was the High-Flux™ tube. For $R_R > 2$, the heat transfer coefficients on the enhanced tubes was 6–12 times that on the plain tube. The High-Flux™ tube provided the highest performance.

Nakayama et al. [1982] measured evaporation of R-11 (101 kPa) on a 300-mm-long vertical plate containing the Thermoexcel-E™ geometry (Figure 11.12d) and a grooved plate (1.1 mm high, 0.4 mm thick, 0.8 mm pitch). The plates were operated over a range of flow rates. The Thermoexcel-E™ geometry provided higher performance than for pool boiling, which decreased to the plain tube value at $q = 100$ kW-m². The performance remained high until dryout of the liquid. The performance of the finned plate approached that of the Thermoexcel-E™ plate at $G_l = 0.1$ kg/s-m, but decreased either side of this maximum. The Takahashi et al. [1990] tests were performed using R-22 on the outer surface of a 12-tube bundle, 0.79 m long with water flow in the tubes. The Thermoexcel-E™ boiling geometry

(Figure 11.12c) comprised the outer tube surface, and the inner surface is shown in Figure 9.14b. They consciously sought to work with $R_R \leq 1.2$. Their evaporator was part of a refrigeration system containing a scroll compressor, which could pump the unevaporated refrigerant. Some dewetting was observed.

Three studies of thin film evaporation on vertical tubes containing axial fins are reported: Rifert et al. [1975], Mailen [1980], and Grimley et al. [1987]. The surface used by both Rifert and Grimley et al. is similar to that in Figure 13.12. Using water, Rifert found 100% enhancement for fins 0.7 mm high at 0.7 mm pitch. However, Mailen's extruded aluminum tube, similar to that in Figure 12.12, showed little performance improvement, relative to a plain tube. Note that the enhancement mechanism of the Figure 12.12 tube is significantly different for condensation than for thin film evaporation. In condensation, surface tension force pulls the condensate into the valley, yielding very thin films, which are beneficial for condensation. However, this is not beneficial for evaporation. Evaporation requires a mechanism to spread the film onto the fin surface. Hence, surface tension force tends to dewet the Figure 12.12 tube, when used with thin film evaporation.

13.9 CONCLUSIONS

Convective vaporization includes the combined effects of nucleate boiling and convective evaporation. Two heat transfer modes may exist: shell-side boiling (as used in flooded evaporators or kettle reboilers) or thin film evaporation. The relative contributions of nucleate boiling or convective evaporation depend on the surface geometry. The geometry may be enhanced to promote either component.

The enhanced tubes discussed in Chapter 11 are very effective for shell-side boiling in flooded evaporators and kettle reboilers. Their performance is dominated by nucleate boiling. These tubes have also been shown to provide a high nucleate boiling contribution in thin film evaporators. Thus, nucleation occurs in the liquid film, even though it is quite thin. The enhanced boiling tubes generally a yield convective evaporation component typical of that of a plain tube.

Enhanced tube geometries used for tube-side convective vaporization generally promote the convective evaporation term, rather than nucleate boiling. However, certain high-performance nucleate boiling surfaces should substantially enhance performance via nucleate boiling. The micro-fin internal geometry provides very high enhancement for tube-side vaporization of refrigerants. Its application potential for other fluids is yet to be explored.

Geometries beneficial to surface-tension-drained condensation are generally unfavorable to vaporization. This is because heated films tend to rupture. If the surface cannot be fully wetted, a significant loss of performance potential will exist.

13.10 REFERENCES

Agrawal, K. N., Varma H. K., and Lal, S., 1982. "Pressure Drop During Forced Convection Boiling of R-12 Under Swirl Flow," *Journal of Heat Transfer*, Vol. 104, pp. 758–762.

Agrawal, K. N., Varma, H. K., and Lal, S., 1986. "Heat Transfer During Forced Convection Boiling of R-12 Under Swirl Flow." *Journal of Heat Transfer*, Vol. 108, pp. 567–573.

Akhanda, M. A. R., and James, D. D., 1988. "An Experimental Study of the Relative Effects of Transverse and Longitudinal Ribbing of the Heat Transfer Surface in Forced Convective Boiling," in *Two-Phase Heat Exchanger Symposium*, J. T. Pearson and J. B. Kitto, Jr., Eds., ASME Symposium, Vol. 44, ASME, New York, pp. 83–90.

Antonelli, R., and O'Neill, P.S., 1981. "Design and Application Considerations for Heat Exchangers with Enhanced Boiling Surfaces," Paper read at International Conference on Advances in Heat Exchangers, September, Dubrovnik, Yugoslavia.

Arai, N., Fukushima, T., Arai, A., Nakajima, T., Fujie, K., and Nakayama, Y., 1977. "Heat Transfer Tubes Enhancing Boiling and Condensation in Heat Exchangers of a Refrigerating Machine." *ASHRAE Transactions*, Vol. 83, Part 2, pp. 58–70.

Azer, N. Z., and Sivakumar, V., 1984. "Enhancement of Saturated Boiling Heat Transfer by Internally Tinned Tubes," *ASHRAE Transactions*, 90, Part 1A, pp. 58–73.

Baker, O., 1954. "Simultaneous Flow of Oil and Gas," *Oil and Gas Journal*, Vol. 53, pp. 185–195.

Bennett, D. L., and Chen, J. C., 1980. "Forced Convective Boiling in Vertical Tubes for Saturated Pure Components and Binary Mixtures," *AIChE Journal*, Vol. 26, No. 3, pp. 454–461.

Bergles, A. E., and Chyu, M. C., 1982. "Characteristics of Nucleate Pool Boiling from Porous Metallic Coatings," *Journal of Heat Transfer*, Vol. 104, pp. 279–285.

Bergles, A. E., Fuller, W. D., and Hynek, S. J., 1971. "Dispersed Flow Boiling of Nitrogen with Swirl Flow." *International Journal of Heat and Mass Transfer*, Vol. 14, pp. 1343–1354.

Blatt, T. A., and Alt, R. R., 1963. "The Effects of Twisted-Tape Swirl Generators on the Heat Transfer and Pressure Drop of Boiling Freon-11 and Water, ASME paper 65-WA-42.

Boling, C., Donovan, W. J., and Decker, A. S., 1953. "Heat Transfer of Evaporating Freon with Inner-Fin Tubing." *Refrigerating Engineering*, Vol. 61, pp. 1338–1340, 1384.

Bukin, V. G., Danilova, G. N., and Dyundin, V. A., 1982. "Heat Transfer from Freons in a Film Flowing Over Bundles of Horizontal Tubes That Carry a Porous Coating," *Heat Transfer–Soviet Research*, Vol. 14, No. 2, pp. 98–103.

Campolunghi, F., Cumo, M., Ferrari, G., and Palazzi, G., 1976. "Full Scale Tests and Thermal Design of Once-Through Steam Generators," AIChE paper presented at 16th National Heat Transfer Conference, St. Louis.

Carey, V. P., 1992. *Liquid–Vapor Phase-Change Phenomena*, Hemisphere Publishing Corp., Washington, D.C.

Carey, V. P., and Shah, R. K., 1988. "Design of Compact and Enhanced Heat Exchangers for Liquid-Vapor Phase-Change Applications," in *Two-Phase Flow Heat Exchangers: Thermal Hydraulic Fundamentals and Design*, S. Kakaç, A. E. Bergles and E. O. Fernandes, Eds., NATO ASI Series E., Vol. 143, Dordecht, Kluwer Academic Publishers, pp. 909–968.

Carnavos, T. C., 1980. "Heat Transfer Performance of Internally Finned Tubes in Turbulent Flow," *Heat Transfer Engineering*, Vol. 4, No. 1, pp. 32–37.

Chen, J. C., 1966. "A Correlation for Boiling Heat Transfer to Saturated Fluids in Convective Flow," *Industrial and Engineering Chemistry, Process Design and Development*, Vol. 5, No. 3, pp. 322–329.

Chen, C. C., Loh, J. V., and Westwater, J. W., 1981. "Prediction of Boiling Heat Transfer in a Compact Plate-Fin Heat Exchanger Using the Improved Local Technique," *International Journal of Heat and Mass Transfer*, Vol. 24, pp 1907–1912.

Chen, T-K., Chen, X-Z., and Chen, X-J., 1992. "Boiling Heat Transfer and Frictional Pressure Drop in Internally Ribbed Tubes," in *Multiphase Flow and Heat Transfer: Second International Symposium*, Vol. 1, X-J. Chen, T. N. Veziroğlu, and C. L. Tien, Eds., Hemisphere Publishing Corp., New York, pp. 621–629.

Chisholm, D., 1967. "A Theoretical Basis for the Lockhart–Martinelli Correlation for Two-Phase Flow." *International Journal of Heat and Mass Transfer*, Vol. 10, pp. 1767–1777.

Chun, K. R., and Seban, R. A., 1971. "Heat Transfer to Evaporating Liquid Films," *Journal of Heat Transfer*, Vol. 93, pp. 391–396.

Chyu, M. C., Bergles, A. E., and Mayinger, F., 1982. "Enhancement of Horizontal Tube Spray Film Evaporators," in *Heat Transfer—1982, Proceedings of the 7th International Heat Transfer Conference*, Hemisphere Publishing Corp., Washington, D.C., Vol. 6, pp. 275–280.

Conklin, J. C., and Vineyard, E. A., 1992. "Flow Boiling Enhancement of R-22 and a Nonazeotropic Mixture of R-143a and R-124 Using Perforated Foils," *ASHRAE Transactions*, Vol. 98, Part 2, pp. 402–410.

Conti, R. J., 1978. "Experimental Investigation of Horizontal Tube Ammonia Film Evaporators with Small Temperature Differentials," in *Proceedings of the 5th International Heat Transfer Conference*, Vol. 6, Hemisphere Publishing Corp., Washington, D.C., pp. 161–180.

Cooper, M. G., 1984. "Saturation Nucleate, Pool Boiling—A Simple Correlation," *International Chemical Engineering Symposium Series*, No. 86, pp. 785–792.

Crain, B., Jr., and Bell, K. J., 1973. "Forced Convection Heat Transfer to a Two-Phase Mixture of Water and Steam in a Helical Coil," *AIChE Symposium Series*, Vol. 69, No. 131, pp. 30–36.

Cui, S., Tan, V., and Lu, Y., 1992. "Heat Transfer and Flow Resistance of R-502 Flow Boiling Inside Horizontal ISF Tubes," in *Multiphase Flow and Heat Transfer: Second International Symposium*, Vol. 1, X-J. Chen, T. N. Veziroğlu, and C. L. Tien, Eds., Hemisphere Publishing Corp., New York, pp. 662–670.

Cumo, M., Farello, G. E., Ferrari, G., and Palazzi, G., 1974. "The Influence of Twisted Tapes in Subcritical, Once-Through Vapor Generators in Counter Flow," *Journal of Heat Transfer*, Vol. 96, pp. 365–370.

Czikk, A. M., Gottzmann, C. F., Ragi, E. G., Withers, J. G., and Habdas, E. P., 1970. "Performance of Advanced Heat Transfer Tubes in Refrigerant-Flooded Coolers," *ASHRAE Transactions*, Vol. 76, Part 1, pp. 99–109.

Czikk, A. M., O'Neill, P. S,. and Gottzmann, C. F., 1981. "Nucleate Pool Boiling from Porous Metal Films: Effect of Primary Variables," in *Advances in Enhanced Heat Transfer*, R. L. Webb, T. C. Carnavos, E. F. Park, and K. M. Hostetler, eds., HTD Vol. 18, ASME, New York, pp. 109–122.

Eckels, S. J., and Pate, M. B., 1991. "In-Tube Evaporation and Condensation of Refrigerant–Lubricant Mixtures of HFC-134a and CFC-12," *ASHRAE Transactions*, Vol. 97, Part 2, pp. 62–70.

Eckels, S. J., Pate, M. B., and Bemisderfer, C. H., 1992. "Evaporation Heat Transfer Coefficients for R-22 in Micro-Fin Tubes of Different Configurations," in *Enhanced Heat Transfer*, M. B. Pate and M. K. Jensen, Eds., ASME Symposium, Vol. HTD-Vol. 202, ASME, New York, pp. 117–126.

Fagerholm, N.-E., Kivioja, K., Ghazanfari, A.-R., and Jarvinen, E., 1985. "Using Structured Surfaces to Enhance Heat Transfer in Falling Film Flow," *I.I.F.-I.I.R. Commission E2*, Trondheim (Norway), pp. 273–279.

Fujie, K., Itoh, N., Innami, T., Kimura, H., Nakayama, N., and Yanugidi, T., 1977, "Heat Transfer Pipe," U. S. Patent 4,044,797, assigned to Hitachi, Ltd.

Fujita, Y., Ohta, H., Hidaka, S., and Nishikawa, K., 1986. "Nucleate Boiling Heat Transfer on Horizontal Tubes in Bundles," *Proceedings of the 8th International Heat Transfer Conference*, Vol. 5, pp. 2131–2136.

Gambill, W. R., 1963. "Generalized Prediction of Burnout Heat Flux for Flowing, Subcooled Wetting Liquids," *Chemical Engineering Progress Symposium Series*, Vol. 59, No. 41, pp. 71–87.

Gambill, W. R., 1965. "Subcooled Swirl-Flow Boiling and Burnout with Electrically Heated Twisted Tapes and Zero Wall Flux, *Journal of Heat Transfer*, Vol. 97, p. 342.

Gambill, W. R., Bundy, R. D., and Wansbrough, R. W., 1961. "Heat Transfer, Burnout, and Pressure Drop for Water in Swirl Flow Tubes with Internal Twisted Tapes," *Chemical Engineering Progress Symposium Series*, Vol. 57, No. 31, pp. 127–137.

Grimley, T. A., Mudawwar I. A., and Incropera, F. P., 1987. "Enhancement of Boiling Heat Transfer in Falling Films," *Proceedings of the 1987 ASME-JSME Thermal Engineering Joint Conference*, Vol. 3. pp. 411–418.

Gu, C. B., Chow, L. C., and Beam, J. E., 1989. "Flow Boiling in a Curved Channel," in *Heat Transfer in High Energy/High Heat Flux Applications*, R. J. Goldstein, L. C. Chow, and E. E. Anderson, Eds., ASME Symposium, Volume HTD-Vol. 119, ASME, New York, pp. 25–32.

Gupte, N. S., 1992. "Simulation of Boiling in Flooded Refrigerant Evaporators," Ph.D. Thesis, The Pennsylvania State University.

Gupte, N. S., and Webb, R. L., 1992. "Convective Vaporization of Refrigerants in Tube Banks," *ASHRAE Transactions*, Vol. 98, Part 2, pp. 411–424.

Hewitt, G. F., and Roberts, D. N., 1969. *Studies of Two-Phase Flow Patterns by Simultaneous X- Ray and Flash Photography*, Report AERE-M 2159, Her Majesty's Stationery Office, London.

Hughes, T. G., and Olson, D. R., 1975. "Critical Heat Fluxes for Curved Surfaces during Subcooled Flow Boiling," *Transactions of the CSME*, Vol. 3, No. 3, pp. 122–130.

Ikeuchi, M., Yumikura, T., Fujii, M., and Yamanaka, G., 1984. "Heat-Transfer Characteristics of an Internal Microporous Tube with Refrigerant-22 Under Evaporating Conditions." *ASHRAE Transactions*, Vol. 90, Part 1A, pp. 196–211.

Ishihara, K., Palen, J. W., and Taborek, J., 1980. "Critical Review of Correlation for Predicting Two-Phase Flow Pressure Drop Across Tube Banks," *Heat Transfer Engineering*, Vol. 1, pp. 23–32.

Ito, M., and Kimura, H., 1979. "Boiling Heat Transfer and Pressure Drop in Internal Spiral-Grooved Tubes," *Bulletin of the JSME*, Vol. 22, No. 171, pp. 1251–1257.

Jensen, M. K., 1984. "A Correlation for Predicting the Critical Heat Flux Condition with Twisted-Tape Swirl Generators," *International Journal of Heat and Mass Transfer*, Vol. 27, pp. 2171–2173.

Jensen, M. K., 1985. "An Evaluation of the Effect of Twisted-Tape Swirl Generators in Two-Phase Flow Heat Exchangers," *Heat Transfer Engineering*, Vol. 6, No. 4, pp. 19–30.

Jensen, M. K., and Bensler, H. P., 1986. "Saturated Forced-Convective Boiling Heat Transfer with Twisted-Tape Inserts," *Journal of Heat Transfer*, Vol. 108, pp. 93–99.

Jensen, M. K., and Bergles, A. E., 1981. "Critical Heat Flux in Helically Coiled Tubes," *Journal of Heat Transfer*, Vol. 103, pp. 660–666.

Jensen, M. K., and Hsu J. T. 1987. "A Parametric Study of Boiling Heat Transfer in a Tube Bundle," *Proceedings of the 1987 ASME-JSME Thermal Engineering Joint Conference*, Vol. 3, pp. 133–140.

Jensen, M. K., Trewin, R. R., and Bergles, A. E., 1992. "Crossflow Boiling in Enhanced Tube Bundles," in *Two-Phase Flow in Energy Exchange Systems*, M. S. Sohal and T. J. Rabas, Eds., ASME Symposium Vol. HTD-Vol. 220, ASME, New York, pp. 11–18.

Kitto, J. B., and Wiener, M., 1982. "Effects of Nonuniform Circumferential Heating and Inclination on Critical Heat Flux in Smooth and Ribbed Bore Tubes," in *Proceedings of the 7th International Heat Transfer Conference*, Vol. 4, pp. 303–308.

Kosky, P. G., 1971. "Thin Liquid Films Under Simultaneous Shear and Gravity Flows," *International Journal of Heat and Mass Transfer*, Vol. 14, pp. 1220–1223.

Kubanek, G. R., and Miletti, D. L., 1979. "Evaporative Heat Transfer and Pressure Drop Performance of Internally-Finned Tubes with Refrigerant 22," *Journal of Heat Transfer*, Vol. 101, pp. 447–452.

Kun, L. C., and Czikk, A. M., 1969. "Surface for Boiling Liquids," U. S. Patent 3,454,081 (Reissued 1979, Ref. 30,077), assigned to Union Carbide Corp.

Lavin, J. G., and Young, E. H., 1965. "Heat Transfer to Evaporating Refrigerants in Two-Phase Flow," *AIChE Journal*, Vol. 11, pp. 1124–1132.

Lewis, L. G., and Sather, N. F., 1978. "OTEC Performance Tests on the Union Carbide Flooded Bundle Evaporator," ANL Report ANL-OTEC-PS-1, Argonne National Lab, Chicago, December.

Lorenz, J. J., and Yung, D. T., 1979. "A Note on Combined Boiling and Evaporation of Liquid Films on Horizontal Tubes," *Journal of Heat Transfer*, Vol. 101, pp. 178–180.

Mailen, G. S., 1980. "Experimental Studies of OTEC Heat Transfer Evaporation of Ammonia on Vertical Smooth and Fluted Tubes," *Proceedings of the 7th Ocean Energy Conference*, Paper 12.5 , pp. 1–10.

Mandrusik, G. D., and Carey, V. P., 1989. "Convective Boiling in Vertical Channels with Different Offset Strip Fin Geometries," *Journal of Heat Transfer*, Vol. 111, pp. 156–165.

Mergerlin, F. E., Murphy, R. W., and Bergles, A. E., 1974. "Augmentation of Heat Transfer by Use of Mesh and Brush Inserts," *Journal of Heat Transfer*, Vol. 96, pp. 145–151.

Muller, J., 1986. "Boiling Heat Transfer on Finned Tube Bundles—The Effect of Tube Position and Intertube Spacing," *Proceedings of the 8th International Heat Transfer Conference*, Vol. 4, pp. 2111–2116.

Nakajima, K., and Shiozawa, A., 1975. "An Experimental Study on the Performance of a Flooded Type Evaporator," *Heat Transfer—Japanese Research*, Vol. 4, No. 3, pp. 49–66.

Nakayama, W., Dikoku, D., and Nakajima, T., 1982. "Enhancement of Boiling and Evaporation on Structured Surfaces with Gravity Driven Film Flow," *Proceedings of the 7th International Heat Transfer Conference*, Vol. 6, pp. 409–414.

O'Neill, P. S., Private Communication, February 16, 1981. Linde Division, Union Carbide Corp., Tonawanda, NY.

O'Neill, P. S., King, R. C., and Ragi, E. G., 1980. "Application of High-performance Evaporator Tubing in Refrigeration Systems of Large Olefins Plants," *AIChE Symposium Series*, Vol. 76, No. 199, pp. 289–300.

Owens, W. L., 1978. "Correlation of Thin Film Evaporation Heat Transfer Coefficients for

Horizontal Tubes," *Proceedings of the 5th Ocean Thermal Energy Conversion Conference*, Miami Beach, FL., Vol. 6, pp. 71–89.

Palen, J. W., Yarden, A., and Taborek, J., 1972. "Characteristics of Boiling Outside Large-Scale Horizontal Multitube Bundles," *AIChE Symposium Series*, Vol. 68, No. 118, pp. 50–61.

Palm, B., 1990. "Heat Transfer Augmentation in Flow Boiling by Aid of Perforated Metal Foils," ASME paper 90-WA/HT-10.

Panchal, C. B., France, D. M., and Bell, K. H., 1992. "Experimental Investigation of Single-Phase, Condensation, and Flow Boiling Heat Transfer for a Spirally Fluted Tube," *Heat Transfer Engineering*, Vol. 13, No. 1, pp. 43–52.

Parken, W. H., Fletcher, L. S., Sernas, V., and Han, J. C., 1990. "Heat Transfer Through Falling Evaporation and Boiling on Horizontal Tubes," *Journal of Heat Transfer*, Vol. 112, pp. 744–750.

Pearson, J. F., and Young, E. H., 1970. "Simulated Performance of Refrigerant-22 Boiling Inside of Tubes in a Four Pass Shell and Tube Heat Exchanger," *AIChE Symposium Series*, Vol. 66, No. 102, pp. 164–173.

Pierre, B., 1964. "Flow Resistance with Boiling Refrigerants," *ASHRAE Journal*, Vol. 6, No. 9, pp. 58–65, Vol. 6, No. 10, pp. 73–77.

Polley, G. T., Ralston, T., and Grant, I. R., 1980. "Forced Cross Flow Boiling in an Ideal In-line Tube Bundle," ASME paper 80-HT-46, ASME/AIChE Heat Transfer Conference, Orlando, FL.

Rifert, V. G., Butuzov, A. I., and Belik, D. N., 1975. "Heat Transfer in Vapor Generation in a Falling Film Inside a Vertical Tube with a Finely-Finned Surface," *Heat Transfer—Soviet Research*, Vol. 7, No. 2, pp. 22–25.

Robertson, J. M., 1980. "Review of Boiling, Condensing and Other Aspects of Two-Phase Flow in Plate Fin Heat Exchangers," *Compact Heat Exchangers—History, Technological Advances and Mechanical Design Problems*, HTD Vol. 10, R. K. Shah, C. F. McDonald, and C. P. Howard, Eds., ASME, New York, pp. 17–27.

Schlager, L. M., Pate, M. B., and Bergles, A. E., 1988a. "Performance of Micro-fin Tubes with Refrigerant-22 and Oil Mixtures," *ASHRAE Journal*, November, pp. 17–28.

Schlager, L. M., Bergles, A. E., and Pate, M. B., 1988b. "Evaporation and Condensation of Refrigerant–Oil Mixtures in a Smooth Tube and a Micro-fin Tube," *ASHRAE Transactions*, Vol. 94, Part 1, pp. 149–166.

Schlager, L. M., Pate, M. B., and Bergles, A. E., 1990. "Evaporation and Condensation Heat Transfer and Pressure Drop in Horizontal, 12.7-mm Micro-fin Tubes with Refrigerant 22," *Journal of Heat Transfer*, Vol. 112, pp. 1041–1047.

Schlünder, E. U., and Chawla, J., 1967. "Local Heat Transfer and Pressure Drop for Refrigerants Evaporating in Horizontal, Internally Finned Tubes," *Proceedings of the International Congress on Refrigeration*, paper 2.47.

Schlünder, E. U., and Chawla, J., 1969. "Ortlicher Warmeubergang und Druckabfall bei der der Stromung verdampfender Kaltemittel in innenberippten, waggerechten Rohren," *Kaltetechnik Klimatisierung*, Vol. 21, No. 5, pp. 136–139.

Shinohara, Y., and Tobe, M., 1985. "Development of an Improved Thermofin Tube," *Hitachi Cable Review*, No. 4, pp. 47–50.

Shinohara, Y., Oizumi, K., Itoh, Y., and Hori, M., 1987. "Heat Transfer Tubes with Grooved Inner Surface," U. S. Patent 4,658,892, assigned to Hitachi Cable, Ltd.

Sideman, S., and Levin, A., 1979. "Effect of the Configuration on Heat Transfer to Gravity Driven Films Evaporating on Grooved Tubes," *Desalination*, Vol. 31, pp 7–18.

Steiner, D., and Taborek, J., 1992. "Flow Boiling Heat Transfer of Single Components in Vertical Tubes," *Heat Transfer Engineering*, Vol. 13, No. 2, pp. 43–68.

Swenson, H. S., Carver, J. R., and Szoeke, G., 1962. "The Effects of Nucleate Boiling Versus Film Boiling on Heat Transfer in Power Boiler Tubes," *Journal of Engineering Power*, Vol. 84, pp. 365–371.

Taitel, Y., and Dukler, A. E., 1976. "A Model for Predicting Flow Regime Transitions in Horizontal and Near Horizontal Gas–Liquid Flow," *AIChE Journal*, Vol. 22, pp. 47–55.

Takahashi, K., Daikoku, T., Yasuda, H., Yamashita, T., and Zushi, S., 1990. "The Evaluation of a Falling Film Evaporator in an R-22 Chiller Unit," *ASHRAE Transactions*, Vol. 96, Part 2, pp. 158–163.

Tatsumi, A., Oizumi, K., Hayashi, M., and Ito, M., 1982. "Application of Inner Groove Tubes to Air Conditioners," *Hitachi Review*, Vol. 32, No. 1, pp. 55–60.

Thome, J. R., 1990. *Enhanced Boiling Heat Transfer*, Hemisphere Publishing Corp., New York.

Varma, H. K., Agrawal, K. N., and Bansal, M. L., 1991. "Heat Transfer Augmentation by Coiled Wire Turbulence Promoters in a Horizontal Refrigerant-22 Evaporator." *ASHRAE Transactions*, Vol. 97, Part 1, pp. 359–364.

Watson, G. B., Lee, R. A., and Wiener, M., 1974. "Critical Heat Flux in Inclined and Vertical Smooth and Ribbed Tubes," *Proceedings of the 5th International Heat Transfer Conference*, Vol. 4, pp. 275–279.

Webb, R.L., and Apparao, T. R., 1990. "Performance of Flooded Refrigerant Evaporators with Enhanced Tubes." *Heat Transfer Engineering*, Vol. 11, No. 2, pp. 30–44.

Webb, R. L., and Gupte, N. S., 1992. "A Critical Review of Correlations for Convective Vaporization in Tubes and Tube Banks," *Heat Transfer Engineering*, Vol. 13, No. 3, pp. 58–81.

Webb, R. L., Choi, K.-D., and Apparao, T., 1990. "A Theoretical Model to Predict the Heat Duty and Pressure Drop in Flooded Refrigerant Evaporators." *ASHRAE Transactions*, Vol. 95, Part 1, pp. 326–338.

Wieland, 1991. *Ripple-Fin Tubes,* Wieland-Werke AG brochure TKI-42e(M)-02.91, Ulm, Germany.

Withers, J. G., and Habdas, E. P., 1974. "Heat Transfer Characteristics of Helical Corrugated Tubes for Intube Boiling of Refrigerant R-12," *AIChE Symposium Series*, Vol. 70, No. 138, pp. 98–106.

Yasuda, K., Ohizumi, K., Hori, M., and Kawamata, O., 1990. Development of Condensing Thermofin-HEX-C Tube," *Hitachi Cable Review*, No. 9, pp. 27–30.

Yilmaz, S., Palen, J. W., and Taborek, J., 1981. "Enhanced Boiling Surfaces as Single Tubes and Tube Bundles," in *Advances in Enhanced Heat Transfer*, R. L. Webb, T. C. Carnavos, E. F. Park, and K. M. Hostetler, eds., HTD Vol. 18, ASME, New York, pp. 123–124.

Yoshida, S., Matsunaga, T., and Hong, H. P., 1987. "Heat Transfer to Refrigerants in Horizontal Evaporator Tubes with Internal, Spiral Grooves," in *Proceedings of the 1987 ASME-JSME Thermal Engineering Joint Conference*, Vol. 5, P. J. Marto, Ed., pp. 165–172.

13.11   NOMENCLATURE

A_i	Inside tube surface area, m² or ft²
A_c	Minimum cross-sectional flow area, m² or ft²
Bo	Boiling number ($q/G\lambda$), dimensionless
d_e	Diameter over fins of externally finned tube, m or ft
d_i	Tube inside diameter, or diameter to the base of internal fins or roughness, m or ft
d_o	Tube outside diameter, m or ft
D_c	Diameter of helical coil, m or ft
D_h	Hydraulic diameter of flow passages, $4LA_c/A$; m or ft
e	Fin height, roughness height, or corrugation depth, m or ft
E_h	h/h_p at constant Re, dimensionless
f	Friction factor, dimensionless
F	Two-phase convection multiplier factor, h_{cv}/h_l; dimensionless
g	Acceleration due to gravity, m/s² or ft/s²
G	Mass velocity, kg/m²-s or lbm/ft²-s
h	Heat transfer coefficient: h (total), h_l (liquid phase flowing alone), h_{lo} (total mass rate flowing as a liquid), h_{nbp} (nucleate pool boiling), h_{nb} ($=$ Sh_{nbp}), h_{cv} (two-phase convection, without nucleate boiling); W/m²-K or Btu/hr-ft²-°F
H	Length for 180-degree revolution of twisted tape, m or ft
J	Conversion factor used in Equation 13.14, 778 ft-lbf/Btu
k	Thermal conductivity, W/m-K or Btu/hr-ft-°F
L	Flow length, m or ft
p	Pressure, Pa or lbf/ft²
p	Axial pitch of surface or roughness elements, m or ft
q	Heat flux, W/m²
R_R	Recirculation ratio (ratio of feed rate to evaporation rate), dimensionless
Re	Reynolds number, defined where used; dimensionless
S	Suppression factor, dimensionless
t	Thickness of tube wall fin or twisted tape, m or ft
T	Temperature. T_w (wall), T_s (saturation), °C or °F
u	Average flow velocity, m/s or ft/s
U	Overall heat transfer coefficient, W/m²-K or Btu/hr-ft²-°F
v_l	Specific volume of liquid, m³/kg
W	Fluid mass flow rate ($= GA_c$), kg/s or lbm/s
x	Local vapor quality, dimensionless
X_{tt}	Martinelli parameter [$= (\Delta p_{F,l}/\Delta p_{F,v})^{0.5}$], flow in tubes (Equation 13.8), dimensionless
y	Twist ratio ($= H/d_i = \pi/[2 \tan \alpha]$), dimensionless

Greek Symbols

α	Void fraction, dimensionless
α	Helix angle relative to tube axis [$\tan \alpha = \pi/2y$], radians or degrees

δ	Film thickness, m or ft
Δp	Total pressure drop for two-phase flow, Pa, or lbf/ft²
Δp_F	Frictional pressure drop for two-phase flow. Liquid phase flowing alone ($\Delta p_{F,l}$), two phase friction (Δp_F), Pa or lbf/ft²
ΔT_{ws}	$T_w - T_s$, K or °F
Δx	Vapor quality change, dimensionless
η_f	Fin efficiency or temperature effectiveness of the fin, dimensionless
λ	Latent heat of vaporization, J/kg or Btu/lbm
μ_l	Dynamic viscosity: μ_l (of liquid), μ_v (of vapor), $\mu_{l,w}$ (of water), μ_a (of air)
ν	Kinematic viscosity, m/s² or ft/s²
ρ	Density: ρ_l (of liquid), ρ_v (of vapor), $\rho_{l,w}$ (of water), ρ_a (of air), kg/m³ or lbm/ft³
σ	Surface tension, σ_w (of water), N/m or lbf/ft
ϕ	Severity factor, $(e/d_i)/(p/e)$, used in Figure 13.20; dimensionless
ϕ_l^2	Two-phase friction multiplier, $\Delta p_F/\Delta p_{F,l}$ (Equation 13.9); dimensionless

Subscripts

cr	Critical heat flux
cv	Convection
i	Inside tube
l,w	Of liquid water
l	Liquid phase
nb	Nucleate boiling
p	Plain tube
v	Vapor phase
w	At tube wall

14

CONVECTIVE CONDENSATION

14.1 INTRODUCTION

When condensation occurs in a tube, convective effects occur which do not exist in vapor space condensation. The convective effects are associated with the influence of shear stress on the liquid–vapor interface. Furthermore, if gravity force exceeds vapor shear forces, the liquid phase will tend to stratify in horizontal tubes.

The highest vapor shear effects exist near the inlet end of the tube, where vapor velocity is highest. Section 13.2 discussed (a) fundamentals of two-phase flow in tubes and (b) how heat transfer is influenced by the flow pattern. If the nucleate boiling contribution in convective vaporization were suppressed (e.g., $h_{cv} \gg h_{nb}$), the convective vaporization and convective condensation in a tube would have a great deal in common. This would be particularly true if the wall were wetted in both cases. As noted in Chapter 13, if the film flow is laminar, the heat transfer coefficient is given by $h = k_l / \delta$, where δ is the film thickness. As the vapor quality increases in vaporization, the vapor velocity increases, causing increased vapor shear. In condensation the converse exists. A two-phase flow experiences different "flow patterns" as vapor quality changes along the tube length, as shown in Figure 13.3. The heat transfer coefficient is significantly affected by the flow pattern. Figure 14.1 shows the flow patterns drawn on the Baker flow map (Figure 13.5) experienced by R-12 condensing in a horizontal tube. This figure shows that the flow patterns are very sensitive to mass velocity. The mass velocity is defined as the total flow rate divided by the pipe cross-sectional area. The reader should review Section 13.2 to understand how flow pattern affects the heat transfer coefficient in two-phase heat transfer.

The pressure drop in condensation contains the same components as in vaporization. However, the acceleration term, Δp_a, in Equation 13.6 causes pressure recov-

482

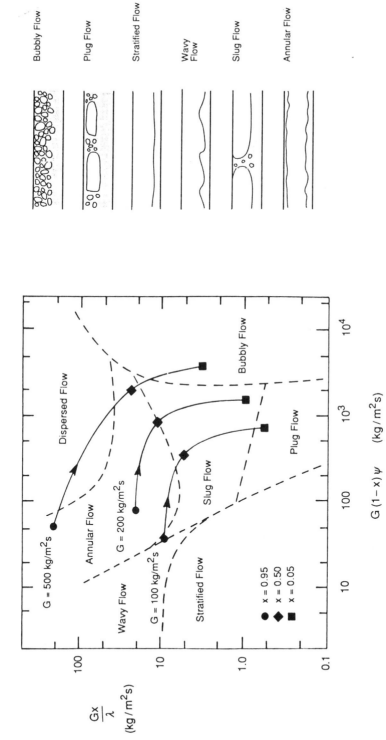

Figure 14.1 Flow patterns for condensation in a horizontal tube drawn on the Baker [1954] flow pattern map. (From Carey [1992].)

483

ery in condensation, whereas the same magnitude of Δp_a adds to the pressure drop in vaporization.

Enhancement should also be beneficial for condensing flows that contain noncondensible gases.

14.2 FORCED CONDENSATION INSIDE TUBES

Enhancement techniques used on the outside of vertical tubes are applicable to condensation inside vertical tubes. At low vapor velocities, enhancement requirements for horizontal tubes are different, because gravity force drains the film transverse to the flow direction. Preferred augmentation techniques for horizontal tubes are yet to be established and commercialized. Those which have been tested include twisted-tape inserts, internal fins, and integral roughness. Internal fins have yielded the highest augmentation levels. However, selection of the preferred geometry must consider the tubing cost and pressure drop.

14.2.1 Internally Finned Tubes

Vrable et al. [1974] and Reisbig [1974] used R-12 in horizontal internally finned tubes. Defining the condensation coefficient in terms of the total surface area, they observed coefficients 20–40% greater than the smooth-tube value for wet vapor. Accounting for surface area increases of 1.5–2.0, the heat conductance hA/L is 2–3 times the smooth tube value. Reisbig found that his full-height fins provided a smaller hA/L than did smooth tubes when condensing superheated R-12. No explanation is offered for this unexpected performance.

Royal and Bergles [1978a] condensed steam in four horizontal internally finned copper tubes, three of which had spiraled fins. The inlet steam was saturated at 4 bar, and all of the vapor was condensed in the test section. A later work by Luu and Bergles [1979] tested the same tubes with R-l13. Their results were generally consistent with the augmentation levels reported by Vrable [1974] and Reisbig [1974] for R-12. The best internal fin geometry provided 20–40% higher heat transfer coefficients (total area basis) than the smooth tubes. Figure 14.2 shows their results, with the tube geometries defined in Table 14.1. The value A/A_p is the internal surface area, relative to a plain tube of the same inside diameter.

The heat transfer coefficients of Figure 14.2a are based on the smooth bore area. The figure also contains data for tubes having twisted-tape inserts, to be discussed later. The internally finned tubes provide significant heat transfer improvement, and considerably higher than provided by the twisted-tapes (tubes B and C). The highest performance is provided by Tube G (16 fins). It provides an enhancement ratio of 2.3 relative to its 73% area increase. The figure shows that the area increase is not the controlling factor. Tube D has approximately the same internal area as tube G, but with twice as many fins of half the height. The enhancement provided by tube D is much less than that of tube G. The pressure drop results (Figure 14.2b) show that

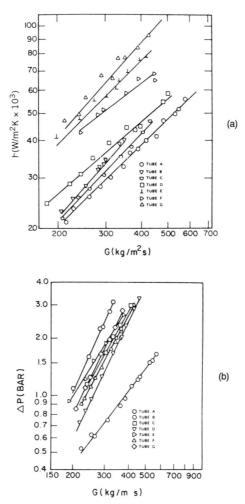

Figure 14.2 Test results of Royal and Bergles [1978a] for steam condensation (4 bar) in horizontal tubes containing internal fins and twisted-tapes. (a) Heat transfer coefficient based on plain tube area. (b) Pressure drop. Geometries defined in Table 14.1. (From Royal and Bergles [1978a].)

the twisted-tapes cause a much larger pressure drop increase than does the heat transfer enhancement.

Royal and Bergles [1978a] correlated their steam condensation data, by modifying the smooth tube correlation of Ackers et al. [1959] to account for geometric parameters. Luu and Bergles [1979] found that the Royal and Bergles [1978a] correlation did not predict their R-113 data well. They developed a modified correlation for the R-113 data, which did not include the steam data. Because of the

Table 14.1   Geometries Tested by Royal and Bergles [1978a]

Code	d_i (mm)	Geometry	e (mm)	n	α	A/A_p (degrees)
A	15.9	Smooth	None	None	0	1
B	15.9	Twisted tape	None	None	18.7	1
C	15.9	Twisted tape	None	None	9.3	1
D	15.9	Internal fin	0.60	32	2.95	1.70
E	12.8	Internal fin	1.74	6	5.25	1.44
F	12.8	Internal fin	1.63	6	0	1.44
G	15.9	Internal fin	1.45	16	3.22	1.73

large fluid property differences of steam and R-113, they conclude that different fin geometries may be preferred for water and refrigerants.

Said and Azer [1983] tested the four internal fin geometries listed in Table 14.2 using R-113. The fins are typically higher and have larger helix angles than those tested by Luu and Bergles [1979]. Because the two investigators performed their tests at different saturation pressures, the results cannot be directly compared. Said and Azer found that the Luu and Bergles [1979] R-113 heat transfer correlation overpredicted their R-113 data. Hence, Said and Azer developed their own empirical correlation. Unfortunately, they did not include the Luu and Bergles [1979] R-113 data. Kaushik and Azer [1988] developed a new empirical power-law correlation for internally finned tubes. This correlation predicts the data of Luu and Bergles [1979], Royal and Bergles [1978a], and Said and Azer [1983] with only modest success. Some points are under- or overpredicted as much as 100%. We conclude that satisfactory correlations to predict the effect of fluid properties and fin geometry on condensation do not exist. Kaushik and Azer [1990] also developed an empirical power-law correlation for pressure drop in internally finned tubes. The correlation predicted 68% of the data points of three investigators within ±40%. Kaushik and Azer [1989] and Sur and Azer [1991] propose analytically based models to predict the condensation coefficient and the pressure drop in internally finned tubes, respectively. The models were not sufficiently validated against existing data to justify recommendation of their use. More work is needed on correlations and theoretically based models.

The selection of internally enhanced tubes for two-phase applications should

Table 14.2   Internal Fin Geometries Tested by Said and Azer [1983]

Code	d_i (mm)	Fin Height (mm)	Number of fins	Helix angle (degrees)	A/A_p
2	13.84	1.58	10	0	1.50
3	17.15	1.80	16	9.7	1.89
4	19.87	1.98	16	12.3	1.76
5	25.38	2.13	16	19.4	1.64

account for the effect of pressure drop on the saturation temperature of the condensing or evaporating fluid. Increased pressure drop, for fixed inlet pressure, will reduce the available logarithmic mean temperature difference (LMTD), as discussed in Chapter 4. Royals and Bergles [1978b] present a performance evaluation analysis of the Royal and Bergles [1978a] test geometries. However, they do not include the effect of pressure drop in their performance evaluation analysis.

Ito and Kimura [1979] and Shinohara and Tobe [1985] describe a tube having very small triangular-shaped fins, which has become known as the "micro-fin" tube. This tube is used in virtually all refrigerant condensers, either air- or water-cooled. Because of the importance of this tube, it is separately described in Section 14.3.

14.2.1.1 Wire-Loop Finned Annulus
Honda et al. [1988] have investigated a different type of internally finned channel than the continuous internal fins previously discussed. Their work is applicable to "tube-in-tube" condensers, which are used in small, water-cooled refrigeration condensers. Condensation occurs in an annulus, with water flow in the inner corrugated tube. Their annulus corrugated copper wire fins were wrapped on the outer surface of a 19.1-mm-diameter tube and were soldered to the tube. Figure 14.3a shows the wire fin geometry, which provided an area enhancement of 3.04. Three outer tube diameters were tested: 24.8, 27.2, and 29.9-mm inside diameter. The local R-113 condensation coefficient was enhanced a factor of 2–13, relative to a plain inner tube at mass velocities between 52 and 201 kg/m^2-s, respectively.

14.2.2 Twisted-Tape Inserts

The previously discussed work of Royal and Bergles [1978a, 1978b] included the evaluation of two twisted-tape geometries for steam. Luu and Bergles [1979] also tested the same twisted-tapes with R-113. These works show that the performance of twisted-tapes is distinctly poorer than that of internally finned tubes (see Figure 14.2). They provide only 30% heat transfer enhancement, but exhibit pressure drops equal to or higher than those of internally finned tubes. Data on R-113 by Said and Azer [1983] generally confirm this conclusion.

Lin et al. [1980] tested the static mixer insert device shown in Figure 7.1c using R-113. This consists of successive axial increments of twisted-tape segments. The insert device successively rotates the flow 180 degrees clockwise, then 180 degrees counterclockwise. The leading edge of each successive element is at 90 degrees to the trailing end of the upstream element. Lin et al. tested elements approximately 38 mm long in 6.35-, 12.7-, and 19.1-mm-diameter tubes. Significantly higher enhancement is obtained with the static mixer than with a twisted-tape. The enhancement was approximately 80% for all tube sizes. However, the pressure drop increased by a factor of four.

Several correlations have been developed for condensation in tubes having twisted tapes. The correlations are based on modifying correlations for condensation in plain, circular tubes. Royal and Bergles [1978a] developed a correlation for their steam data, based on a modification of the smooth tube correlation of Ackers et

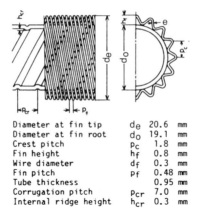

Diameter at fin tip	d_e	20.6	mm
Diameter at fin root	d_o	19.1	mm
Crest pitch	p_c	1.8	mm
Fin height	h_f	0.8	mm
Wire diameter	d_f	0.3	mm
Fin pitch	p_f	0.48	mm
Tube thickness		0.95	mm
Corrugation pitch	p_{cr}	7.0	mm
Internal ridge height	h_{cr}	0.3	mm

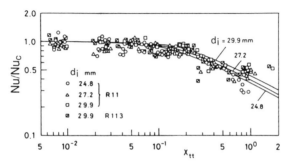

Figure 14.3 (a) Detail of wire finned inner tube tested by Honda et al. [1988]. (b) Correlation of data versus X_{tt}. (From Honda et al. [1988].)

al. [1959]. The modification consisted of defining the Reynolds number in terms of the condensate properties, using the hydraulic diameter, and calculating the velocity as done in the Lopina and Bergles [1969] correlation for single-phase flow. The velocity used is the resultant of the average axial velocity and tangential velocity at the wall, assuming that the tape induces solid body rotation. The correlation assumed that the tape acted as a fin, with no contact resistance, which is probably optimistic. Luu and Bergles [1980] abandoned the Royal and Bergles [1978a] correlation for their R-113 data. Rather, they chose to use a modification of the Boyko and Kruzhilin [1967] plain tube correlation. The Reynolds number was defined by the hydraulic diameter and the average swirling velocity (rather than that at the wall). Luu and Bergles [1980] showed that most of their R-113 data were correlated within ±30%. However, Said and Azer [1983] state that the Luu and Bergles correlation underpredicted their R-113 data by 0–30%. They developed their own heat transfer correlation based on a modification of the Shah [1979] plain tube correlation. They also developed a pressure drop correlation, based on modifying plain tube correlations.

14.2.3   Roughness

The work of Nicol and Medwell on the effect of roughness for vapor space conden-sation (Chapter 12) provided motivation for investigating roughness for forced condensation inside tubes. Luu and Bergles [1979] found that two-dimensional repeated-rib roughness ($e/d = 0.013$, $p/e = 9.7$, and $e/d = 0.021$, $p/e = 20.8$) increased the R-113 condensing coefficient more than 100%. Both roughness sizes provided approximately equal augmentation levels. For the same augmentation level, roughness may be of greater interest than the relatively high internal fins discussed in Section 14.2.1, since roughness requires less tube material.

In a U.S. patent, Fenner and Ragi [1979] describe a method for applying a "sand-grain"-type roughness inside tubes. The favored roughness consists of a single layer of metal particles bonded to the surface and covering approximately 50% of the projected surface. The spaced metal particles provide extended surface for high vapor qualities and cause turbulence in the condensate film at lower vapor qualities. Data are presented for R-12 condensing in a 14.5-mm-diameter tube having a 50% area density of $e/d_i = 0.031$ particles. This tube yielded enhancement levels (smooth tube area basis) of 2.4 for low exit qualities (0.25–0.60) and 4.0 for high exit qualities. The accompanying pressure drops were only 68% and 105% larger than the corresponding smooth tube values. It would appear that this roughness geometry can provide the same enhancement as the Section 14.2.1 internally finned tubes with smaller pressure drop and using less additive material content.

Shinohara and Tobe [1985] provide R-22 condensation data on the tubes shown in Figure 13.21a. Table 14.3 compares the dimensions and performance of a 16-mm-outside-diameter TFIN-CR tube with a plain and several other geometries for the standard Figure 13.21b corrugated tube (CORG). The geometry parameters of the corrugated tube are the roughness height (e) and roughness spacing (p). The table also contains performance data on a micro-fin tube (TFIN) and five-leg star fin insert (Figure 13.10). The heat transfer coefficients are based on the outer diameter (16 mm). The micro-grooves in the TFIN tubes are 0.2 mm, and the helix angle is 8 degrees. The 5.7-m-long test section was operated at 35°C. Note that the enhance-ment level of the TFIN tube is not as high as that listed in Table 14.4.

Table 14.3   Tube Comparison for R-22 Condensation at $G = 100$ kg/hr

Geometry	e_i (mm)	p (mm)	h (kW/m²)	Δp (kPa)	h/h_p	$\Delta p/\Delta p_p$
Plain	0	NA	1.28	7.80	1.00	1.00
Star insert	0	NA	2.62	15.7	2.04	2.01
CORG	0.7	8.0	2.03	14.2	1.59	1.82
TFIN	0	NA	2.06	7.80	1.61	1.00
TFIN-CR	0.7	8.0	3.29	14.2	2.57	1.82

[a]NA, not applicable

Table 14.4 Chronological Improvements in the Hitachi Thermofin™ Tube

Year	Geometry	e	p/e	α	β	n	A/A_p	Wt/Wt_p	h/h_p
1977	Original	0.15	2.14	25	90	65	1.28	1.22	1.8
1985	EX	0.20	2.32	18	53	60	1.51	1.19	2.4
1988	HEX	0.20	2.32	18	40	60	1.60	1.19	2.5
1989	HEX-C	0.25	2.32	30	40	60	1.73	1.28	3.1

14.2.4 Wire Coil Inserts

Wang [1987] investigated wire coil inserts for tube-side condensation with stratified flow. He developed a theoretical model to predict the condensation coefficient. The model assumes that surface tension force at the base of the wire pulls the condensate to the base of the wire, where it is gravity-drained to the liquid level in the tube. This is the same concept described by Thomas [1967] and illustrated in Figure 12.15. They show that

$$\frac{h}{h_p} = \frac{0.75(1 + F_1)^{1/4}(p - e)}{p}$$

(14.1)

where $F_1 = 4\sigma d_i/\rho_l g(p - e)^2 r_s$. The F_1 parameter defines the strength of the surface tension force, relative to gravity force. r_s is the radius of the condensate film at the base of the wire. Data were obtained with R-12 condensing in an 8.4-mm-inside-diameter tube using two different wire coil inserts: $e = 0.5$ mm, $p = 2.6$ mm, and $e = 0.3$ mm, $p = 1.5$ mm. Near the exit of the tube, where the flow was stratified, they obtained 40% enhancement. They also ran tests with the wire coil insert in the full circuit length and obtained an average enhancement of approximately 35%, with $\Delta p/\Delta p_p \cong 10$. At high vapor qualities, the wire coil acts as a roughness. Equation 14.1 correlated their data within ±20%.

14.2.5 Coiled Tubes and Return Bends

Brdlik and Kakabaev [1964] and Miroploskii and Kurbanmukhamedov [1975] have studied steam condensation in helical tube coils. Brdlik and Kakabaev [1964] show that the local condensation coefficient is given by

$$\frac{hd_i}{k_l} = \frac{388}{1 - x}\left(\frac{d_i G}{\mu_l}\right)^{0.15}\left(\frac{D_c}{d_i}\right)^{-0.54}$$

(14.2)

where d_i is the tube diameter and D_c is the coil diameter. Traviss and Rohsenow [1973] measured the local condensation coefficients downstream from a return bend for R-12 condensing in a 0.80-mm-diameter tube. The effect of the return bend is negligible when averaged over a length of 90 tube diameters.

14.3 MICRO-FIN TUBE

This tube was introduced in Chapter 13, and is illustrated in Figure 13.7d. Virtually all air-cooled residential air conditioners use this tube. This includes central air conditioners and window units. It is also frequently used in automotive refrigerant condensers, which use aluminum tubes. Some manufacturers use the same tube for enhancement of both evaporation and condensation. Hence, the descriptive information presented in Section 13.3.2 is equally applicable to the condensation version and will not be repeated here. The tube is made in diameters between 4 and 16 mm. The micro-fin tube provides a R-22 condensation coefficient approximately 100% higher than that of a plain tube. Table 14.4 is a repeat of Table 13.1, except the last column shows the condensing coefficient relative to a plain tube (h/h_p) operated at the same mass velocity. Figure 13.12, from Yasuda et al. [1990], shows the cross-section fin geometries associated with Table 14.4.

The condensation tube (HEX-C) differs from the evaporator tube (HEX) only in the helix angle. Figure 14.4 shows the R-22 condensing coefficients of the 9.52-mm-diameter Table 14.4 tubes as a function of mass velocity. Table 14.4 shows that the condensing coefficient of the HEX-C tube is 3.1 times that of a plain tube at the same mass velocity. The improved versions have higher fins (e) and a smaller fin

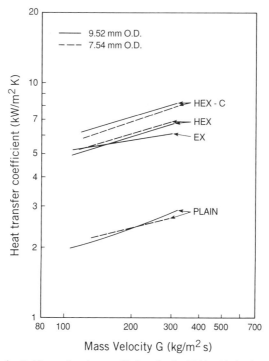

Figure 14.4   R-22 condensing coefficient in the Table 14.4 micro-fin tubes.

included angle (β). Work by Shinorara and Tobe [1985] for the Thermofin-EX™ tube shows that the condensing coefficient gradually increases with helix angle (α), from 7 degrees, up to 30 degrees. The preferred helix angle for condensation is 30 degrees.

The pressure drop for condensation should be slightly less than for vaporization, since the momentum contribution reduces pressure drop. The pressure drop for evaporation was shown in Figure 13.15.

No detailed work has been done to explain the condensation enhancement mechanism in the micro-fin tube. In addition to vapor shear, the author proposes that surface tension forces (see Chapter 12) are influential. Surface tension pulls the condensate from the fin tips into the drainage channel at the base of the fins. No correlations have been published to predict the performance of the tube as a function of the flow, fluid properties, and geometric variables.

There is no uniform agreement on the optimum helix angle. Khanpara et al. [1987] measured R-113 evaporation in 9.5-mm-outside-diameter micro-fin tubes made by eight different manufacturers. There were significant differences in the cross-section fin profiles, as shown in Figure 14.5. All tubes, except Tube 6, had 60–70 fins, 0.15- 0.19-mm fin height, and 20- to 25-degree helix angle. Tube 6 had 0.1-mm fin height and 9 degree helix angle. At $x = 0.5$, the enhancement ratio of the eight tubes varied from 1.6 to 2.2 at $G = 200$ kg/m²-s, and from 1.8 to 3.3 at $G = 575$ kg/m²-s. The Shinohara et al. [1987] patent shows that the R-22 condensation coefficient in a 9.5-mm-outside-diameter tube increases approximately 20% as the helix angle increases from 10 to 35 degrees. Schlager et al. [1990c] tested three 12.7-mm-outside-diameter micro-fin tubes having different helix angles 15, 18, and 25 degrees) with R-22. However, their tubes also had different fin heights ($0.15 \le e_i \le 0.3$ mm) and pitch, so they could not define the effect of specific geometry factors on the performance differences. The highest condensation coefficient was provided by the tube having the highest fin height and 18-degree helix angle.

No theoretical models or correlations (either empirical or rationally based) have been proposed for the micro-fin tube. This is surprising, in view of the importance of the tube.

Schlager et al. [1988, 1989] and Eckels and Pate [1991] report the effect of oil on

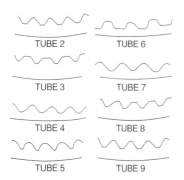

Figure 14.5 Cross sections of the micro-fin tubes tested by Khanpara et al. [1987]. (From Khanpara et al. [1987].)

the performance of R-22, R-134, and R-12, respectively. Oil slightly reduces the condensation coefficient for typically used oil concentrations.

14.4 AUTOMOTIVE CONDENSERS

Automotive air conditioners frequently use a flat tube as illustrated in Figure 14.6a. Use of a flat tube, rather than a round tube, provides the air-side pressure drop. A flat tube presents less projected frontal area to the air stream, and hence will reduce the air-side pressure drop. The tubes contain membrane webs between the flat surfaces for pressure containment. Initially, these tubes contained a smooth inner surface as shown in Figure 14.6b. Based on the success of the round micro-fin tube, it is logical to apply micro-fins inside a flat extruded aluminum tube. Such an extruded aluminum tube with axial micro-fin grooves is shown in Figure 14.6c. Ohara [1983] describes an extruded tube with a corrugated insert brazed to the top and bottom walls of the extruded tube passages (Figure 14.6d). Rather than using an extruded tube, Figure 14.6e shows a welded tube with corrugated insert brazed to the tube wall.

If surface tension forces promote condensation from the fins for round micro-fin tubes, the same phenomena should occur for the Figure 14.6c aluminum tube with micro-grooves. This is illustrated in Figure 14.6f. The role of surface tension force is different in the Figure 14.6c and Figure 14.5e tubes. This is because the Figure 14.6d and 14.6e tubes do not have sharp fin tips, on which a surface tension pressure gradient may act. However, surface tension may act to pull the condensate into the acute corners, where the corrugated insert joins with the base tube.

The automotive industry has used small-diameter, flat tubes in radiators for many years. Use of flat tubes for condensers is a recent trend that began in the late 1970s using extruded aluminum tubes with membrane webs. The condenser operates at a much higher pressure than does a radiator. Hence, the membrane webs illustrated in the Figure 14.6 tubes are required for an automotive condenser.

The flat tubes are used with air-side fins. If the minor diameter of a flat tube is reduced, the air-side pressure drop will be reduced. This has been known for many years, and trends in both the refrigeration and automotive industries have been toward smaller tube diameters. When the tube diameter is reduced, more refrigerant circuits are required. Figure 14.7 shows two methods of circuiting automotive refrigerant condensers. Figure 14.7a uses a one-pass, serpentine construction, with the refrigerant supplied directly to the flat tube. Figure 14.7b from Hoshino et al. [1991] shows such a condenser designed for three refrigerant passes, as illustrated in Figure 14.7c. The condenser has inlet and outlet headers. The headers consist of round aluminum tube. Figure 14.7d shows the method of dividing the flow at the manifold headers to provide multipasses on the refrigerant side. Figure 14.7c shows that the number of tubes in parallel are reduced with each subsequent pass. This maintains high vapor velocity in each pass. A key advantage of the parallel flow, multipass design is that it allows use of a thinner tube (smaller minor tube diameter), since only part of the total refrigerant flow rate is carried by each tube. Use of a

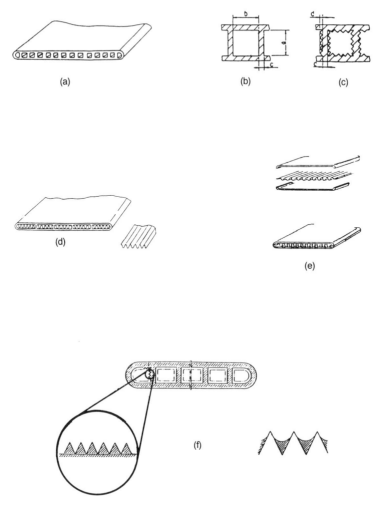

Figure 14.6 Aluminum tubes used in automotive, refrigerant condensers. (a) Extruded tube with membrane partitions. (b) Detail of Figure 14.6a tube, (c) Detail of tube having micro-grooves. (d) Extruded tube with corrugated insert. (e) Tube made by brazing a corrugated strip in a tube formed from flat strip and seam-welded. (f) Illustration of surface-tension-drained condensation on the tips of the microgrooves.

thinner tube results in lower air-side pressure drop. The extruded aluminum tube used in automotive condensers is typically 3.0-mm minor tube diameter. The 180-degree bends required by the Figure 14.7a serpentine arrangement impose a minimum fin height requirement, to satisfy the minimum allowable bend radius. This limitation does not exist in the Figure 14.7b design.

A natural consequence of such thinner extruded tubes used with the parallel flow is reduced hydraulic diameter on the tube side. No data have been reported for condensation in extruded aluminum tubes of the type shown in Figure 14.6.

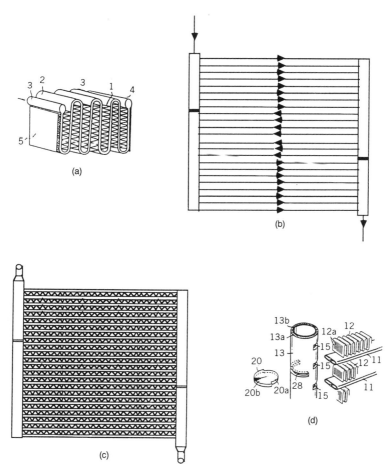

Figure 14.7   (a) Serpentine refrigerant circuiting method. (b) Parallel flow method of circuiting. (c) Illustration of Figure 14.7b condenser having three refrigerant circuits. (d) Method of dividing the refrigerant flow at headers to provide multipass circuiting.

14.5 PLATE-TYPE HEAT EXCHANGERS

Figure 1.16 shows the plates used in common plate-type heat exchangers. Although the plate-type exchanger was developed for liquid–liquid applications, it is now frequently for condensation and vaporization. Figure 14.8a shows the cross section of a plate heat exchanger developed for condensation, with heat rejection to a single-phase coolant. Condensation occurs in the "A" channels, with the coolant in the "B" channels. The recommended corrugation pitch (p) and height (e) are $1 \leq p \leq 2$ mm and $0.3 \leq e \leq 0.6$ mm, respectively. Surface tension pulls the condensate into the crevices formed by the intersection of the corrugation and the adjacent flat plate. This maintains a thin film on the flat plate region between corrugations. Figure 14.8b shows the overall heat transfer coefficient for ammonia and R-114 as a

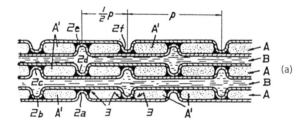

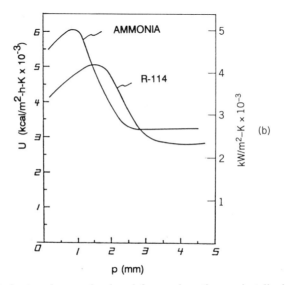

Figure 14.8   Plate heat exchanger developed for condensation against liquid coolant by Uehara and Sumitomo [1985]. (a) End view showing plate configuration with condensing vapor (A) and coolant (B) streams. (b) Measured overall heat transfer coefficient for condensation of ammonia and R-114.

function of the corrugation pitch (p). The figure shows that the optimum corrugation pitch is approximately 1.0 mm for ammonia and 1.5 mm for R-114. The condensing coefficient significantly decreases for pitches greater than 3.0 mm. This is because there are fewer channels to carry the condensate, so thicker condensate film exist on the flat plate regions.

14.6   NONCONDENSIBLE GASES

When noncondensibles are present, an additional thermal resistance is introduced in the gas at the vapor–liquid interface. Mixing in the gas film will substantially reduce this thermal resistance. The maintenance of high vapor velocities, or special surface geometries, which promote a higher heat transfer coefficient in this gas film will substantially alleviate the performance deterioration due to noncondensibles.

Enhancement should be beneficial when noncondensibles are present. This possibility was investigated by Chang and Spencer [1971], who condensed R-12 in a vertical annulus geometry. Noncondensibles cause a mass transfer resistance at the liquid–vapor interface and depress the liquid–vapor interface temperature. The O-rings (diameter unspecified) were spaced at 25-m axial pitch on the inner annulus condensing surface. Mixing of this gas boundary layer by the O-ring roughness should reduce this resistance, increase the interface temperature, and increase the condensation coefficient. Figure 14.9 shows the dimensionless condensing coefficient of R-12 with noncondensible gas in the annulus with and without O-rings. As shown in Figure 14.9, the O-rings cause considerable enhancement at low Reynolds (Re_d) number, but not at high Reynolds numbers. The entering vapor velocity increases with increasing Re_d. The lack of enhancement at high Re_d is believed to be due to surface roll waves which naturally occur because of an interaction between the light noncondensible layer and the heavy condensing vapor. As the molecular weight of the noncondensibles approaches that of the condensing vapor, enhancement due to the O-rings decreases. Chang and Spencer found negligible enhancement for condensation of pure R-12.

The enhancement concept discussed in Section 15.2.2 is directly applicable to the noncondensible gas problem. This concept (Figure 15.2) consists of a transverse-rib roughness, which is displaced from the tube wall and mixes the vapor boundary layer. This is probably a better approach for dealing with noncondensibles than the wall-attached Chang and Spencer O-rings.

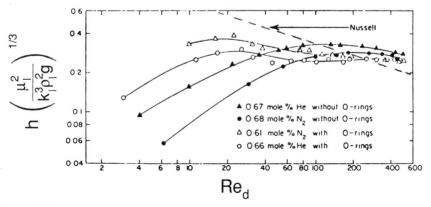

Figure 14.9 Effect of spaced ring roughness on condensation of an R-12/air mixture for vertical flow in an annulus. (From Chang and Spencer [1971].)

14.7 PREDICTIVE METHODS FOR CIRCULAR TUBES

Predictive equations have been developed for several of the enhancement techniques. These equations are based on modifications to condensation in plain tubes. Schlager et al. [1990a] provide a literature survey of correlations for condensation inside tubes. The remarks and warnings concerning predictive equations for vaporization in tubes given in Section 13.5 also apply here. Only heat transfer correlations are presented here. Refer to the references for the accompanying pressure drop correlations.

14.7.1 High Internal Fins

The first correlation for internal fins was developed by Vrable et al. [1974] based on their R-12 data. They proposed an empirical correlation of their R-12 data for one internal fin geometry.

The Royal and Bergles [1978a] correlation of their steam data in four internally finned tubes is based on modifying the Ackers et al. [1959] correlation for smooth tubes. The correlation is

$$\frac{hD_h}{k} = 0.0265 \left(\frac{G_e D_h}{\mu_l} \right)^{0.8} \mathrm{Pr}_l^{0.33} \left[160 \left(\frac{e^2}{sD_h} \right)^{1.91} + 1 \right] \qquad (14.3)$$

The modification is the term in square brackets, which includes the fin geometry parameters and is strictly empirical. Because the data were taken only for steam at 1.0 bar, there is no guarantee that it will apply to other pressures or fluids. It is likely that surface tension drainage effects occur on the fin tips and fin base, as discussed in Chapter 12, which are not accounted for in Equation 14.3. Note that the equation does not include fin efficiency.

Luu and Bergles [1980] found that Equation 14.3 did not predict their R-113 data very well. They proceeded to develop another correlation for the R-113 data, which did not include the steam data. This correlation was based on an empirical correction to the Boyko and Kruzhilin [1967] correlation for smooth tubes. This correlation should have the same limitations as for the steam correlation.

Kaushik and Azer [1988] developed a correlation based on steam, R-113, and R-11 data. The correlated data included the steam data of Royal and Bergles [1978a], the R-113 data of Luu and Bergles [1980] and of Said and Azer [1983], and the R-11 data of Venkatesh [1984]. The correlation is a strictly empirical power-law regression type given by

$$\frac{hd_i}{k_l} = C \left(\frac{G_e d_i}{\mu_l} \right)^{0.507} \left(\frac{\Delta x d_i}{L} \right)^{0.198} \left(\frac{p}{p_{cr}} \right)^{-0.14} F_1^{n_1} F_2^{n_2} \qquad (14.4)$$

where F_1 and F_2 are geometric parameters (see Nomenclature) used by Carnavos [1980] to correlate his single-phase flow data. For $F_1 \leq 1.4$, $n_1 = 0.874$ and $n_2 =$

−0.814. For $F_2 > 1.4$, $n_1 = 4.742$ and $n_2 = 0.0$. Again, this empirical correlation does not account for probable surface tension drainage effects or fin efficiency. However, it is probably the most general of those presented. The correlation predicted 71% of the data points within ±30%.

14.7.2 Wire-Loop Internal Fins

Honda et al. [1988] developed a rationally based correlation for the friction and heat transfer in an annulus containing wire-loop fins discussed in Section 14.2.1. The heat transfer correlation is shown in Figure 14.3b, which shows the heat transfer results for R-11 and R-113 plotted in the form Nu/Nu_c versus the Martinelli parameter, X_{tt}. The Nusselt number is defined in terms of the wire diameter, $Nu_e = he/\mu_l$. The normalizing parameter in Figure 14.3b is defined as

$$Nu_c = (Nu_s^4 + Nu_v^4)^{1/4} \qquad (14.5)$$

where Nu_s and Nu_v are calculated by correlations for the surface-tension- and vapor-shear-controlled regimes, respectively. These correlations are given by Honda et al. [1988]. Figure 14.3b shows that the vapor-shear-controlled regime prevails for $X_{tt} \leq 0.1$. In the surface-tension-controlled regime, condensation occurs on the wire fins and is pulled to the fin root, and then it is gravity-drained from the tube. In the vapor-shear-controlled regime, the model treats condensation on the wire fins similar to shear-controlled condensation normal to a cylinder. This is a very comprehensive model, and is well justified by theoretical reasoning.

Kaushik and Azer [1989] worked to develop an analytically based model for condensation in internally finned tubes. This model includes surface tension drainage effects. This rationally based model predicted the steam data of Royal and Bergles [1978a] within ±30. However, it overpredicts the R-113 data of Said and Azer [1983]. Sur and Azer [1991] developed an analytical model to predict the pressure drop in internally finned tubes.

14.7.3 Twisted Tapes

The Royal and Bergles [1978a] correlation for steam condensation in tubes is based on modifying the Ackers et al. [1959] correlation for smooth tubes. The correlation assumes no thermal contact resistance of the tape at the tube wall. The correlation is

$$\frac{hD_h}{k_l} = 0.0265 \left(\frac{F_t G_e D_h}{\mu_l} \right)^{0.8} Pr_l^{0.4} F_{tt} \qquad (14.6)$$

The F_t in the Reynolds term accounts for the velocity of swirling flow at the wall and is given by

$$F_t = (1 + \tan^2\alpha)^{1/2} = [1 + (\pi/2y)^2]^{1/2} \qquad (14.7)$$

The term F_{tt} accounts for the fin efficiency of the tape and is given by

$$F_{tt} = \frac{(\pi d_i - 2t_t + \eta_f d_i)}{\pi d_i} \qquad (14.8)$$

where η_f is the fin efficiency. Their steam data for two tapes were correlated within $\pm 30\%$. Because Royal and Bergles [1978a] state that their tape did not have good thermal contact with the tube wall, their assumption of zero contact resistance in the correlation raises questions concerning validity of the correlation. It is doubtful that typical twisted-tapes will have perfect thermal contact with the tube wall. Assuming no heat transfer from the tape, one would set $\eta_f = 0$ in Equation 14.8.

Luu and Bergles [1980] correlate their R-113 data using a modification of the Boyko and Kruzhilin [1967] smooth tube correlation. The modification is similar to that of Royal and Bergles [1978a], except they assume that only 50% of the tape thickness is in contact with the wall. Again, most of the data were correlated within $\pm 30\%$.

Said and Azer [1983] found that the Luu and Bergles [1980] correlation underpredicted their R-113 data. Said and Azer [1983] proposed a different correlation, based on the steam data of Royal and Bergles [1978a] and the R-113 data of Luu and Bergles [1980] and of Said and Azer [1983]. This empirical correlation is

$$\frac{hd_i}{k_l} = 0.023 \mathrm{Re}_{d,l}^{0.8} \mathrm{Pr}_l^{0.4} (0.665 + 1.86 p_r^{-0.38}) \left[1 + 0.019 \left(\frac{\mathrm{Re}_{d,l}}{y} \right)^{0.26} \right] \qquad (14.9)$$

Equation 14.9 is a modification of the Dittus–Boelter equation for turbulent, single-phase heat transfer in tubes. The first modifying parameter in parentheses accounts for pressure, and the term in square brackets accounts for the tape twist. The Reynolds number is based on the nominal tube diameter, rather than the hydraulic diameter; it is defined by $\mathrm{Re}_d = d_i G / \mu_l$. Equation 14.9 predicted all three data sets within $\pm 30\%$. Note that the correlation makes no assumptions concerning the fin efficiency of the tape. This correlation is recommended because of its ability to predict data for several fluids and because an assumption concerning the fin efficiency is not required.

14.7.4 Roughness

Luu and Bergles [1980] provide an interesting correlation of their helical-rib roughness R-113 data. The correlation is based on the single-phase roughness correlation of Webb et al. [1971] discussed in Chapter 9. They correlated their data on two roughness geometries within $\pm 30\%$. The reader should consult Luu and Bergles [1980] for use of the correlation.

14.7.5 Micro-fins

No serious work has been done to develop a correlation to predict the heat transfer or pressure drop in the micro-fin tube.

14.8 CONCLUSIONS

Clearly, the micro-fin tube provides the highest heat transfer performance and the lowest pressure drop of the competing internal enhancements. However, data have not been reported for fluids other than the refrigerants. The twisted-tape is a relatively low enhancement device that may be of value only to increase capacity of an existing system.

Rationally based semiempirical correlations have been developed for the twisted-tape. However, the correlations for the high internal fins discussed in Section 14.2.1 are largely empirical, and do not seem to predict other than the developers' data very well. Surprisingly little progress has been made in predictive methods for the micro-fin tube.

14.9 REFERENCES

Ackers, W. W., Deans, H. A., and Crosser, O. K., 1959. "Condensing Heat Transfer within Horizontal Tubes," *Chemical Engineering Progress Symposium Series*, Vol. 55, No. 29, pp. 1711–1176.

Baker, O., 1954. "Simultaneous Flow of Oil and Gas," *Oil and Gas Journal*, Vol. 53, pp. 185–195.

Boyko, L. D., and Kruzhilin, G. N., 1967. "Heat Transfer and Hydraulic Resistance during Condensation of Steam in a Horizontal Tube and in a Bundle of Tubes," *International Journal of Heat and Mass Transfer*, Vol. 10, pp. 361–373.

Brdlik, P. M., and Kakabaev, A., 1964. "An Experimental Investigation of the Condensation of Steam in Coils," *International Chemical Engineering*, Vol. 2, pp. 216–239.

Carey, V. P., 1992. *Liquid–Vapor Phase-Change Phenomena*," Hemisphere Publishing Corp., Washington, D.C.

Carnavos, T. C., 1980. "Heat Transfer Performance of Internally Finned Tubes in Turbulent Flow," *Heat Transfer Engineering*, Vol. 4, No. 1, pp. 32–37.

Chang, K. I., and Spencer, D. L., 1971. "Effect of Regularly Spaced Surface Ridges on Film Condensation Heat Transfer Coefficients for Condensation in the Presence of Noncondensible Gas," *International Journal of Heat and Mass Transfer*, 14, pp. 502–505.

Eckels, S. J., and Pate, M. B., 1991. "In-tube Evaporation and Condensation of Refrigerant–Lubricant Mixtures of HFC-134a and CFC-12," *ASHRAE Transactions*, Vol. 97, Part 2, pp. 62–70.

Fenner, G. W., and Ragi, E., 1979. "Enhanced Tube Inner Surface Heat Transfer Device and Method," U.S. Patent 4,154,293.

Honda, H., Nozu, S., Matsuoka, Y, and Aomi, T., 1988. "Condensation of Refrigerants R-11 and R-113 in Horizontal Annuli with an Enhanced Inner Tube," *Proceedings of the 1st World Conference on Experimental Heat Transfer, Fluid Mechanics and Thermodynamics*, R. K. Shah, E. N. Ganic, and K. T. Yang, Eds., Elsevier Science Publishers, New York, pp. 1069–1076.

Hoshino, R., Sasaki, H., and Yasutake, K., 1991. "Condenser for Use in Car Cooling System," U. S. Patent 5,025,855, assigned to Showa Aluminum Co., Japan.

Ito, M. and Kimura, H., 1979. "Boiling Heat Transfer and Pressure Drop in Internal Spiral-Grooved Tubes." *Bulletin of the JSME*, Vol. 22, No. 171, pp. 1251–1257.

Kaushik, N., and Azer, N. Z., 1988. "A General Heat Transfer Correlation for Condensation Inside Internally Finned Tubes," *ASHRAE Transactions*, Vol. 94, Part 2, pp. 261–279.

Kaushik, N., and Azer, N. Z., 1989. "An Analytical Heat Transfer Prediction Model for Condensation Inside Longitudinally Internally Finned Tubes," *ASHRAE Transactions*, Vol. 95, Part 2, pp. 516–523.

Kaushik, N., and Azer, N. Z., 1990. "A General Pressure Drop Correlation for Condensation Inside Internally Finned Tubes," *ASHRAE Transactions*, Vol. 96, Part 1, pp. 242–255.

Khanpara, J. C., Pate, M. B., and Bergles, A. E., 1987. Local Evaporation Heat Transfer in a Smooth Tube and a Micro-fin Tube Using Refrigerants 22 and 113," in *Boiling and Condensation in Heat Transfer Equipment*, E. G. Ragi, Ed., HTD Vol. 85, ASME, New York, pp. 31–39.

Lin, S. T., Azer, N. S., and Fan, L. T., 1980. "Heat Transfer and Pressure Drop During Condensation Inside Horizontal Tubes with Static Mixer Inserts," *ASHRAE Transactions*, Vol. 86, Part 2, pp. 649–651.

Lopina, R. F., and Bergles, A. E., 1969. "Heat Transfer and Pressure Drop in Tape-Generated Swirl Flow of Single-Phase Water," *Journal of Heat Transfer*, Vol. 91, pp. 434–442.

Luu, M., and Bergles, A. E., 1979. "Experimental Study of the Augmentation of the In-tube Condensation of R-113," *ASHRAE Transactions*, Vol. 85, Part 2, pp. 132–146.

Luu, M., and Bergles, A. E., 1980. "Enhancement of Horizontal In-tube Condensation of R-113," *ASHRAE Transactions*, Vol. 85, Part 2, pp. 293–312.

Miroploskii, Z. L., and Kurbanmukhamedov, A., 1975. "Heat Transfer with Condensation of Steam within Coils," *Thermal Engineering*, No. 5, pp. 111–114.

Ohara, 1983. "Heat Exchanger," Japanese Patent 58–221390, assigned to Nippondenso Co.

Reisbig, R. L., 1974. "Condensing Heat Transfer Augmentation Inside Splined Tubes," *AIAA/ASME Thermophysics Conference*, Boston, Paper 74-HT-7.

Royal, J. H., and Bergles, A. E., 1978a. "Augmentation of Horizontal In-tube Condensation by Means of Twisted Tape Inserts and Internally Finned Tubes," *Journal of Heat Transfer*, Vol. 100, pp. 17–24.

Royal, J. H., and Bergles, A. E., 1978b. "Pressure Drop and Performance Evaluation of Augmented Tube Condensation," *Proceedings of the 6th International Heat Transfer Conference*, Toronto, Vol. 2, pp. 459–464.

Said, S. A., and Azer, N. Z., 1983. "Heat Transfer and Pressure Drop During Condensation Inside Horizontal Finned Tubes," *ASHRAE Transactions*, Vol. 89, Part 1, pp. 114–134.

Schlager, L. M., Pate, M. B., and Bergles, A. E., 1988. "Performance of Micro-fin Tubes with Refrigerant-22 and Oil Mixtures," *ASHRAE Journal*, November, pp. 17–28.

Schlager, L. M., Pate, M. B., and Bergles, A. E., 1989. "A Comparison of 150 and 300 SUS Oil Effects on Refrigerant Evaporation and Condensation in a Smooth Tube and a Micro-fin Tube, *ASHRAE Transactions*, Vol. 95, Part 1, pp. 387–397.

Schlager, L. M., Pate, M. B., and Bergles, A. E., 1990a. "Performance Predictions of Refrigerant–Oil Mixtures in Smooth and Internally Finned Tubes—Part 1: Literature Review," *ASHRAE Transactions*, Vol. 96, Part 1, pp. 160–169.

Schlager, L. M., Pate, M. B., and Bergles, A. E., 1990b. "Performance Predictions of Refrigerant–Oil Mixtures in Smooth and Internally Finned Tubes—Part 2: Design Predictions," *ASHRAE Transactions*, Vol. 96, Part 1, pp. 170–182.

Schlager, L. M., Pate, M. B., and Bergles, A. E., 1990c. "Evaporation and Condensation Heat Transfer and Pressure Drop in Horizontal, 12.7-mm Micro-fin Tubes with Refrigerant 22," *Journal of Heat Transfer*, Vol. 112, pp. 1041–1047.

Shah, M. M., 1979. "A General Correlation for Heat Transfer During Film Condensation Inside Pipes," *International Journal of Heat and Mass Transfer*, Vol. 22, pp. 547–556.

Shinohara, Y., and Tobe, M., 1985. "Development of an Improved Thermofin Tube," *Hitachi Cable Review*, No. 4, pp. 47–50.

Shinohara, Y., Oizumi, K., Itoh, Y., and Hori, M., 1987. "Heat Transfer Tubes with Grooved Inner Surface," U. S. Patent 4,658,892.

Sur, B., and Azer, N. Z., 1991. "An Analytical Pressure Drop Prediction Model for Condensation Inside Longitudinally Internally Finned Tubes," *ASHRAE Transactions*, Vol. 97, Part 2, pp. 54–61.

Thomas, D. G., 1967. "Enhancement of Film Condensation Rates on Vertical Tubes by Vertical Wires," *Industrial and Engineering Chemistry Fundamentals*, Vol. 6, pp. 97–102.

Traviss, D. P., and Rohsenow, W. M., 1973. "The Influence of Return Bends on the Downstream Pressure Drop and Condensation in Tubes," *ASHRAE Transactions*, Vol. 79, Part 1, pp. 129–137.

Uehara, H., and Sumitomo, H., 1985. "Condenser," U. S. Patent 4,492,268 assigned to Iisaka Works, Ltd.

Venkatesh, K., 1984. "Augmentation of Condensation Heat Transfer of R-11 by Internally Finned Tubes," M.S. Thesis, Department of Mechanical Engineering, Kansas State University.

Vrable, D. A., Yang, W. J., and Clark, J. A., 1974. "Condensation of Refrigerant-12 Inside Horizontal Tubes with Internal Axial Fins," *Heat Transfer 1974, 5th International Heat Transfer Conference*, Vol. 3, pp. 250–254.

Wang, W., 1987. "The Enhancement of Condensation Heat Transfer for Stratified Flow in a Horizontal Tube with Inserted Coil," in *Heat Transfer Science and Technology*, B-X. Wang, Ed., Hemisphere Publishing Corp., New York, pp. 805–811.

Webb, R. L., Eckert, E. R. G., and Goldstein, R. J., 1971. "Heat Transfer and Friction in Tubes with Repeated-Rib Roughness," *International Journal of Heat and Mass Transfer*, Vol. 14, pp. 601–617.

Yasuda, K., Ohizumi, K., Hori, M., and Kawamata, O., 1990. "Development of Condensing "Thermofin-HEX-C Tube," *Hitachi Cable Review*, No. 9, pp. 27–30.

14.10 NOMENCLATURE

A	Heat transfer surface, m_2 or ft^2
A_c	Cross-sectional flow area, m^2 or ft^2
d_i	Tube inside diameter, or diameter to the base of internal fins or roughness; m or ft
d_{im}	internal diameter if enhancement material is uniformly returned to wall thickness, m or ft
d_o	Tube outside diameter, m or ft
D_c	Diameter of helix, m or ft

D_h	Tube hydraulic diameter, m or ft
e	Roughness height, corrugation depth, or wire diameter, m or ft
F_1	Geometric parameter used in Equation 14.4, $[(d_i/d_{im})(1 - 2e/d_i)]^2$, dimensionless
F_2	Geometric parameter used in Equation 14.4, $d_i D_h/d_{im}^2$, dimensionless
G	Mass velocity in tube: G_v (of vapor component), G_l (of liquid component); kg/m²-s or lbm/ft²-s
G_e	$G[(1 - x_{av}) + x_{av}(\rho_l/\rho_v)^{1/2}]$, kg/m²-s or lbm/ft²-s
h	Heat transfer coefficient, W/m²-K or Btu/hr-ft²-°F
h_{cv}	Convective component of two-phase heat transfer coefficient, W/m²-K or Btu/hr-ft-°F
h_{nb}	Nucleate boiling component for convective vaporization, W/m²-K or Btu/hr-ft²-F
h_s	Heat transfer coefficient on smooth surface, W/m²-K or Btu/hr-ft²-F
H	Axial distance for one fin (or flute, or twisted-tape), 180-degree revolution, m or ft
L	Length of vertical condensing surface; tube length, m or ft
LMTD	Logarithmic mean temperature difference, K or °F
n	Number of flutes or internal fins in tube, dimensionless
N	Number of horizontal tube rows in depth, dimensionless
p	Pitch of enhancement surface elements, m or ft
p	Fluid pressure, kPa or lbf/ft²
p_{cr}	Critical fluid pressure, kPa or lbf/ft²
p_r	Reduced pressure, p/p_{cr}; dimensionless
Δp	Pressure drop, kPa or lbf/ft²
P	Pumping power, W or HP
Nu	Nusselt number, hd/k_l; dimensionless
Nu_c	Nusselt number defined by Equation 14.5, dimensionless
Nu_e	Nusselt number based on wire diameter, he/μ_l; dimensionless
Pr	Prandtl number, dimensionless
q	Heat flux, W/m² or Btu/hr-ft²
Q	Heat transfer rate, W or Btu/hr
Re	Reynolds number, $\text{Re}_d\ (=Gd_i/\mu_l)$, $\text{Re}_{Dh}\ (=GD_h/\mu_l)$, $\text{Re}_{d,l}\ [=G(1 - x)d_i/\mu_l]$, dimensionless
s	Spacing between fins, m or ft
U	Overall heat transfer coefficient
u	Frictional shear velocity $(\tau/\rho)^{1/2}$, m/s or ft/s
Wt	Tube weight, kg or lbm
x	Vapor quality, x_{av} (average in tube), dimensionless
X_{tt}	Defined by Equation 13.8, dimensionless
y	Twist ratio $= H/d_i = \pi/(2 \tan \alpha)$, dimensionless

Greek Symbols

α	Helix angle, measured from tube axis ($\alpha = \tan^{-1} \pi d_i/H$), radians
β	Included angle of fin cross section, radians

δ	Condensate film thickness, m or ft
η_f	Fin efficiency, dimensionless
μ	Dynamic viscosity: μ_l (of liquid), μ_v (of vapor) kg/m-s or lbm/s-ft
ρ	Density: ρ_l (of liquid), ρ_v (of vapor), kg/m³ or lbm/ft³
τ_o	Wall shear stress based on pressure drop, kg/m² or lbf/ft²
ϕ	Severity factor, $(e/D)/(p/e)$, dimensionless

Subscripts

i	Designates inner surface of tube
o	Designates outer surface of tube
p	Plain tube
s	Smooth surface

Unsubscripted variables refer to enhanced tube.

15

ENHANCEMENT USING
ELECTRIC FIELDS

15.1   INTRODUCTION

Considerable work has been done using electric fields to enhance heat transfer. Ohadi [1991] provides a survey of the technology in lay terms. Yabe et al. [1987] and Yabe [1991] provide an excellent discussion of the fundamentals of EHD. Yabe [1991] provides particular focus on enhancement of condensation and boiling. The application of an electric field to a dielectric fluid will impose a body force (F_e) on the fluid. This body force adds a term to the momentum equation and influences the fluid motion. It is important that the fluid be a dielectric—that is, not conduct an electric current. The body force (F_e) used in the Navier–Stokes equations is described by the change of Helmholtz free energy for virtual work with the energy stored in the fluid by the electric fluid. As described by Panofsky and Phillips [1962], application of electric field strength E to a constant temperature dielectric fluid of permittivity ϵ, and density ρ results in F_e given by

$$F_e = \rho_c E - 0.5 E^2 \nabla \epsilon + 0.5 \nabla \left[\rho E^2 \left(\frac{\partial \epsilon}{\partial \rho} \right) \right] \qquad (15.1)$$

The electric field strength will vary with distance from the electrode and depends on the electrode design. The physical significance of the three terms in Equation 15.1 is as follows:

1. The first term is the "electrophoretic" (Coulomb) force acting on the net free charge of electric field space charge density, ρ_c. The direction of the force depends on the relative polarities of the free charges and the electric field.

2. The second term is the "di-electrophoretic" force produced by the spatial change of permittivity, ϵ (also called the dielectric constant). In two-phase flow, this force arises from the difference in permittivity of the vapor and liquid phases.

3. The third term is called the "electrostriction" force and is caused by inhomogeneity of the electric field strength. This force is dependent on inhomogeneities in the electric field strength and may be likened to an "electrical pressure."

The electric current is given by

$$i = \rho_c u + \sigma_e E \qquad (15.2)$$

where u is the fluid velocity and σ_e is the electrical conductivity of the fluid. In steady state, without convection, the electric charge density is given by

$$\rho_c = \frac{\epsilon}{\sigma_e} E \nabla \sigma_e \qquad (15.3)$$

Thus, the electric charges are generated by the gradient to the electrical conductivity. Because the electrical conductivity of liquids is temperature-dependent, Equation 15.3 shows that temperature gradients in the fluid generates the electric charge. This makes the Coulomb force (first term of Equation 15.1) an active force.

The momentum and energy equations are coupled through the temperature dependence of the permittivity and thermal conductivity of the fluid. Cooper [1992] notes that analytical solution of the electrohydrodynamic (EHD) coupled momentum and energy equations is not possible, except for the simplest configurations. Rather, data are correlated, based on the predicted electric field strength distribution at the heat transfer surface on the basis of uniform electrical fluid properties.

The sum of the second and third terms of Equation 15.1 gives the force exerted on dielectric fluids. For polarization of dielectric fluids, the net charge is zero. However, the force on the polarized charges formed in the stronger field zone is greater than the force on the charges in the weaker field zone. Thus, the resultant force (the sum of the forces exerted on each polarized charge) moves fluid elements to the stronger electric field region. Yabe [1991] notes that the effect of the magnetic field generated by the current is negligible compared with the pressure generated by the electric field.

Table 15.1, from Ohadi [1991] summarizes enhancement levels obtained by various investigators for natural convection, laminar flow forced convection, boiling, and condensation.

Table 15.1 EHD Heat Transfer Enhancement in Heat Exchangers

Source	Maximum Enhancement (%)	Test Fluid	Wall/ Electrode Configuration	Process
Fernandez and Poulter [1987]	2,300	Transformer oil	Tube–wire	Forced convection
Ohadi et al. [1991]	320	Air	Tube–wire/rod	Forced convection
Levy [1964]	140	Silicone oil	Tube–wire	Forced convection
Yabe and Maki [1988]	10,000	96% R-113	Plate–ring	Natural convection
Cooper [1990]	1,300	90% R-113 10% oil	Tube–wire mesh	Pool boiling
Uemura et al. [1990]	1,400	R-113	Plate–wire mesh	Film boiling
Bologa et al. [1987]	2,000	Diethylether, R-113, hexane	Plate–plate	Film condensation
Sunada et al. [1991]	400; 600	R-113; R-123	Vertical wall– plate	Condensation
Ohadi et al. [1992]	480	R-123	Tube–wire	Boiling

Source: Ohadi [1991].

15.2 ELECTRODE DESIGN AND PLACEMENT

The design and location of the electrodes is dependent on whether the flow is single-phase or two-phase, as well as on the thermophysical and electrical properties of the fluid. The electrodes must be positioned such that the electric field force aids the hydrodynamic forces, particularly at the heat transfer surface. For separated two-phase flows, the permittivity of the liquid and vapors will be distinctly different. Hence $\Delta\epsilon$ exhibits a singularity at a liquid–vapor interface. Jones [1978] shows that the fluid component having the highest ϵ will move to the region of highest field strength, E. A liquid drop will move toward the electrode, and a bubble will move toward the region of lower field strength.

Placement of the electrodes is different for tube-side or shell-side heat transfer enhancement. Consider heat transfer to a fluid flowing inside a tube. Figure 15.1 shows a wire electrode in the center of a tube. If the fluid inside the tube has constant permittivity (ϵ), the field strength distribution for this configuration is given by

$$E = \frac{V}{r}\left[\ln\left(\frac{R_2}{R_1}\right)\right]^{-1} \tag{15.4}$$

where V is the applied voltage, r is the radial distance from the tube axis, and R_1 and R_2 are defined in Figure 15.1. The field strength decays with radial distance from the central electrode wire. If a liquid–gas mixture flows in the tube, the liquid phase

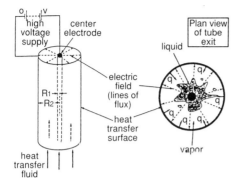

Figure 15.1 Wire electrode in center of tube with two-phase flow. (From Cooper [1992].)

will have the highest permittivity. Hence, the liquid will move toward the central electrode, leaving vapor at the tube wall. This will enhance condensation, but not evaporation, which requires liquid at the tube wall. Condensation enhancement would occur, because the electrode pulls condensate away from the condensing surface. Yabe et al. [1992] used a perforated sheet wrapped in a cylindrical form (5-mm diameter with 1- to 2-mm-diameter holes) as the electrode for enhancement of tube-side convective evaporation of refrigerants. Figure 15.2 shows electrode designs used by Cooper [1992] for enhancement of shell-side boiling in a tube bundle. Figure 15.3 illustrates electrode designs used by Yabe et al. [1987] for enhancement of condensation on the outer surface of vertical tubes. The Figure 15.3 electrodes produce a nonuniform electric field. Both AC and DC voltages provide enhancement.

The enhancement effect will be increased by increasing the applied voltage. The electrical breakdown strength of the liquid and vapor must be known and should not be exceeded. As noted by Cooper [1992], careful design of the solid insulation

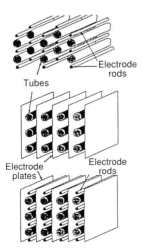

Figure 15.2 Electrode designs used by Cooper [1992] for enhancement of shell-side boiling in a tube bundle. (From Cooper [1992].)

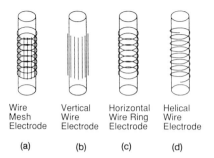

Wire
Mesh
Electrode

(a)

Vertical
Wire
Electrode

(b)

Horizontal
Wire Ring
Electrode

(c)

Helical
Wire
Electrode

(d)

Figure 15.3 Electrode designs used by Yabe et al. [1987] for enhancement of condensation on the outer surface of vertical tubes. (From Yabe et al. [1987].)

system is required. Poor design may result in electrical breakdown, which would make the EHD enhancement system useless. The insulation material must be compatible with the heat transfer media. Inert dielectrics, such as PTFE and proprietary epoxy composites, are candidates. Cooper [1992] notes that high electric stress exists near small radii of curvature on solids, so electric breakdown may occur at these locations. Electrode and insulation surfaces should be rounded, and conducting rods of large radius should be used to reduce the local electrical field strength. Design aspects of the insulation system are discussed in texts on power transformers and switch-gear, such as Kuffel and Zaengl [1984].

Voltages up to 25 kV are typically used in EHD enhancement. Although high voltages are used, the electric current is quite small (e.g., 1 mA) and thus the power consumption is small. To place these voltages in perspective, automotive spark plugs and electronic air cleaners operate at approximate voltages of 25 kV and 10 kV, respectively. The power is supplied by a relatively low kVA power supply consisting of a high-voltage transformer operating at very small current. A device, such as an automotive spark plug, may be modified to bring the high-voltage supply conductor through the heat exchanger wall.

The use of very small currents minimizes physical hazards. All high-voltage surfaces are completely contained within the heat exchanger, which is electrically insulated from the shell and the tubes. Provided that adequate insulation is used, no danger of electric shock should exist.

15.3 SINGLE-PHASE FLUIDS

For single-phase fluids, the permittivity (ϵ) is approximately constant and the second and third terms of Equation 15.1 are negligible. Thus, for single-phase flows, the first term of Equation 15.1 can cause EHD-induced fluid motion known as the ionic (or Corona) wind effect. Ions are produced close to the surface of a wire anode. The Coulomb force on the ions causes them to move to the cathode surface. Interaction of the Corona wind with the main flow produces mixing by secondary flows. The greatest enhancement occurs for natural convection and low-Reynolds-number laminar flows. As the main flow velocity increases into the turbulent regime, the secondary flow speed is negligible compared to the strength of the turbulent eddies. Table 15.1 identifies several studies for natural convection and laminar flow forced convection.

Data for forced convection enhancement are illustrated by the data of Ohadi et al. [1991a]. They used single and double axial (two axial wires spaced at 4.0 mm) electrodes to measure enhancement of air flow in a 32-mm-inside-diameter, 305-mm-long tube. Figure 15.4 shows that the two-electrode design provided a modestly higher enhancement than did the single-electrode design. A maximum enhancement of 3.5 was obtained at 7.5 kV using the two-electrode system at $Re_d = 3000$. Note that the two-electrode design provided only 20% enhancement at $Re_d = 10,000$. For 7.75 kV, the current flow was 385 μA, which results in 2.8-W power expenditure. Hence, the power consumption is minimal. Ohadi, et al. [1991b] used axial wire electrodes in a double-pipe heat exchanger with air flow in both streams. They provide test results for EHD enhancement of the tube side, the shell side, and both fluids. Their results show almost total decay of the enhancement for $Re_d > 4000$. However, for shell-side and tube-side Reynolds numbers of 500 and 2000, respectively, the overall heat transfer coefficient was enhanced a factor of four. Nelson et al. [1991] studied EHD enhancement of air flow in a circular tube using a central, axial wire electrode. Their results show that the friction factor increase is generally comparable with the heat transfer enhancement.

Fernandez and Poulter [1987] worked with transformer oil flowing on the annulus side of a double-pipe heat exchanger. The electric field was applied across the

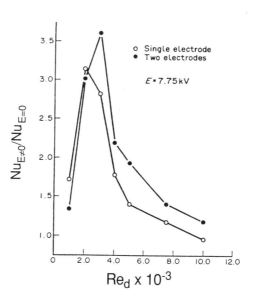

Figure 15.4 Enhancement provided by axial single- and double-wire electrodes in a 38.4-mm-diameter tube. (From Ohadi et al. [1991b].)

annular spacing. Although the heat transfer coefficient was enhanced a factor of 20 (Table 15.1), the oil pressure drop was increased by a factor of only 3.0.

Working in a 20-mm-high rectangular channel, Ishiguro et al. [1991] placed axial wire electrodes spaced at 20-mm pitch, 5 mm from the wall. A mixture consisting of 96% R-113 and 4% ethanol was used as the working fluid. The authors sought to disturb the controlling thermal resistance in the boundary layer at the wall. As expected, the widthwise distribution of the local Nu occurred at the position closest to the wire electrodes. A maximum Nu enhancement of 23 occurred for laminar flow. A ninefold enhancement was measured at $Re_{Dh} = 9100$ for 10-kV applied voltage. At $Re_{Dh} = 27,300$, the enhancement at 10 kV was 2.8.

15.4 CONDENSATION

Early work was performed by Velkoff and Miller [1965], who used a screen-type electrode, which enhanced R-113 condensation on a vertical plate by 150%. Recent work by Yabe [1991] and Sunada et al. [1991] provides fundamental understanding of the concept. Working with Yabe, Sunada et al. [1991] condensed R-113 and R-123 on a vertical plate with electric field strengths between 5 and 8 MV/m. The electrode consisted of a transparent glass plate coated with a 500-Å-thick, electrically conducting InO_3 film. Figure 15.5 shows the local condensation coefficient versus field strength for R-113 condensation at 48°C. The electrode was placed 6.0 mm from the surface, and the condensing plate was cooled by silicone oil. No enhancement was observed until the electrode voltage was increased to 4 MV/m (24-kV electrode voltage). The test results for no electric field are shown by the data

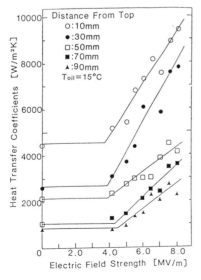

Figure 15.5 Heat transfer enhancement versus electric field strength for R-113 condensing at 48°C on a vertical plate. (From Sunada et al. [1991].)

points for zero field strength. The average enhancement on the lower part of the plate exceeds 300% for 8 MV/m.

Above 4 MV/m, the character of the condensate film radically changed, as illustrated in Figure 15.6. The condensate formed in surface drops, similar to those observed in dropwise condensation. Yabe calls this "pseudo-dropwise" condensation. R-113 is not amenable to dropwise condensation without an electric field, because its low surface tension causes surface wetting. The high enhancement results from the thin film thickness between the drops.

Use of a screen electrode, as shown in Figure 15.3, provides a second enhancement phenomenon known as *liquid jetting*, which is illustrated in Figure 15.7. Spaced electrodes near the surface cause a jet flow of condensate away from the surface (EHD liquid extraction), which thins the condensate film. This removed 95% of the condensate from the surface and resulted in a 2.8-factor enhancement level.

Finally, Yabe [1991] combined the "liquid jetting" and "pseudo-dropwise" phenomena in condensation on a vertical tube. This was done using a helical wire

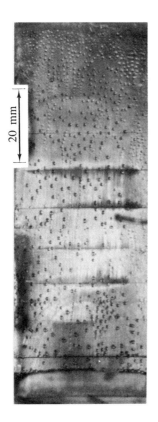

Figure 15.6 Photograph of EHD "pseudo-dropwise condensation" of R-113 at 48.7 K on a vertical plate observed by Sunada et al. [1991]. Electric field strength was 7.5 MV/m, and condensing temperature was 48.7°C. (From Sunada et al. [1991].)

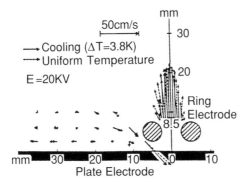

Figure 15.7   Illustration of EHD liquid jet phenomenon. (From Yabe [1991].)

electrode (extraction mode) and a perforated, curved plate (pseudo-dropwise mode). The R-113 condensation coefficient was increased by a factor of 4.5.

Yamashita et al. [1991] proceeded to develop a vertical tube condenser containing 102 19-mm-diameter, 1.4-m-long tubes. Tests were performed condensing perfluorohexane C_6F_{14}. The Figure 15.3a electrode design was used, which promoted both liquid extraction and pseudo-dropwise condensation. The authors obtained a maximum heat transfer enhancement ratio of approximately 6.0, relative to that for zero electric field (for the same heat flux). Tests were also performed with R-114. For a total condensation rate of 60 kW, the electric power consumption by the electric fields at 18-kV voltage was less than 1.4 W! Figure 15.8 compares test results on the 102-tube vertical condenser with data for single horizontal and vertical, mechanically enhanced tubes. For a condensate film Reynolds number greater

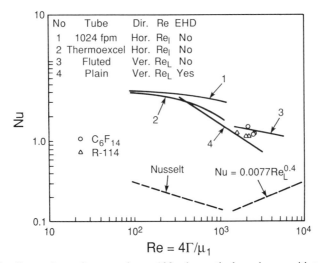

Figure 15.8   Comparison of test results on 102-tube vertical condenser with test results on single horizontal mechanically enhanced tubes.

than 1000, the EHD enhancement is higher than that for vertical fluted tubes. The dashed lines at the bottom of the figure show the laminar and turbulent equations for film condensation on a vertical plate. Typically, a vertical tube will operate with higher Re_L ($4\Gamma_L/\mu_l$) than the Re_l ($4\Gamma/\mu_l$) for a horizontal tube, so the comparison is somewhat inconsistent. The vertical EHD enhanced tube operates at a higher film Reynolds number than do the horizontal mechanically enhanced tubes. Although EHD phenomena offer interesting possibilities, they have not shown enhancement significantly superior to that of high-performance horizontal enhanced tubes (e.g., the Figure 12.6a Thermoexcel-C™). The data do not suggest that the vertical EHD enhanced tube will be superior to horizontal mechanically enhanced tubes for tube bundle applications.

15.5 NUCLEATE BOILING

EHD also provides dramatic enhancement of nucleate boiling. Figure 15.9 shows Yabe's [1991] test results on a plain horizontal tube using a 90%/10% mixture (by weight) of R-11 and ethanol. The Figure 15.3d helical wire electrode design was used with the electrodes placed 5 mm from the boiling surface.. The performance is also compared to that of the High-Flux™ porous surfaces and the Thermoexcel-E™ mechanically enhanced surface geometries (see Chapter 11) both operating without EHD enhancement. The performance increases with increasing applied voltage, and is as high as or higher than that provided by the mechanically enhanced surfaces. Although it is unclear in the Yabe [1991] publication, it appears that the porous surface and the Thermoexcel-E™ data are for pure R-11.

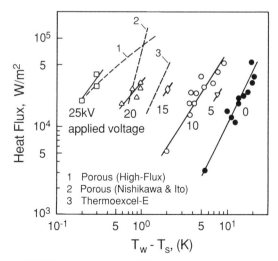

Figure 15.9 Yabe's [1991] test results for EHD boiling enhancement on a plain horizontal tube using a 90%/10% mixture (by weight) of R-114 and ethanol. Also shown are data for boiling on mechanically enhanced tubes without EHD enhancement.

The proposed mechanism for EHD enhancement on a plain surface is as follows. The EHD body force pushes the vapor bubbles away from the electrode. The greatest force occurs on the part of the bubble closest to the electrode. The bubble is pushed toward the heating surface (away from the electrode), and the radial components of the body force cause violent movement of the bubbles on the heating surface. This enhances thin film evaporation at the base of the bubble and promotes breakup of large bubbles into a greater number of smaller-diameter bubbles. This theory was substantiated by Ogata and Yabe [1991], who observed the effect of an electric field on a single air bubble in a silicone oil/ethyl alcohol mixture. They explain that the bubble breakup can be explained by the Taylor instability in the electric field on the liquid–gas interface, whose shape is deformed by the field.

Boiling mixtures have a smaller heat transfer coefficient than do pure fluids. Yabe theorized that the violent mixing at the heater surface should be beneficial for enhancement of mixture boiling. Figure 15.9 supports this proposal. This mechanism is supported by the visualization studies of Ohadi et al. [1992], who boiled oil–refrigerant mixtures using R-11 and R-123. The high-speed films showed that EHD force causes the number of bubbles to increase and causes the diameter of the bubbles to decrease.

Cooper [1992] states that the electrode should be designed to promote a high degree of nonuniformity of the electric field at the surface. Figure 15.10 shows Cooper's [1990] results on a standard integral-fin tube using R-114. A nonuniform electric field was generated by a coaxial electrode. Figure 15.10 shows that significant hysteresis exists with no electric field (0 kV). The 0-kV data are for increasing (curve A) and decreasing (curve B) heat flux, respectively. However, application of the electric field totally eliminated hysteresis. Cooper argues that EHD forces activate nucleation sites at lower superheat than are required without an electric field. After activation, the forces provided by the electric field further enhance

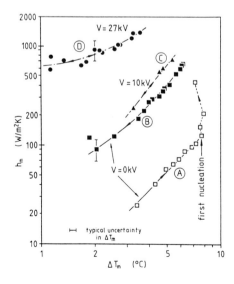

Figure 15.10 R-114 boiling at 21.5°C on a 1428-fin/m horizontal, integral-fin tube tested by Cooper [1990]. (From Cooper [1990].)

nucleate boiling. This is shown by comparing the decreasing heat flux data at 0 kV (curve B) with the EHD enhanced data.

The above results have been corroborated by Damianidis et al. [1992], who tested a plain tube, an integral-fin tube, and a 1420-fin/m, sawtooth finned tube (designed for condensation enhancement) using R-114. The electric field increased the boiling coefficient for all three tube geometries. The authors also provide data on a nine tube horizontal tube bundle.

Ogata et al. [1992] tested R-123 boiling on a horizontal bundle of 50 plain tubes using six axial wire electrodes spaced at 3.0 mm from each tube. They show that the performance of the tube bundle is competitive with that of a single tube. An economic analysis of EHD enhancement is also provided.

Yabe et al. [1992] studied convective vaporization of R-123 and R-134a inside a circular tube. A three-factor enhancement was obtained over a wide range of vapor qualities. More dramatic enhancement is obtained in nucleate pool boiling.

15.6 CONCLUSIONS

EHD enhancement of boiling and condensing processes appears to offer exciting new possibilities. Data taken to date suggest that the performance of both plain and mechanically enhanced tube geometries is increased.

However, the case is not yet proved that EHD enhancement will lead to higher enhancement than may be obtained from well-designed mechanically enhanced surface geometries. More data are required using the same fluids and operating conditions for the EHD and mechanical enhancements.

15.7 REFERENCES

Bologa, M. K., Savin, I. K., and Didovsky, A. B., 1987. "Electric-Field-Induced Enhancement of Vapor Condensation Heat Transfer in the Presence of a Non-condensible Gas," *International Journal of Heat and Mass Transfer*, Vol. 30, No. 8, pp. 1577–1585.

Cooper, P. 1990. "EHD Enhancement of Nucleate Boiling," *Journal of Heat Transfer*, Vol. 112, pp. 458–464.

Cooper, P., 1992. "Practical Design Aspects of EHD Heat Transfer Enhancement in Evaporators," *ASHRAE Transactions*, Vol. 98, Part 2, pp. 445–454.

Damianidis, C., Karayinnis. T., Al-Dadah, R. K., James, R. W., Collins, M. W., and Allen, P. H. G., 1992. "EHD Boiling Enhancement in Shell-and-Tube Evaporators and Its Application to Refrigeration Plants," *ASHRAE Transactions*, Vol. 98, Part 2, pp. 462–473.

Fernandez, J., and Poulter, R., 1987. "Radial Mass Flow in Electrohydrodynamically-Enhanced Forced Heat Transfer in Tubes," *International Journal of Heat and Mass Transfer*, Vol. 30, pp. 2125–2136.

Ishiguro, H., Nagata, S., Yabe, A., and Nariai, 1991. "Augmentation of Forced-Convection Heat Transfer by Applying Electric Fields to Disturb Flow Near a Wall," *Heat Transfer Science and Technology*, B-X. Wang, Ed., Hemisphere Publishing Corp., New York, pp. 25–31.

Jones, T. B., 1978. "Electohydrodynamically Enhanced Heat Transfer in Liquids—A Review," *Advances in Enhanced Heat Transfer*, Vol. 14, pp. 107–148.

Kuffel, E., and Zaengl, W. S., 1984. *High-Voltage Engineering*, Pergamon, Oxford.

Levy, E., 1964. *"Effects of Electrostatic Fields on Forced Convection Heat Transfer,"* M.S. thesis, Massachusetts Institute of Technology, Cambridge, MA.

Nelson, D. A., Ohadi, M. M., Zia, S., and Whipple, R. L., 1991. "Electrostatic Effects on Heat Transfer and Pressure Drop in Cylindrical Geometries," *Heat Transfer Science and Technology*, B-X. Wang, Ed., Hemisphere Publishing Corp., New York, pp. 33–39.

Ogata, J., Iwafuji, Y., Shimada, Y., and Yamaziki, T., 1992. "Boiling Heat Transfer Enhancement in Tube-Bundle Evaporator Utilizing Electric Field Effects," *ASHRAE Transactions*, Vol. 98, Part 2, pp. 435–444.

Ogata, J., and Yabe, A., 1991. "Augmentation of Nucleate Boiling Heat Transfer by Applying Electric Fields: EHD Behavior of Boiling Bubble," in *Heat Transfer Science and Technology*, B-X. Wang, Ed., Hemisphere Publishing Corp., New York, pp. 41–46.

Ohadi, M. M., 1991. "Heat Transfer Enhancement in Heat Exchangers," *ASHRAE Journal*, Vol. 33, No. 12, pp. 42–50.

Ohadi, M. M., Nelson, D. A., and Zia, S., 1991a. "Heat Transfer Enhancement of Laminar and Turbulent Pipe Flows Via Corona Discharge," *International Journal of Heat and Mass Transfer*, Vol. 34, pp. 1175–1187.

Ohadi, M., Sharaf, N., and Nelson, D., 1991b. "Electrohydrodynamic Enhancement of Heat Transfer in a Shell-and-Tube Heat Exchanger" *Experimental Heat Transfer*, Vol. 4, No. 1, pp. 19–39

Ohadi, M., Faani, M., Papar, R., Radermacher, R., and Ng, T., 1992. "EHD Heat Transfer Enhancement of Shell-Side Boiling Heat Transfer Coefficients of R-123/Oil Mixture," *ASHRAE Transactions*, Vol. 98, Part 2, pp. 427–434.

Panofsky, W., and Phillips, M., 1962. *Classical Electricity and Magnetism*, 2nd edition, Addison–Wesley, Reading, MA, pp. 107–116.

Sunada, K., Yabe, A., Taketani, T., and Yoshizawa, Y., 1991. "Experimental Study of EHD Pseudo-dropwise Condensation," *Proceedings of the Third ASME/JSME Thermal Engineering Conference*, Vol. 3, pp. 47–53.

Uemura, M., Nishiio, S., and Tanasawa, I., 1990. "Enhancement of Pool Boiling Heat Transfer by Static Electric Field," *Proceedings of the Ninth International Heat Transfer Conference*, Jerusalem, Israel, Vol. 4, pp. 69–74.

Velkoff, H. R., and Miller, T. H., 1965. "Condensation of Vapor on a Vertical Plate with a Transverse Electrostatic Field," *Journal of Heat Transfer*, Vol. 87, pp. 197–201.

Yabe, A. 1991. "Active Heat Transfer Enhancement by Applying Electric Fields," *Proceedings of the Third ASME/JSME Thermal Engineering Conference*, Vol. 3, pp. xv–xxiii.

Yabe, A., and Maki, N., 1988. "Augmentation of Convective and Boiling Heat Transfer by Applying an Electro-hydrodynamical Liquid Jet," *International Journal of Heat and Mass Transfer*, Vol. 31, No. 2, pp. 407–417.

Yabe, A., Taketani, T., Kikuchi, K., Mori, Y., and Hijikata, K., 1987. "Augmentation of Condensation Heat Transfer Around Vertical Cooled Tubes Provided with Helical Wire Electrodes by Applying Nonuniform Electric Fields," in *Heat Transfer Science and Technology*, Bu-Xuan Wang, Ed., Hemisphere Publishing Corp., Washington, D.C., pp. 812–819.

Yabe, A., Taketani, T., Maki, H., Takahashi, K., and Nakadai, Y., 1992. "Experimental

Study of Electrohydrodynamically (EHD) Enhanced Evaporator for Nonazeotropic Mixtures," *ASHRAE Transactions*, Vol. 98, Part 2, pp. 812–819.

Yamashita, K., Kumagai, M., Sekita, S., Yabe, A., Taketani, T., and Kikuchi, K., 1991. "Heat Transfer Characteristics of an EHD Condenser," *Proceedings of the Third ASME/JSME Joint Thermal Engineering Conference*, Reno, Nevada. pp. 61–67.

15.8 NOMENCLATURE

d_i	Plain tube inside diameter, or diameter to the base of internal fins or roughness, m
d_o	Tube outside diameter, fin root diameter for a finned tube, m
E	Electric field strength, V/m or V/ft
F_e	Electric field body force, N
h	Heat transfer coefficient, W/m²-K or Btu/hr-ft²-°F
i	Electric current, amperes
L	Tube length, m or ft
Nu	Nusselt number, dimensionless
p	Pressure, Pa or lbf/ft²
R	Radius of tube or electrode, m or ft
Re_d	Reynolds number based on the tube diameter = Gd/μ, $d = d_i$ for flow inside tube and $d = d_o$ for flow outside tube, dimensionless
Re_{Dh}	Reynolds number based on the hydraulic diameter = GD_h/μ, dimensionless
Re_l	Condensate Reynolds number $(4\Gamma/\mu_l)$, dimensionless
Re_L	Condensate Reynolds number leaving vertical tube $(4\Gamma_L/\mu_l)$, dimensionless
r	Radial coordinate, m or ft
T	Temperature, °C or K
t	Time, s
u	Fluid velocity, m/s or ft/s
V	Electric potential, V
W	Condensate flow rate leaving tube, kg/s or lbm/s

Greek Symbols

Γ	Condensate flow rate, per unit tube length, leaving horizontal tube; kg/s-m or lbm/s-ft
Γ_L	Condensate flow rate, per unit plate width, leaving vertical plate; kg/s-m or lbm/s-ft
Δ	Gradient, dimensionless
ϵ	Fluid permittivity (dielectric constant of fluid), dimensionless
μ	Fluid viscosity, kg/m-s or lbm/ft-s
ρ	Fluid density, kg/m³ or lbm/ft³
ρ_c	Electric field space charge density, C/m³ or C/ft³
σ_e	Electrical conductivity, A/V-m or A/V-ft

16

SIMULTANEOUS HEAT
AND MASS TRANSFER

16.1 INTRODUCTION

There are a number of convective heat transfer processes that also involve convective mass transfer. Typically, these processes involve two-phase heat transfer involving either condensation or evaporation of mixtures, one of which may be inert at the process conditions. Examples of evaporation processes include vaporization of binary fluids, air humidification, and desorption of gases from liquid mixtures. Heat/mass transfer in a cooling tower is an excellent example of air humidification. Examples of condensation processes include condensation of mixtures, or condensation with noncondensible gases, air dehumidification, and absorption of vapor in a liquid film. The cooling of moist air involves both heat and mass transfer. These problems involve simultaneous heat and mass transfer. A mass transfer resistance exists, because the active component must diffuse through the possibly inert component. Enhancement techniques offer excellent possibilities for performance improvement.

The processes differ depending whether the mass transfer resistance exists in the gas phase, the liquid phase, or both. Important examples of each type of problem are presented in the following sections.

Many applicable processes are within the domain of chemical engineering heat transfer, which involve separation of mixtures. We will not intimately delve into this complex area of transport phenomena of mixtures. Rather, our intent is to illustrate how enhancement may be applied to such processes.

16.2   MASS TRANSFER RESISTANCE IN THE GAS PHASE

This class of problems involves either (a) condensation onto a liquid film or (b) evaporation from a liquid film. In many problems the inert component is air, although it may be any inert component.

16.2.1   Condensation with Noncondensible Gases

Figure 16.1 shows the cross section of a tube wall with a condensing vapor on one side, and cooling water on the other side. The curves in the condensing region show the temperature and pressure distributions. The dashed line profiles indicate conditions that would exist if a noncondensible gas were not present. The solid line shows the situation with noncondensible gas. For a pure vapor, there is no temperature drop between the bulk vapor and the liquid–vapor interface, where condensation occurs; thus, the interface temperature (T_i) is equal to the saturation temperature (T_s) at the system pressure. With noncondensible gas present, the value of T_i is reduced, yielding a smaller value of $T_i - T_s$. The $T_i - T_s$ is reduced for two reasons:

1. Because the condensing vapor exists at its partial pressure, its saturation temperature is reduced.
2. Because the condensing vapor must diffuse through the vapor–gas mixture, the vapor partial pressure at the interface is below that of the bulk mixture. This reduces the saturation temperature at the interface.

The first reduction of $T_i - T_s$ is directly related to the volume concentration of

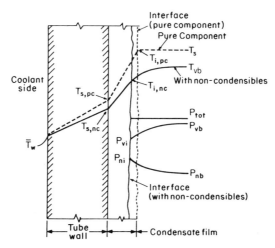

Figure 16.1   Illustration of temperature and pressure profiles for condensation of pure vapor (*solid lines*) and with noncondensible gas (*dashed line*), (From Webb and Wanniarachchi [1980].)

the noncondensible gas. However, the second reduction of $T_i - T_s$ is inversely proportional to the convective heat transfer coefficient of the gas–vapor mixture. High vapor velocity, or enhancement, will reduce the temperature drop across the gas boundary layer.

Because the vapor is saturated at the liquid–vapor interface, the vapor in the bulk mixture at T_{vb} is superheated. Because $T_{vb} > T_i$, there is a sensible heat flow though the gas to the liquid–vapor interface, where condensation occurs. Neglecting subcooling of the condensate, the heat transfer rate to the coolant is the sum of the sensible and latent portions. The heat transfer rate from the vapor mixture to the interface is given by

$$q = h_g \alpha_s (T_{vb} - T_i) + K_p i_{gv} (p_{vb} - p_{vi}) \qquad (16.1)$$

The first term of Equation 16.1 is for sensible heat transfer from the vapor, and the second term is the latent heat transfer. The α_s term is the Ackermann correction factor, which accounts for the effect of the mass flux on the sensible heat transfer rate. The gas phase mass transfer coefficient (K_p) may be obtained from the heat–mass transfer analogy,

$$\frac{h_g}{c_p} \mathrm{Pr}^{2/3} = K_p p M \mathrm{Sc}^{2/3} \qquad (16.2)$$

where h_g is for single-phase flow of vapor in the geometry of interest. Use of Equation 16.2 to obtain K_p assumes that h_g for the condensation situation is the same as for all gas flow. Roughness on the liquid film may act to slightly enhance h_g for the condensation situation.

Equation 16.2 shows that K_p is directly proportional to the gas-phase heat transfer coefficient (h_g). Hence, techniques applicable to enhancement of the gas-phase heat transfer coefficient will also be effective in enhancing the mass transfer coefficient. Note that the enhancement is required at the gas–liquid interface, rather than at the pipe wall. Therefore, selection of the enhancement should account for the effect of the liquid film thickness.

Section 14.5 described an enhancement technique that increased the h_g term for shell-side condensation in an axial flow condenser. This consisted of a two-dimensional rib roughness on the tube outer surface. The roughness increased the gas-phase heat transfer coefficient and, hence, the mass transfer coefficient. Any enhancement technique that is effective for convective heat transfer to gases should be beneficial to convective condensation with noncondensible gases.

Webb [1991] discusses the heat–mass transfer analogy (Equation 16.2) and gives a standard nomenclature for mass transfer processes. The nomenclature of Webb [1991] is used here.

16.2.2 Evaporation into Air

Evaporation from a water film into an air stream is an important process and is used in several engineering applications. Perhaps the most important is the cooling tower.

Other applications include the humidifier, evaporative fluid coolers, and evaporative condensers. The heat transfer process involves evaporation from a water film into a flowing air stream. The controlling mass transfer resistance is in the gas boundary layer. We will discuss how enhancement may be applied to enhance the gas-phase mass transfer coefficient.

16.2.2.1 Cooling Tower Packings.
Cooling tower packings are good candidates for enhancement. In this case, water evaporated at the liquid film surface is transferred by convective heat and mass transfer to the moist air, which flows over the surface. Figure 1.17 shows typical film packings used in a cooling tower. Both packings provide enhancement for the gas flow. Equation 16.1 is applicable to the cooling tower problem, as discussed by Webb [1988]. However, the latent heat term is typically written in terms of the specific humidity (W). Then, Equation 16.1 becomes

$$q = h_g\alpha_s(T_{vb} - T_i) + K_W i_{gv}(W_b - W_i) \tag{16.3}$$

Equation 16.3 is frequently expressed in terms of the enthalpy driving potential, which involves use of the heat–mass transfer analogy to express h_g in terms of K_W. The relation between K_W, K_P, and h_g is given by Webb [1991] and is

$$K_W = pM(1 - y)^2 K_p = \frac{h_g}{c_p}(1 - y)^2 \left(\frac{Pr}{Sc}\right)^{2/3} \tag{16.4}$$

Equation 16.4 shows that a surface geometry that provides high h_g will also provide a high mass transfer coefficient. Substitution of h_g in terms of K_W into Equation 16.3, along with use of approximations, allows one to combine the sensible and latent terms. The result is a driving potential defined in terms of the moist air enthalpy (i). This derivation is shown by Webb [1988]. Thus

$$q = K_W(i_b - i_i) \tag{16.5}$$

16.2.2.2 PEC Example 16.1.
The air flow through the Figure 1.17a ceramic cooling tower packing is similar to that which occurs in the offset-strip-fin geometry (Figure 5.3). Hence, one may use correlations for the offset strip fin to estimate the mass transfer coefficient in the ceramic cooling tower packing.

Assume that air flows at 3.8 m/s and 32°C in the ceramic cooling tower packing. Predict the mass transfer coefficient using the heat–mass transfer analogy. The ceramic block is 150 mm high and has 45.0-mm square cells with 6.4-mm membrane thickness. Using the heat–mass transfer analogy, the mass transfer coefficient is $K_W \simeq h_g/c_p(Pr/Sc)^{2/3}$, where h_g is the heat transfer coefficient for air flow in the packing. The Reynolds number, based on the cell length, is given by

$$Re_L = \frac{L_p u}{\nu} = \frac{0.15 \times 3.8}{16.2E{-}6} = 35{,}185 \tag{16.6}$$

Using the offset strip fin correlation (Equation 5.9), which uses Re_{Dh} (= 10,555), gives a j factor 0.0028. Then, $h_g = j(\rho u c_p Pr^{-2/3}) = 23.9$ W/m^2-K. Using the heat–mass transfer analogy (Equation 16.4), one obtains

$$K_W = \frac{h_g}{c_p}\left(\frac{Pr}{Sc}\right)^{2/3} = \frac{23.9}{1007}\left(\frac{0.7}{0.6}\right)^{2/3} = 0.0263 \text{ kg/s-m}^2 \qquad (16.7)$$

16.2.2.3 Enhancement for Evaporation into an Air Stream.

Webb and Perez-Blanco [1986] used gas-phase enhancement for a process involving evaporation of water into an air–steam mixture flowing inside a vertical tube. Figure 16.2 shows a water film draining down the inside surface of a round vertical tube, with moist air in counterflow. They placed a $p/e = 10$ transverse-rib roughness at the edge of the gas boundary layer to provide enhancement of h_g. The roughness was displaced from the wall a distance equal to the calculated liquid film thickness, so that the ribs were not contained in the liquid film. The heat transfer coefficient of the rib roughness was calculated using the correlation of Webb et al. [1970] described in Section 9.3.1. The wire of diameter e was chosen to operate at roughness Reynolds number (Equation 9.2) of $eu^*/\nu_v = 20$, as recommended in Section 9.6.4. They obtained a 38% enhancement of the gas-phase heat transfer and mass transfer coefficients.

16.2.3 Dehumidifying Finned-Tube Heat Exchangers

Finned-tube heat exchangers used for cooling of air will experience moisture condensation on the fin surface, if the fin surface temperature is below the dewpoint temperature. This is typical for refrigerant evaporators. Equation 16.1 (or Equation 16.3 or 16.5) applies to this process. Hence, a surface geometry having a high h_g should also enhance the mass transfer coefficient and provide high performance

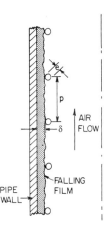

Figure 16.2 Transverse-rib concept applied to enhance mass transfer in the vapor boundary layer. (From Webb and Perez-Blanco [1986].)

under wet conditions. Moisture condensation on the fin surface provides a naturally occurring enhancement.

Senshu et al. [1981] measured the performance of the Figure 6.7 convex louver finned tube geometry, under dehumidifying conditions for which moisture condensation occurred on the fin surface. Their tests were run using 5°C water flow in the tubes. The Figure 6.7 surface geometry provides high enhancement for dry air flow. The heat transfer rate for the wetted surface condition was then predicted in terms of the enthalpy driving potential (e.g., Equation 16.5). In this case the mass transfer is from the bulk air to the interface, so the equation is written as

$$q = K_W(i_b - i_i) \qquad (16.8)$$

The authors used their measured h_g (obtained without moisture condensation) and the heat–mass transfer analogy relation (Equation 16.4) to obtain K_W. For their steam–air mixture, they used the acceptable approximation, $(1 - y)^2(\text{Pr}/\text{Sc})^{2/3} \simeq 1.0$. Thus, $K_W = h_g/c_p$. The predicted K_W was compared with the experimental value obtained using Equation 16.8. Equation 16.8 requires evaluation of the moist air enthalpy at the fin surface temperature. Senshu et al. [1981] developed an equation for this purpose. They showed that their K_W obtained from h_g/c_p was within ±5% of the measured value. The good prediction of K_W using the heat–mass transfer analogy shows that the draining condensate did not bridge the fin louvers or substantially alter the air-flow pattern over the louvers.

The above theory is applicable, provided that the surface is fully wetted. Or, it is applicable to the portion of the surface that is wetted. Because the air must be cooled to the dewpoint temperature, it is probable that a fraction of the entering air region of the heat exchanger will not be wetted.

16.2.4   Water Film Enhancement of Finned Tube Exchanger

In Section 16.2.3, heat was transferred from the air stream to the fin surface, with condensation occurring at the liquid–vapor interface. The reverse situation— evaporation from a wetted surface into the air stream—is of interest here. The concept involves spraying water on the finned tube heat exchanger. The intent is to cause the water to flow as a film on the fin surface, with water evaporation from the film surface. Unless good surface wetting is achieved, the concept will not work well. Tests have been done by spraying water on operating heat exchangers. Performance data are obtained for both the dry and the wetted conditions. Quantification of the degree of surface wetting and drainage characteristics of the water film typically has not been investigated. Among the earliest work to measure the effect of water spray on heat exchanger performance was by Yang and Clark [1974], who sprayed a water mist on the face of an automotive radiator. They found only a 10% increase of performance. Apparently, the water did not wet the fins.

Hudina and Sommer [1988] used mist cooling to enhance two different plate finned tube heat exchangers (Figure 6.1a) having plain fins. The 1.04-m² frontal area heat exchangers were tested in a wind tunnel using upstream water spray nozzles. The heat exchanger dimensional data are listed in Table 16.1. The dimensions differ only in the in-tube pitch.

Figure 16.3 shows their heat transfer and pressure drop test results for fixed water spray rate, tube-side flow rate (7.3 kg/s), 50% relative humidity, 293 K water spray temperature, and inlet air temperatures between 298 and 303 K. Figure 16.3a shows that the enhancement level decreases with increasing air mass velocity. Figure 16.3a shows that an enhancement level of 5.0 was obtained for the Fo geometry at 1.0-kg/m^2-s air mass velocity. Figure 16.3b shows that the pressure drop increase is less than 15% for the Fs geometry. However, a much larger pressure drop enhancement occurs for the Fo geometry at $G_a < 3$ kg/m^2-s. Note that the Fo geometry was operated at a 2.6 times higher water spray rate than for the Fs geometry. Figure 16.4 shows the effect of water spray rate on the heat transfer enhancement level. The enhancement level increased with increasing water spray rate.

Associated bench scale experiments by Sommer [1984] showed that a uniform, thin film does not exist over the fin surface. At low ITD, $(T_{w1} - T_{a1})$, Sommer showed that the sprayed water is collected on the fins near the air inlet in the form of large droplets, which eventually bridge the fin spacing. The water will drain downward by gravity. As the ITD (or air velocity) is increased, water bridging will be less, but the atomized water may pass through the heat exchanger as fine entrained droplets. Because of its high contact angle, water will not wet the fin surface unless the surface is moderately oxidized or specially treated.

Kreid et al. [1983b] tested three different fin-and-tube geometries (Figure 16.5) in a wind tunnel. The dimensions of the aluminum fin heat exchangers are given in Table 16.2 and were 0.61 m wide and 1.8 m high. The heat exchangers were oriented with the fins in the vertical direction to facilitate water drainage. The water was supplied by what is called the "deluge" method. This involves flowing the water onto the heat exchanger from a supply manifold above the heat exchanger. They used a water deluge rate approximately 10 times the amount evaporated. Surface B provided the highest performance for a fixed air pressure drop. The heat transfer rate of sprayed surface B was 2–5 times that for operation without water spray. The water did not flow as a film on surfaces A and C; rather it broke up into droplets, which significantly increased the air pressure drop. They did not precisely determine to what degree the water wetted the entire fin surface as a film. Kreid et al. [1978] and Kreid et al. [1983a] developed an analytical model, based on use of

Table 16.1   Geometries Tested by Hudina and Sommer [1988]

Geometry code:	Fs	Fo
Tube diameter (d_o):	17.3	17.3
Transverse tube pitch (S_t):	60	57
Longitudinal tube pitch (S_l):	25	45
Number of tube rows:	6	6
Fin pitch:	0.028	0.028
Fin thickness:	0.30	0.30

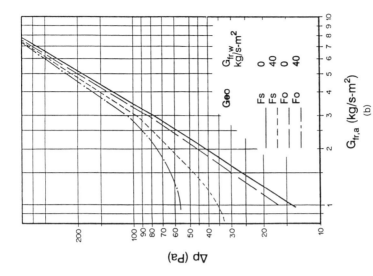

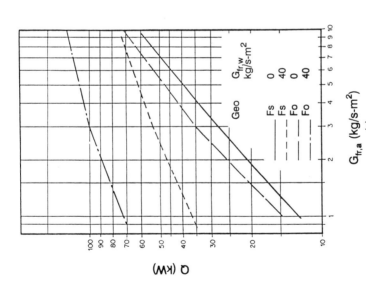

Figure 16.3 Test results of Hudina and Sommer [1988] for spray enhancement of the Table 16.1 finned tube heat exchangers. (a) Heat transfer enhancement. (b) Pressure drop enhancement. (From Hudina and Sommer [1988].)

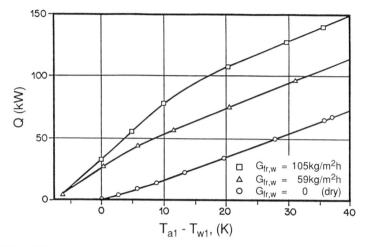

Figure 16.4   Effect of water spray rate on heat transfer enhancement for Fs finned tube geometry of Table 16.1 tested by Hudina and Sommer [1988]. (From Hudina and Sommer [1988].)

Equation 16.3. The model, which assumes that the surface is fully wetted, moderately predicted the results under wet conditions. These studies show that, with a good mechanism for wetting, the heat transfer performance of a finned tube heat exchanger can be significantly increased. The Kreid et al. [1983b] tests were part of a program described by Johnson et al. [1981] to develop a water enhanced dry cooling tower.

Bentley et al. [1978] performed a bench scale laboratory investigation of a vertical plate having spaced, gravity-drained water channels. They did not attempt to wet the entire surface. Rather, they sought to enhance the dry surface performance by evaporation from the surface of spaced, discrete water channels. The water channels occupied about 5% of the total area. Their tests showed that the heat transfer by evaporation varied between 20% and 40% of the total heat transfer. Their aluminum plate provided enhancement for dry heat transfer by use of repeated rib roughness, whose channels also drained water. However, they obtained only 20% improvement in dry heat transfer over a plain surface at the same fan power. This dry enhancement level is quite small, relative to currently used dry tower surfaces. They did not address practical means to supply and distribute the water to finned tube banks.

Successful evaporative enhancement of a dry cooling tower fin surface requires that a significant fraction of the surface be wetted, and that the dry air heat transfer coefficient be enhanced over that for a plain surface. Furthermore, the surface must be continuously wetted during the evaporative enhancement period, or mineral deposits and corrosion will occur. The excess water drained from the surface should prevent accumulation of corrosion forming deposits, provided that dewetting of the surface does not occur. The particular challenge is how to obtain a high air-side heat

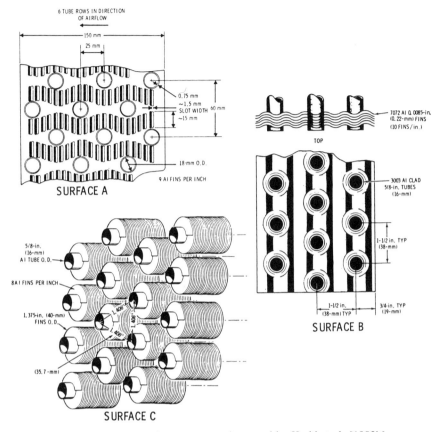

Figure 16.5 Heat exchanger geometries tested by Kreid et al. [1983b].

Table 16.2   Geometries Tested by Kried et al. [1983b]

Parameter	Geometry		
	A	B	C
$A_o/A_{fr}N$ (m²/m²)	13.88	27.05	14.70
A_c/A_{fr} (m²/m²)	0.495	0.534	0.509
S_t (mm)	25.0	38.0	30.93
S_l (mm)	60.0	38.0	35.7
D_h (mm)	3.87	2.96	3.18
d_o (mm)	18.0	16.0	16.0
t_f (mm)	0.43	0.22	0.20
fins/m (m⁻¹)	354	393	315
N	6	3	4
d_i (mm)	20.2	13.4	13.4

transfer coefficient, and to continuously wet a significant fraction of the surface. In order to maintain a continuously wetted surface, it is necessary to flow an excess amount of water than that evaporated. Not all enhanced fin geometries are amenable to surface wetting and water drainage. The Figure 16.5 surface A interrupted strip fins would prevent lateral spreading of a water film, which enters from the front in a spray. The concave channel elements of the Figure 16.5 surface B wavy fins should provide natural drainage channels for a water film introduced at the top of the heat exchanger. In this case, the entire fin surface would not be wetted. The water–air interface should exist in the concave fin channels.

Commercial coatings exist that reduce the contact angle of water on a surface and that promote surface wetting. Use of such coatings will reduce the amount of excess water that must be applied to the surface.

16.3   CONTROLLING RESISTANCE IN LIQUID PHASE

Examples of this problem are the absorption process, or condensation of mixtures. In the absorption process, a vapor (e.g., steam) is condensed at the surface of a liquid film, and the condensed liquid is absorbed in a binary solution (e.g., ammonia-water). If the condensing vapor is a pure component, there is no mass transfer resistance in the vapor phase. The mass transfer resistance exists in the liquid film. Enhancement is needed in the liquid film, which has both heat and mass transfer resistances. One should determine whether the limiting resistance is heat or mass transfer. In the present problem illustrated, the mass transfer resistance is expected to be the limiting resistance. The primary mass transfer resistance exists at the liquid–vapor interface, not at the tube wall. Hence, reduction of the mass transfer resistance requires mixing of the liquid film at the liquid–vapor interface. Mixing of the liquid film at the wall will be beneficial to the heat transfer resistance. An extended surface would benefit the thermal resistance, but probably not the mass transfer resistance.

This situation also applies to problems involving boiling of mixtures. Mixtures have a lower heat transfer coefficient than do pure components, because of the added mass transfer resistance in the liquid phase.

16.4   SIGNIFICANT RESISTANCE IN BOTH PHASES

If the condensing vapor of Section 16.3 contained a noncondensible gas, mass transfer resistance would exist in both phases. Mass transfer resistance will exist in both phases for the boiling or condensation of mixtures. The previously discussed approaches to reducing the mass transfer resistance for noncondensible gases are applicable. Hence, mixing is required in both the liquid and vapor phases.

Distillation columns involve two-phase flow and separation of mixtures in either tray towers or packed columns. The packing in Figure 1.17b is an enhanced mass transfer device for packed towers. It causes a tortuous flow path for the gas phase and causes gas-phase mixing. Thin liquid films exist on the packing.

16.5 CONCLUSIONS

Significant potential exists for heat transfer enhancement in simultaneous heat–mass transfer processes. One must first determine which phase requires enhancement—the gas phase, the liquid phase, or possibly both. The controlling mass transfer resistance is in the gas-phase for processes involving (a) humidification or dehumidification of steam–air mixtures and (b) condensation with noncondensible gases. Use of the heat–mass transfer analogy provides a powerful tool to evaluate enhanced heat transfer surfaces for application to gas-phase mass transfer.

Typical problems involve heat and mass transfer between a gas phase and a liquid film. Cooling towers currently use enhanced surface geometries to reduce the gas-phase resistance.

Significant potential exists for water film enhancement in fin-and-tube heat exchangers used for heat rejection. For evaporation of the liquid film, it is very important that the surface be fully wetted. Surface geometries that promote, or allow, full surface wetting are not easy to obtain. However, problems in surface wetting (and corrosion) have prevented their successful commercial exploitation. Commercial coatings that promote surface wetting of a water film are available and offer potential for this application.

Applications involving mass transfer in both the liquid and the gas-phases are yet to be seriously addressed. Further developments in water film enhanced heat transfer to gases are expected.

16.6 REFERENCES

Bentley, J. M., Snyder, T. K., Glicksman, L. R., and Rohsenow, W. M., 1978. "An Experimental Study of a Unique Wet/Dry Surface for Cooling Towers," *Journal of Heat Transfer*, Vol. 100, pp. 520–526.

Hudina, M., and Sommer, A., 1988. "Heat Transfer and Pressure Drop Measurements on Tube and Fin Heat Exchangers," in *Proceedings of the 1st World Conference on Experimental Heat Transfer, Fluid Mechanics and Thermodynamics*, R. K. Shah, E. N. Ganic, and K. T. Yang, Eds., Elsevier Science Publishers, New York, pp. 1393–1400.

Johnson, B. M., Bartz, J. A., Alleman, R. T., Fricke, H. D., Price, R. E., and McIlroy, 1981. "Development of an Advanced Concept of Dry/Wet Cooling for Power Plants," Battelle Pacific Northwest Laboratories Report BN-SA-1296. Also presented at the American Power Conference, April 27–29, Chicago, IL.

Kreid, D. K., Johnson, B. M., and Faletti, D. W., 1978. "Approximate Analysis of Heat Transfer from the Surface of a Wet Finned Heat Exchanger," ASME paper 78-HT-26.

Kreid, D.K., Hauser, S. G., and Johnson, B. M., 1983a. "Investigation of Combined Heat and Mass Transfer from a Wet Heat Exchanger, Part 1—Analytical Formulation," *Proceedings of the ASME-JSME Joint Thermal Engineering Conference*, Vol 1, pp. 517–524.

Kreid, D. K., Hauser, S. G., and Johnson, B. M., 1983b. "Investigation of Combined Heat and Mass Transfer from a Wet Heat Exchanger, Part 2—Experimental Results," *Proceedings of the ASME-JSME Joint Thermal Engineering Conference*, Vol 1, pp. 525–534.

Senshju, T., Hatada, T., and Ishibane, K., 1981. "Heat and Mass Transfer Performance of Air Coolers Under Wet Conditions," *ASHRAE Transactions*, Vol. 87, Part 2, pp. 109–115.

Sommer, A., 1984. "Wasserverteilung auf den Lamellen besprühter FORGO-GLATT-Wärmetauscher, Beobachtungen and einem Plexiglas-Aluminum Modell," Report EIR-TM-23-84-09, Würenlingen.

Webb, R. L., 1988. "A Critical Evaluation of Cooling Tower Design Methodology," in *Heat Transfer Equipment Design*, R. K. Shah, E. C. Subbarao and R. A. Mashelkar, Eds., Hemisphere Publishing Corp., Washington, D.C., pp. 547–558.

Webb, R. L., 1991. "Standard Nomenclature for Mass Transfer Processes," *ASHRAE Transactions*, Vol. 97, Part 2, pp. 114–118.

Webb, R. L., Eckert, E. R. G., and Goldstein, R. J., 1970. "Heat Transfer and Friction in Tubes with Repeated-Rib Roughness, "*International Journal of Heat and Mass Transfer*, Vol. 14, pp. 601–617.

Webb, R. L., and Perez-Blanco, H., 1986. "Enhancement of Combined Heat and Mass Transfer in a Vertical Tube Heat and Mass Exchanger," *Journal of Heat Transfer*, Vol. 108, pp. 70–75.

Webb, R. L., Wanniarachchi, A. S., and Rudy, T. M., 1980. "The Effect of Non-condensible Gases on the Performance of an R-11 Centrifugal Water Chiller Condenser," *ASHRAE Transactions.*, Vol. 86, Part 2, pp. 170–184.

Yang, W-J., and Clark, D. W., 1974. "Spray Cooling of Air-Cooled Compact Heat Exchangers," *International Journal of Heat and Mass Transfer*, pp. 311–317.

16.7 NOMENCLATURE

a	Ackermann correction factor ($q_{lat}c_p/i_{gv}h_g$), dimensionless
A	Heat transfer surface area on one side of a direct transfer type exchanger, m^2 or ft^2
A_c	Flow cross-sectional area in minimum flow area, m^2 or ft^2
A_{fr}	Air-flow frontal area, m^2 or ft^2
A_o	Total air-side heat transfer surface area, m^2 or ft^2
c_p	Specific heat of fluid at constant pressure, J/kg-K or Btu/lbm-°F
d_o	Tube outside diameter, fin root diameter for a finned tube, m or ft
D	Diffusion coefficient, m^2/s or ft^2/s
D_h	Hydraulic diameter of flow passages, $4LA_c/A$; m or ft
e	Fin height or roughness height, m or ft
G	Mass velocity based on the minimum flow area: G_a (air), G_w (water), kg/m^2-s or lbm/ft^2-s
G_{fr}	Mass velocity based on flow frontal area: $G_{fr,a}$ (air), $G_{fr,w}$ (water), kg/m^2-s or lbm/ft^2-s
h_g	Heat transfer coefficient, W/m^2-K or Btu/hr-ft^2-°F
i	Enthalpy of steam–air mixture, kJ/kg or Btu/lbm

i_{gv}	Enthalpy of saturated vapor, kJ/kg or Btu/lbm
ITD	Temperature difference between entering fluid temperatures, K or °F
j	Colburn factor ($= \text{StPr}^{2/3}$), dimensionless
K_p	Mass transfer coefficient based on partial pressure driving potential, kg/s-m^2 or lbm/s-ft^2
K_W	Mass transfer coefficient based on specific humidity or enthalpy driving potential, kg/s-m^2 or lbm/s-ft^2
L	Flow length, m or ft
L_p	Strip flow length of OSF or louver pitch of louver fin, m or ft
M	Molecular weight of mixture, kg/kmol or lbm/lbmol
N	Number of tube rows in the flow direction, or number of tubes in heat exchanger; dimensionless
p	Pressure, Pa or lbf/ft^2
p	Axial spacing between roughness elements, m or ft
q	Heat flux, W/m^2 or Btu/hr-ft^2
q_{lat}	Latent component of heat transfer, W/m^2 or Btu/hr-ft^2-°F
Q	Heat transfer rate in the exchanger, W or Btu/hr
Pr	Prandtl number ($= c_p\mu/k$), dimensionless
Re_{ph}	Reynolds number based on hydraulic diameter ($= GD_h/\mu$), dimensionless
Sc	Schmidt number, ν/D; dimensionless
S_l	Longitudinal tube pitch, m or ft
S_t	Transverse tube pitch, m or ft
St	Stanton number ($= h/Gc_p$), dimensionless
T	Temperature: T_s (saturated), T_w (wall), T_{vb} (of bulk vapor), T_i (interface), T_{av}(average), T_{a1} (entering air), T_{w1} (entering water); K or °F
u	Local fluid velocity, m/s or ft/s
W	Fluid mass flow rate ($= \rho u_m A_c$), kg/s or lbm/s
W	Specific humidity, dimensionless
y	Mass fraction of diffusing component in vapor, dimensionless

Greek Letters

α	$(1 - e^{-1})/a$, dimensionless
δ	Liquid film thickness, m
μ	Fluid dynamic viscosity coefficient, Pa-s or lbm/s-ft
ν	Kinematic viscosity, m/s^2 or ft/s^2
ρ	Fluid density, kg/m^3 or lbm/ft^3

Subscripts

$a1$	Entering air
b	In bulk vapor
i	At liquid–gas interface
v	Vapor
$w1$	Entering water

17

ADDITIVES FOR GASES AND LIQUIDS

17.1 INTRODUCTION

Additives for *liquids* include solid particles or gas bubbles in single-phase flows and liquid trace additives for boiling systems. Additives for *gases* are liquid droplets or solid particles, either dilute-phase (gas–solid suspensions) or dense-phase (packed beds and fluidized beds).

17.2 ADDITIVES FOR SINGLE-PHASE LIQUIDS

17.2.1 Solid Particles

Kofanov [1964] performed a very detailed study of solid particle additives, for flow in a circular tube, which spanned $4000 \leq \mathrm{Re}_d \leq 200{,}000$. Figure 17.1a shows his heat transfer correlation, which includes 18 data sets from five authors. The data shown in Figure 17.1a are as follows:

1. Kofanov [1964]: 1, pure water; 2, water–chalk; 3, water–coal; 4, water–sand; 5, water–aluminum; 6, water–iron
2. Orr and Dallavalle [1954]: 7, water–clay; 8, water–copper; 9, water–glass; 10, water–graphite; 11, water–aluminum; 12, ethylene glycol–graphite; 13, water–aluminum; 14, water
3. Bonilla et al. [1953]: 15, water–chalk
4. Salamone and Newman [1955]: 16, water–copper; 17, water–sand
5. Miller and Moulton [1956]: 18, kerosene–graphite

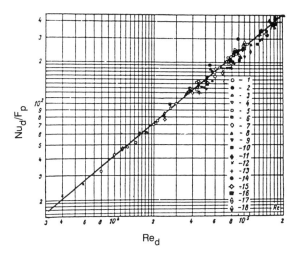

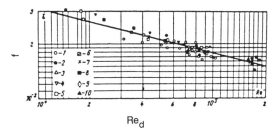

Figure 17.1 Correlations of Kofanov [1964] for solid particles in liquid. (a) Heat transfer correlation. (b) Friction factor correlation. See text for data sources. (From Kofanov [1964].)

The correlation shown in Figure 17.1 is given by

$$\mathrm{Nu}_d = 0.026\mathrm{Re}_d^{0.8}\mathrm{Pr}^{0.4}F_p \tag{17.1}$$

where F_p is a property group defined by

$$F_p = \left(\frac{x_v}{1 - x_v}\right)^{0.15}\left(\frac{\rho}{\rho_p}\right)^{0.15}\left(\frac{c_p}{c_{pp}}\right)^{0.15}\left(\frac{d_i}{d_p}\right)^{0.02} \tag{17.2}$$

The fluid properties used in the Nusselt, Reynolds, and Prandtl numbers are those of the solid–liquid suspension. The properties of the solid-liquid mixture (ρ_m, $c_{p,m}$, μ_m, and k_m) are calculated as follows: The specific heat and the thermal conductivity are calculated on the basis of particle mass fraction (x_m). The suspension density is calculated on the basis of the particle volume fraction (x_v). The suspension viscosity is calculated using the volume fraction by the equation

$$\mu_m = \mu(1 + 2.5x_v + 7.17x_v^2 + x_v^3) \tag{17.3}$$

where μ is the viscosity of the liquid, in which the particles are suspended. Kofanov obtained good correlation of the friction factor of the flowing suspension using the Blasius equation

$$f = 0.079\text{Re}_d^{-0.25} \tag{17.4}$$

where Re_d is defined as specified above. The correlated data are shown in Figure 17.1b.

One should be aware that suspensions of very small, low-thermal-conductivity particles can cause fouling at low velocity. This possibility should exist for particles, such as clay or chalk. Such fouling would be caused by particulate fouling, which is discussed in Chapter 10.

More recent work on suspensions is reported by Watkins et al. [1976], who worked with plastic beads in laminar oil flow. They observed a maximum enhancement of 40%.

17.2.2 PEC Example 17.1

Calculate the enhancement provided by 15% volume fraction of 0.05-mm-diameter glass beads in water flowing in a 20-mm-diameter tube at 2.0 m/s and 20°C.

Solution. First calculate the heat transfer coefficient for pure water (h_o). The Reynolds number is 39,840. Using the Dittus–Boelter equation, $h_o = 7192$. Next calculate the properties of the suspension and use Equation 17.1 to calculate h. We obtain $F_p = 0.6517$, $\text{Re}_{d,m} = 30,600$, $\text{Pr}_m = 5.14$, and $h = 4102$. Hence, the enhancement ratio is $h/h_o = 4102/7192 = 0.57$. The suspension of glass beads actually reduces the heat transfer coefficient by 43%, compared to pure water flow. Note that the F_p term significantly contributes to the reduction. Furthermore, the Prandtl number of the suspension is 5.14, as compared to 6.99 for water.

Examination of Equations 17.1 and 17.2 show that a high particle concentration, and high values of ρ_p/ρ, $c_{p,p}/c_p$, and k_p/k are required to obtain $h/h_o > 1$. Hence, we conclude that typical solid–liquid suspensions will give little, if any, enhancement.

17.2.3 Gas Bubbles

Bubbling a gas through a stationary liquid simulates the conditions that occur in nucleate boiling on a surface. Enhancement will occur because of the liquid agitation on the surface, caused by the vapor bubbles. Tamiri and Nishikawa [1976] injected air into water (or ethylene glycol) heated on a vertical plate. The air was injected at the base of the plate. They measured enhancement levels up to 400%. Kenning and Kao [1972] injected air into turbulent flow of water in a tube, and they measured 50% enhancement level.

17.2.4 Suspensions in Dilute Polymer Solutions

Considerable work has been done on flowing suspensions that reduce fluid friction. Long fibers damp fluid turbulence and reduce the turbulent energy dissipation. For example, Lee et al. [1974] used asbestos fibers at 200 and 800 ppm in an aqueous dilute polymer solution (150 ppm Separan AP-30). For the asbestos fibers in water, no drag reduction was achieved. The asbestos fiber-polymer solution gave 50–64% drag reduction in the turbulent regime.

Moyls and Sabersky [1975] investigated the effects of dilute suspensions of asbestos fibers (300 ppm) in an aqueous solution containing a polymer (50 ppm Polyox). For $10,000 < Re_d < 200,000$, the Nusselt number of the polymer solution (without fibers) was only 20% as high as that of pure water. When the asbestos fibers were added, heat transfer was improved. However, the heat transfer coefficient was far below that of pure water. Hence, we conclude that both suspensions in dilute polymer solutions reduce both heat transfer and fluid friction.

17.3 ADDITIVES FOR SINGLE-PHASE GASES

17.3.1 Solid Additives

Two situations exist, where solid additives are quite important. The first is flow of gas–solid suspensions in ducts. A number of papers have been written on this subject over the last 30 years. Our discussion will focus on heat transfer enhancement and heat transfer correlations for suspended particles in ducts. The principal mechanism for enhancement is increased capacity of the flowing media at a given Re_d. The heat transferred from the heated surface to the particle is usually assumed to be negligible.

The second situation is fluidized beds, which is a high interest topic. This involves heat transfer between a bundle of horizontal (or vertical) tubes and a fluidized gas–solid media. We will not attempt to survey the extensive publications on fluidized beds, which are reviewed by Gabor and Botterill [1985]. In this discussion attention is directed toward enhancement of gas-to-wall heat transfer, rather than understanding the gas-to-particle heat transfer mechanism.

Typical particulate sizes are in the range of 20–600 μm, and the solid–gas particle "loading ratio" (G/G_o) spans $1 < G/G_o < 15$. Figure 17.2, from Bergles et al. [1976], shows the enhancement ratio (h/h_o) for various gas–solid suspensions flowing inside a tube. Enhancement ratios (h/h_o) as high as 3.5 are seen. However, several of the suspensions give very low enhancement.

Several empirical correlations have been developed for gas–solid suspensions flowing inside tubes, none of which appear to be totally satisfactory. Furchi et al.

Curve	Re_d	Particle	d_p (μm)	Gas	d_p/d_i
A	18,000	Glass	60	Air	0.0011
B	18,000	Glass	120	Air	0.0027
C	19,000	Sand	230	Air	0.0060
D	19,000	Sand	80	Air	0.0021
E	15,000	Graphite	65	Air	0.0085
F	53,000	Zinc	40	Air	0.0005
G	53,000	Zinc	40	Air	0.0008
H		Graphite			
I	53,000	Zinc	40	Air	0.016
J	15,000	Al$_2$O$_3$	65	Air	0.0085
K	13,500	Glass	30	Air	0.0016
L	13,500	Glass	200	Air	0.0111

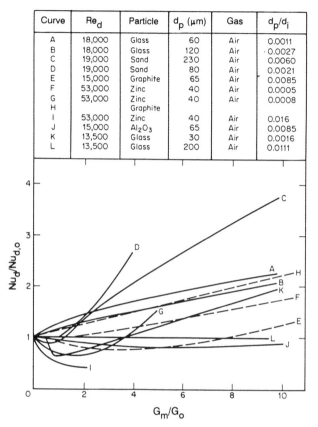

Figure 17.2 Enhancement ratio h/h_o versus G/G_o loading ratio for various solid particle mixtures. (From Bergles et al. [1976].)

[1988] present recent experimental information and correlations for the heat transfer coefficient of gas–solid suspensions. Figure 17.3 shows the correlated data of Furchi et al. [1988] and three other investigators. The correlation shown in the figure was developed by Sadek [1972]. The dimensional correlation [(d_i, d_p (mm), C_p (m^{-3})] may be written in terms of the enhancement ratio (h/h_o):

$$h/h_o = 1 + 0.20(C_p d_p^2 d_i)^{1.19} \qquad (17.5)$$

Figure 17.3 shows that Equation 17.5 does not do a very good job correlating the data, which spans $17 < d_i < 102$ mm, $20 < d_p < 600$ μm, $4000 < Re_d < 80,000$, and $0 < G_m/G_g < 300$. The correlation shows that h/h_o increases with particle concentration and particle diameter. Notice that the correlation does not include the thermal conductivity of the particles, which may not be valid for high thermal conductivity particles. Contrary to Equation 17.5, the Furchi et al [1988] data show that h/h_o increases with decreasing particle size.

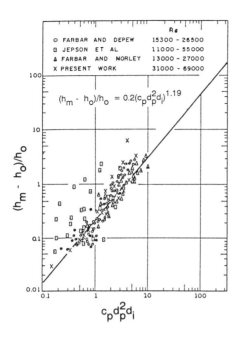

$$(h_m - h_o)/h_o = 0.2(c_p d_p^2 d_i)^{1.19}$$

	R_e
O FARBAR AND DEPEW	15300 - 26500
⊟ JEPSON ET AL	11000 - 55000
▲ FARBAR AND MORLEY	13000 - 27000
X PRESENT WORK	31000 - 69000

$(h_m - h_o)/h_o$ (vertical axis)

$c_p d_p^2 d_i$ (horizontal axis)

Figure 17.3 Correlation of gas–solid suspension heat transfer data for turbulent flow in tubes. (From Furchi et al. [1988].)

Gabor and Botterill [1985] give heat transfer correlations for horizontal bundles of plain and finned tubes. Several studies have been performed to measure the advantages offered by finned and rough surfaces in fluidized beds. Use of an enhanced surface is considered to be "compound enhancement," because two enhancement techniques are combined. Petrie et al. [1968] used finned tubes in a horizontal fluidized bed and examined the effect of fin pitch for plain tubes and for 9.5-mm-high fins. For 199 fins/m, they measured a gas-side enhancement ratio (E_{ho} = $\eta hA/h_s A_s$) of 1.7, compared to an area increase of 6.8. Because the heat transfer coefficient in the fluidized bed is so high, it operates at low fin efficiency, which reduced the enhancement level. Bartel and Genetti [1973] measured the heat transfer coefficients for horizontal bundles of steel tubes having radial, plain aluminum fins. They tested 15.9-mm-diameter tubes having 0-, 3.18-, 7.1-, 12.8-, and 16.6-mm fin height. Krause and Peters [1983] also tested steel segmented finned tubes in a fluidized bed. Their tubes had d_o = 19.2 mm, 300 fins/m, and 0.76-mm fin thickness. Three fin heights were tested (4.76, 8.33, and 11.11 mm) using 0.21- and then 0.43-μm-diameter particles. Table 17.1 summarizes their experimental results, which are presented as ratios, relative to the plain tube bundle. The highest enhancement ratio ($hA/h_s A_s$) of 5.82 was obtained with the largest particle size and the highest fins. However, the lowest fins gave the best performance for the smaller particle size.

Chen and Withers [1978] tested vertical 19-mm-diameter integral-fin tubes (cf. Figure 6.1c) in a 140-mm-diameter, tube using glass particles (d_p = 0.13, 0.25, and 0.6 mm). Table 17.2 shows their results for d_p = 0.25 mm, presented as ratios, relative to the plain tube. Table 17.2 shows that the highest $\eta h/h_s$ was obtained from

Table 17.1   Segmented Finned Tube Performance in Horizontal Fluidized-Bed Bundle ($G/G_{mf} \cong 1.2$)

e_o (mm)	d_p = 0.21 mm			d_p = 0.43 mm		
	η_f	h/h_s	hA/h_sA_s	η_f	h/h_s	hA/h_sA_s
4.76	0.78	0.88	3.74	0.81	0.89	3.80
8.33	0.65	0.72	4.93	0.64	0.81	5.57
11.11	0.66	0.34	3.03	0.68	0.65	5.82

the 354-fin/m tube having 1.57-mm fin height. Furthermore, their data showed that the E_{ho} does not increase with decreasing particle size. Rather, they found that the maximum E_{ho} occurred when $3 < s/d_p < 4$, where s is the fin spacing.

Grenwal and Saxena [1979] investigated tubes having a closely spaced V-thread or knurled roughness in a horizontal fluidized bed. The roughness height was 1.07 mm or less. The best roughness provided 40% higher heat transfer coefficient than did the plain tube.

17.3.2   Liquid Additives

Liquid additives generally refer to adding water droplets to the air stream. The water wets the heat transfer surface, providing evaporation from the water film surface into the air stream. Enhancement provided by an evaporating water film on a heat transfer surface is discussed in Section 16.2.3. Very high enhancement can be provided, if the surface can be fully wetted. Using a water spray onto a wedge, Thomas and Sunderland [1970] measured an enhancement (E_{ho}) of 20 by adding 5% water to the air stream.

Moderate enhancement can also be obtained without wetting the surface. Under this condition, the upstream water mist would cool the incoming air to its wet-bulb temperature, and the water droplets would exchange heat with the air in transit through the heat exchanger. Greater enhancement would be obtained for water mist temperature below the air temperature. Bhatti and Savery [1975] analyze the problem of suspended water droplets in a boundary layer flow over a flat plate. Kosky [1976] and Nishikawa and Takase [1979] experimentally and analytically study the problem of a mist flow normal to a circular cylinder.

Table 17.2   Performance of Vertical Integral-Fin Tubes in Fluidized-Bed (Glass particles, d_p = 0.25 mm)

E_{ho}	Fins/m	e_o (mm)	A/A_s	$\eta h/h_s$
0.60	197	3.18	2.22	0.70
1.50	354	1.57	1.88	1.00
2.30	433	3.18	3.87	0.90
1.80	748	3.06	3.06	0.80

17.4 ADDITIVES FOR BOILING

As discussed by Thome [1990], the pool boiling heat transfer coefficient of a mixture is less than that of either pure component. In spite of this understanding, work has been done using minute amounts of a volatile additive, which show that the boiling coefficient may be modestly increased. Examples of this poorly understood phenomenon are provided by Lowery and Westwater [1959] and by Gannett and Williams [1971].

Furthermore, it appears that a trace additive may also provide a significant increase of the critical heat flux (CHF) in pool boiling, as shown by van Wijk et al. [1956] and by van Stralen [1959]. van Stralen [1959] measured a 240% increase of the CHF for 3–5% concentration of 1-penthanol in water. The optimum concentration is a function of boiling pressure. Bergles and Scarola [1966] found that addition of 1-penthanol reduced the CHF for subcooled boiling.

17.5 ADDITIVES FOR CONDENSATION

The objective of such additives is to promote dropwise condensation. Typically, the heat transfer coefficient for dropwise condensation is much higher than that for filmwise condensation. Additives which establish the existence of dropwise condensation are called dropwise condensation "promoters." Griffith [1985] surveys the technology of dropwise condensation and discusses various promoters which have been found for steam (water) condensation. A satisfactory promoter must cause the surface to become hydrophobic to the condensate. This means that the contact angle of the condensate must be increased to approach 90 degrees. Hence, the condensate will not spread on the surface but will exist as discrete droplets. The droplets agglomerate and run off the surface. This is possible for high-surface-tension fluids such as water. However, low-surface-tension fluids (e.g., organics, alcohols, and refrigerants) have such a low surface tension that no promoter has been found that will make the surface hydrophobic to the condensate.

Table 2 of Griffith [1985] lists various promoters that work for steam condensation. However, contamination of the surface, or depletion of the promoter, will cause cessation of dropwise condensation. The desired promoter is also a function of the base surface material. To date, practical concerns have prevented the commercial use of such promoters for steam condensation. Tanasawa [1978] discusses practical concerns for commercial use of dropwise condensation promoters.

The only known promoters that are indefinitely durable are certain coatings on the base surface. These are gold and Teflon. Gold has obvious cost problems. Although Teflon effectively promotes dropwise condensation, the coating also adds a significant thermal resistance that substantially diminishes the benefits of the coating. Depew and Reisbig [1964] found that a 1.27-μm-thick Teflon coating, which promoted dropwise condensation, will increase the condensing coefficient 100% over the filmwise condensation coefficient on a 12.7-mm-diameter horizontal tube.

17.6   CONCLUSIONS

Typical solid–liquid suspensions will provide little, if any, enhancement. However, significant enhancement can be obtained from solid–gas suspensions. The significant enhancement provided by solid-gas suspensions provides justification for the fluidized bed heat exchanger. Finned surfaces can be used in the fluidized bed to further increase the heat transfer coefficient.

Aqueous, dilute polymer additives will reduce both the heat transfer coefficient and friction factor. Use of a rough surface will recover some of the heat transfer reduction, but at the expense of increased friction.

Adding a minute amount of a volatile fluid may moderately enhance the nucleate boiling coefficient. However, this phenomenon is poorly understood, since mixtures typically reduce the heat transfer coefficient.

Additives can promote high-performance dropwise condensation for water or other high-surface-tension fluids. However, long-term surface contamination, or depletion of the additive, will likely reduce the performance. Additives that promote dropwise condensation for low-surface-tension fluids have not been found.

17.7   REFERENCES

Bartel, W. J., and Genetti, W. E., 1973. "Heat Transfer from a Horizontal Bundle of Bare and Finned Tubes in an Air Fluidized Bed," *AIChE Symposium Series*, Vol. 69, No. 128, pp. 85–92.

Bergles, A. E., and Scarola, L. S., 1966. "Effect of a Volatile Additive on the Critical Heat Flux for Surface Boiling of Water in Tubes," *Chemical Engineering Science*, Vol. 21, pp. 721–723.

Bergles, A. E., Junkhan, G. H., and Hagge, J. K., 1976. "Advanced Cooling Systems for Agricultural and Industrial Machines," SAE paper 751183.

Bhatti, M. S., and Savery, S. W., 1975. "Augmentation of Heat Transfer in a Laminar External Gas Boundary Layer by the Vaporization of Suspended Droplets," *Journal of Heat Transfer*, Vol. 97, pp. 179–184.

Bonilla, C. F., Cervi, A., Jr., Colven, T. J., Jr., and Wang, S. J., 1953. "Heat Transfer to Slurries in Pipe, Chalk, and Water in Turbulent Flow," *AIChE Symposium Series*, Vol. 49, No. 5, pp. 127–134.

Chen, J. C., and Withers, J. G., 1978. "An Experimental Study of Heat Transfer from Plain and Finned Tubes in Fluidized Beds," *AIChE Symposium Series*, Vol. 74, No. 174, pp. 327–333.

Depew, C. A. and Kramer, T. J., 1973. "Heat Transfer to Flowing Gas-Solid Mixtures," in *Advances in Heat Transfer*, Vol. 9, J. P. Hartnett and T. F. Irvine, eds., Academic Press, New York, pp. 113–180.

Depew, C. A., and Reisbig, R. L., 1964. "Vapor Condensation on a Horizontal Tube Using Teflon to Promote Dropwise Condensation," *Industrial Engineering Chemistry, Processing, Design, and Development*, Vol. 11, pp. 365–369.

Furchi, J. C. L, Goldstein, L., Lombardi, G., and Mohseni, M. 1988. "Heat Transfer Coefficients in Flowing Gas–Solid Suspensions," *AIChE Symposium Series*, Vol. 84, No. 263, pp. 26–30.

Gabor, J. D., and Botterill, J. S. M., 1985. "Heat Transfer in Fluidized and Packed Beds," Chapter 6 in *Handbook of Heat Transfer Applications*, McGraw–Hill, New York.

Gannett, H. J., Jr., and Williams, M. C., 1971. "Pool Boiling in Dilute Nonaqueous Polymer Solutions," *International Journal of Heat Mass Transfer*, Vol. 11, pp. 1001–1005.

Grenwal, N. S., and Saxena, S. C., 1979. "Effect of Surface Roughness on Heat Transfer from Horizontal Immersed Tubes in a Fluidized Bed," *Journal of Heat Transfer*, Vol. 101, pp. 397–403.

Griffith, P., 1985. "Condensation, Part 2—Dropwise Condensation," Chapter 11 in *Handbook of Heat Transfer Applications*, McGraw–Hill, New York.

Kenning, D. B. R., and Kao, Y. S., 1972. "Convective Heat Transfer to Water Containing Bubbles: Enhancement Not Dependent on Thermocapillarity," *International Journal of Heat Mass Transfer*, Vol. 15, pp. 1709–1718.

Kofanov, V. I., 1964. "Heat Transfer and Hydraulic Resistance in Flowing Liquid Suspensions in Piping," *International Chemical Engineering*, Vol. 4, No. 3, pp. 426–430.

Kosky, P. G., 1976. "Heat Transfer to Saturated Mist Flowing Normally to a Heated Cylinder," *International Journal of Heat Mass Transfer*, Vol. 19, pp. 539–543.

Krause, W. B., and Peters, A. R., 1983. "Heat Transfer from Horizontal Serrated Finned Tubes in an Air-Fluidized Bed of Uniformly Sized Particles," *Journal of Heat Transfer*, Vol. 105, pp. 319–324.

Lee, W. K., Vaseleski, R. C., and Metzner, A. B., 1974. "Turbulent Drag Reduction in Polymeric Solutions Containing Suspended Fibers," *AIChE Journal*, Vol. 20, pp. 128–133.

Lowery, A. J., Jr., and Westwater, J. W., 1959. "Heat Transfer to Boiling Methanol—Effect of Added Agents," *Industrial Engineering Chemistry*, Vol. 49, pp. 1445–1448.

Miller, A. P., and R. W. Moulton, 1956. "Heat Transfer to Liquid–Solid Suspensions in Turbulent Flow in Pipes," *The Trend in Engineering*, April, pp. 15–21.

Moyls, A. L., and Sabersky, R. H., 1975. "Heat Transfer to Dilute Asbestos Dispersions in Smooth and Rough Tubes," *Letters in Heat Mass Transfer*, Vol. 2, pp. 293–302.

Nishikawa, N., and Takase, H., 1979. "Effects of Particle Size and Temperature Difference on Mist Flow over a Heated Circular Cylinder," *Journal of Heat Transfer*, Vol. 101, pp. 705–711.

Orr, C., and Dallavalle, J. M., 1954. "Heat Transfer Properties of Liquid–Solid Suspensions," *Chemical Engineering Progress Symposium Series*, Vol. 50, No. 9, pp 29–45.

Petrie, J. C., Freeby, J. A., and Buckham, J. A., 1968. "In-Bed Heat Exchangers," *Chemical Engineering Progress*, July, pp. 45–51.

Sadek, S. E., 1972. "Heat Transfer to Air–Solids Suspensions in Turbulent Flow," *Industrial Engineering Chemistry—Processing Design and Development*, Vol. 11, pp. 133–135.

Salamone, J. J., and Newman, M., 1955. "Heat Transfer Design Characteristics-Water Suspensions of Solids," *Industrial Engineering Chemistry*, Vol. 47, No. 2, pp. 283–288.

Tamari, M., and Nishikawa, K., 1976. "The Stirring Effect of Bubbles upon the Heat Transfer to Liquids," *Heat Transfer—Japanese Research*, Vol. 5, No. 2, pp. 31–44.

Tanasawa, I., 1978. "Dropwise Condensation: The Way to Practical Applications," *Proceedings of the 6th International Heat Transfer Conference*, Vol. 6, pp. 393–405.

Thomas, W. C., and Sunderland, J. E., 1970. "Heat Transfer between a Plane Surface and Air Containing Water Droplets," *Industrial Engineering Chemistry, Fundamentals*, Vol. 9, pp. 368–374.

Thome, J. R., 1990. *Enhanced Boiling Heat Transfer*, Hemisphere Publishing Corp., Washington, D.C.

van Stralen, S. J. D., 1959. "Heat Transfer to Boiling Binary Liquid Mixtures," *British Chemical Engineering*, Vol. 4, Part I, pp. 8–17; Vol. 4, Part II, pp. 78–82.

van Wijk, W. R., Vos, A. S., and van Stralen, S. J. D., 1956. "Heat Transfer to Boiling Binary Liquid Mixtures," *Chemical Engineering Science*, Vol. 5, pp. 68–80.

Watkins, R. W., Robertson, C. R., and Acrivos, A., 1976. "Entrance Region Heat Transfer in Flowing Suspensions," *International Journal of Heat Mass Transfer*, Vol. 19, pp. 693–695.

17.8   NOMENCLATURE

C_p Particle volume concentration, m^{-3} or ft^{-3}
d_i Tube inside diameter, m or ft
d_p Particle diameter, m or ft
e_o External fin height, m or ft
E_{ho} Heat transfer enhancement ratio ($hA/h_s A_s$) relative to plain surface, dimensionless
G Mass velocity, kg/s-m^2 or lbm/s-ft^2
G_{mf} Minimum fluidization mass velocity, kg/s-m^2 or lbm/s-ft^2
h Heat transfer coefficient, W/m^2-K or Btu/hr-ft^2-°F
k Thermal conductivity, W/m-K or Btu/hr-ft-F
Nu_d Nusselt number based on tube diameter, dimensionless
Pr   Prandtl number, dimensionless
Re_d Reynolds number based on tube diameter, dimensionless
s Fin spacing, m or ft
x_m Particle–fluid mass fraction ($= x_v \rho_m/\rho_p$), dimensionless
x_v Particle–fluid volume fraction, dimensionless

Greek Symbols

η_f Fin efficiency, dimensionless
μ Viscosity, kg/s-m or lbm/s-ft^2
ρ Density, kg/m^3 or lbm/ft^3

Subscripts

m Solid–gas or solid–liquid mixture
o Flow without solids
p Particles
s Smooth or plain surface

PROBLEM SUPPLEMENT

1.1 Benefits of Enhancement A existing four-row finned tube refrigerant evaporator has 551 fins/m on 12.7 mm diameter tubes located on 31.8 mm equilateral triangular pitch. The fins are plain and the tube inside surface is smooth. The air and refrigerant-side heat transfer coefficients are 400 and 3000 W/m²-K, respectively. The heat exchanger has $A_o/A_{fr}N = 21.6$ m and $A_i/A_{fr}N = 1.19$, where N is the number of tube rows. If the Figure 1.3e air-side surface geometry is employed, the air-side heat transfer coefficient will be increased 80%. To obtain the same UA value, how many tube rows would be required? [Ans. 0.83N]

It is also possible to enhance the refrigerant-side using the Figure 1.15 microfin tube, which will provide 120% increased refrigerant heat transfer coefficient. Would use of the tube side enhancement, retaining the plain fins on the air-side, provide comparable size reduction benefits to those obtained in part a? [Ans. 0.61N]

3.1 Simple Single-Phase PEC A smooth tube has $f_s = 0.079 \, Re^{-0.25}$ and $j_s = f_s/2$. An internally roughened tube provides $j = 0.079 \, Re^{-0.25}$ and has $f = 0.11 \, Re^{-0.2}$. Assuming all of the thermal resistance is on the tube side, calculate G/G_s, A/A_s and h/h_s using Case VG-1 for $Re_s = 20,000$. What is A/A_s? [Ans. $G/G_s = 0.937$, $A/A_s = 0.525$, $h/h_s = 1.90$]

Now consider Case FN-1. Outline the procedure you would use to obtain G/G_s and A/A_s. Will A/A_s be greater or smaller than for Case VG-1? Why?

3.2 Complex Single-Phase PEC Reconsider the rough tube of Problem 3.1 for a case, in which $h_{i,s} = 4000$ kW/m²-K and 75% of the total thermal resistance is on the inside of the smooth tube. For simplicity, assume $B_s = B = 1$, ignore the metal wall and fouling resistances. Calculate G/G_s and A/A_s for Case VG-1. Will A/A_s be greater or smaller than for Case VG-1? Why? [Ans. Smaller]

4.1 Two-Phase PEC Consider steam condensing in an annulus against water inside the tubes. Water enters the tube at 1.0 kg/s and 300 K. Steam enters the annulus at 330 K for which dp/dT = 22.94 kPa/K. The water-side heat transfer coefficient is 10,000 W/m²-K. If plain tubes are used, the steam condensation coefficient is 5,000 W/m²-K and the steam pressure drop is 50 kPa. Because the steam side has the controlling thermal resistance, a tube having steam side enhancement is to be applied. This tube provides 2.5 times higher condensing coefficient, and the steam pressure gradient is also increased 2.5 times. Calculate the tube length reduction for $Q/Q_s = 1$ with the same operating conditions. For simplicity, neglect the effect of the steam pressure drop. [Ans. $L/L_s = 0.6$]

If you use $L/L_s = 0.6$, and account for the effect of steam pressure drop, what would be the resulting value of Q/Q_s? [Ans. $Q/Q_s = 0.98$]

5.1 Compare Louver and OSF Fin Geometries Compare the performance of the offset strip fin (OSF) with the louver fin using PEC FN-2. Both fin geometries have 472 fins/m, 7.62 mm fin height, 2.03 mm louver pitch, 0.10 fin thickness, and $D_h = 3.18$ mm. The louver geometry has 26 degree louver angle and $L_L/H = 0.8$. Compare for air flow at at 26.7 °C (80 °F) with 8.94 m/s (29.3 ft/s) air frontal velocity. Calculate the j and f factors using the Davenport correlations (Eqs. 5.16–5.18) for the louver fin, and the Manglik and Bergles power law correlations (Eqs. 5.6–5.9) for the OSF geometry. Calculate the A_{osf}/A_{louv} and P_{osf}/P_{louv} where A is the fin surface area. [Ans. $A_{osf}/A_{louv} = 1.03$ and $P_{osf}/P_{louv} = 0.66$]

5.2 Effect of Strip Length on OSF Performance Consider an offset-strip fin (OSF) having 787 fins/m, 0.025 mm fin thickness, and 9.52 mm fin height. Air enters the fin array at 5.1 m/s frontal velocity ($v = 15.9 \times 10\text{-}6$ m²/sec, $\rho = 1.16$ kg/m³). Compare the performance of 3.18 mm (1/8 in) and 12.7 mm (1/2 in) strip lengths using the simple Kays model to predict j and f. Let subscripts "1" and "2" refer to the 3.18 and 12.7 mm strip lengths, respectively. Compute j and f for each surface. (Ans. $j_1 = 0.929 \times 10\text{-}2$, $j_2 = 1.86 \times 10\text{-}2$, $f_1 = 2.74 \times 10\text{-}2$, $f_2 = 7.24 \times 10\text{-}2$.)

Calculate $(j/f)_2/(j/f)_1$ and the fraction of the friction factor due to form drag. [Ans. $(j/f)_2/(j/f)_1 = 1.32$. For $L_p = 3.18$ mm, 48.6% is form drag.]

If hA = constant, calculate A_2/A_1 and P_2/P_1 (Ans. $A_2/A_1 = 0.5$ and $P_2/P_1 = 1.32$). Draw conclusions regarding the performance and acceptability of the two surface geometries.

5.3 Joshi and Webb Model of OSF Use the analytically based model of Joshi and Webb [1987] to predict the performance of the offset strip fin (OSF) geometry, whose j and f vs. Reynolds number characteristic is shown in Figure 10-58 of Kays and London [1984]. The Kays and London definition of Reynolds number is Re = $D_{h,KL}G/v$, where $D_{h,KL} = 2s/(1 + \alpha + \delta)$ with G based on flow area $A_c = s(b - t)$. The Figure Kays and London geometry has $L_p = 2.8$ mm, t = 0.10 mm, $\alpha = 0.1837$, $\delta = 0.036$, $\gamma = 0.111$, and $D_{h,KL} = 1.495$ mm. The j and f vs Re characteristic is the same as that of the OSF geometry shown on Figure 5.4.

Predict the j and f values for air flow at 27 C with Re = 500 and 3,000 and compare with the experimental values. Assume 100% fin efficiency. At Re = 500, Figure 5.4 yields j = 0.0207 and f = 0.088, and at Re = 3000, j = 0.0095 and f = 0.0398.

For the Joshi and Webb model, use Equations 5.13 and 5.14 supported by the equations given in Table 5.1. As shown in Table 5.1, Joshi and Webb define D_h and Re_{Dh} differently from Kays and London. Hence, the Kays and London Reynolds numbers must be converted using the definitions given at the bottom of Table 5.1. For the Joshi and Webb model, $D_h = 1.333$ mm, and $D_{sh} = 1.540$ mm. [Ans. For Kays and London Re $= 500$, $Re_{Dh} = 600$, $Re_s = 610$, $Re_{sh} = 515.2$, and $L_s^+ = 0.00253$, $j_p = 0.01805$ and $f_p = 0.0681$].

5.4 Performance of OSF Variants Figures 5.4, 6.6, and 6.13 show the performance of three closely related heat transfer enhancement concepts. At abscissa values corresponding to the same air velocity, the heat transfer enhancement ratio (j/j_p), relative to the plain fin geometry (j_p), are Fig. 5.4 $(j/j_p = 2.5)$, Fig. 6.6 $(j/j_p = 1.8)$, and Fig. 6.13 $(j/j_p = 1.4)$. Explain why the enhancement ratio successively decreases for Figures 6.6 and 6.13. Assume that the strip width is the same for all of the enhancements.

6.1 Circular Finned Tubes Predict j and f for the five circular finned tube bank geometries taken from Figure 10-89 of Kays and London [1984] shown in Figure P6.1. All geometries have 19.66 mm diameter tubes, 356 fins/m, and 0.305 mm fin thickness. The j and f vs. Re_{Dh} curves are fit by the equations $j = C_j(Re_{Dh}/1000)^{-m}$ and $f = C_f(Re_{Dh}/1000)^{-n}$, where the constants and exponents are given in the table. The term β is the total surface area per unit heat-exchanger volume. Assume 27 C (80 F) air at 3.05 m/s (600 ft/min) approaches the tube bank. Use the Briggs and young correlation (Eq. 6.10) for heat transfer, and the Robinson and Briggs correlation (Eq. 6.11) for pressure drop. Compare the predicted values with the experimental values shown on the graph. Draw conclusions regarding the ability of the correlations to predict the data. [Ans. The last two lines in the table shows the ratio of the predicted-to-experimental values]

Item	a	b	c	d	e
S_d (mm)	48.51	51.05	56.39	40.13	42.93
D_h (mm)	5.131	8.179	13.59	4.846	6.426
$\sigma = A_c/A_{fr}$	0.455	0.572	0.688	0.537	0.572
$\beta = A/V$ (m²/m³)	354	279	203	443	354
C_j	0.00925	0.0105	0.0128	0.0090	0.0099
m	0.387	0.343	0.373	0.362	0.362
C_f	0.0365	0.0520	0.0715	0.0400	0.0506
n	0.316	0.278	0.264	0.309	0.275
j_p/j_{exp}	1.09	1.08	1.09	1.10	1.07
f_p/f_{exp}	1.06	1.17	1.34	1.72	1.19

6.2 PEC Analysis of Circular Finned Tubes Which of the five tube-bank layouts described in Problem 6.1 would you choose for a heat exchanger application? Assume fixed heat duty (Q), inlet temperature difference (ΔT_i) and pumping power (P). Use geometry "a" as the reference surface, with 27 C (80F) air at 3.05 m/s (600 ft/min) approaching the tube bank. Assume counterflow performance with

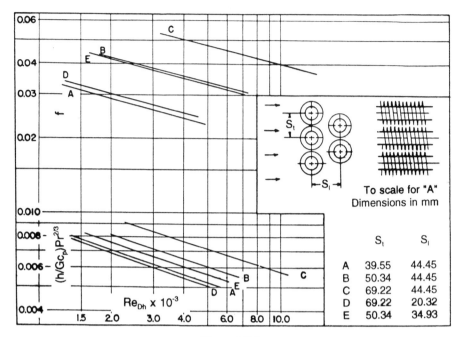

Figure P6.1

$C_{min}/C_{max} = 0$, and geometry "a" designed for $\epsilon = 0.6$ (heat exchanger effectiveness). Consider only the air-side performance in your PEC analysis.

Use the Performance Evaluation Criteria (PEC) VG-1 shown in Table 3.1 to justify your choice. Also calculate $A_{fr}/A_{fr,r}$, A/A_r and L/L_r, where L is the air-flow depth. [Ans. See table below]

	a	b	c	d	e
A/A_r	1.0	1.08	1.18	1.06	1.09
$A_{fr}/A_{fr,r}$	1.0	0.90	0.83	0.89	0.92
L/L_r	1.0	1.54	2.54	1.09	1.23

Calculate A/A_r for geometry "d" using PEC case FN-1 of Table 3.1. Compare the calculated A/A_r with the result obtained in part a. Explain why the results for case FN-1 are greatly different from the value calculated for case VG-1. Hint: You must account for the effect of reduced flow rate on the required hA. [Ans. $A/A_r = 2.60$]

7.1 Twisted-Tape with Laminar Flow Consider heat transfer at constant heat flux to fully developed laminar flow inside a plain tube at Re = 1000 for two cases: Water at 60 C (140 F) with Pr 3 and oil at 100 C (212 F) with Pr = 276. The tube contains a twisted-tape insert with having y = 2.5 and t = 0. Calculate h/h_p and f/f_p for each fluid, where subscript "p" refers to a plain tube with no tape insert.

Comment on the enhancement ratio (h/h_p) and the "efficiency index", $(h/h_p)/(f/f_p)$ for each case. For the plain tube, $Nu = 4.36$ and $f = 16/Re$. For the twisted tape, use the Hong and Bergles [1976] correlation for Nu, and the Manglik and Bergles [1992a] correlation for the friction factor. [Ans. $Nu_p = 4.36$, $Nu = 23.4$ (water), $Nu = 113.8$ (oil), $f_p = 0.016$, $f/f_p = 6.17$] Would h/h_p increase of decrease if the oil were cooled?

7.2 Manglik and Bergles Correlation Repeat the heat transfer calculation for Problem 7.1 for the case of constant wall temperature. Assume $L/d_i = 150$ and use the Manglik and Bergles correlation (Eqs. 7.2 and 7.8). How do the results differ for heating as compared to cooling?

8.1 Laminar Flow In Internal-Fin Tubes Evaluate the heat transfer enhancement for laminar flow in internally finned tubes, based on the numerical results shown in Table 8.1. Assume a constant wall temperature boundary condition, and that the fin height is limited to $e/d_i = 0.1$. What is the preferred number of fins to obtain the highest enhancement level? What is the magnitude of h/h_p for the chosen geometry? What is the efficiency index? [Ans. $n_f = 8$, $\eta = 0.68$]

Compare your results with the experimental data of Marner and Bergles [1985] for $n_f = 12$ and $e/d_i = 0.084$ as shown on Figure 7.9. Comment on the observed differences.

8.2 Entrance Length for Laminar Flow Oil having the properties of Problem 8.3 flows on the tube-side of an oil cooler having internally finned tubes ($d_i = 17$ mm, $e/d_i = 0.10$, $n_f = 16$) at $Re_d = 1000$. The tube is 1.5 m long giving $L/d_i = 88$. Assuming constant wall temperature, what is the thermal entrance length for this flow? [Ans. 13.2 m].

Would you expect the fully developed flow solutions shown in Table 8.1 to be applicable for estimating the average heat transfer coefficient in this tube?

8.3 Internal-Fin Tube Heat Exchanger Evaluate the benefits of an internal-fin tube design, which provides the same heat duty, tube-side pressure drop, and flow rate as a smooth tube design (Case VG-1 of Table 3.1). The heat-exchanger has 19.1 mm (3/4 in) outside diameter, 0.35 mm (0.035 in) wall copper tubes on a triangular pitch with 25.4 mm (1.0 in) tube pitch. The smooth tube exchanger has 136 tubes, 305 mm (12.2 in) diameter shell (D_s) with four tube side passes. Oil flows on the tube-side at 3.05 m/s (10 ft/s) and is cooled from 98.9-to-87.8 °C (210-to-190°F). Cooling water flows on the segmentally baffled shell-side. The water enters at 32.2 °C (90 °F) and exits at 54.4 °C (130 °F) . The shell-side heat transfer coefficient is $h_o = 4825$ W/m^2-K (850 Btu/hr-ft^2-F). Assume $R_f = 0$ on both the tube and shell sides. The oil properties at 93 °C (200 °F are: $\rho = 865$ kg/m^3 (54 Ibm/ft^3), $c_p = 2.13$ kJ/kg-K (0.51 Btu/lbm-F), $\nu = 4.3E$-6 m^2/s (4.6 × 10^{-5} ft^2/sec), $k = 01.28$ W/m-K (0.074 Btu/hr-ft-F), and $Pr = 62$.

First, evaluate the potential for internal-fin tubes (IFT) by calculating R_i/R_{tot} (internal/total thermal resistance of the smooth-tube exchanger). Will reduction of the tube-side resistance be beneficial? [Ans. $R_i/R_{tot} = 0.83$]

Select a candidate IFT geometry based on your study of Figure 8.10. Do you concur that 16 fins, × 1.5 mm high × 0.50 mm thick with 30 degree helix angle is a reasonable choice?

Calculate the h and f vs. Re characteristics of the IFT using the Carnavos [1980] correlation (Equations 8.5 and 8.6). Convert the h and f based on the total area to values based on the nominal area using Equations 8.9 and 8.10 for use in the PEC equations.

After calculating A/A_s, calculate L/L_s (H-X length ratio), and V/V_s (tube material volume). Draw conclusions regarding the value of IFT for this application. [Ans. $\beta_s = 0.207$, G/G_s 0.908, $A/A_s = 1.04$, $N/N_s = 1.174$, $L/L_s = 0.435$]

9.1 Sand-Grain Roughness A heat exchanger uses 15.9 mm (0.625 in) inside diameter tubes with 27 °C (80° F) water having $\nu = 863$ E-6 m²/s (0.929 E-6 ft²/sec) at 2.44 m/s (8 ft/sec) velocity. What is the largest roughness size (e) that will act as hydraulically smooth? Hint: figure 9.4b shows that the hydraulically smooth line is approached at $\log_{10}e^+ = 0.7$. [Ans. e = 0.037 mm]

What heat transfer enhancement (h/h_s) would you expect at this value of e? [Ans. None]

9.2 Rough Rectangular Channel Consider a 5:1 rectangular channel having a sand-grain wall roughness. How would you use Equation 9.4 to calculate the friction factor for the fully rough condition?

9.3 Two-Dimensional Rib Roughness Two-dimensional rib-roughness (p/e = 10, $\alpha = 90$ degrees) is to be used to enhance heat transfer on the tube-side of a water chiller evaporator, in which water flows at 3.05 m/s (10 ft/sec) and 7.2 °C (45 °F) in 16.0 mm (0.63 in) inside diameter tubes. The evaporator tubes presently use a smooth tube inner surface. The heat conductance on the outer tube surface is $h_o A_o/L$ = 865 W/m-K (500 Btu/h-ft-F) and the fouling resistance is zero.

The objective is to select an appropriate roughness height (e) and to satisfy the VG-1 PEC of Table 3.1. As suggested on page 258, select $e^+ = 14$ as the rough tube design criterion. Obtain $\bar{g}(e^+) = 10.8$ and $B(e^+) = 4.5$ at $e^+ = 14$ from Figures 9.8 and 9.9, respectively. Because the required G/G_s is unknown, assume u = 2.6 m/s (8.5 ft/s) tube-side velocity. Use Equation 9.31 to solve for the roughness height (e) and the friction factor (f). Then use Equation 9.14 to solve for St. [Ans. $e/d_i = 0.00563$, f = 0.014, St = 1.716E-3]

Calculate L/L_s and P/P_s for $Q/Q_s = 1$. [Ans. $L/L_s = 0.706$, $P/P_s = 1.06$].

Because $P/P_s > 1$, G/G_s must be reduced to meet the $P/P_s = 1$ constraint. Estimate the G/G_s required to meet the $P/P_s = 1$ constraint. [Ans. $G/G_s \simeq 0.833$ and u = 2.54 m/s]

9.4 Optimum Internal Roughness Before deciding to use the internal fin tube design evaluated in Problem 8.3, it may be advisable to consider an internally roughened tube. Calculate the material savings for the VG-1 case offered by two-dimensional rib roughness ($\alpha = 90$, p/e = 10, t/e = 0.5) assuming $R_{fi} = 0$. Following the guidance of Webb and Eckert [1972] discussed on page 258, select e^+ = 20 as the design criterion. Figures 9.8 and 9.9 yield $\bar{g}(20) = 11.0$ and $B(20) =$ 4.2. Since f and St are functions of e/D and Re, the PEC equation for A/A_s implicitly contains two independent variables, e/d_i and Re. Therefore a range of possible e/d_i exist. Use the iterative solution procedure outlined in Webb and Eckert [1972]. The solution will converge on the "optimum" e/d_i and the G/G_s needed to satisfy the VG-1 PEC equation. [Ans. $e/d_i = 0.0176$, $G/G_s = 0.81$, $A/A_s = 0.535$, $N/N_s = 1.234$, $L/L_s = 0.434$]

11.1 Superheat Required for Bubble Existence Consider a vapor bubble of radius $R = 0.75$ mm in a uniformly superheated liquid. Calculate the superheat required ($T_f - T_s$) for existence of the bubble at the following conditions: (a) Saturated water 300, 100 and 10 °C, [Ans. 0.38, 0.043, 2.4 K], (b) Saturated R-11 at 7 °C [Ans. 0.026 K] The surface tension of water at 10, 100 and 300 °C is 0.0742, 0.05878 and 0.01439 N/m, respectively. The R-11 surface tension at 7 °C is 0.021 N/m.

11.2 Effect of Cavity Shape on Required Superheat Consider the cavity shapes shown in Figure 11.18a, b and d. All have the same cavity radius (r_c) and depth, H $= 4r_c$. Perform the following analysis for liquid contact angles (θ) of 15 and 90 degrees: (1) Will any of the cavities trap air when the surface is flooded with the liquids having the two contact angles? [Ans. $\theta = 15°$ (no, no, yes), $\theta = 90°$ (yes, yes, yes) (2) Sketch the shape of the liquid-vapor interface within the cavities for the two liquids, (3) Draw a graph of $(T_f - T_s)m/2\sigma$ vs. y/r for the cavities, (4) At what position in the cavity (y/r_c) is the greatest superheat required for existence of the vapor-liquid interface?, (5) Will any of the cavities allow any subcooling of the liquid? [Yes, cavity C].

11.3 Effect of Cavity Shape on Incipient Boiling You plan to develop an enhanced surface geometry for boiling R-11 at 4.4 °C (40 °F) on 19.1 mm (3/4 in) diameter copper tubes. The R-11 properties at 4.4 °C are: $\nu_l = 0.381$ m²/s, $\sigma = 0.0205$ N/m, $Pr_l = 5.4$, $k_l = 0.0935$ W/m-K, $m = dp/dT = 1994$ Pa/K, $\beta = 1.5E$-3 K⁻¹ (volume expansion coefficient).

Assume the boiling will occur in conical cavities (Fig. 11.18a) having $r_c = 0.025$ mm (0.001 in), and $2\phi = 30°$. Calculate the liquid superheat ($T_l - T_s$) required for a bubble of radius r_c to exist. [Ans. 0.86 K]

Calculate $T_w - T_s$ required for incipient boiling. For this condition, required $T_l - T_s$ at a distance $y = 1.5\ r_c$ above the heating surface to equal $2\sigma/mr_c$. Assume that the temperature profile in the liquid is linear. Heat will be transferred from the surface by natural convection at the incipient boiling condition. For natural convection, assume $Nu_d = 0.5\ (Gr_dPr)^{1/4}$, where Nu and Gr use the tube diameter for their characteristic dimension. The thermal boundary layer thickness is calculated by $\delta = k_l/h$. Calculate the temperature at the top of the bubble assuming a linear temperature profile in the liquid. [Ans. 0.86 K]

11.4 Effect of Cavity Shape Continuing with Problem 11.3, consider the effect of the cavity shape and contact angle on the value of $T_l - T_s$ required for existence of the vapor nucleus. The contact angle (θ) of R-11 on copper is approximately 10 degrees (or less). If a conical cavity having $2\phi = 30$ degrees is employed, what value of $T_l - T_s$ is required for a nucleus to exist at $x/l = 1/4$? Compare your calculated $T_l - T_s$ with the $T_w - T_s$ of Part b of Problem 11.3. [Ans. 3.2 K]

Assume a re-entrant cavity (Figure 11.18c) having $r_c = 0.025$ mm (0.001 in), $2\phi = 120$ and $y_l = 2r_c$. What $T_l - T_s$ is required at $x/H = 0.75$ (H = cavity depth)? [Ans. 0.11 K]

11.5 Porous Boiling Surface Consider the Code C surface in Figure 11.10b, whose physical characteristics are given in Table 11.2. Predict $(T_w - T_s)$ at $q = 20$ kW/m^2 for R-11 at 1.0 atm using the Nishikawa correlation (Equation 11.33). Compare your results with the experimental value (0.79 K). The physical properties of R-11 at 1.0 atm are: $\sigma = 0.0188$ N/m, $\lambda = 0.18E6$ J/kg-K, $\mu_v = 10.5E-6$ kg/m-s, $k_l = 0.091$ W/m-K, $\rho_l = 1479.3$ kg/m^3, $\rho_v = 5.87$ kg/m^3, m = dp/dt = 1.015kPa/K. The thermal conductivity of the copper matrix is $k_m = 197$ W/m-K. [Ans. 0.26 K]

12.1 Surface Tension Drained Condensation Adamek [1981] defined a family of profile shapes, which are discussed in Section 12.4.2. The condensate interface geometry for these profiles is given by $r_o/r + s/S_m = 1$. Integrate this equation over $0 < \theta < \theta_m$ using $d\theta = ds/r$ to obtain Equation 12.21.

One may derive the equation for the condensation coefficient following a procedure similar to that used by Nusselt for a gravity drained vertical plate. Defining $\kappa' = d(1/r)/ds$, show that

$$\frac{3v_l\kappa\Delta T}{\sigma\lambda} = \delta \frac{d}{ds}(\delta^3\kappa')$$

Solve the differential equation to obtain Equation 12.23.

12.2 The Bond Number The Adamek analysis used to obtain Equation 12.23 assumes that surface tension forces are much greater than the gravity force. Consider steam condensing at 60 °C on a profile having $\zeta = 0$ and $S_m = 1.0$ mm. Calculate the Bond number at $s = S_m$ using Equation 12.25. At 60 °C, $\sigma = 66E-3$ N/m and $\rho_l = 984$ kg/m^3. [Ans. Bo = 0.094]

12.3 Condensation on Vertical Fluted Surface The Figure 12.18 fluted surface concept is to be applied to condensation of ammonia at 10 °C with $\Delta T_{vs} = 2.22$ °C on a 1.52 m high vertical plate having 236 fins/m and the Gregorig profile ($\zeta = 2$) with $S_m = 1.52$ mm. The ammonia properties are given in Problem 12.4. Neglecting the condensation in the concave drainage channels, calculate the condensation coefficient based on the projected surface area and compare with the value for a smooth plate (h_s) [Ans. h = 133.8 kW/m^2-K, h/h_s = 25.9]

12.4 Condensation on Integral-Fin Tube Ammonia condenses at 50 F on a horizontal integral-fin tube having 748 fin/m (19 fins/in), $d_e = 19.1$ mm (0.75 in) and $d_o = 15.88$ mm (0.625 in) with $\Delta T_{vs} = 4$ F. The fins are trapezoidal shape having $t_b = 0.38$ mm (0.015 in) and $t_t = 0.23$ mm (0.009 in). Compute the condensing coefficient for $\eta_f = 1.0$ using: (1) The Beatty and Katz gravity drained model., and (2) The approximate model given in Section 12.5.3 assuming $\zeta = -0.8$ model. Assume rectangular fins ($t_{av} = 0.30$ mm) for calculation of c_b. What fraction of the surface is condensate flooded (c_b)? The ammonia properties at 8.9 °C are $\rho_l = 626$ kg/m^3, 0.531 W/m-K, 0.368 m^2/s, $v_l = 1.426E-2$ ft^2/hr, $\sigma = 0.0236$ N/m, $\lambda = 1226$ kJ/kg. [Ans. $A_o/L = 0.166$ m^2/m, h = 19.7 kW/m^2-K (B&K model), h = 21.1 kW/m^2-K (surface tension model), $c_b = 0.43$.]

INDEX

WIDENER UNIVERSITY
WOLFGRAM
LIBRARY
CHESTER, PA.